The Environment in a Magnet

Applications of NMR Techniques to Environmental Problems

New Developments in NMR

Editor-in-chief:
William S. Price, Western Sydney University, Australia

Series editors:
Bruce Balcom, *University of New Brunswick, Canada*
István Furó, *Industrial NMR Centre at KTH, Sweden*
Masatsune Kainosho, *Tokyo Metropolitan University, Japan*
Maili Liu, *Chinese Academy of Sciences, Wuhan, China*

Editorial Advisors:
Kavita Dorai, *IISER, India*
Danuta Kruk, *University of Warmia and Mazury in Olsztyn, Poland*

Titles in the series:

1: Contemporary Computer-Assisted Approaches to Molecular Structure Elucidation
2: New Applications of NMR in Drug Discovery and Development
3: Advances in Biological Solid-State NMR
4: Hyperpolarized Xenon-129 Magnetic Resonance: Concepts, Production, Techniques and Applications
5: Mobile NMR and MRI: Developments and Applications
6: Gas Phase NMR
7: Magnetic Resonance Technology: Hardware and System Component Design
8: Biophysics and Biochemistry of Cartilage by NMR and MRI
9: Diffusion NMR of Confined Systems: Fluid Transport in Porous Solids and Heterogeneous Materials
10: NMR in Glycoscience and Glycotechnology
11: Fast NMR Data Acquisition: Beyond the Fourier Transform
12: Cross-relaxation and Cross-correlation Parameters in NMR: Molecular Approaches
13: Contrast Agents for MRI: Experimental Methods
14: NMR-based Metabolomics
15: Modern Methods in Solid-state NMR: A Practitioner's Guide
16: Paramagnetism in Experimental Biomolecular NMR
17: Optimizing NMR Methods for Structure Elucidation: Characterizing Natural Products and Other Organic Compounds
18: Field-cycling NMR Relaxometry: Instrumentation, Model Theories and Applications
19: Hybrid MR-PET Imaging of the Brain: Systems, Methods and Applications
20: NMR Methods for Characterization of Synthetic and Natural Polymers
21: In-cell NMR Spectroscopy: From Molecular Sciences to Cell Biology

22: Long-lived Nuclear Spin Order: Theory and Applications
23: NMR and MRI of Gels
24: Advanced Diffusion Encoding Methods in MRI
25: NMR and MRI of Electrochemical Energy Storage Materials and Devices
26: NMR Spectroscopy for Probing Functional Dynamics at Biological Interfaces
27: Fast 2D Solution-state NMR: Concepts and Applications
28: Practical NMR for Oil and Gas Exploration
29: Biophysics of Molecular Chaperones: Function, Mechanisms and Client Protein Interactions
30: Integrated Structural Biology
31: Residual Dipolar Couplings: Principles and Applications
32: The Environment in a Magnet: Applications of NMR Techniques to Environmental Problems

How to obtain future titles on publication:
A standing order plan is available for this series. A standing order will bring delivery of each new volume immediately on publication.

For further information please contact:
Book Sales Department, Royal Society of Chemistry, Thomas Graham House, Science Park, Milton Road, Cambridge, CB4 0WF, UK
Telephone: +44 (0)1223 420066, Fax: +44 (0)1223 420247
Email: booksales@rsc.org
Visit our website at books.rsc.org

The Environment in a Magnet
Applications of NMR Techniques to Environmental Problems

Edited by

Pellegrino Conte
University of Palermo, Italy
Email: pellegrino.conte@unipa.it

Delia Francesca Chillura Martino
University of Palermo, Italy
Email: delia.chilluramartino@unipa.it

and

Paolo Lo Meo
University of Palermo, Italy
Email: paolo.lomeo@unipa.it

ROYAL SOCIETY OF CHEMISTRY

New Developments in NMR No. 32

Print ISBN: 978-1-83916-734-8
PDF ISBN: 978-1-83767-125-0
EPUB ISBN: 978-1-83767-126-7
Print ISSN: 2044-253X
Electronic ISSN: 2044-2548

A catalogue record for this book is available from the British Library

The Royal Society of Chemistry is a charity, registered in England and Wales, Number 207890, and a company incorporated in England by Royal Charter (Registered No. RC000524), registered office: Burlington House, Piccadilly, London W1J 0BA, UK, Telephone: +44 (0)20 7437 8656.

For further information see our web site at www.rsc.org

Printed in the United Kingdom by CPI Group (UK) Ltd, Croydon, CR0 4YY, UK

Preface

A few years ago, just before the pandemic occurred in 2020, the three of us started a collaboration, joining our different fields of expertise. Although all of us are chemists, we developed different experiences since we graduated in chemistry and concluded our PhD studies. Paolo is a physical organic chemist whose main research activity deals with the synthesis and characterization of nanosponges. He teaches the theory and applications of different spectroscopic techniques in organic chemistry, among which NMR plays a key role. Delia is a physical chemist whose primary research interest is in the structural characterization of materials, with a focus on soft matter. She teaches the theory and application of various spectroscopies, among which solid-state NMR techniques with applications in cultural heritage play a pivotal role. Additionally, she is involved in developing innovative teaching strategies. Rino (that is Pellegrino) is a soil chemist teaching soil chemistry, while his research activity is focused on the development of new NMR techniques to be applied in the evaluation of soil quality and the characterization of environmentally relevant materials, among which biochar is one of the most important.

After around thirty years from the start of our activities, once our career and role reached a "steady state", we were looking for new occasions to reestablish our enthusiasm. Therefore, we had the idea to join all we know to find new possibilities for improving our research and expanding our teaching possibilities. In this view, we have developed different projects by which we are trying to understand how to obtain new environmentally sustainable materials based, *e.g.*, on the combination of nanosponges and biochars to be used in agriculture and on the transformation of production waste, with the aim to reduce the environmental impact of the most common poorly degradable plastics. At the same time, we aimed to achieve new

New Developments in NMR No. 32
The Environment in a Magnet: Applications of NMR Techniques to Environmental Problems
Edited by Pellegrino Conte, Delia Francesca Chillura Martino and Paolo Lo Meo

Published by the Royal Society of Chemistry, www.rsc.org

water and soil remediation systems or even to develop new films for food and cultural heritage protection.

The combination of our teaching experiences made us understand that, at least in Italy, where we work, the NMR world's wideness is underrated. For example, NMR is not considered in any chemistry courses in agriculture studies, while only spectroscopy is taught in the chemistry departments. Moreover, magnetic resonance for imaging seems to be significant only in the medicine courses, while no information at all is provided about fast field cycling (FFC) NMR relaxometry. The latter can be considered the experimental side of the computational chemistry investigating molecular dynamics. Therefore, we decided to project a book where the "*classic*" NMR spectroscopy, traditionally viewed as the main technique – the sovereign one! – in aiding the molecular structure determination of organic compounds, is analyzed together with other techniques (such as solid-state NMR, FFC NMR relaxometry, MRI and so on), which have been perceived for a long time as niche methods, despite their increasing popularity and their great potential, specifically in addressing problems such as the characterization of porous materials and many complex systems. Fortunately, a more and more widespread consideration and consciousness of the problems related to complex and interdependent systems is likely to impose, in the next few decades, a profound revision of the educational perspectives and a significant innovation in the teaching of spectroscopic characterization methods as well. It is our intention that this book will provide a scholarly overview of the most recent scientific perspectives and handy and reliable guidance for exploring the field of otherwise "exoteric" work instruments, useful for students, self-updating of professionals, technicians, and interested people.

The result of what we had in mind at the beginning of this adventure is the "*object*" that you, readers, hold in your hands. The three of us know that this book may not satisfy all the readers, as for someone there is something missing, while for someone else something can be written too simplistically. We three are the only ones responsible for the possible inaccuracies and simplistic things you may find herein.

This book is intended for non-specialized people having very different backgrounds. Were we successful? Were we not? This judgment is up to our readers. We hope this book can contribute to understanding the complexity and fascination of a set of techniques all grouped under the name of "*Nuclear Magnetic Resonance*".

Pellegrino Conte
Delia Francesca Chillura Martino
Paolo Lo Meo

Contents

New Developments in NMR No. 32
The Environment in a Magnet: Applications of NMR Techniques to Environmental Problems
Edited by Pellegrino Conte, Delia Francesca Chillura Martino and Paolo Lo Meo

Published by the Royal Society of Chemistry, www.rsc.org

CHAPTER 1

The Meaning of Pollution and the Powerfulness of NMR Techniques

P. CONTE,[a] D. F. CHILLURA MARTINO*[b] AND P. LO MEO[b]

[a] Department of Agricultural, Agronomical and Food Sciences (SAAF), University of Palermo, v.le delle Scienze ed. 4, 90128 Palermo, Italy;
[b] Department of Biological, Chemical and Pharmaceutical Sciences and Technologies (STEBICEF), University of Palermo, v.le delle Scienze ed. 17 "S. Cannizzaro", 90128 Palermo, Italy
*Email: delia.chilluramartino@unipa.it

"*Hostemque nostrum comprime ne polluantur corpora*"
(from a Christian hymn attributed to St. Ambrose of Milan, IV century AD)

1.1 Introduction: Pollution and Its Multifaceted Aspects

Pollution is a Latin-derived word, tracing back to the term *lutum* (mud) and, through it, to the Indo-European root **leu-* (filth, dirty).[1] It holds in itself the sense of dirtiness and impurity, both in a physical and in a ritual or moral sense. Hence, bishop Ambrose of Milan (339–397 AD) could pray in pathetic words: "*...Thou [Lord] smash our Enemy, so our bodies will be not polluted*". There are a couple of interesting ideas, in our opinion, hidden in

New Developments in NMR No. 32
The Environment in a Magnet: Applications of NMR Techniques to Environmental Problems
Edited by Pellegrino Conte, Delia Francesca Chillura Martino and Paolo Lo Meo

Published by the Royal Society of Chemistry, www.rsc.org

the short Ambrose passage that fit quite well for introducing the topics covered in the present book:

(i) Pollution is *evil*. It treacherously affects the quality of the environment we live in. In a more or less apparent way, it introduces harmful substances into air, water, and soils, which pass into the food chain, and poison and harm animal and vegetal populations. This ultimately results in the reduction of biodiversity, in damage to wildlife environments, forests and oceans, and in the present global warming problems everyone is perfectly aware of.
(ii) Pollution is an *enemy*. It threatens our lives, affecting life quality and human health, reducing the availability of safe food and water resources, and the salubrity of air, and causing directly or indirectly diseases and pathologies. Thus, large amounts of human and economic resources have to be devoted to repairing the damage that pollution has caused and is causing, to recovering soils and water bodies, and to preventing possible future problems, in an extenuating everlasting war against time, and against human short sight and stupidity, as well. However, fighting against an enemy requires weapons: NMR is a weapon.

This book aims at presenting a general overview of how various NMR techniques can be profitably exploited in monitoring the analytical parameters of waters and soils and in aiding the design of smart solutions for solving environmental problems.

It is important to stress that environmental monitoring and remediation are multifaceted problems, for dealing with which a change of paradigm in scientists' mind is somehow required, with the transition from a simply reductionist towards a more holistic and multidisciplinary approach.[2–4] Physicists, chemists, biologists, and ecologists have all to co-operate, crossing their own competencies, skills, and knowledge. This is necessary because the environment is a *complex* entity. As an example, Figure 1.1 reports a simple scheme where four different environmental compartments intersect to form a fifth compartment indicated as soil. The hydrosphere is the environmental compartment referring to all the forms of water present on and in the planet Earth; the lithosphere refers to the environmental compartment containing the rocks; the atmosphere is the term concerning the gas phase (mainly oxygen and nitrogen) held around the planet by gravitational forces; the biosphere comprises all the living carbon-based organisms occupying all the ecological niches on the Earth.

The graph in Figure 1.1 highlights that all the five environmental compartments are related to each other. This means that polluting one of them causes pollution of all the others. Just as a simple example, misuse of any type of agrochemical implies soil pollution. However, due to the direct contact between soil and water, agrochemicals and their residues can translocate from the former towards the latter. Moreover, agrochemical residues can be highly volatile. Therefore, they can also pollute the atmosphere. Finally, once

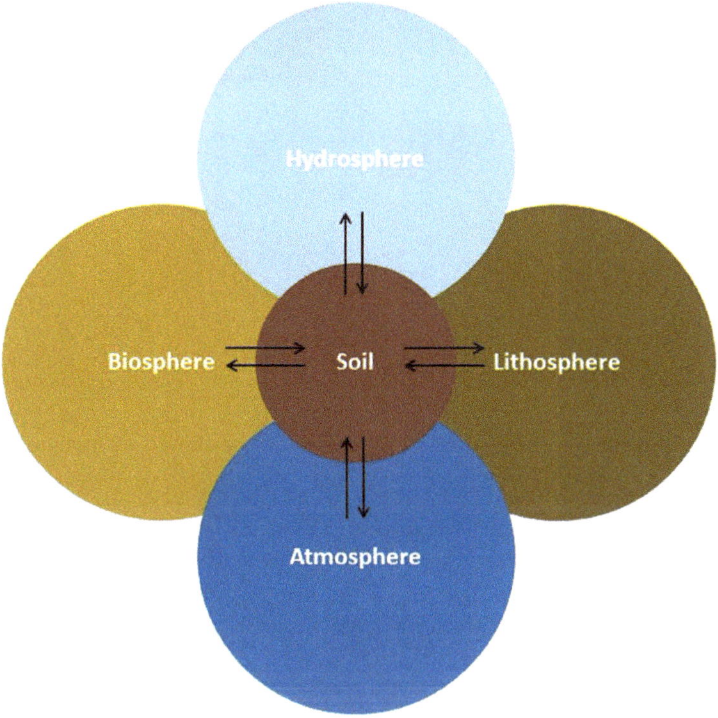

Figure 1.1 Example of interactions between the different environmental compartments. See the main text for the definition of each compartment.

pollutants enter drinking water or any other food system, they are ingested by humans with all the consequences on human health.

According to what has been outlined above, the interrelations among environmental compartments reveal a multi-layered reality, where complex interactions and interdependences occur, even at molecular, cellular, physiological, and ecological levels. At the same time, projecting a preservation or remediation intervention in an environmental context similarly requires a comprehensive evaluation of all the aspects previously mentioned and of how the different counterparts of such a complex system interact.

Figure 1.2 shows a simple scheme summarizing the sequence of operations that must be accounted for before a remediation is applied.

The first three points referring to the "collection of historical data", "hydrological investigations", and "chemical and microbiological analyses" answer the first of the few questions that must be satisfied to design a remediation approach: is the environmental site intensely polluted? Indeed, in order to establish the pollution extent, one has to know which kinds of activities have been carried out on the polluted site. Hence, for facilitating the hydrogeological, chemical, and biological investigations applied to assess present pollutants quantitatively/qualitatively, a historical reconstruction of

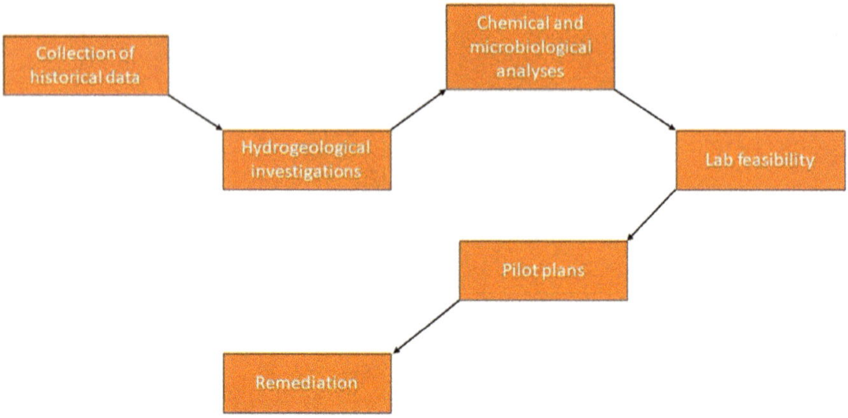

Figure 1.2 Summary of the operations that must be done to project an environmental remediation.

the anthropic actions must be retrieved. The next question is: how urgent site's remediation is? The answer to the latter question requires an evaluation of the remediation aims. As a matter of fact, the more the pollution affects human life quality, the more urgent the remediation is. For instance, if pollutants affect groundwater quality that is addressed to human feed, remediation is much more urgent than in the case of the achievement of a golf field! Following the scheme in Figure 1.2, the feasibility of a remediation intervention requires a preliminary lab-scale assessment. This involves working on a limited amount of samples (*i.e.*, from a few mg up to a few grams) somehow modelling the real system. Then, engineers must upgrade the lab-scale feasibility to the pilot-scale one. This means working with medium sized amounts of samples (*i.e.*, from kg to quintals) strictly resembling the real environmental system. Ultimately, everything is ready for the real large-scale action. The latter can be accomplished either *in situ* (*i.e.*, the environmental compartment is de-polluted without moving it from its original place) or *ex situ* (*i.e.*, the de-pollution is achieved by removing the contaminated system from its original site and carried out in a separate plant), depending on the practical and economical conveniences.[5,6]

Such a titanic task requires the availability of more and more sophisticated, versatile, sensitive, and powerful analytical tools, able to provide quick, accurate and complete qualitative and quantitative information. It is in this context that modern NMR techniques find their placement, providing a set of smart tools for diverse tasks.

1.2 Modern NMR in the Context of Environmental Analysis and Remediation

Modern NMR is indeed a collection of different techniques, by which various aspects and features of the behaviour of both organic and inorganic matter can be elucidated, going from molecular structure to molecular dynamics, imaging analysis and so on. Moreover, NMR techniques have largely benefitted from the enormous progresses in instrumental technologies that have been continuously developed in the last few decades.[7–12] More and more powerful cryomagnets have been designed and produced, as well as high sensitivity multichannel probes and cryoprobes. This has allowed a large and widespread availability of high-resolution instruments. At the same time, solid-state probes have become increasingly diffused. In parallel, more and more sophisticated pulse sequences have been developed and implemented. Therefore, as long as environmental problems are concerned, NMR can provide diverse useful tools for different tasks.

The main use of 1D and 2D NMR spectroscopy in organic chemistry is undoubtedly constituted by the determination of molecular structure. In combination with mass spectrometry,[13–15] "traditional" NMR represents an essential tool for the structural elucidation of organic contaminants and their metabolites or degradation by-products.[16] From this standpoint, hyphenated techniques such as liquid chromatography (LC) combined with NMR can provide a powerful and versatile analytical tool.[15,17,18] Certainly, the need to detect at the same time several low-concentration substances in complex matrices has stimulated the development of various high-sensitivity techniques, such as SABRE,[19] CIDNP,[20] and NUS.[21] It is important to stress that NMR can be profitably used in both targeted and untargeted methods, *i.e.*, it can be used for both identifying particular marker species or retrieving an entire metabolic or microconstituent fingerprint[22–24] (that can be subjected, in turn, to suitable multivariate analysis[25]). Indeed, chemical fingerprinting is an excellent way to trace back the origin of some contaminations.[26–28] Metabolomics analysis by NMR[29–32] can also provide important information on the fate of contaminant species (in particular under sub-lethal conditions) when interacting with living species, in order to elucidate their toxicity from a biochemical standpoint.[33] Furthermore, NMR also allows the detection and study of heavy metal contaminants.[34] Last, but not least, the use of NMR is essential for the characterization of materials devoted to environmental remediation applications such as sorbent polymers[35–38] or biochar.[39]

An important strength point of NMR is its versatility in relation to the nature and the constitution of the sample to be analysed. For long time, it has been kept for granted that NMR spectra could be acquired only from non-viscous liquid systems. However, the magic angle spinning (MAS) technique has enabled the acquisition of good quality spectra[40] also from solid or semi-solid samples. More recent developments, such as CMP-NMR,[41] have made possible studying even heterogeneous systems.

Beyond chemical characterization, NMR provides useful tools even for a physical characterization of real systems. In particular, relaxometry (*i.e.*, the study of both longitudinal and transversal relaxation of excited nuclei) allows access to very useful information on the dynamics of molecular systems.[42–44] In particular, motions of liquids within porous media can be monitored and studied by the FFC technique,[45–47] which in turn allows retrieval of useful information on the textural features of the porous medium itself.[48,49] Hence, relaxometric techniques have been proven as a valuable tool for the physico-chemical characterization of systems such as soils[50–52] or sorbents,[39,49,53] thereby providing an essential piece of information for their treatment or possible employment. Besides, FFC NMR finds useful applications in the characterization of foods,[54–57] in monitoring degradation processes,[58] and in studying paramagnetic species.[59–61] As long as environmental problems are concerned, relaxometric techniques have been profitably exploited not only to assess the quality of soils,[52] but also to study the presence of contaminants[62] or even to verify the quality of marine life.[63] Finally, magnetic resonance imaging (MRI) techniques provide a further powerful tool for the study of environmental systems. In particular, the characteristics and the interactions occurring within the rhizosphere can be adequately investigated, even in combination with other techniques.[64,65] Indeed, MRI allows monitoring not only roots and the surrounding water content, but also the possible accumulation of solutes.[66,67]

1.3 Future Perspectives

In the previous sections, the potential uses of NMR techniques for addressing environmental remediation and relevant issues have been briefly summarized. As Simpson *et al.* have evidenced in a recent seminal paper,[68] it can be envisaged that NMR will play a more and more important role in future research concerning environmental problems. At present, the main drawbacks to cope with include the lower sensitivity in comparison to mass spectrometry techniques, the cost, portability, and maintenance of instrumentation, as well as the spread of necessary knowledge and skills. Noticeably, the progress in computer science, the development of artificial intelligence, and the increasing capability of miniaturization will increase the powerfulness of NMR in environmental monitoring, enable the ability of NMR to evaluate human diseases in more detail, and enable early diagnoses.

In general, NMR techniques may become central in environmental issues, provided that more and more information is provided at different levels. Indeed, one of the main problems in extending the application of NMR outside the academic community is the lack of professionals which, in contrast, are more abundant when dealing with chromatography. In other words, even because of the costs and/or the needed quantum physics knowledge, NMR techniques are less used than other more widespread and conceptually simpler analytical methods.

It is the authors' belief that this book will contribute to opening researchers' minds in envisaging the potential and the challenges NMR offers in this context.

References

1. J. P. Mallory and D. Q. Adams, *The Oxford introduction to proto-Indo-European and the proto-Indo-European world*, Oxford University Press on Demand, 2006.
2. H. Andersen, *Endeavour*, 2001, **25**, 153–156.
3. R. Levi-Montalcini and P. Calissano, *BMC Neurosci.*, 2006, **7**, S1.
4. A. Fardet and E. Rock, *Healthcare*, 2015, **3**, 1054–1063.
5. H. I. Gomes, C. Dias-Ferreira and A. B. Ribeiro, *Sci. Total Environ.*, 2013, **445–446**, 237–260.
6. A. Hussain, F. Rehman, H. Rafeeq, M. Waqas, A. Asghar, N. Afsheen, A. Rahdar, M. Bilal and H. M. N. Iqbal, *Chemosphere*, 2022, **289**, 133252.
7. A. G. Webb, in *Annual Reports on NMR Spectroscopy*, ed. G. A. Webb, Academic Press, 2006, vol. 58, pp. 1–50.
8. Y. Yanagisawa, M. Hamada, K. Hashi and H. Maeda, *Supercond. Sci. Technol.*, 2022, **35**, 044006.
9. T. Castaing-Cordier, D. Bouillaud, J. Farjon and P. Giraudeau, *Annual Reports on NMR Spectroscopy*, 2021, vol. 103, pp. 191–258.
10. R. W. Martin, J. E. Kelly and J. I. Kelz, *J. Struct. Biol.*, 2019, **206**, 73–89.
11. S. C. Chiliveri, A. J. Robertson, Y. Shen, D. A. Torchia and A. Bax, *Chem. Rev.*, 2022, **122**, 9307–9330.
12. E. W. McFarland and B. R. Rosen, *CardioVasc. Interv. Radiol.*, 1986, **8**, 238–250.
13. L. M. Labine and M. J. Simpson, *Curr. Opin. Environ. Sci. Health*, 2020, **15**, 7–15.
14. S. Faixo, N. Gehin, S. Balayssac, V. Gilard, S. Mazeghrane, M. Haddad, G. Gaval, E. Paul and J.-C. Garrigues, *Anal. Chim. Acta*, 2021, **1152**, 338284.
15. O. Corcoran and M. Spraul, *Drug Discovery Today*, 2003, **8**, 624–631.
16. B. Yuan, D. H. Lysak, R. Soong, A. Haddad, A. Hisatsune, A. Moser, S. Golotvin, D. Argyropoulos, A. J. Simpson and D. C. G. Muir, *Environ. Sci. Technol. Lett.*, 2020, **7**, 496–503.
17. A. Pukalskas, T. A. van Beek and P. de Waard, *J. Chromatogr. A*, 2005, **1074**, 81–88.
18. S. H. Hansen, A. G. Jensen, C. Cornett, I. Bjørnsdottir, S. Taylor, B. Wright and I. D. Wilson, *Anal. Chem.*, 1999, **71**, 5235–5241.
19. P. M. Richardson, A. J. Parrott, O. Semenova, A. Nordon, S. B. Duckett and M. E. Halse, *Analyst*, 2018, **143**, 3442–3450.
20. G. J. Janssen, P. Bielytskyi, D. G. Artiukhin, J. Neugebauer, H. J. M. de Groot, J. Matysik and A. Alia, *Sci. Rep.*, 2018, **8**, 17853.
21. F. Delaglio, G. S. Walker, K. A. Farley, R. Sharma, J. C. Hoch, L. W. Arbogast, R. G. Brinson and J. P. Marino, *Am. Pharm. Rev.*, 2017, **20**, 339681.
22. A. P. Sobolev, F. Thomas, J. Donarski, C. Ingallina, S. Circi, F. C. Marincola, D. Capitani and L. Mannina, *Trends Food Sci. Technol.*, 2019, **91**, 347–353.
23. I. Bertini, S. Cacciatore, B. V. Jensen, J. V. Schou, J. S. Johansen, M. Kruhøffer, C. Luchinat, D. L. Nielsen and P. Turano, *Cancer Res.*, 2012, **72**, 356–364.

24. P. Krishnan, N. J. Kruger and R. G. Ratcliffe, *J. Exp. Bot.*, 2004, **56**, 255–265.
25. Y. B. Monakhova, T. Kuballa and D. Lachenmeier, *J. Anal. Chem.*, 2013, **68**, 755–766.
26. E. Shumilina, T. Johansen, C. Glasner, J. Robert, T. Rustad and A. Dikiy, *Waste Biomass Valorization*, 2020, **11**, 1349–1357.
27. J. Lu, A. C. Chang and L. Wu, *Environ. Pollut.*, 2004, **132**, 365–374.
28. R. M. Duarte, J. T. Matos and A. C. Duarte, *Appl. Sci.*, 2021, **11**, 2539.
29. A.-H. Emwas, R. Roy, R. T. McKay, L. Tenori, E. Saccenti, G. N. Gowda, D. Raftery, F. Alahmari, L. Jaremko and M. Jaremko, *Metabolites*, 2019, **9**, 123.
30. A. Smolinska, L. Blanchet, L. M. Buydens and S. S. Wijmenga, *Anal. Chim. Acta*, 2012, **750**, 82–97.
31. A. Vignoli, V. Ghini, G. Meoni, C. Licari, P. G. Takis, L. Tenori, P. Turano and C. Luchinat, *Angew. Chem., Int. Ed.*, 2019, **58**, 968–994.
32. J. L. Markley, R. Brüschweiler, A. S. Edison, H. R. Eghbalnia, R. Powers, D. Raftery and D. S. Wishart, *Curr. Opin. Biotechnol.*, 2017, **43**, 34–40.
33. M. C. Celander, *Aquat. Toxicol.*, 2011, **105**, 72–77.
34. L. Cruz, J.-M. Savariault, C. Morais, C. Fernandez, J. P. D. Jesus and J. Rocha, *J. Solid State Chem.*, 2004, **177**, 1549–1558.
35. G. Crini, *Environ. Chem. Lett.*, 2021, **19**, 2383–2403.
36. M. Russo, M. L. Saladino, D. Chillura Martino, P. Lo Meo and R. Noto, *RSC Adv*, 2016, **6**, 49941–49953.
37. A. Spinella, M. Russo, A. D. Vincenzo, D. Chillura Martino and P. Lo Meo, *Beilstein J. Org. Chem.*, 2018, **14**, 1498–1507.
38. A. Di Vincenzo, A. Palumbo Piccionello, A. Spinella, D. Chillura Martino, M. Russo and P. Lo Meo, *Beilstein J. Org. Chem.*, 2019, **15**, 633–641.
39. P. Conte, R. Bertani, P. Sgarbossa, P. Bambina, H. P. Schmidt, R. Raga, G. Lo Papa, D. F. Chillura Martino and P. Lo Meo, *Agronomy*, 2021, **11**, 615.
40. A. J. Simpson, M. J. Simpson, W. L. Kingery, B. A. Lefebvre, A. Moser, A. J. Williams, M. Kvasha and B. P. Kelleher, *Langmuir*, 2006, **22**, 4498–4503.
41. D. Courtier-Murias, H. Farooq, H. Masoom, A. Botana, R. Soong, J. G. Longstaffe, M. J. Simpson, W. E. Maas, M. Fey and B. Andrew, *J. Magn. Reson.*, 2012, **217**, 61–76.
42. R. K. Wangsness and F. Bloch, *Phys. Rev.*, 1953, **89**, 728–739.
43. N. Bloembergen, E. M. Purcell and R. V. Pound, *Phys. Rev.*, 1948, **73**, 679–712.
44. E. Anoardo, G. Galli and G. Ferrante, *Appl. Magn. Reson.*, 2001, **20**, 365–404.
45. R. Kimmich, *Field-cycling NMR relaxometry: instrumentation, model theories and applications*, Royal Society of Chemistry, 2018.
46. R. M. Steele, J.-P. Korb, G. Ferrante and S. Bubici, *Magn. Reson. Chem.*, 2016, **54**, 502–509.
47. J. Ward-Williams and L. F. Gladden, *Magn. Reson. Imaging*, 2019, **56**, 57–62.
48. D. Faux, R. Kogon, V. Bortolotti and P. McDonald, *Molecules*, 2019, **24**, 3688.

49. P. Lo Meo, F. Mundo, S. Terranova, P. Conte and D. Chillura Martino, *J. Phys. Chem. B*, 2020, **124**, 1847–1857.
50. P. Conte and V. Ferro, *Hydrol. Processes*, 2018, **32**, 93–101.
51. P. Conte and V. Ferro, *Magn. Reson. Chem.*, 2020, **58**, 41–50.
52. P. Conte and P. Lo Meo, *Agronomy*, 2020, **10**, 1040.
53. P. Lo Meo, S. Terranova, A. Di Vincenzo, D. Chillura Martino and P. Conte, *Anal. Chem.*, 2021, **93**, 8553–8558.
54. E. G. Ates, V. Domenici, M. Florek-Wojciechowska, A. Gradišek, D. Kruk, N. Maltar-Strmečki, M. Oztop, E. B. Ozvural and A.-L. Rollet, *Trends Food Sci. Technol.*, 2021, **110**, 513–524.
55. P. Conte, L. Cinquanta, P. Lo Meo, F. Mazza, A. Micalizzi and O. Corona, *Food Res. Int.*, 2020, 109845.
56. P. Conte, V. Mineo, S. Bubici, C. De Pasquale, F. Aboud, A. Maccotta, D. Planeta and G. Alonzo, *Anal. Bioanal. Chem.*, 2011, **400**, 1443–1450.
57. F. Aboud, P. Conte, C. D. Pasquale, A. Sinacori and G. Alonzo, *J. ApiProd. ApiMed. Sci.*, 2011, 164–173.
58. E. M. Erro, C. C. Fraenza, L. Gerbino and E. Anoardo, *Mol. Phys.*, 2019, **117**, 983–989.
59. P. Conte, V. Marsala, C. De Pasquale, S. Bubici, M. Valagussa, A. Pozzi and G. Alonzo, *GCB Bioenergy*, 2013, **5**, 116–121.
60. R. Zairov, S. Pizzanelli, A. P. Dovzhenko, I. Nizameev, A. Orekhov, N. Arkharova, S. N. Podyachev, S. Sudakova, A. R. Mustafina and L. Calucci, *J. Phys. Chem. C*, 2020, **124**, 4320–4329.
61. D. Faux, R. Kogon, V. Bortolotti, and P. McDonald, *Molecules*, 2019, **24**, 3688.
62. S. A. Hedberg, R. J. Knight, A. L. MacKay and K. P. Whittall, *Water Resour. Res.*, 1993, **29**, 1163–1170.
63. P. Fantazzini, S. Mengoli, S. Evangelisti, L. Pasquini, M. Mariani, L. Brizi, S. Goffredo, E. Caroselli, F. Prada and G. Falini, *Environ. Sci. Technol.*, 2013, **47**, 12679–12686.
64. S. Haber-Pohlmeier, C. Tötzke, E. Lehmann, N. Kardjilov, A. Pohlmeier and S. Oswald, *Vadose Zone J.*, 2019, **18**, 1–11.
65. J. S. MacFall, G. A. Johnson and P. J. Kramer, *New Phytol.*, 1991, **119**, 551–560.
66. A. Perelman, N. Lazarovitch, J. Vanderborght and A. Pohlmeier, *Plant Soil*, 2020, **454**, 171–185.
67. A. Moradi, S. Oswald, J. Massner, K. P. Pruessmann, B. Robinson and R. Schulin, *Eur. J. Soil Sci.*, 2008, **59**, 476–485.
68. A. J. Simpson, M. J. Simpson and R. Soong, *Anal. Chem.*, 2018, **90**, 628–639.

CHAPTER 2

The Importance of NMR as a Discovery Tool

K. DOWNEY,[†a] K. RONDA,[†a] K. STEINER,[a] K. JOBST,[b] S. KLEYWEGT,[c] M. J. SIMPSON[a] AND A. J. SIMPSON*[a]

[a] Environmental NMR Centre, University of Toronto Scarborough, Toronto, ON, Canada; [b] Memorial University of Newfoundland, St. John's, NL, Canada; [c] Technical Assessment and Standards Development Branch, Ontario Ministry of the Environment, Conservation and Parks, Toronto, ON, Canada

*Email: andre.simpson@utoronto.ca

2.1 Introduction

NMR is invaluable in identifying critical novel contaminants when mass spectrometry (MS) alone is insufficient. Mass spectrometry is the workhorse of contaminant monitoring with excellent limits of detection and targeted capabilities. However, the limitation of MS is its ability to identify isomeric structures from scratch and without libraries, a feat which is essential for novel compound discovery when standards are not available. The potential and need for NMR as a contaminant discovery tool is best demonstrated through a few important real-world examples.

The first example of NMR use in contaminant discovery is that of Donald Taves, a physician at the University of Rochester who, in 1976, found fluorinated organic compounds in his own blood using ^{19}F NMR.[1] Fluorinated compounds have been present in commercial products such as textiles and

[†]Joint first authors; both authors contributed equally.

New Developments in NMR No. 32
The Environment in a Magnet: Applications of NMR Techniques to Environmental Problems
Edited by Pellegrino Conte, Delia Francesca Chillura Martino and Paolo Lo Meo

Published by the Royal Society of Chemistry, www.rsc.org

surfactants since the 1940s,[2] and upon roughly identifying the compound as perfluorooctanoic acid (PFOA, or something structurally similar), Taves presumed that these consumer products were the source of PFOA in his blood.[2] The presence of PFOA and other per- and polyfluorinated compounds in the environment was only "discovered" by mass spectrometry nearly two decades later.[3] Numerous studies were subsequently conducted, finding that not only are perfluorinated compounds ubiquitous in the environment,[4] but they are persistent, exhibit toxic properties at relatively low doses, and bioaccumulate in organisms.[2] The presence of PFOA, perfluorooctanesulfonic acid (PFOS), and others has also been found within human serum at concentrations over 30 ppb,[5] which is over 10 times as much in some wild animals.[6] Additionally, comprehensive studies suggest that ppb levels of perfluorinated compounds are present in most or all humans today.[2] Considering that perfluorinated compounds pose health risks, such as increased occurrence of certain cancers and attention-deficit hyperactivity disorder (ADHD), changes in normal hormone and cholesterol levels, and reproductive issues (such as altered female and male fertility, preeclampsia, and birth defects),[7] strict regulations surrounding perfluorinated compound production and use have been implemented in some countries.[2] In regions that have increased restrictions regarding perfluorinated compounds, levels in humans have decreased,[7] which begs the question: *if NMR had been employed more routinely in contaminant discovery in the 1970s, could these fluorinated compounds have been identified and phased out earlier, thereby minimizing human and environmental health impacts?*

A second example that shows the power of NMR as a tool in environmental research is that of the 2021 study looking into the cause of acute mortality in coho salmon (*Oncorhynchus kisutch*).[8] Here, a quinone transformation product of *N*-(1,3-dimethylbutyl)-*N*′-phenyl-*p*-phenylenediamine (6PPD) was identified as the compound responsible for the elevated mortality rates in the salmon. Experiments done by ultra-performance liquid chromatography–high resolution mass spectrometry (UPLC–HRMS) indicated that a $C_{18}H_{22}N_2O_2$ by-product, formed during the ozonation of 6PPD, was present. While other analytical techniques provide limited information about the exact structure of contaminants, the use of 2D NMR experiments allowed the exact structure to be identified as the novel compound 6PPD-quinone.[8] These results contradict previous work by liquid chromatography and mass spectrometry, which suggested that the structure of the 6PPD ozonation by-product was a dinitrone, a belief that persisted for decades.[8] Upon identification of the contaminant, extensive toxicity testing confirmed that 6PPD-quinone was in fact responsible for the elevated mortality rates in coho salmon in the US Pacific Northwest.[8]

These results are critical in terms of protecting environmental health. 6PPD has been added to tires since the mid-1960s to prevent the rubber from cracking over time. Tire rubber deposits 6PPD on roadways where it then leaches into aquatic systems along with its transformation products.[8] Since 6PPD is used globally as a tire rubber antioxidant, the leaching of the highly toxic by-product 6PPD-quinone may be affecting aquatic organisms on a much larger scale.[8] 6PPD-quinone has an LC50 of 95 ng L^{-1} in coho salmon,[9] and a single tire has enough parent chemical to theoretically kill 1 billion salmon. It is estimated that the chemical kills between 40 and 90% of wild salmon spawning in urban creeks[10] and has been found to be toxic to a range of species, as well as prevalent in the fraction of air particles that can enter human lungs.[11] Within a year of the discovery of 6PPD-quinone, discussions in the United States Congress began, and acts to ban the parent chemical in California and the European Union were proposed. However, the question remains: *why is NMR not used more in environmental research, and if it was, could 6PPD-quinone have been discovered decades earlier?*

The unrivaled ability of NMR spectroscopy to provide information regarding the connectivity of atoms in a molecule means that it can be used for investigations into complex mixtures when other methods alone are inadequate.[12] For example, chlorinated paraffins (CPs) can exist as thousands of different isomers, between which MS cannot distinguish. However, by analyzing these complex mixtures with both MS and NMR, a great deal of information can be obtained.[12] In one study, the large number of isomers present in a mixture meant that the NMR spectra recorded had a great deal of overlap, making the assignment of specific isomers impossible.[12] However, using NMR databases, which were generated from all the isomers using neural network spectral simulation, allowed the most likely isomers to be reported.[12] These results show that using NMR and MS as complementary tools can make the analysis of highly complex mixtures much more manageable.

This chapter aims to summarize and discuss the various contexts in which NMR spectroscopy has been a useful complementary tool in compound discovery to date, namely in environmental research, but also as a tool in the identification of transformation products, as well as in compound discovery for metabolomics and food science. The chapter concludes by discussing the use of other NMR-active nuclei in compound discovery and by postulating future directions. One key goal of this chapter is to demonstrate that NMR has a critical and complementary role in environmental molecular discovery and needs to be more routinely employed by the next generation of researchers.

2.2 Environmental Discovery

Due to its unparalleled ability to solve structures *de novo* (without the need for libraries), NMR spectroscopy is an ideal tool for discovery-based research. Recent advances have allowed improved sensitivity and solvent suppression, meaning that NMR can now be applied to increasingly complex and environmentally relevant samples, which often have complex matrices composed of mixtures of unknown compounds. The following sections will describe recent applications of NMR spectroscopy in environmental research.

2.2.1 Water

NMR has proved to be a useful discovery tool in various aquatic systems. Water suppression techniques can be used to attenuate the large, broad water signal that is present in these samples, allowing the identification of organic signatures that are present at low concentrations.[13,14] This has been applied to the characterization of substances present in fresh and ocean waters, as well as wastewater samples.

Dissolved organic matter (DOM) is known to significantly impact global systems such as nutrient cycling and the fate and transport of contaminants.[15] Therefore, improving methods for the analysis and comparison of DOM has far-reaching implications.

Techniques such as solid-phase extraction can be used prior to analysis by NMR in order to concentrate the DOM portion of a sample. One study used this technique to analyze the DOM composition of water samples collected from the South Atlantic Ocean. When analyzing these extracts with both ^{1}H and ^{13}C NMR experiments, differences in the composition of DOM could be observed between samples collected at different depths.[16] For instance, ^{1}H NMR analysis showed that the concentrations of oxygenated aliphatics decrease with increasing depth in the South Atlantic Ocean.[16] Conversely, groups such as methyl esters were found to increase in concentration as the sample depth increased, with the lowest concentrations being detected at the ocean surface.[16] Similarly, ^{13}C NMR data indicated that functional groups such as carboxylic acids and ketones increase in concentration with increasing depth.[16] 2D NMR experiments run on DOM that was isolated by solid-phase extraction were used to detect differences in the molecular signatures of samples collected at varying depths.[16] Resin columns can also be used along with pre-concentration processes to isolate DOM fractions for analysis by NMR.[17] Using these methods, the distribution of polar and non-polar DOM was investigated in samples collected from the Great Salt Lake.[17] Cross-polarization magic angle spinning (CP-MAS) NMR was found to be a useful tool for the characterization of the isolated fractions in this study. Results indicated that 56% of the DOM in the Great Salt Lake is composed of polar compounds, of which *N*-acetyl hydroxy carboxylic acids were found to be a significant contributor.[17] It was determined that alicyclic ring structures made up a major component of the remaining, non-polar, DOM.[17] Using NMR experiments in conjunction with complementary techniques such as mass spectrometry allowed models to be generated for the composition of the different fractions of DOM in the Great Salt Lake.[17]

In some cases, the DOM of a natural water sample can even be analyzed by NMR without the need for pre-concentration.[15] For example, Lam and Simpson characterized the DOM of unaltered water samples. The samples were collected from various sites in order to provide a robust representation of global water sources.[15] Using direct analysis by ^{1}H NMR spectroscopy, it was determined that the DOM present in ocean waters is primarily composed of carbohydrates, while the DOM from rivers and lakes was found to be very similar in composition to that of wetland samples.[15] It should be noted that the low concentration of DOM in natural samples such as these means that analysis without preconcentration requires a greater number of scans, and as such longer experiment times.[13,15] Rainwater samples have also been analyzed by NMR and differences in the composition of DOM investigated between seasons.[13] The results of ^{1}H NMR analysis suggest that the samples collected in the spring had a higher percent composition of carbohydrates but lower contributions from saturated fatty acids compared to winter rains.[13] These results provide insight into air mass trajectory and may provide valuable information for investigations into the origins of dissolved organic carbon present in the atmosphere.[13]

The insight provided by NMR into the composition and behaviour of DOM is hugely useful in understanding carbon cycling, as well as in the investigation of paleoclimatic events,[15] and NMR can even be applied to the characterization of uncommon samples, such as pore water and samples collected from ice cores.[15] For instance, pores in sea ice have been studied using pulse gradient spin echo experiments to quantify sea ice brine. This was done by examining protons in the liquid phase inside ice pores, allowing an improved understanding of the structure of brine pores.[18] Additional investigations into DOM have been applied to glacial ice.[13,15,19] Looking at unaltered samples collected from glaciers in various locations, NMR spectroscopy was used to characterize DOM. Over 90% of the DOM in these complex samples was characterized by NMR.[13] Simple ^{1}H NMR experiments found that the majority of the DOM in glacial ice is composed of small molecules, which are likely microbially derived.[19]

In addition to characterization studies, NMR provides useful information for research focused on identifying the origins of a specific compound in the environment. For example, the humic fractions of sinking particulate matter collected from two different locations were characterized by NMR. Despite one site being fed primarily by terrestrial sources and the other by aquatic sources, solid state NMR experiments showed significant similarities in their organic matter compositions.[20] The samples used in this study were collected from the Saguenay Fjord and the St. Lawrence Estuary, both at depths of 20 m. In both cases, NMR characterization indicated that sinking particulate matter is primarily composed of peptides, aliphatics, aromatics, and carbohydrates. This information allowed the authors to conclude that, despite the differences in the origins of the sinking particulate matter, bacterial activity was a significant contributor to both.[20]

A great deal of work has been done using NMR to identify and quantify contaminants present in water originating from different sources, including industrial, agricultural, and domestic.[14,21] Following a spill of perfluorinated surfactants in Etobicoke Creek (Toronto, Ontario), NMR was used to determine the concentrations of surfactants in the contaminated water.[22] In this study, ^{19}F NMR experiments were used for the analysis of the water and calibration curves were used to determine the total concentrations of perfluorinated surfactants present. Results indicated that the concentrations ranged from below the limit of detection (10 $\mu g L^{-1}$) to 17 000 $\mu g L^{-1}$. In recent work, the same group used noise reduction algorithms to improve ^{19}F NMR sensitivity, permitting detection limits down to 389 $pg L^{-1}$ in rainwater.[23]

Chemical fingerprinting is one common method by which the origins of organic matter can be identified. This principle has been applied to groundwater DOM, wherein a nitrate-contaminated groundwater sample collected from the Chino Basin dairy area was studied.[24] The results of NMR fingerprinting allowed researchers to conclude that nitrate contamination originated primarily from dairy waste disposal rather than the application of N-containing fertilizers.[24] Similarly, research aiming to identify contaminants that originate from dairy wastewater is significant in improving the sustainability of the agri-food sector.[21] 2D NMR experiments can be used to provide additional information about the metabolomic composition of dairy wastewater. However, simple 1D NMR experiments have proven sufficient for the characterization of much of the organic matter present in these samples.[21] Using NMR spectroscopy in addition to more conventional methods, such as total organic carbon (TOC) and chemical oxygen demand (COD), can provide a more complete picture of the total organic matter present, in that conventional methods alone cannot account for the contributions of the volatile components of organic matter due to the requirement for sample processing.[21] Information gained by applying NMR to samples of this nature can allow effective recommendations regarding wastewater recycling.[21]

In addition to experiments that identify and quantify specific components of wastewater samples, NMR has been used for the investigation of different processes, such as wastewater treatment, reaction monitoring, and contaminant interactions,[14] all of which will be discussed further in the following sections. An extensive review of the work that has been done on wastewater from different sources has been written by Anaraki *et al.*[14] This review describes investigations into the various treatment phases and includes discussions of different NMR experiments that can be used to identify previously unknown compounds. Additionally, techniques that can be used to avoid issues such as line broadening, which become a concern in complex samples like wastewater, are discussed.[14] Interested parties are encouraged to refer to this review for additional information on wastewater research done by NMR.

2.2.2 Soil

Soil chemistry is another area of environmental research in which NMR is highly applicable. Topics such as soil organic matter (SOM) composition, microbial processes, and the implications of various agricultural practices continue to be investigated in this way.

Agricultural practices can have a significant impact on the health and composition of soils. As such, research into the effects of these practices can provide valuable information that contributes to improving sustainable farming. One specific approach is to investigate the long-term effects that fertilizer application can have on soil organic matter (SOM).[25] This is especially important since SOM is a significant contributor to the global carbon cycle, as well as to the overall fertility of soil. For investigations of this nature, ^{13}C NMR is ideally suited. Using ^{13}C CP-MAS NMR, the major components of soil organic carbon (SOC) present in different soils can be identified, and the effects of fertilizer applications determined.[25] One study found that the long-term application of fertilizers on agricultural soil can result in reduced activity of lignin-degrading enzymes.[25] However, the findings of the same study suggest that, in some cases, fertilizer applications may result in reduced SOC mineralization and therefore an increase in the stability of SOC, especially if the applied fertilizer is inorganic.[25]

Additionally, NMR has served as a useful tool in determining the extent of microbial contributions to SOM.[26] One investigation looked specifically at prairie and mixed aspen-dominated forest soils. The results of diffusion-edited NMR spectroscopy indicate that more than 50% of the total high molecular weight species in the A and B horizons can be attributed to microbial biomass.[26] Based on the C/N ratio found by NMR, it was determined that over 80% of the total organic N in the humic fraction of soil may be microbially derived. This suggests that previous estimates may have significantly underestimated microbial contributions to SOM.[26]

One area of interest in recent years has been the identification and quantification of humic substances in soils.[27–29] Before NMR had been applied to soil analysis, it had not been possible to identify one of the major components of soil. This component had traditionally been referred to as ‘humic substances’ and was believed to be composed of complex compounds of high molecular weight. However, the introduction of 2D NMR experiments to the analysis of SOM ultimately allowed the identification of this significant soil component. It was determined that what had previously been thought of as complicated macromolecules was actually a complex mixture of plant and microbially derived biopolymers that are present in various stages of decomposition.[28,29] Later, it was discovered that this component of soil is primarily composed of carbohydrates, peptides, aliphatic species, peptidoglycan, and lignin.[29]

NMR also represents an ideal tool for monitoring how complex systems, such as soil, are altered by changes in the environment. Climate change, for example, is predicted to have a significant effect on soil composition. Studies on soil warming provide insight into how soil will be affected by the predicted changes in global temperatures. One such study used soils from a mixed temperature forest to identify changes in SOM composition with rising temperature. The analysis of the warmed soil by NMR indicated an increase in alkyl groups, which authors suggest is likely from plant cuticles in the soil.[30] This was accompanied by a decrease in methoxy groups. Despite increases in the input of plant matter, carbohydrate concentrations decreased, likely due to increased degradation rates.[30] The results of this study indicate that future changes in SOM with increasing temperature will likely result from changes in decomposition rates.[30]

2.2.3 Atmospheric

As previously mentioned, improvements in water suppression techniques and overall NMR sensitivity have allowed the analysis of compounds present at very low concentrations. This has opened up opportunities for the analysis of atmospheric samples, an area of interest due to the potential impacts that air pollutants can have on both human and environmental health.[31,32]

Similar to the analysis of rainwater discussed above, information about the organic content in the atmosphere can be investigated by analyzing aerosols or atmospheric water. One study looked at fog water, which was collected from Fresno, California[33] and lyophilized to pre-concentrate the samples prior to analysis by ^{13}C CP-MAS NMR. It was found that the most prominent components in the fog water collected in 2003 and 2004 were volatiles and hydrophilic acids. Additionally, significant contributions from biological sources, such as proteins and lipids derived from bacteria, were found.[33]

To further examine the fate of volatile and semi-volatile organic compounds in the environment, work has been done to examine the composition of urban films. These thin films form on surfaces exposed to the environment as primary and secondary products of emissions accumulate.[34,35] Simpson *et al.* collected urban film samples from Toronto, Ontario and used a variety of NMR techniques to examine the compounds present in various phases.[35] This analysis found that carbohydrates and aliphatics made up a major portion of the total carbon content present in urban films, with each contributing about 35% of the total carbon content. In addition to these groups, aromatics were found to make up approximately 20% of the total carbon content, with the remaining 10% from carbonyl groups.[35] In addition, polybutadiene was found (up to a few percent by weight of the total organic material), likely derived from car tire residues in the heavily urbanized environment. This study clearly demonstrates the ability of NMR to characterize complex samples containing volatile, semi-volatile and polymeric materials.

Related research into the composition of DOM in atmospheric aerosols has used 1D and 2D NMR experiments to identify various organic compounds of low molecular weight in addition to long-chain acids and mono- and dicarboxylic acids.[31,36] A comparison of aerosols collected near Aveiro, Portugal during both the spring and winter seasons showed changes in the composition of DOM between the different seasons.[36] Methoxyphenols that originated as lignin were observed in the winter samples, but not in the spring.[36] A second study examined aerosols collected in pastures and rainforests in Rondônia, Brazil during various seasons. Using ^{1}H NMR experiments, it was possible to identify various functional groups, including aliphatics, carboxylic acids, and aromatic groups.[37] A similar investigation into quantifying the components of urban atmospheric aerosols by Duarte *et al.* detected several O- and N-containing aliphatic groups that are suggested to have originated from the burning of fossil fuels. The stage of decomposition observed for these compounds suggests that they were transported over a long distance.[32] The results of these three studies demonstrate how NMR can be used to characterize atmospheric DOM and identify the origins of atmospheric contaminants based on their NMR fingerprints.[32,36,37]

Samples originating from biological sources represent an additional area of interest in terms of atmosphere-based NMR studies. In 2016 and 2017, air samples collected in Munich were analyzed by NMR and used to create algorithms for pollen count predictions.[38] By determining the concentrations of metabolites present in air samples, it was possible to develop algorithms to identify pollen types. This was done with accuracies of 74%, 82% and 93% for tree, grass and weed pollens, respectively. These results indicate that NMR can be used successfully in the determination of pollen counts and can provide valuable information regarding pollen types.[38]

Similarly, aerosols generated by algal blooms have been investigated by NMR for the purpose of identifying the toxins present as well as for comparing differences between phases.[39] The composition of these aerosols was especially interesting due to the known impacts that algal blooms can have on human health. The authors of this study used ^{1}H NMR spectroscopy to show that the composition of the organic matter in the water suspensions was the same as that in the aerosols generated.[39] These results indicate that there is a selective transfer of organic compounds between marine suspensions and atmospheric environments.[39]

2.2.4 Partitioning

In addition to the investigations intended to identify, quantify, or compare pollutants in different states, NMR has proved very useful for studying interactions, sorption, and partitioning mechanisms, to which techniques such as HR-MAS NMR, sometimes called "gel phase NMR", are commonly applied. This technique works based on the principle that components that are swollen in their natural state or by a solvent that is added can be visualized by NMR. The approach uses magic angle spinning[40] to narrow the NMR lineshape, while the solvent introduces enough molecular dynamics to permit high resolution ^{1}H NMR experiments on soluble, swollen materials or materials at the liquid interface. As such, this has proven to be very useful for analyzing interactions between various phases.

For example, some studies have examined the method of interaction between organic compounds and the mineral components of soils. A study in 2006 looked at the sorption interactions between organic matter and montmorillonite surfaces. Using HR-MAS NMR, it was found that, compared to carbohydrates and amino acids, aliphatic components are the most likely to sorb to clay surfaces.[41] The results of this study clearly illustrate how NMR spectroscopy can be used to improve the understanding of sorption mechanisms as well as elucidate the resulting organo-clay structures.[41] In some cases, research looking into binding mechanisms can be aided through the use of isotope labelling.[42]

Since HR-MAS NMR allows the analysis of processes occurring at the interface of two phases, it has been applied to fields related to improving agri-food sustainability. This is because it has the potential to study herbicide and pesticide interactions between both the solids and solutions present in soils. An example of this is a study where interactions at the solid–solution interface were observed following the application of the herbicide trifluralin. Here, fatty acids, aliphatic esters, and other O-containing groups were identified at the solid–solution interface of a soil.[43] By examining differences between the NMR spectra of the free *versus* the bound trifluralin, it was possible to understand the types of interactions present between the herbicide and the soil.[43] Further studies using saturation transfer difference (STD) NMR have been conducted to understand the binding mechanisms between agrochemicals and soil organic matter[44,45] or whole soils.[46] More recent work applied a comprehensive approach to monitor the binding, orientation and phase transfer of contaminants as they were sequestered into a whole soil.[47]

Furthermore, interactions between agriculturally relevant compounds and plant tissue can be investigated by NMR. For example, Sachleben *et al.* investigated the interactions between pyrene and cuticular matter in tomato plants.[48] Variable temperature NMR experiments were used to observe differences in plant materials, and it was found that cutin and cutan behave differently at different temperatures.[48] This observation is significant due to the implications it has on sorption properties. Additionally, 2D NMR experiments indicated that sorbed pyrene is spatially very close to aliphatics in the cutin.[48]

2.2.5 Others

Many metabolomics studies have also been performed using NMR. As a keystone species, *Daphnia magna* (freshwater fleas) are common test organisms in these types of toxicity and metabolomics studies. Their small size means that they can be placed in standard NMR tubes with room to swim around. As a result, when a flow system is used to provide oxygen and food, it is possible to keep the organisms alive for the duration of a study (*i.e.*, up to ~2 days), allowing real-time observation of metabolic changes.[13,49] These studies provide information related to the sub-lethal toxic effects of specific contaminants and provide valuable insight into the effects surrounding the release of contaminants into the environment.

Plant matter represents another biological sample that can be analyzed by NMR. Advances in NMR such as magic angle spinning (MAS) have allowed the investigation of heterogeneous samples. For instance, Miglietta and Lamanna detected carotenoids, including β-carotene, in raw vegetables.[50] The use of mobile NMR probes also opens up the opportunity for analyzing complex environmental samples. A mobile probe was used by Casieri *et al.* to determine the water content in wood samples without destroying the sample itself. Using this method, water from different parts of the plant could be separated to determine their relative contribution to the overall water content.[51] This method gave comparable results to those obtained by more traditional gravimetric methods.

2.3 Transformation Products

2.3.1 Environmental

2.3.1.1 Soil

With an ever-increasing global population, there has been growing pressure for agricultural productivity, often without due consideration of the environmental implications surrounding these practices. The complex matrices associated with the impacted environmental systems mean that studying effects and interactions with these introduced substances is not trivial. Despite this, NMR is an ideal tool for looking at these interactions as well as the degradation processes that occur. One such study looked specifically at chlordecone, a pesticide commonly used in the French West Indies.[52] While organochlorine pesticides such as this have been banned for years, they still persist in the environment.[52] As such, understanding the biodegradation of these compounds is of critical interest. While techniques such as GC–MS and LC–HRMS can provide valuable information regarding the number of different transformation products that form as chlordecone breaks down, they provide little to no insight into the exact structures of these products.[52] The detailed information NMR can provide about the connectivity of atoms in a molecule makes it ideally suited for the identification of previously unknown transformation products. Using both 1D and 2D NMR experiments, Chevallier *et al.* were able to determine the exact structures of chlordecone's transformation products following biodegradation in a laboratory setting.[52]

A similar study by Seger and Maciel looked at the degradation and adsorption properties of transformation products formed when chlorpyrifos, an organothiophosphate pesticide, breaks down in soils.[53] This study used ^{31}P NMR experiments to examine changes over time. These experiments indicate that the transformation products of chlorpyrifos sorb more strongly to clay compared to the parent compound.[53] This illustrates how heteronuclear NMR experiments can allow a simple and effective analysis of transformation processes and products as well as their interactions in complex environmental systems.

The cycling of nutrients and their transformation processes is critically important in soils. For example, soil phosphorus cycles through various organic and inorganic forms through processes in soils such as mineralization and immobilization.[54] The cycling of P through the mountain soils of the Alay Range, Kyrgyzstan, was investigated using sequential extractions and ^{31}P NMR to look at the alkali soluble forms.[54] Samples from both pastures and forests were analyzed for comparison. The results of the ^{31}P NMR analysis indicated that the forms of P in both pasture and forest samples were identical, but that concentrations were consistently higher in the pasture soils.[54] By examining the relative concentrations of soil P forms, the effects of specific processes on the nutrient's form and availability can be studied, and the effects of deforestation on soil pools can be determined by noting changes in the ratio of monoester to diester P in forest and pasture soils.[54] It was found that the monoester-to-diester ratio is much higher in pasture soils than in forest soils, suggesting that deforestation can lead to decreased labile organic P species.[54]

2.3.1.2 Water

The transformation processes of natural organic matter in both marine and terrestrial environments are of great interest for both carbon cycling and water dynamics.[55] Solid-state ^{13}C NMR studies can be used in addition to other tools, such as mixing models, to provide a comprehensive picture of the composition of organic matter and how it changes during decomposition.[56] Using the combined NMR-modeling approach as a complementary tool to more traditional methods, NMR can provide additional insight into the breakdown and changes that occur as natural organic matter degrades in both terrestrial and marine settings.[56]

Plant-derived compounds make up a significant portion of the organic matter present in both terrestrial and aquatic environments. Several studies have used NMR as a tool for improving the understanding of how these compounds break down over time and the products that form. For instance, the biomass derived from aquatic plants in Tai Lake, China was found to be primarily composed of carbohydrates, meaning that it is highly labile.[57] These results indicate that ^{13}C and ^{31}P NMR experiments can be used to better understand the forms and sources of important nutrients in aquatic systems, as well as provide additional insight into biogeochemical cycles.[57]

Additionally, NMR can be used to provide valuable information about the transformation products and processes that are involved in the breakdown of environmental contaminants. Commonly used as a secondary explosive, 2,4,6-trinitrotoluene (TNT) has been found to be a significant pollutant in both soil and groundwater and its degradation processes were therefore investigated.[58] Two known degradation processes of TNT include oxidative and reductive degradation as a result of exposure to sunlight, and these processes can occur in both solution and solid states.[59] Thorn used ^{13}C and ^{15}N NMR experiments to investigate the decomposition of TNT in aqueous environments, which led to the determination of various transformation products.[59] The results of solid-state analysis also found these transformation products, but a greater number of products were observed in the aqueous samples.[59] This provided new information related to the breakdown processes of TNT in the environment.

Other studies on the environmental impacts of TNT have been conducted using hyphenated LC–NMR systems. One specific study by Markus Godejohann[60] examined groundwater collected from a former ammunition site. Using a hyphenated LC–NMR system, it was possible to detect novel transformation products that form as a result of both photolytic and microbial degradation of TNT. The exact structures of these novel compounds were not conclusively determined; however, this study does illustrate the importance of hyphenated techniques as these compounds were not detectable by either HPLC or MS alone.[60] A second study by Godejohann looked at identifying the contaminants and transformation products present in the groundwater collected at a former ammunition destruction site.[60] The use of MS in the analysis of this complex matrix detected the presence of two unknown compounds; however, hundreds of potential compounds were proposed as a match for these unknowns. When using ^{1}H NMR, the exact structures were identified.[60] These examples go to show the importance of hyphenated NMR systems for environmental research and highlight the benefits of using NMR and MS as complementary techniques.

2.3.1.3 Atmospheric

NMR has also been a useful tool in atmospheric studies when identifying transformation products of common contaminants. This type of research has been done on fluoropolymers, which are widely used and are known to degrade when under thermal stress. The result of this process is the formation of gaseous fluorinated species that persist in the atmosphere.[61] As such, fluorinated acids and their fluoropolymer precursors have been investigated using ^{19}F NMR experiments.[61] Using both 1D and 2D ^{19}F NMR experiments, the processes involved in fluoropolymer degradation were studied, and many previously unknown by-products were identified.[61] This shows that NMR is ideally suited for the identification of compounds that are either overlooked or incorrectly assigned by other techniques.

Since coal soot is often formed as a result of the condensation of polyaromatic hydrocarbons (PAHs), many of which are known to be carcinogens, the processes involved in its formation have been the subject of investigation. Using ^{13}C NMR to analyze coal soot samples, novel information about the mechanisms involved in its formation was obtained.[62] From this analysis, it was determined that, in some cases, the aromatic ring systems in soot may contain 80 to 90 carbons, though these formation processes were found to have some dependence on temperature. Additionally, it was noted that, before a growth in cluster size, samples lost aliphatic and O-containing groups.[62]

Secondary organic aerosols (SOA) are known to pose significant risks to human health and impact climate and visibility.[63] For these reasons, investigations into the exact composition as well as mechanisms of formation have become an increasingly important area of research. In order to better understand the partitioning behaviour of SOA into the particulate phase, some studies have focused primarily on identifying the types of functional groups present.[64,65] Moretti *et al.* used ^{1}H NMR experiments to analyze for total carbonyl groups.[64] Related studies have shifted their focus to examine the mechanism of formation of SOA. Using 2D NMR experiments, Maksymiuk *et al.* were able to gain additional insight into the mechanisms involved in SOA generating reactions between ozone and gaseous limonene. By examining HSQC spectra, it was confirmed that the terminal unsaturation present in limonene is depleted when the ozone content is low.[63] A combination of 1D and 2D NMR experiments were used in this study to confirm that it is the double bond within the limonene ring that is oxidized in the gas phase when in an environment with low concentrations of NO_x, whereas this process occurs heterogeneously when there is an excess of ozone.[63]

2.3.1.4 Wastewater Treatment

Wastewater treatment is an interesting research area for investigations into the breakdown of contaminants. Using ^{13}C and ^{1}H NMR experiments, the composition of wastewater before and after treatment has been examined.[66,67] Solakyildrim *et al.* used NMR spectroscopy to examine the degradation of personal care products and pharmaceuticals during wastewater treatment with chlorinated species.[68] This study looked specifically at the transformation products of salicylic acid, naproxen, diclofenac, gemfibrozil, bisphenol A, and *tert*-octylphenol. Ultra-performance liquid chromatography coupled with quadrupole time-of-flight mass spectrometry (UPLC–QTOF-MS) was used to determine the number of chlorinated transformation products and NMR was used to elucidate the structures. This emphasized the valuable information that can be provided by NMR in terms of identifying the types of secondary contaminants that may be present in the environment.[68] In another study, Dignac *et al.* investigated the processes involved in the treatment of wastewater using activated sludge.[66] Following treatment, it was found that branched alkyl chains increased compared to long, straight chains and that polymethylene carbons were removed.[66] Both the treated and untreated water samples were found to be dominated by carbohydrates, but differences between the treated and untreated water samples were not significant.[66] In contrast, Piterina *et al.* examined the changes that occur when domestic sludge is treated with autothermal aerobic digestion. In this case, the NMR results showed that long-chain aliphatics were efficiently removed by microbial populations and an increase in O-alkyl carbons was observed.[67] In addition to these findings, the authors were able to observe the different stages of decomposition and the changes that occurred in each stage.[67]

2.3.1.5 Miscellaneous

Metabolomics studies can also be used to examine the transformation of nutrients in living systems. Using small organisms, such as aquatic bacteria, as the subject of study means that *in vivo* NMR can be used to analyze real-time metabolomic changes. One such study found that the breakdown of glucose by bacteria resulted in the formation of twenty-two different transformation products falling into the general groups of amino acids, dipeptides, organic acids, nucleosides, nucleobases, and amino alcohols.[69] Investigations of this nature typically involve isotope labelling and are not limited to bacterial populations.[13,49,69,70]

Many pollutants can be introduced into the environment *via* wastewater if the treatment processes do not sufficiently remove them. One such compound is carbamazepine (CBZ), a pharmaceutical sometimes used to treat epilepsy. Since CBZ does not adsorb to the sludge used in wastewater treatment processes, it is not removed and is therefore released into the environment, where it can undergo photo-transformations to form epoxides and hydroxylation products.[71] Since effective wastewater treatment involves the removal of both the parent product and its transformation products, it is essential to understand how pollutants will transform. For this reason, Kråkstrom *et al.* looked at the process and products of CBZ ozonation. While the majority of this study was done using MS, uncertainties remained in the actual products formed through these processes. As such, ^{1}H and ^{13}C NMR experiments were used to piece together the exact structures of the products formed.[71] In this way, a never-before documented transformation product of CBZ was identified as 2,2′-azanediyldibenzaldehyde.[71] These results show that when NMR and MS are used as complementary tools, the identification of novel compounds is possible.

2.3.2 Reaction Monitoring

In addition to applications in the identification of novel compounds, NMR has the potential to be used in reaction monitoring. One major area of interest for this is in tracking the progress of catalytic cycles. NMR has been used to identify products, examine mechanisms and even monitor reaction kinetics.[72–77] Some studies have examined catalyst activation processes using NMR to identify the mechanisms involved in transformations between pre-catalysts, active catalysts, and resting forms.[74,76,78] Others have applied *in situ* NMR to examine the exact mechanism of the catalytic cycle as well as identify the products formed.[76] Flow systems can be used to allow real-time analysis of the changes that occur throughout a catalytic cycle,[79,80] or for systems with very fast reaction rates, stopped-flow systems can be used to provide insight into the intermediates and allow analysis of the kinetics involved.[73] Another option for the identification of intermediates in a catalytic cycle is to use a DOSY experiment at low temperatures as this will separate the smaller intermediate compounds from those that are bound to the catalyst itself.[80] The application of NMR to catalytic systems opens up the opportunity to improve the efficiency and sustainability of catalytic processes and demonstrates how NMR can be applied to complex mixtures such as those found in the environment.

2.4 Nontargeted Environmental Metabolomics

Metabolomics is a rapidly growing field focused on the comprehensive analysis of endogenous and exogenous low molecular mass metabolites in biological systems. Metabolomics studies are broadly categorized into two categories: targeted analysis (*i.e.*, the investigation of defined groups of metabolites) and nontargeted analysis (*i.e.*, characterization of metabolomes in a non-biased manner).[81,82] The primary analytical techniques used in metabolomics studies are NMR, LC–MS and GC–MS, each with their own advantages and limitations; the enhanced sensitivity and selectivity of MS make it ideal for targeted metabolomics studies, especially when working with low concentrations, but the non-selective nature of NMR makes it highly advantageous in nontargeted and compound discovery experiments.

Due to its ability to analyze samples in a non-destructive, non-invasive, and non-selective manner, NMR can be applied to *in vivo* and *in vitro* studies on cells, tissues, and entire organisms.[81] In fact, NMR can provide real-time *in vivo* data on whole, intact samples. Heteronuclear NMR experiments can also be used to detect various biologically relevant compounds, which expands the overall coverage of the metabolome by permitting the analysis of different metabolite classes. Ultimately, however, what makes NMR a standout discovery tool is its ability to identify novel compounds without requiring the use of a database. While databases can be useful complementary tools, they lack the ability to identify novel metabolites and transformation products.

2.4.1 Aquatic Organisms

Discovering and understanding biochemicals and their roles in complex biological pathways is a challenging and daunting task that multi-dimensional NMR experiments can help solve. Nontargeted metabolomics studies involving aquatic organisms are critical for many research areas, including toxicology, ecological monitoring and human disease modeling.[42,83] Green algae and *Daphnia magna* are amongst the most common models in environmental toxicology research as they both play essential roles in maintaining balance within aquatic ecosystems.[84,85] These types of toxicology studies often involve exposing the organisms to sub-lethal concentrations of contaminants and monitoring the subsequent changes in their metabolic profiles.[86] For example, one study used NMR to investigate the effects of triclosan, carbamazepine, and ibuprofen on the metabolomes of *D. magna*, and discovered novel bioindicators which could significantly improve the efficiency and reliability of future toxicity testing.[87] Similar studies have looked at the effect of perfluorooctane sulfonate (PFOS), a common manufacturing additive and notorious pollutant, on the metabolome of *D. magna*,[86] finding that toxic effects were heavily dependent on PFOS concentrations. Moreover, Kovacevic *et al.* monitored the metabolic responses of *D. magna* when exposed to contaminants in either the presence or absence of DOM,[88] finding that the impact of DOM on toxicity depends on the contaminant. Overall, these studies illustrate how NMR-based metabolomics can be used as a non-biased discovery tool.

Spectral overlap can be especially problematic for 1D NMR spectra of complex environmental samples, such as those used in *in vivo* studies.[89] A combination of 1D and multidimensional NMR experiments can greatly increase resolution and enable unambiguous assignments of metabolites. A recent study by Anaraki *et al.* showed how 1D, 2D and 3D ^{1}H and ^{13}C NMR *in vivo* and *ex vivo* experiments could be used to assign 60 metabolites in *D. magna.*[89] An *in vivo* HSQC showing the metabolomic assignments of ^{13}C-labelled *D. magna* is shown in Figure 2.1.

In addition to *D. magna*, research by Akhter *et al.* shows how solution-state NMR can help characterize the metabolic profiles of aquatic vegetation such as whole intact algal cells.[90] The authors successfully identified 43 and 49 metabolites in the species *Chlamydomonas reinhardtii* and *Chlamydomonas vulgaris*, respectively. Solution-state NMR is ideal for identifying metabolites in algae since most biologically relevant metabolites are dissolved within cells' cytoplasm. Solid-state NMR and HR-MAS can provide complementary data to create a more comprehensive metabolic profile.[91–93] Metabolomics studies like those discussed here have provided insight into the impacts of toxins such as heavy metals on algal growth.[84]

One of the exceptional benefits of NMR is its ability to analyze samples in various phases. However, analysis of individual phases may not necessarily reflect the true nature of a system, making comprehensive multiphase (CMP) NMR an exceptionally powerful discovery tool. CMP NMR has the ability to study all phases (*i.e.*, solutions, gels/swollen materials and true solids) in heterogeneous systems.[94] This enables *in vivo* studies on invertebrates where not only can the metabolites be studied, but all the swollen materials (tissue) and solid material (exoskeleton and shell) can be analyzed as well.[94]

Although NMR is a powerful discovery tool, its ability to carry out *in vivo* experiments on living organisms is limited to relatively small organisms (at least when vertical bore NMR spectrometers are considered, as opposed to horizontal MRIs). Organisms such as *D. magna* are ideal for 5 mm NMR probes as they often reach a size of only 1–3 mm. Copper toxicity in large abalone was studied *in vivo* using a horizontal MRI system in early work,[95] but in general organisms such as fish are too large to fit inside vertical NMR systems. As such their metabolomics studies are commonly restricted to biofluids (urine, plasma, *etc.*). For instance, Tsentalovich *et al.* analyzed eleven tissue samples from freshwater zander fish (*Sander lucioperca*) organs and assigned over 65 metabolites.[96] Nontargeted NMR-based metabolomics of fish has actually been a focus of marine and freshwater research for many

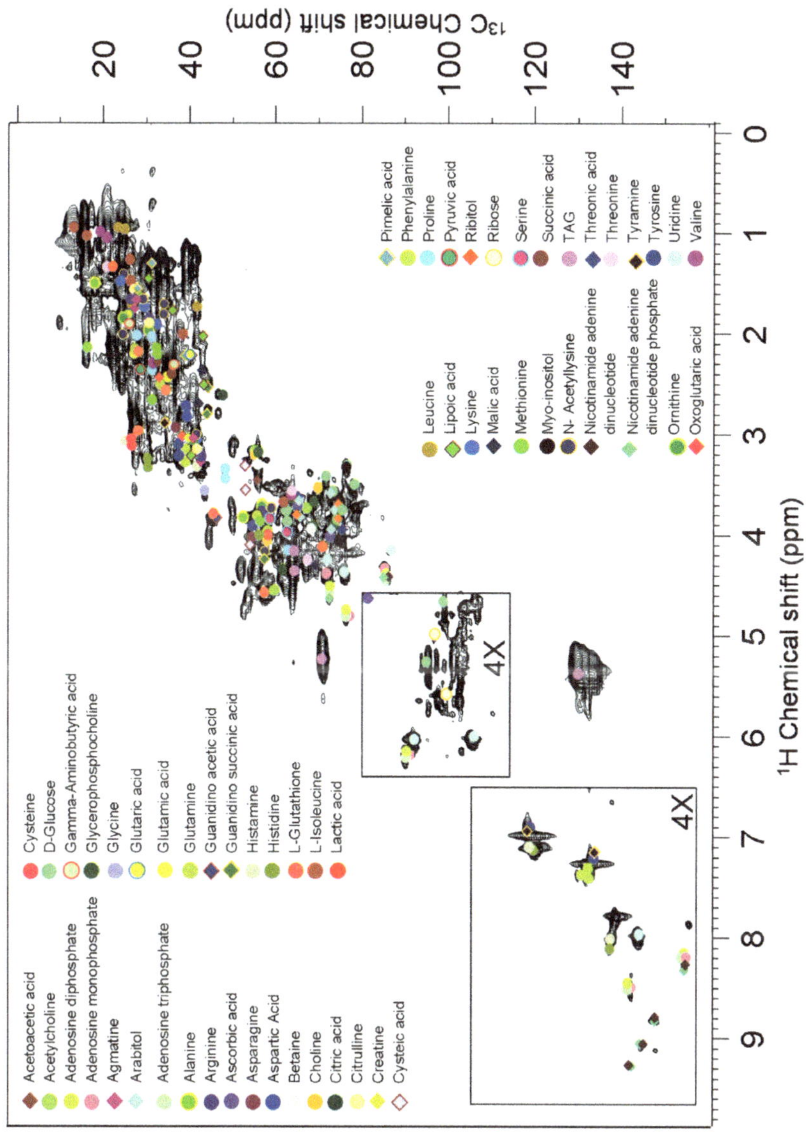

Figure 2.1 HSQC spectrum of ^{13}C-labelled *D. magna* (*in vivo*) with metabolite assignments. Reproduced from ref. 89 with permission from the Royal Society of Chemistry.

years; some species of particular interest include salmon, fathead minnows, trout, and zebrafish.[97–100] Since fish play a vital role not only in aquatic ecosystems, but also in global and local economies, it is crucially important to understand how contaminants impact their health, growth and reproductive systems. To date, NMR has been widely employed to characterize fish metabolomes after exposure to contaminants such as microplastics, pharmaceuticals and synthetic hormones.[101–103] A particularly interesting study by Martinovic-Weigelt *et al.* even identified the metabolites responsible for territorial male behaviours, demonstrating the ability of NMR metabolomics to help understand behavioural changes in stressed organisms.[104] Other research has investigated the impact of feed composition on farm-raised rainbow trout, concluding that plant-based diets may cause an imbalance in essential amino acids.[99]

2.4.2 Terrestrial Vegetation

Plants require highly complex signaling pathways to adapt to their ever-changing environments, and although many plants share biochemical and cellular similarities, there is a great deal of diversity within the kingdom. Therefore, plant metabolomics strives to better understand the various ways by which plants adapt to external stressors at the molecular level. With ongoing food shortages, climate change, and the arms race between plants and pathogens, NMR-based plant metabolomics is of particular interest.

Research often focuses on enhancing pathogen and drought resistance, as well as nutritional value *via* genetic modification (GM) of crops; NMR is ideally suited to help differentiate between original and modified organisms. A study by Le Gall *et al.* compared the metabolomes of GM and nonmodified tomatoes, and ^{1}H NMR results showed over 20 metabolic changes.[105] Such research demonstrates that NMR has the potential to identify the unintended effects that GM can have on a crop. This is a critically important feat in that the genetic modification of crops is a growing field. Similar experiments have been conducted on other crops, such as lettuce, rice, peas, and potatoes,[106–109] all of which conclude that NMR can identify minute metabolome differences between GM and non-GM crops. Other studies have investigated the impacts of drought and shifting water regimes on various crops,[110] many of which demonstrated that various crops can alter their metabolic pathways in response to differing soil moisture conditions. For example, research by Mdlalose *et al.* showed that potatoes grown under extreme conditions (drought, flood) exhibit elevated levels of valine and leucine,[110] which are both essential for plant defense and pathogen resistance. Furthermore, NMR has also been used to investigate secondary metabolites, which are not directly involved in growth or development but instead often aid in pathogen resistance.[111] For instance, research by Copley *et al.* showed how ^{1}H NMR can be utilized to discover soybean metabolic fluctuations in response to the *Rhizoctonia solani* AG1-IA phytopathogen at different life stages.[112] These results helped identify candidate genes that could be used to make future GM soybeans more pathogen resistant. Similar experiments have also been conducted on other crops (*e.g.*, lemons) to improve agricultural sustainability.[113]

Lastly, CMP NMR has been applied to monitor structural and metabolic changes, as well as molecular flux, within developing wheat seeds.[114] This experiment perfectly demonstrated the power of NMR as a discovery tool, since a wide range of metabolites were identified in the 1D and 2D spectra; research like this provides a foundation for better understanding crop germination and growth. In other work, CMP NMR was used to study the genetic manipulation in *Arabidopsis*,[115] and NMR has also been used for many years to better understand how external conditions impact plant growth, such as how changes in plant growth regulators affect strawberry metabolomes.[116]

2.5 Food Science

Although chromatography–mass spectrometry methods are mainstream analytical techniques used in food science and quality control, NMR spectroscopy is becoming a more commonplace tool for food analysis. Its nontargeted nature and simple sample preparation make it ideal for direct analysis of food and beverage samples. To date, it has been used in the food industry for safety assessment (*i.e.*, identifying adulterants), investigating potential contamination of food from packaging, and in food fraud and authentication, thereby becoming a well-established field in its own right.

2.5.1 Food Safety

One of the most high-profile food safety emergencies was that of the 2008 melamine crisis in China, wherein infant milk powders were adulterated with melamine. Since melamine is high in nitrogen, adding it to the product decreased the production cost while increasing the apparent amount of protein (determined by N content) present in the diluted powders. However, melamine posed a serious health hazard and caused kidney problems in nearly 300 000 children, with effects being fatal to some.[117] Since analysis of milk was based mainly on targeted techniques, adulteration with melamine went unnoticed for many months. However, Lachenmeier *et al.* show that nontargeted NMR analysis of milk would have enabled immediate identification of melamine and argue that NMR should play a larger role in food quality assessment to prevent or minimize a similar crisis from occurring in the future.[117] Figure 2.2 shows the ^{1}H NMR spectrum of a melamine-contaminated sample compared to reference milk samples, illustrating the ease with which NMR can identify the abnormal presence of melamine (*i.e.*, the peak at 5.93 ppm). Lachenmeier *et al.* also found that melamine was quantifiable down to the low parts-per-million range.[117]

Contaminant discovery using NMR has been equally effective in nontargeted analysis of poorly monitored or unregulated consumables and

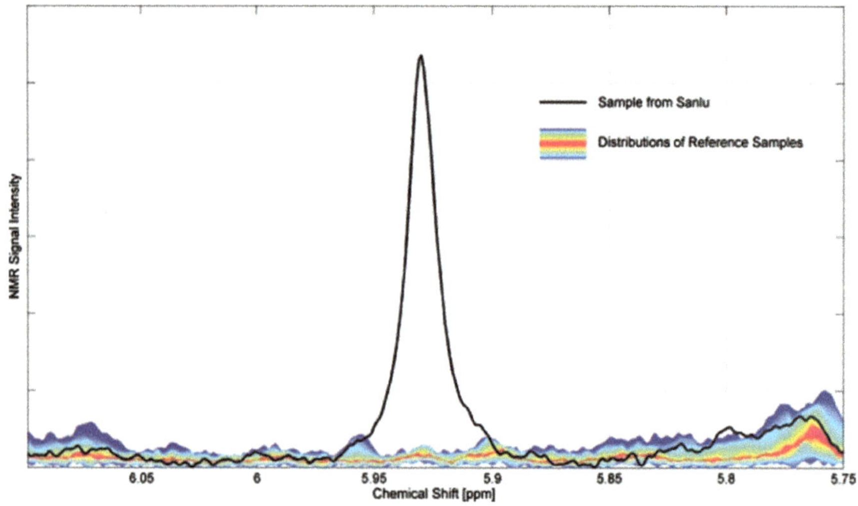

Figure 2.2 The ^{1}H NMR spectra of reference milk powder samples as well as the melamine-contaminated sample (black spectrum), showing the characteristic melamine peak at ~5.93 ppm. Reproduced from ref. 117 with permission from American Chemical Society, Copyright 2009.

supplements. For instance, Monakhova *et al.* used ^{1}H NMR and principal component analysis (PCA) to detect and identify hazardous compounds in both commercial and unregulated alcoholic beverages from 13 different countries.[118] Using both water and ethanol suppression techniques, hazardous compounds such as diethyl phthalate and polyhexamethylene guanidine, as well as dangerously high levels of methanol and ethyl carbamate were identified in a handful of samples.[118] This study thereby highlights the need for nontargeted NMR analysis, as new or unknown adulterants in beverages could easily go undetected using traditional, targeted analysis.

Moreover, supplements in the diet industry are often poorly regulated and can pose serious health hazards, but NMR could play a valuable role in nontargeted analysis and safety assessment. Yun *et al.* analyzed an illegal weight loss supplement and discovered a novel compound (now termed chlorosipentramine) that was analogous to a common weight loss drug.[119] Little information was known regarding its potential side effects, but other analogues had shown to be addictive and cause cardio-excitatory and psychostimulatory issues.[119] The unknown compound was isolated and analyzed by HPLC, LC–MS, IR, and NMR. NMR spectroscopy, however, offered the most information regarding structure elucidation, and using ^{1}H and ^{13}C 1D and 2D (DEPT, COSY, HMBC, and HSQC) experiments, they proposed a chemical structure. To confirm the unknown's identity, they synthesized the proposed compound and compared its NMR spectra to those of the unknown.[119] This ultimately demonstrates how NMR can aid in discovering novel, potentially dangerous, compounds in poorly regulated supplements.

Similarly, Moon Han *et al.* used NMR to help discover a novel phosphodiesterase-5 inhibitor (propoxyphenyl thiosildenafil) in an herbal supplement for male sexual enhancement.[120] This compound has not been studied thoroughly and the safety and health risks are therefore unknown. In another study, Häberli *et al.* used NMR to characterize an unknown compound in an illegal erectile dysfunction drug.[121] The authors do not believe that the unknown compound was intentionally added, but instead was likely a result of drug incompatibility and an unanticipated reaction between the drug's contents and active pharmaceutical ingredient.[121] Since these drugs are not regulated, they could pose serious health risks to consumers, but NMR could be used to investigate their safety in a nontargeted approach.

Furthermore, NMR was used to identify three unknown chromatophores in a Japanese red food dye, one of which is not considered an "acceptable" food dye in the country and another being a product from an unanticipated chemical reaction within the dye.[122] In addition, Nishiyama *et al.* used NMR to identify an isomer of sorbic acid (*i.e.*, a common food preservative) in products containing sorbic acid.[123] Prior to this, HPLC data had shown an unknown peak in the chromatograms, but the compound corresponding to this peak could only be identified using NMR (in conjunction with GC–MS and GC–FTIR).

Concerns related to food safety extend beyond poor regulation of ingredients, as food contamination from packaging can pose significant risks to human health. For instance, Kawamura *et al.* used NMR spectroscopy to identify 13 compounds in polystyrene packaging that could leach into the food they contain and pose health risks to the consumer.[124] Similarly, the plastic lining inside metal food cans can pose health risks if compounds leach from the plastic into the food. Since the traditional coating material bisphenol A (BPA) is a suspected endocrine disruptor, there has been a push for development of safer polymeric linings.[125] However, prior to widespread implementation of a new coating, it must be rigorously tested for possible leaching of compounds into foodstuffs. Therefore, Paseiro-Cerrato *et al.* investigated leaching from polyester coatings and found that 29 different compounds could migrate from the lining.[125] Although the presence of unknowns was detected using chromatography–mass spectrometry systems, unambiguous identification and characterization of the unknowns could only be achieved using NMR spectroscopy (^{1}H and ^{13}C NMR). Ultimately, NMR spectroscopy plays a vital role in the discovery of potentially hazardous compounds in foodstuffs. Its unparalleled structure elucidation capacities, as well as its nontargeted nature and lack of sample preparation requirements, make it an ideal tool for food safety assessment and contaminant discovery.

2.5.2 Food Authentication

In addition to its role in food safety assessment, NMR spectroscopy has also been instrumental in detecting and identifying adulteration of higher value products with those of lower value. For instance, nontargeted 1D and 2D ^{1}H and ^{13}C NMR have been used to investigate adulteration of goat milk (high value) with bovine and soy milks (lower value).[126] This was done by using the different metabolic profiles of the individual milk products to differentiate between pure and adulterated milks.[126] Similarly, paprika (an expensive spice) is often adulterated with lower value compounds. Using ^{1}H NMR, Horn *et al.* analyzed paprikas (extracted in methanol) and found azorubine, ponceau 4R, and beetroot and sumac powders in several samples.[127] Vegetable oils are another commonly adulterated product, and Nascimento *et al.* used ^{1}H and ^{13}C NMR to investigate the purity of 23 oils from Brazilian shops by comparing their NMR spectra to those of pure oils.[128] They found that NMR was particularly adept at identifying occurrences of omega-3-spiked oils. Furthermore, as honey is becoming increasingly scarce due to climate change and decreases in honeybee populations, honey supplies are dwindling and it is thus becoming more valuable.[129,130] Honey is consequently becoming prone to adulteration, and Cagliani *et al.* used 1D ^{1}H, TOCSY, and HSQC NMR to identify adulterants such as inulin, corn syrup, malt syrup, and invert sugar.[129] This illustrates how NMR could play a critical role in discovering adulteration in honey screening and quality assurance. The adulteration of fruit juices, though generally considered a less valuable product, can be done easily and remains a practice that often goes unnoticed by targeted analysis. However, using 1D ^{1}H and JRES NMR, Spraul *et al.* found that adulteration of these products can be detected by comparing experimental spectra to databases of pure/regulated juices.[131] Lastly, coffee beans and grounds are susceptible to food fraud, so Burton *et al.* used quantitative ^{1}H NMR to identify 12 compounds present in coffee beans to differentiate between 100% arabica blends and those adulterated with lower value robusta beans.[132] Although many of the above examples of food adulteration use comparison with spectral databases, NMR still provides a more concrete investigation of food adulteration, offering the ability to elucidate the structures of unknown/unanticipated compounds if desired. Structure elucidation can become time-consuming, and for routine analysis in the food industry, simply detecting adulteration is often all that is required for food fraud evaluations. However, when outlier compounds arise or pose safety risks, NMR offers unparalleled structure elucidation capabilities.

The last major discovery application of NMR in food science is in food authentication and confirming the origins of a given food. This is most common in wine analysis and verifying the geographic origin, grape variety, and age of the wine, as they vary considerably in price.[133,134] Godelmann *et al.* used nontargeted 1D ^{1}H, JRES, and NOESY NMR (with water and ethanol suppression) to fingerprint 600 samples of wine from different regions in Germany.[134] They were able to identify methanol, organic acids, acetone, and amino acids, and used this information to determine the vineyard location, grape type, and age of the wines with near or above 90% accuracy in these three categories.[134] Similarly, Palmioli *et al.* used 1D ^{1}H,

TOCSY, and HSQC NMR to differentiate between craft and industrial beer varieties based on their metabolite profiles.[135] NMR in food authentication has equally been used for analysis of produce. Klare *et al.* used nontargeted 1D ^{1}H NMR to differentiate between and trace the geographic origins of asparagus from 5 countries and 13 regions in Germany.[136] By identifying marker compounds in the different samples, they were able to accurately predict the geographic origin of 90% of the samples.[136] Similarly, Dimitrakopoulou *et al.* used 1D ^{1}H and JRES NMR to differentiate between currants grown in 5 different regions of Greece based on nontargeted analysis of their metabolites.[137] Greek olives are also differentiable based on where they were grown in Greece, as shown by Beteinakis *et al.* who analyzed olive extracts in various solvents with ^{1}H NMR to determine their region of cultivation.[138] Lastly, Longobardi *et al.* used 1D ^{1}H NMR to determine whether lentil samples came from Canadian or Italian crops by comparing the experimental spectra of lentil samples to the spectra of known Canadian or Italian lentils.[139] Ultimately, these examples highlight the unparalleled strength of NMR spectroscopy as a tool in food authentication and discovery of food metabolites characteristic of a particular place of origin.

2.5.3 Other Applications

Lastly, NMR spectroscopy has been utilized in a handful of other contexts, such as how different foods affect the human body. For example, pulses (*i.e.*, chickpeas, lentils, *etc.*) have time and again shown to have positive effects on overall human health and longevity.[140] However, the biological reason behind these health benefits has not been well-characterized to date. Therefore, Madrid-Gambin *et al.* used 1D ^{1}H and HSQC NMR to identify potential biomarkers of pulse consumption in urine and serum samples as a first step in helping deconvolute the effects of pulses on biological processes.[140] Upon comparison with spectral databases, they found metabolites such as trigonelline, 3-methylhistidine, and dimethylglycine, among others to be promising biomarkers.[140] Although this was but a starting point, identification of metabolite markers could not have been achieved as confidently or as simply with other analytical instrumentation. Köbler *et al.* used nontargeted ^{1}H and ^{13}C NMR to investigate potential biomarker compounds in pine nuts that cause pine nut syndrome (*i.e.*, a phenomenon wherein the sense of taste is altered after consuming certain pine nuts).[141] The cause of pine nut syndrome has been the topic of much debate over the past couple of decades, but researchers have been unable to determine the compound(s) responsible for this effect.[141] Although Köbler *et al.* were also unable to identify a specific compound in the pine nuts, the developed NMR protocol is robust and could be used in future studies.

Ultimately, NMR spectroscopy has proven to be an irreplaceable tool in food science, as its nontargeted nature has been of use in food safety assessments (*i.e.*, in identifying unknown contaminants in foods), in food fraud and authentication, as well as in food metabolomics. In addition to its structure elucidation abilities, NMR spectroscopy requires little sample preparation, resulting in higher throughput, and produces little waste, making it a more informative, rapid, and environmentally-friendly analytical tool in food science.

2.6 Heteronuclear NMR

Due to generally broad lineshape and poor resolution, NMR of nuclei other than ^{1}H and ^{13}C is not often used for structure elucidation or for contaminant discovery. However, there are certain nuclei and contexts in which heteronuclear NMR has shown to be an informative and complementary tool in compound discovery.

2.6.1 Fluorine (^{19}F)

The second-most sensitive nucleus (after ^{1}H, ignoring ^{3}H which is radioactive) ^{19}F is spin $\frac{1}{2}$ and is 100% abundant, making it ideal to study by NMR.[14] Interestingly, it is estimated that ~20% of pharmaceuticals contain fluorine[142] and perfluorinated chemicals often used as surfactants, firefighting foams, and water proofing and non-stick agents are ubiquitous in the environment and are linked to harmful health effects in humans and animals.[143]

Tront and Saunders applied ^{19}F NMR to track and monitor the fate and transformation of 4-chloro-2-fluorophenol in duckweed.[144] To do this, duckweed was exposed to varying concentrations of halogenated phenol and solution-state ^{19}F NMR was used not only to monitor the amount absorbed, but also to identify metabolites in the duckweed.[144] This demonstrates that ^{19}F NMR could play an important role in environmental modeling of the fate, transformation, and bioavailability of fluorinated contaminants like perfluoroalkyl and polyfluoroalkyl pollutants. In related work, it was found that PFCs bound irreversibly to human serum albumin in human blood and bound so strongly that fatty acids and other drugs could no longer bind.[145]

In addition, ^{19}F NMR has been used to characterize novel reaction mechanisms and products. For instance, metal fluorides (MgF_3^- or AlF_4^-) have been used as substitute ions (*i.e.* "labels") for the PO_3^- ion in phosphoryl transfer reactions to study and characterize transition state geometry.[146] Jin *et al.* used this method and solution-state ^{19}F NMR to identify the geometry of transition state protein complexes (trigonal bipyramidal in their case).[146] Moreover, Saloutin *et al.* used ^{19}F NMR (as well as ^{1}H NMR and mass spectrometry) to elucidate the structures of products from the chemical reaction of perfluoropentene-2 with ethylenediamine, diethylenetriamine, and triethylenetetramine, which were unknown prior to this study.[147] Similarly, Lux and Karaghiosoff used ^{19}F NMR (as well as ^{31}P NMR) to identify the reduction products of CF_3PCl_2 as $(CF_3P)_3$, $(CF_3P)_4$, and $(CF_3P)_5$, of which

$(CF_3P)_3$ had not previously been observed.[148] Ultimately, these examples demonstrate that, despite the uncommon occurrence of fluorine in compounds of interest, ^{19}F NMR has played a vital role in the discovery of new metabolites and reaction products, and these capabilities can translate to the identification of transformation products in the environment.

2.6.2 Phosphorus (^{31}P)

Phosphorus (^{31}P) is another spin $\frac{1}{2}$ nucleus that can be highly useful in compound discovery. With a natural abundance of 100%, and many compounds containing only 1 phosphorus atom, ^{31}P spectra generally have only one peak,[149] thereby simplifying detection and spectral interpretation.

Like with ^{19}F NMR, ^{31}P NMR has been used as a discovery tool in elucidating the structures and geometry of previously unknown reaction products. For instance, products from the reaction of vinylidenediphosphonic acid with various amino acids were characterized by ^{31}P NMR.[150] The formation of novel 1,2,3-triphosphetenes was also discovered using ^{31}P NMR,[151] as was the production of the new benzo-substituted dioxaphosphonines.[152] This technique was further employed to characterize a reaction product as an 8-membered ring[153] and used in another study to characterize both a hitherto unknown reaction product and the intermediate cation.[154] The versatility and unmatched abilities of NMR spectroscopy are highlighted in a study by Gusarova *et al.*, in which they determined the coordination of the phosphorus centre in bis{[2-(4-*tert*-butyl)phen]ethyl}phosphine sulfide.[155]

^{31}P NMR has perhaps been most prolifically used as a discovery tool in environmental research, especially in relation to the characterization of phosphorus forms in different ecosystems. For instance, Stebbins *et al.* used ^{31}P (and ^{27}Al) NMR to study the mineral stishovite, in which trace amounts of phosphorus can be substituted for silicon atoms in the structure. Using solid-state NMR, they successfully determined that the phosphorus atoms in the mineral were octahedrally coordinated.[156] However, more commonly studied are the forms of phosphorus present in freshwater ecosystems and soils. Phosphorus is essential for soil fertility and is an important plant nutrient, but in excess, phosphorus is also a causal factor in eutrophication of freshwater systems. Therefore, characterization of phosphorus forms is critical for predicting the fate, mobility, and transformation of phosphorus in the environment, an important component of environmental management. El-Rifai *et al.* used ^{31}P NMR as a complementary tool to mass spectrometry to identify the forms of phosphorus in Florida wetlands, which act as phosphorus sinks.[157] Since the forms of phosphorus in these wetlands were previously unknown, there was little understanding of the stability and potential for remobilization of the phosphorus, which could result in eutrophication. Similarly, peatlands act as phosphorus sinks, but have historically been drained for agriculture. To restore these peatlands, they must be rewet, but again, this could mobilize the sequestered phosphorus.

Therefore, Negassa *et al.* used ^{31}P NMR (in conjunction with X-ray absorption near-edge structure [XANES] spectroscopy) to identify and quantify the dominant form of phosphorus in peatland soils in Northern Germany during the draining and rewetting process.[158] In lakes, phosphorus compounds settle on and/or sorb to suspended particles and accumulate in sediments, and the tendency of the phosphorus to be released (and potentially cause eutrophication) depends on their dominant forms. Therefore, Shinohara *et al.* identified the forms of phosphorus on suspended particles and in sediments in a eutrophic lake in Japan by analyzing extracts with solution-state ^{31}P NMR spectroscopy.[159] To confirm the identities of the unknowns, samples were spiked with the proposed compounds and the resulting NMR spectra were compared to the original sample spectra. The use of ^{31}P NMR spectroscopy has also contributed to the discovery and identification of phosphorus-containing compounds in various soils, such as alpine,[160] glacial till,[161] and semi-arid arable soils,[162] as well as in both cultivated and uncultivated soils to assess changes in phosphorus forms with changing land use.[163] Since phosphorus is an important soil nutrient, its forms were identified by Tchienkoua *et al.* in soils used to cultivate either food crops, eucalyptus, or tea and they observed slight differences in the proportions of phosphorus forms depending on soil use.[164] ^{31}P NMR has also been used as a discovery tool in other environmental applications. For instance, Koukol *et al.* used ^{31}P to study forms of phosphorus in spruce needles before, during, and after decomposition by different strains of fungus.[165] Since plant matter decomposition is crucial for forest health and phosphorus cycling, the insight provided by NMR is highly useful for ecosystem management. As such, unambiguous characterization of metabolites from nontargeted analysis like this would be complementary to other instruments in compound identification. Lastly, Ghotra *et al.* used ^{31}P as a discovery tool to determine the forms of phosphorus in oat and barley β-glucans, which are important proteins in plant cell walls. Although they were previously assumed to be neutral, Ghotra *et al.* used ^{31}P NMR to discover that the phosphorus present in β-glucans mainly existed as phosphate ions, thereby suggesting the proteins were charged.[166]

In addition to characterizing reaction products and environmental samples, ^{31}P has been used in other miscellaneous contexts. For instance, ^{31}P NMR has been used in medicine to analyze the urine of patients taking cyclophosphamide (a chemotherapy drug) for two consecutive days.[167] Joqueviel *et al.* found known metabolites and some intact cyclophosphamide, but also discovered 7 unknown metabolites that were not anticipated nor present in the blank urine spectra.[167] This suggests that the unknowns were products of cyclophosphamide metabolism and provides valuable information contributing to a better understanding of the drug's mode of action, as well as potential side effects. Structure elucidation of proteins in the bacterium *Helicobacter pylori* has greatly benefitted from the application of ^{31}P NMR as Petzold *et al.* developed and used a 2D ^{31}P, ^{1}H-COSY experiment to identify and quantify the phospholipids in *H. pylori* membranes.[168] Finally, even solid-state ^{31}P NMR has been used to aid in compound discovery and identification; Schäfer *et al.* used solid-state ^{31}P NMR to identify flame retardants in epoxy thermosets (*i.e.*, materials that do not melt after hardening).[169]

Ultimately, although ^{31}P NMR spectroscopy is not used as routinely as ^{1}H and ^{13}C NMR spectroscopy, it can be particularly helpful in compound discovery, especially in the characterization of unknown forms of phosphorus in environmental compartments.

2.6.3 Nitrogen (^{15}N)

Nitrogen has two naturally occurring isotopes: ^{14}N and ^{15}N. Although ^{14}N is most abundant (~99.63%), it is spin 1, which, although NMR active, exhibits broad lines that reduce the applicability and amount of information that can be obtained from a ^{14}N NMR spectrum.[14] However, ^{15}N is spin $\frac{1}{2}$ and results in more informative spectra, but is less sensitive than ^{1}H or ^{13}C NMR due in part to its low natural abundance (~0.37%).[14]

Similar to ^{31}P, ^{15}N NMR has been used for compound discovery and characterization of reaction products. For instance, Ringler and Hänisch used ^{15}N NMR (as well as ^{1}H NMR, ^{13}C NMR, mass spectrometry and IR spectroscopy) to determine the chemical structures of novel stibano amines using a new reaction scheme.[170] Similarly, Adam *et al.* used ^{15}N NMR to aid in the structure elucidation of products generated in a novel reaction scheme for producing *N*-oxides from 1,2,4,5-tetrazines.[171] Lastly, Schwarz *et al.* synthesized new potential prodrugs (*para*-substituted benzamidoximes and benzamidinium salts) and used 1D ^{15}N NMR (as well as ^{1}H and ^{13}C NMR) to characterize their final structures.[172] Ultimately, without ^{15}N NMR spectroscopy, unambiguous characterization of reaction products would be highly difficult, but NMR's nontargeted nature enables confident compound identification, making it an important complementary tool in compound discovery in the chemical industry.

As nitrogen is an important nutrient for plants and other organisms, characterizing forms of nitrogen in soils is essential for assessing and predicting ecosystem health. Therefore, ^{15}N NMR has also been used in several environmental applications. To characterize nitrogen forms in SOM, one usually performs a series of soil extraction procedures.[173] However, with solid-state NMR, intact soil samples can be studied rather than individual extracts. Therefore, VanZomeren *et al.* used solid-state ^{15}N NMR to analyze intact soils and determine the main forms of nitrogen present.[173] Furthermore, due to the critical importance of plant decomposition in global nutrient cycles, specific plant matter transformation processes and products are of great scientific interest. Kelleher *et al.* used HR-MAS NMR of ^{15}N-enriched plants to identify transformation products during decomposition.[174] They employed 2D ^{1}H–^{15}N NMR (along with 1D experiments and ^{13}C NMR) and found that when comparing different portions of the same plant, decomposition pathways and rates were very similar and proceeded at comparable rates; additionally, the recalcitrant compounds remaining at the end of decomposition were very similar between the 2 plant types. These results show that NMR can be used as a tool for investigations into the mechanisms of various transformation processes as well as in kinetics studies. Using NMR as a discovery tool in this context was particularly beneficial in that, with HR-MAS, an intact plant sample can be analyzed and not destroyed during analysis, resulting in more reliable and representative results.[174] In analysis of wastewater, Dignac *et al.* used 1D solid-state ^{15}N NMR (along with 1D solid ^{13}C and solution ^{1}H NMR) to characterize the organic matter remaining at the end of wastewater treatment (*i.e.*, refractory organic matter) and observed that nitrogen was primarily found in amide groups.[175]

Overall, despite the low natural abundance of ^{15}N and its low sensitivity in NMR, it has proven to be a useful complementary tool in compound discovery, particularly in reaction product and environmental sample characterization.

2.6.4 Vanadium (^{51}V)

Vanadium has two isotopes, ^{51}V and ^{50}V, but ^{50}V is only 0.25% naturally abundant and is slightly radioactive with a half-life of 1.5×10^{17} years. In contrast, ^{51}V has a natural abundance of 99.75%, is spin 7/2 (*i.e.*, NMR active), and has a relatively sharp lineshape (though still broader than ^{1}H or ^{13}C).[14] Despite the reasonable sensitivity of ^{51}V NMR, it is seldom employed, in part because when in a 0, +2 or +4 oxidation state, it becomes paramagnetic and thus no longer NMR active.[14] It has, however, been used in a handful of contexts to aid in compound discovery.

Vanadium has recently been declared an emerging contaminant by the Environmental Protection Agency,[176] and its dominant forms, behaviour, and fate in the environment must be characterized for proper environmental management. Therefore, Wen *et al.* monitored the transformation of four common forms of vanadium under varying pH levels in aqueous solutions, and compounds identified by mass spectrometry were confirmed by ^{51}V NMR analysis.[176] Furthermore, since different forms (*i.e.*, oxidation state and coordination) of vanadium have shown to be teratogenic, carcinogenic, and mutagenic, characterizing vanadium species in the environment is critical for human and environmental health. Lu *et al.* thus used ^{51}V NMR to monitor how the major form of vanadium in aqueous solutions of organic matter changes depending on factors such as pH.[177]

In chemical research, ^{51}V NMR has been used to monitor reaction pathways and identify reaction products. Rana *et al.* used ^{51}V NMR (as well as UV-Vis and Raman spectroscopies) to elucidate the structure of an intermediate product involved in decarbonylative halogenation.[178] Furthermore, the structures of novel vanadium compounds have been elucidated with the aid of ^{51}V NMR spectroscopy. In 1994, Cornman *et al.* synthesized the first vanadium(III)–amide complex and used ^{51}V NMR (as well as ^{1}H NMR, elemental analysis, and X-ray spectroscopy) to elucidate its structure.[179]

Similarly, Zandvakili *et al.* used ^{51}V NMR (as well as elemental analysis and other spectroscopic analyses) to characterize a new oxo-vanadium dimer complex,[180] and Tricot *et al.* used ^{51}V and ^{31}P solid-state NMR (as well as X-ray diffraction) to characterize the novel compound $Pb_4(VO_2)(PO_4)_3$.[181]

Overall, despite its uncommon use, ^{51}V NMR can play a vital role in compound discovery and structure elucidation in certain niche contexts.

2.6.5 Other Metals

A variety of other metals have been used as NMR nuclei to aid in compound discovery and identification, especially in reaction product characterization. Although use of the following metals is rare, these show that metal NMR can be helpful in compound identification.

Of lead's four stable isotopes (^{204}Pb, ^{206}Pb, ^{207}Pb, and ^{208}Pb), ^{207}Pb is the only spin $\frac{1}{2}$ nucleus, making it NMR active, though it is still less than 1% as sensitive as ^{1}H.[14] Nevertheless, ^{207}Pb NMR has been used in some contexts for discovery and characterization of synthesized complexes. For instance, Sharma *et al.* synthesized the complexes [Et_3Pb{ON=ClAr}] [where R = H or Me and Ar = 2-C_5H_4N, 2-C_4H_3O or 2-C_4H_3S], and based on the ^{207}Pb, ^{1}H, ^{13}C, and ^{119}Sn NMR spectra, they concluded that the complex was tetrahedral in geometry (with Pb as the central atom).[182] In another study, Morsali and Mahjoub used NMR spectroscopy (^{207}Pb, ^{1}H, ^{13}C) in conjunction with elemental analysis and IR spectroscopy to characterize a novel Pb(II) complex with 3 ligands.[183] Pu *et al.* also synthesized a novel group of lead complexes (metalloplumbylenes) and used ^{207}Pb, ^{1}H, ^{13}C NMR along with other forms of spectroscopy to discover that the complexes had bent geometry (*i.e.*, a Pb centre with 2 ligands).[184] Lastly, Cruz *et al.* synthesized a novel group of compounds (tin–lead niobates) and characterized them with solid-state ^{207}Pb, ^{119}Sn, and ^{93}Nb MAS NMR spectroscopy.[185]

Cadmium NMR has also been utilized in a few contexts, and although it has 8 isotopes, the most frequently used isotope is ^{113}Cd as it is spin $\frac{1}{2}$ and is the most sensitive.[14] Although cadmium is a relatively uncommon metal, its chemical shift values are highly dependent upon factors such as geometry and coordination, and chemical shifts can even be used to identify ligands bound to a cadmium centre.[186] Therefore, Chung and Moon used solution-state ^{113}Cd NMR to identify cadmium carboxylate complexes, an important feat in that carboxylates are ubiquitous in the environment.[186] In this study, they looked at various factors that could affect complex formation, such as ligand properties and steric effects, among others.[186] Solid-state ^{113}Cd NMR has even been used to investigate the ability of cadmium to incorporate into perovskite lattices, a mechanism and structure that were previously unknown.[187] Interactions of cadmium and organic matter have also been studied. Larive *et al.* investigated how ^{113}Cd interacts with humic substances based on chemical shift values and found that $^{113}Cd^{2+}$ primarily binds *via* oxygen atoms (often carboxyl groups) in the humic substances. They also used ^{113}Cd NMR to determine the rates and extent of complexation of ^{113}Cd to humic substances.[188] Otto *et al.* expanded on this work using ^{113}Cd NMR and relaxation times (*i.e.*, spin–lattice/T_1 and spin–spin/T_2) to characterize the interactions of ^{113}Cd and fulvic acid.[189] Their findings agreed that $^{113}Cd^{2+}$ mainly interacts with organic matter *via* the oxygen atoms in carboxylate functional groups. Hertkorn *et al.* further investigated the interactions of ^{113}Cd with natural organic matter (NOM) at varying pH values and the concentrations of Cd^{2+} and NOM. Results agreed with previous work in that Cd^{2+} was shown to interact primarily with oxygen sites in the NOM, but data also suggested that, in alkaline solutions, Cd^{2+} interacts with nitrogen atoms at binding sites as well.[190] Moreover, Simpson used ^{113}Cd diffusion ordered NMR spectroscopy (DOSY) to analyze Cd^{2+} interactions with fulvic acids. Upon comparing the ^{113}Cd 1D and DOSY spectra of a $CdCl_2$ solution before and after the addition of fulvic acids (see Figure 2.3), Simpson found that cadmium binds reversibly.[191]

Lastly, aluminum NMR shows potential to be used in environmental discovery, as it is present in the environment both naturally and from anthropogenic inputs, though the rare use of aluminum NMR can be attributed to the fact that ^{27}Al is a spin 5/2 quadrupole, leading to broad signals in NMR spectra.[14] However, Goudarzi used ^{27}Al NMR to identify the formation of five aluminate compounds in different solutions based on the signals appearing in the ^{27}Al NMR spectra.[192] Due to its abundance in numerous minerals, ^{27}Al NMR has also been used to characterize its form(s) in various geological settings. For instance, Kim used ^{27}Al NMR to characterize the forms and amounts of Al minerals in acid mine drainage precipitates.[193] Fitzgerald *et al.* also used ^{27}Al NMR to study the mineral kaolinite before and after dealumination (*i.e.*, HCl extractions intended to remove the aluminum) and successfully observed concentration changes. They additionally used ^{29}Si NMR to characterize changes in Si chemical environments before and after removal of the aluminum.[194]

Silicon, another ubiquitous element in minerals, has 3 stable isotopes: ^{28}Si, ^{29}Si, and ^{30}Si. Although it is only 4.67% abundant, the ^{29}Si isotope is used in NMR as it is spin $\frac{1}{2}$ and has a large spectral width (~1000 ppm).[14] ^{29}Si NMR has seldom been used in compound discovery, but it has been used extensively in the characterization of the geometry of silicon minerals and complexes. Schmücker *et al.* used ^{29}Si NMR to investigate the 3D geometry of silicon in the mineral mullite,[195] and Herreros *et al.* found that the mineral sodalite is formed *via* a 5-coordinated silicon intermediate.[196] Furthermore, Stebbins and Kanzaki used ^{29}Si NMR to elucidate the structures and coordination of silicon nuclei in calcium and magnesium silicates.[197]

Lastly, mercury (especially in its methylmercury form) is an important contaminant that can pose serious risks to human and environmental health. Mercury has one spin $\frac{1}{2}$ isotope, ^{199}Hg, and it is 16.9% abundant with a large spectral width of approximately 4300 ppm, making it fairly receptive to NMR. Although it has not been studied extensively in an environmental

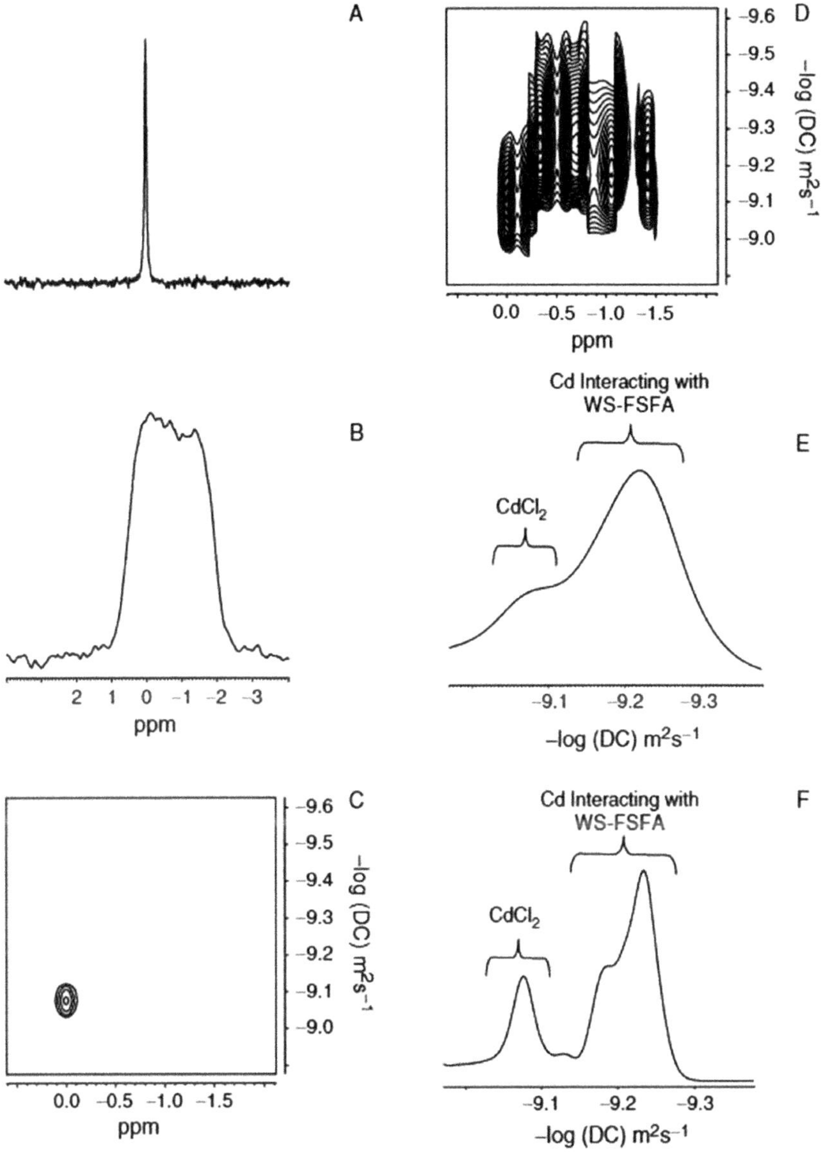

Figure 2.3 The 1D [113]Cd spectra of $CdCl_2$ before (A) and after (B) the addition of fulvic acids, the [113]Cd DOSY spectra of $CdCl_2$ before (C) and after (D) the addition of fulvic acids, and the F_1 diffusion projections of (D) with a linewidth factor of 0 (E) and 0.3 (F). Reproduced from ref. 191 with permission from John Wiley & Sons, Copyright 2002.

context, it has been used to characterize the structure of complexes with thiacrown ligands,[198,199] suggesting that [199]Hg NMR has untapped potential in environmental research.

Thus, although metal NMR is relatively uncommon as a discovery tool due to sensitivity and lineshape issues with many nuclei (as well as the rarity of some metals in the environment), it has been useful in product characterization in chemical research.

2.7 Conclusion

2.7.1 Potential of Low-field and Portable NMR as a Discovery Tool

A major obstacle to the widespread use of NMR as a complementary discovery tool is its inherent inaccessibility. Most NMR spectrometers used in research and industry today are deemed "high-field" due to the strong magnetic fields produced by their cryogenically cooled electromagnets. However, although sensitivity increases with magnetic field strength, high-field NMR spectrometers are also very expensive (both upfront and over time), are large in size (often standing over six feet tall) and require considerable maintenance as well as expertise for operation. Conversely, "low-field" NMR spectrometers (*i.e.*, those with proton resonance frequencies of less than ~100 MHz) use permanent magnets, thereby requiring only a power source for operation. Low-field NMR spectrometers are also much cheaper, small in size (can fit on a benchtop), require little maintenance, and are relatively simple to operate, thus making them more physically and financially accessible.

Due to their weaker magnetic fields, low-field NMR spectrometers are less sensitive than high-field systems, and complex samples often exhibit considerable spectral overlap. However, low-field NMR has already been utilized successfully in numerous research contexts to aid in compound discovery. For example, low-field NMR has been used to characterize previously unknown chemical compositions of nonaqueous phase liquids, which are mixtures of organic solvents that are released into the environment as waste from the chemical industry.[200] Low-field NMR has also been used to identify adulterants and active pharmaceutical ingredients in dietary supplements and drugs. For example, Pagès *et al.* identified adulterant compounds in dietary supplements using a 60 MHz NMR spectrometer,[201] and Antonides *et al.* compared the 60 MHz ^{1}H spectra of illegal drugs to a spectral database to identify the active pharmaceutical(s) present in each.[202] In order to aid with spectral dispersion, Assemat *et al.* used diffusion-ordered NMR spectroscopy (DOSY) at 43 MHz to separate and identify different pharmaceuticals in commercially available mixtures (see Figure 2.4).[203]

Lastly, low-field NMR has also been used to aid in discovery of adulterants and authentication in the food industry. For instance, Defernez *et al.* identified a compound (16-*O*-methylcafestol) that is an indicator of arabica coffee beans with a 1-step extraction as the only sample preparation (60 MHz).[204] In addition, Jakes *et al.* used 60 MHz NMR to distinguish between beef and horse meat based on identification of characteristic triglyceride peaks in the

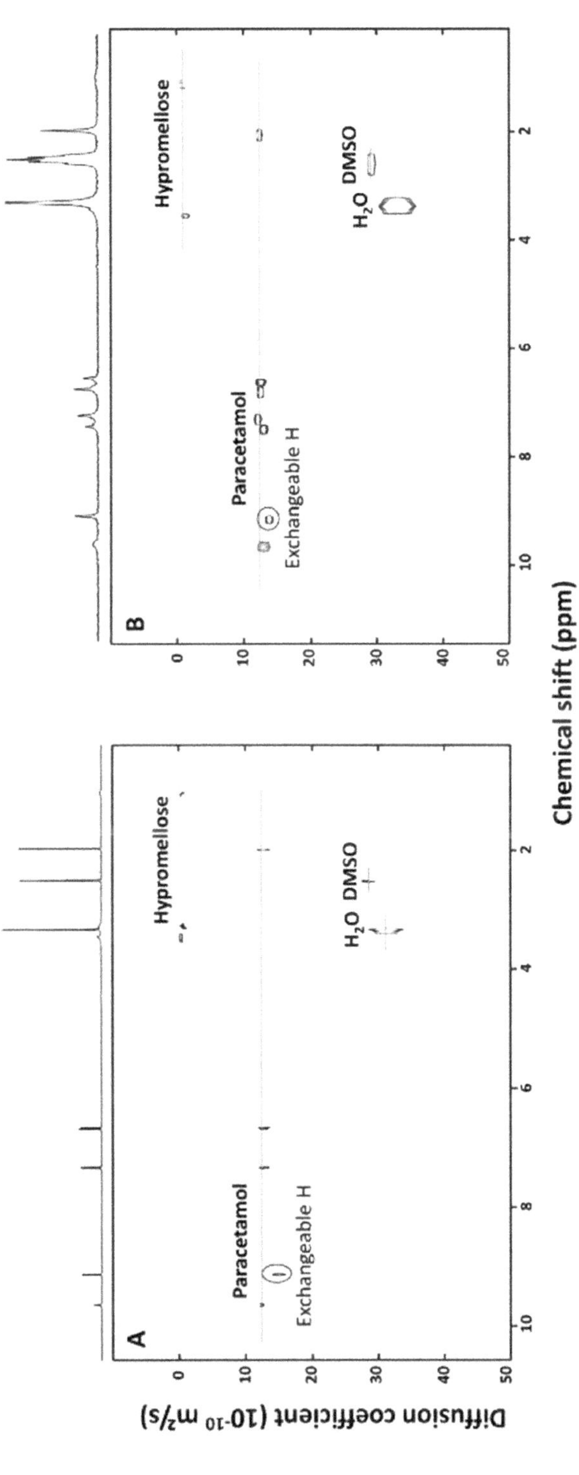

Figure 2.4 The DOSY NMR spectrum of a pharmaceutical mixture recorded at (A) high field strength (500 MHz) and at (B) low field strength (43 MHz). Reproduced from ref. 203 with permission from Elsevier, Copyright 2018.

two meats.[205] These are just a few examples of the numerous studies that show how low-field NMR can be successfully employed to identify adulteration of edible oils.[206–209] Most recently, low-field ^{19}F NMR has been used to analyze perfluorooctanoic acid as it is broken down, and although it could be used to determine the structures of some transformation products, not all breakdown products' structures could be elucidated.[210]

Although these are just a few examples of low-field NMR use in compound discovery, they demonstrate the immense potential for future use of low-field NMR. For instance, low-field NMR could be used on-site to identify hazardous or unexpected transformation products (or by-products) in chemical industry effluent prior to its release into the environment. Low-field NMR could also be used for groundwater quality monitoring and rapid identification of contamination.

Furthermore, low-field NMR relaxometry (*i.e.*, time-domain NMR wherein relaxation data are used in analysis) and fast field cycling (FFC) NMR have played a vital role in porosity and hydrological analysis of environmental media. In FFC NMR, the magnetic field "cycles" through 3 stages: the first is polarization (wherein nuclei are polarized to increase sensitivity), the second is relaxation (wherein the nuclei are allowed to relax), and the third is acquisition. In this low-field relaxometry technique, the length and intensity of the magnetic field applied during the relaxation stage are changed over time, which enables the determination of the longitudinal (or spin–lattice) relaxation times (T_1) of the components in the sample.[211] The T_1 relaxation rates depend on several factors and indicate the types of movement occurring in the sample (*e.g.*, tumbling, diffusion, *etc.*). This can thereby inform the chemical and/or physical properties of a sample, such as porosity or interactions that are occurring.[211] The porosity of a material can be assessed by FFC NMR in that longer relaxation times generally indicate that water is more mobile, whereas shorter relaxation times suggest that water is confined to small pores and is thus less mobile. With this knowledge, Maccotta *et al.* used FFC NMR to study marsh sediment cores and characterize physical–chemical properties at varying depths, finding that porosity varies with depth and is related to sediment formation over time.[212] FFC NMR has also been used to analyze the soils of 4 recently (over the previous 60 years) grown forests, which had been afforested using different plant types. Laudicina *et al.* found that the plant cover affected soil chemistry and hydrology, which was closely related to the soluble organic content of each soil.[213] FFC NMR has further been used to analyze biochar produced from various sources, such as wood chips from different tree species,[214,215] as well as winemaking waste,[215] and to characterize biochar porosity and the nature of the water–biochar interface. This type of research could be integral to predicting the behaviour of biochar-amended soils, an important feat as biochar continues to gain traction in environmental research. Furthermore, FFC NMR has been employed to evaluate novel methods for cellulose degradation. As biofuels have begun gaining popularity as an alternative to fossil fuels, cellulose has been identified as a potential source of glucose, which is needed in fermentation, and production of bioethanol. However, cellulose is a recalcitrant polymer and methods need to be developed to decompose it into glucose units. Therefore, FFC has been utilized to track the degradation of cellulose over time in phosphoric acid at 25 °C[216] and at 80 °C,[217] as the phosphoric acid protons relax quickly when bound to cellulose, and slowly when in the bulk solvent. FFC NMR has additionally been used to study plant materials, such as the investigation of leaf molecular dynamics,[218] as well as the study of the tissue of a wild fruit, wherein the presence of hydrophilic (*e.g.*, sugars) and hydrophobic (*e.g.*, lipids in cell membranes) components could be identified based on longitudinal relaxation times.[211] Lastly, FFC NMR has been used in other applications, such as characterizing organic matter adsorption to different minerals,[219] as well as investigating how water interacts with TiO_2 minerals, which are used for many purposes like photocatalysis and solar energy production.[220] Ultimately, NMR relaxometry and field cycling techniques have shown to be very promising tools that help better understand the physical and chemical properties of various environmental media.

With developments in low-field NMR, portable spectrometers may soon become available, allowing on-site, real-time chemical analysis in nearly any location, thereby further expanding the accessibility and potential of NMR. With portable low-field NMR spectrometers, identification of pollutants in contaminated sites would become possible and enable more accurate assessments of their behaviour, transformation, and transport in the environment. Portable NMR spectrometers could also be taken to remote locations, making ground-breaking research possible, such as the investigation of climate change impacts on environmental health at the North and South Poles.

2.7.2 Hyphenated NMR

Some research has also been conducted using hyphenated and parallel NMR systems, which can increase the quality and quantity of information gained from the analysis of complex samples. The most common systems used thus far consist of liquid chromatography coupled with NMR spectroscopy.[13,221–223] Coupled LC–NMR and MS was used by Hansen *et al.* for drug discovery and to identify metabolites in plant extracts.[223] The combination of these different analytical techniques allowed the detection of various known compounds as well as the characterization of one novel compound.[222] Other studies have also applied these systems to the characterization of plant extracts. For example, an HPLC–DAD–SPE–NMR system was used to analyze commercial rosemary extract, allowing characterization of the five main active compounds.[221] A similar study applied LC–NMR and MS to the analysis of plant extracts. In this study, the structures of various compounds were confirmed, and the structure of a previously unreported compound elucidated.[223]

The benefits of hyphenated systems, such as those discussed here, present future opportunities for improved resolution when analyzing complex systems with much overlap.[13,222] Some preliminary environmental studies have been done with hyphenated LC–SPE–NMR systems, one of which looked at identifying the components of non-living natural organic matter that are present in aquatic or soil environments,[224] allowing the identification of various components of the complex mixture and indicating that future environmental studies may greatly benefit from the information provided by these techniques.[224] These results suggest that the development of truly hyphenated systems would allow improved sensitivity and separation in complex environmental samples.

While NMR has yet to reach its full potential in environmental research, the examples discussed here illustrate that it has unique information to offer related to contaminant discovery, monitoring transformation processes and improving the understanding of living systems. As such, future environmental research should implement NMR spectroscopy more routinely. Ultimately, NMR spectroscopy is the premiere tool for structure elucidation and is highly complementary to mass spectrometry-based detection. When used in combination, researchers have the best of both worlds with the sensitivity of mass spectrometry (providing unparalleled targeted analysis and molecular formula information), whereas NMR maps out molecular connectivity for discovery of the unknown and unexpected.

Acknowledgements

A. J. S would like to thank the Natural Sciences and Engineering Research Council of Canada (NSERC) [Alliance (ALLRP 549399), Alliance (ALLRP 555452) and Discovery Programs (RGPIN-2019-04165)], the Canada Foundation for Innovation (CFI), the Ontario Ministry of Research and Innovation (MRI), the Krembil Foundation for providing funding, and the Government of Ontario for an Early Researcher Award.

References

1. D. R. Taves, W. S. Grey and W. S. Brey, *Toxicol. Appl. Pharmacol.*, 1976, **37**, 120.
2. A. B. Lindstrom, M. J. Strynar and E. L. Libelo, *Environ. Sci. Technol.*, 2011, **45**, 7954.
3. R. Renner, *Anal. Chem.*, 2005, 15.
4. J. P. Giesy and K. Kannan, *Environ. Sci. Technol.*, 2001, **35**, 1339.
5. K. Kannan, S. Corsolini, J. Falandysz, G. Fillmann, K. S. Kumar, B. G. Loganathan, M. A. Mohd, J. Olivero, N. Van Wouwe, J. H. Yang and K. M. Aldous, *Environ. Sci. Technol.*, 2004, **38**, 4489.
6. A. J. Simpson, *Magn. Reson. Chem.*, 2015, **53**, 633.
7. G. Webster, *Potential human health effects of perfluorinated chemicals (PFCs)*, Vancouver, 2010.

8. Z. Tian, H. Zhao, K. T. Peter, M. Gonzalez, J. Wetzel, C. Wu, X. Hu, J. Prat, E. Mudrock, R. Hettinger, A. E. Cortina, R. G. Biswas, F. Vinicius, C. Kock, R. Soong, A. Jenne, B. Du, F. Hou, H. He, R. Lundeen, A. Gilbreath, R. Sutton, N. L. Scholz, J. W. Davis, M. C. Dodd, A. Simpson, J. K. Mcintyre and E. P. Kolodziej, *Science*, 2020, **371**, 185.
9. Z. Tian, M. Gonzalez, C. A. Rideout, H. N. Zhao, X. Hu, J. Wetzel, E. Mudrock, C. A. James, J. K. McIntyre and E. P. Kolodziej, *Environ. Sci. Technol. Lett.*, 2022, **9**, 140.
10. H. Mahoney, F. C. Da, S. Junior, C. Roberts, M. Schultz, X. Ji, J. Alcaraz, D. Montgomery, S. Selinger, J. K. Challis, J. P. Giesy, L. Weber, D. Janz, S. Wiseman, M. Hecker and M. Brinkmann, *Environ. Sci. Technol. Lett.*, 2022, **9**, 765.
11. W. Wang, G. Cao, J. Zhang, P. Wu, Y. Chen, Z. Chen, Z. Qi, R. Li, C. Dong and Z. Cai, *Environ. Sci. Technol.*, 2022, **56**, 10629.
12. B. Yuan, D. H. Lysak, R. Soong, A. Haddad, A. Hisatsune, A. Moser, S. Golotvin, D. Argyropoulos, A. J. Simpson and D. C. G. Muir, *Environ. Sci. Technol. Lett.*, 2020, 7, 496.
13. A. J. Simpson, M. J. Simpson and R. Soong, *Environ. Sci. Technol.*, 2012, **46**, 11488.
14. M. T. Anaraki, D. H. Lysak, K. Downey, F. V. C. Kock, X. You, R. D. Majumdar, A. Barison, L. M. Lião, A. G. Ferreira, V. Decker, B. Goerling, M. Spraul, M. Godejohann, P. A. Helm, S. Kleywegt, K. Jobst, R. Soong, M. J. Simpson and A. J. Simpson, *Prog. Nucl. Magn. Reson. Spectrosc.*, 2021, **126–127**, 121.
15. B. Lam and A. J. Simpson, *Analyst*, 2008, **133**, 263.
16. N. Hertkorn, M. Harir, B. P. Koch, B. Michalke and P. Schmitt-Kopplin, *Biogeosciences*, 2013, **10**, 1583.
17. J. A. Leenheer, T. I. Noyes, C. E. Rostad and M. L. Davisson, *Biogeochemistry*, 2004, **69**, 125.
18. P. T. Callaghan, C. D. Eccles, T. G. Haskell, P. J. Langhorne and J. D. Seymour, *J. Magn. Reson.*, 1998, **133**, 148.
19. B. G. Pautler, G. C. Woods, A. Dubnick, A. J. Simpson, M. J. Sharp, S. J. Fitzsimons and M. J. Simpson, *Environ. Sci. Technol.*, 2012, **46**, 3753.
20. J. D. Mao, L. Tremblay, J. P. Gagné, S. Kohl, J. Rice and K. Schmidt-Rohr, *Geochim. Cosmochim. Acta*, 2007, **71**, 5483.
21. E. Shumilina, T. Johansen, C. Glasner, J. Robert, T. Rustad and A. Dikiy, *Waste Biomass Valorization*, 2020, **11**, 1349.
22. C. A. Moody, Wai Chi Kwan, J. W. Martin, D. C. G. Muir and S. A. Mabury, *Anal. Chem.*, 2001, **73**, 2200.
23. J. R. Gauthier and S. A. Mabury, *Anal. Chem.*, 2022, **94**, 3278.
24. J. Lu, A. C. Chang and L. Wu, *Environ. Pollut.*, 2004, **132**, 365.
25. A. Mustafa, X. Hu, S. Atizaz Ali Shah, M. Mohsin Abrar, A. Akbar Maitlo, K. Ali Kubar, Q. Saeed, M. Kamran, M. Naveed, W. Boren, S. Nan and X. Minggang, *J. Soils Sediments*, 2021, **21**, 2483.
26. A. J. Simpson, M. J. Simpson, E. Smith and B. P. Kelleher, *Environ. Sci. Technol.*, 2007, **41**, 8070.

27. J.-D. Mao, W.-G. Hu, K. Schmidt-Rohr, G. Davies, E. A. Ghabbour and B. Xing, *Soil Sci. Soc. Am. J.*, 2000, **64**, 873.
28. B. P. Kelleher and A. J. Simpson, *Environ. Sci. Technol.*, 2006, **40**, 4605.
29. A. J. Simpson, G. Song, E. Smith, B. Lam, E. H. Novotny and M. H. B. Hayes, *Environ. Sci. Technol.*, 2007, **41**, 876.
30. X. Feng, A. J. Simpson, K. P. Wilson, D. D. Williams and M. J. Simpson, *Nat. Geosci.*, 2008, **1**, 836.
31. Y. Suzuki, M. Kawakami and K. Akasaka, *Environ. Sci. Technol.*, 2001, **35**, 2656.
32. R. M. B. O. Duarte, P. Duan, J. Mao, W. Chu, A. C. Duarte and K. Schmidt-Rohr, *Atmos. Environ.*, 2020, **230**, 117503.
33. P. Herckes, J. A. Leenheer and J. L. Collett, *Environ. Sci. Technol.*, 2007, **41**, 393.
34. N. L. Law and M. L. Diamond, *Pergamon Chemosphere*, 1998, **36**, 2607.
35. A. J. Simpson, B. Lam, M. L. Diamond, D. J. Donaldson, B. A. Lefebvre, A. Q. Moser, A. J. Williams, N. I. Larin and M. P. Kvasha, *Chemosphere*, 2006, **63**, 142.
36. R. M. B. O. Duarte, A. M. S. Silva and A. C. Duarte, *Environ. Sci. Technol.*, 2008, **42**, 8224.
37. B. Graham, O. L. Mayol-Bracero, P. Guyon, G. C. Roberts, S. Decesari, M. C. Facchini, P. Artaxo, W. Maenhaut, P. Köll and M. O. Andreae, *J. Geophys. Res.: Atmos.*, 2002, **107**, 1.
38. L. J. Klimczak, C. Ebner von Eschenbach, P. M. Thompson, J. T. M. Buters and G. A. Mueller, *Atmos. Environ.*, 2020, **243**, 117746.
39. N. I. Medina-Pérez, M. Dall'osto, S. Decesari, M. Paglione, E. Moyano and E. Berdalet, *Environ. Sci. Technol.*, 2021, **55**, 468.
40. O. Beckonert, M. Coen, H. C. Keun, Y. Wang, T. M. D. Ebbels, E. Holmes, J. C. Lindon and J. K. Nicholson, *Nat. Protoc.*, 2010, **5**, 1019.
41. A. J. Simpson, M. J. Simpson, W. L. Kingery, B. A. Lefebvre, A. Moser, A. J. Williams, M. Kvasha and B. P. Kelleher, *Langmuir*, 2006, **22**, 4498.
42. D. Lane, Y. Liaghati Mobarhan, R. Soong, P. Ning, W. Bermel, M. Tabatabaei Anaraki, B. Wu, H. Heumann, M. Gundy, H. Boenisch, T. Y. Jeong, V. Kovacevic, M. J. Simpson and A. J. Simpson, *Anal. Chem.*, 2019, **91**, 15000.
43. A. J. Simpson, W. L. Kingery, D. R. Shaw, M. Spraul, E. Humpfer and P. Dvortsak, *Environ. Sci. Technol.*, 2001, **35**, 3321.
44. J. G. Longstaffe and A. J. Simpson, *Environ. Toxicol. Chem.*, 2011, **30**, 1745.
45. A. Shirzadi, M. J. Simpson, Y. Xu and A. J. Simpson, *Environ. Sci. Technol.*, 2008, **42**, 1084.
46. A. Shirzadi, M. J. Simpson, R. Kumar, A. J. Baer, Y. Xu and A. J. Simpson, *Environ. Sci. Technol.*, 2008, **42**, 5514.
47. H. Masoom, D. Courtier-Murias, R. Soong, W. E. Maas, M. Fey, R. Kumar, M. Monette, H. J. Stronks, M. J. Simpson and A. J. Simpson, *Environ. Sci. Technol.*, 2015, **49**, 13983.
48. J. R. Sachleben, B. Chefetz, A. Deshmukh and P. G. Hatcher, *Environ. Sci. Technol.*, 2004, **38**, 4369.

49. A. Bunescu, J. Garric, B. Vollat, E. Canet-Soulas, D. Graveron-Demilly and F. Fauvelle, *Mol. BioSyst.*, 2009, **6**, 121.
50. M. L. Miglietta and R. Lamanna, *Magn. Reson. Chem.*, 2006, **44**, 675.
51. C. Casieri, L. Senni, M. Romagnoli, U. Santamaria and F. de Luca, *J. Magn. Reson.*, 2004, **171**, 364.
52. M. L. Chevallier, O. Della-Negra, S. Chaussonnerie, A. Barbance, D. Muselet, F. Lagarde, E. Darii, E. Ugarte, E. Lescop, N. Fonknechten, J. Weissenbach, T. Woignier, J. F. Gallard, S. Vuilleumier, G. Imfeld, D. le Paslier and P. L. Saaidi, *Environ. Sci. Technol.*, 2019, **53**, 6133.
53. M. R. Seger and G. E. Maciel, *Environ. Sci. Technol.*, 2006, **40**, 791.
54. M.-B. Turrión, B. Glaser, D. Solomon, A. Ni and W. Zech, *Biol. Fertil. Soils*, 2000, **31**, 134.
55. P. J. Hernes and R. Benner, *Deep Sea Res., Part I*, 2002, **49**, 2119.
56. J. A. Baldock, C. A. Masiello, Y. Gelinas and J. I. Hedges, *Mar. Chem.*, 2004, **92**, 39.
57. S. Liu, Y. Zhu, F. Wu, W. Meng, H. Wang, Z. He, W. Guo, F. Song and J. P. Giesy, *Environ. Sci. Pollut. Res.*, 2017, **24**, 1880.
58. C. M. Kao, B. H. Lin, S. C. Chen, S. F. Wei, C. C. Chen, C. L. Yao and C. C. Chien, *Biorem. J.*, 2016, **20**, 165.
59. K. A. Thorn, *PLoS One*, 2019, **14**, e0224112.
60. M. Godejohann, *NMR Spectroscopy: A Versatile Tool for Environmental Research*, John Wiley & Sons Inc, Chichester, 2014, vol. 1, pp. 49–68.
61. D. A. Ellis, J. W. Martin, D. C. G. Muir and S. A. Mabury, *Analyst*, 2003, **128**, 756.
62. M. S. Solum, A. F. Sarofim, R. J. Pugmire, T. H. Fletcher and H. Zhang, *Energy Fuels*, 2001, **15**, 961.
63. C. S. Maksymiuk, C. Gayahtri, R. R. Gil and N. M. Donahue, *Phys. Chem. Chem. Phys.*, 2009, **11**, 7810.
64. F. Moretti, E. Tagliavini, S. Decesari, M. C. Facchini, M. Rinaldi and S. Fuzzi, *Environ. Sci. Technol.*, 2008, **42**, 4844.
65. F. Cavalli, M. C. Facchini, S. Decesari, L. Emblico, M. Mircea, N. R. Jensen and S. Fuzzi, *Atmos. Chem. Phys.*, 2006, **6**, 993.
66. M.-F. Dignac, P. Ginestet, A. Bruchet, J.-M. Audic, S. Derenne and C. Largeau, *Water Sci. Technol.*, 2001, **43**, 51.
67. A. V. Piterina, J. Barlett and J. T. Pembroke, *Int. J. Environ. Res. Public Health*, 2009, **6**, 2288.
68. K. Solakyildirim, D. N. Bulloch and C. K. Larive, *Magn. Reson. Chem.*, 2014, **52**, 310.
69. M. Uchimiya, Y. Tsuboi, K. Ito, Y. Date and J. Kikuchi, *Metabolites*, 2017, 7, 52.
70. J. G. Bundy, M. P. Davey and M. R. Viant, *Metabolomics*, 2009, **5**, 3.
71. M. Kråkström, S. Saeid, P. Tolvanen, N. Kumar, T. Salmi, L. Kronberg and P. Eklund, *Environ. Sci. Pollut. Res.*, 2020, **27**, 23258.
72. M. P. Robinson and G. C. Lloyd-Jones, *ACS Catal.*, 2018, **8**, 7484.
73. P. A. Cox, M. Reid, A. G. Leach, A. D. Campbell, E. J. King and G. C. Lloyd-Jones, *J. Am. Chem. Soc.*, 2017, **139**, 13156.

74. Y. Ji, R. E. Plata, C. S. Regens, M. Hay, M. Schmidt, T. Razler, Y. Qiu, P. Geng, Y. Hsiao, T. Rosner, M. D. Eastgate and D. G. Blackmond, *J. Am. Chem. Soc.*, 2015, **137**, 13272.
75. D. B. G. Berry, A. Codina, I. Clegg, C. L. Lyall, J. P. Lowe and U. Hintermair, *Faraday Discuss.*, 2019, **220**, 45.
76. E. C. Keske, T. H. West and G. C. Lloyd-Jones, *ACS Catal.*, 2018, **8**, 8932.
77. C. Mclaughlin, A. M. Z. Slawin and A. D. Smith, *Angew. Chem., Int. Ed.*, 2019, **131**, 15255.
78. M. Butters, J. N. Harvey, J. Jover, A. J. J. Lennox, G. C. Lloyd-Jones and P. M. Murray, *Angew. Chem., Int. Ed.*, 2010, **49**, 5156.
79. K. Singh, E. Danieli and B. Blümich, *Anal. Bioanal. Chem.*, 2017, **409**, 7223.
80. Y. Qiao, W. Ge, L. Jia, X. Hou, Y. Wang and C. M. Pedersen, *Chem. Commun.*, 2016, **52**, 11418.
81. L. D. Roberts, A. L. Souza, R. E. Gerszten and C. B. Clish, *Current Protocols in Molecular Biology*, John Wiley & Sons Inc., Chichester, 2012, vol. 1, pp. 30.2.1–30.2.24.
82. B. P. Lankadurai, E. G. Nagato and M. J. Simpson, *Environ. Rev.*, 2013, **21**, 180.
83. A. S. Edison, R. D. Hall, C. Junot, P. D. Karp, I. J. Kurland, R. Mistrik, L. K. Reed, K. Saito, R. M. Salek, C. Steinbeck, L. W. Sumner and M. R. Viant, *Metabolites*, 2016, **6**, 8.
84. D. Martin-Creuzburg, A. Wacker and E. von Elert, *Oecologia*, 2005, **144**, 362.
85. J. Stevenson, *J. Phycol.*, 2014, **50**, 437.
86. M. N. Kariuki, E. G. Nagato, B. P. Lankadurai, A. J. Simpson and M. J. Simpson, *Metabolites*, 2017, **7**, 15.
87. V. Kovacevic, A. J. Simpson and M. J. Simpson, *Comp. Biochem. Physiol., Part D: Genomics Proteomics*, 2016, **19**, 199.
88. V. Kovacevic, A. J. Simpson and M. J. Simpson, *Ecotoxicol. Environ. Saf.*, 2018, **164**, 189.
89. M. T. Anaraki, D. H. Lysak, R. Soong, M. J. Simpson, M. Spraul, W. Bermel, H. Heumann, M. Gundy, H. Boenisch and A. J. Simpson, *Analyst*, 2020, **145**, 5787.
90. M. Akhter, R. Dutta Majumdar, B. Fortier-McGill, R. Soong, Y. Liaghati-Mobarhan, M. Simpson, G. Arhonditsis, S. Schmidt, H. Heumann and A. J. Simpson, *Anal. Bioanal. Chem.*, 2016, **408**, 4357.
91. A. A. Arnold, J. P. Bourgouin, B. Genard, D. E. Warschawski, R. Tremblay and I. Marcotte, *J. Biomol. NMR*, 2018, **70**, 123.
92. N. Merkley and R. T. Syvitski, *J. Appl. Phycol.*, 2012, **24**, 535.
93. M. S. Chauton, T. R. Størseth and G. Johnsen, *J. Appl. Phycol.*, 2003, **15**, 533.
94. Y. L. Mobarhan, B. Fortier-McGill, R. Soong, W. E. Maas, M. Fey, M. Monette, H. J. Stronks, S. Schmidt, H. Heumann, W. Norwood and A. J. Simpson, *Chem. Sci.*, 2016, **7**, 4856.
95. M. R. Viant, J. H. Walton, P. L. TenBrook and R. S. Tjeerdema, *Aquat. Toxicol.*, 2002, **57**, 139.

96. Y. P. Tsentalovich, E. A. Zelentsova, L. V Yanshole, V. V Yanshole and I. M. Odud, *Sci. Rep.*, 2020, **10**, 17128.
97. P. F. Liu, Y. Du, L. Meng, X. Li and Y. Liu, *Fish Shellfish Immunol.*, 2016, **58**, 292.
98. T. W. Collette, D. M. Skelton, J. M. Davis, J. E. Cavallin, K. M. Jensen, M. D. Kahl, D. L. Villeneuve, G. T. Ankley, D. Martinović-Weigelt and D. R. Ekman, *Comp. Biochem. Physiol., Part D: Genomics Proteomics*, 2016, **19**, 190.
99. C. Deborde, B. M. Hounoum, A. Moing, M. Maucourt, D. Jacob, G. Corraze, F. Médale and B. Fauconneau, *J. Nutr. Sci.*, 2021, **10**, 1.
100. M. Y. Mushtaq, R. M. Marçal, D. L. Champagne, F. van der Kooy, R. Verpoorte and Y. H. Choi, *Planta Med.*, 2014, **80**, 1227.
101. N. Zitouni, T. Cappello, O. Missawi, I. Boughattas, G. de Marco, S. Belbekhouche, M. Mokni, V. Alphonse, H. Guerbej, N. Bousserrhine and M. Banni, *Sci. Total Environ.*, 2022, **838**, 155872.
102. L. M. Samuelsson, B. Björlenius, L. Förlin and D. G. J. Larsson, *Environ. Sci. Technol.*, 2011, **45**, 1703.
103. L. M. Samuelsson, L. Förlin, G. Karlsson, M. Adolfsson-Erici and D. G. J. Larsson, *Aquat. Toxicol.*, 2006, **78**, 341.
104. D. Martinovic-Weigelt, D. R. Ekman, D. L. Villeneuve, C. M. James and Q. Teng, *PLoS One*, 2012, 7, 46579.
105. G. le Gall, I. J. Colquhoun, A. L. Davis, G. J. Collins and M. E. Verhoeyen, *J. Agric. Food Chem.*, 2003, **51**, 2447.
106. A. P. Sobolev, D. Capitani, D. Giannino, C. Nicolodi, G. Testone, F. Santoro, G. Frugis, M. A. Iannelli, A. K. Mattoo, E. Brosio, R. Gianferri, I. D'Amico and L. Mannina, *Nutrients*, 2010, **2**, 1.
107. K. Keymanesh, M. H. Darvishi and S. Sardari, *Ric. Sci.*, 2009, **16**, 119.
108. N. Ellis, C. Hattori, J. Cheema, J. Donarski, A. Charlton, M. Dickinson, G. Venditti, P. Kaló, Z. Szabó, G. B. Kiss and C. Domoney, *Front. Plant Sci.*, 2018, **9**, 1022.
109. M. Defernez, Y. M. Gunning, A. J. Parr, L. V. T. Shepherd, H. V. Davies and I. J. Colquhoun, *J. Agric. Food Chem.*, 2004, **52**, 6075.
110. S. P. Mdlalose, M. Raletsena, K. Ntushelo, O. Bodede and D. M. Modise, *Front. Sustainable Food Syst.*, 2022, **6**, 801504.
111. K. A. Leiss, Y. H. Choi, R. Verpoorte and P. G. L. Klinkhamer, *Phytochem. Rev.*, 2011, **10**, 205.
112. T. R. Copley, R. Duggavathi and S. Jabaji, *PLoS One*, 2017, **12**, 1.
113. V. Muccilli, A. Vitale, L. Sheng, A. Gentile, N. Cardullo, C. Tringali, C. Oliveri, R. la Rosa, M. di Guardo, S. la Malfa, Z. Deng and G. Distefano, *J. Agric. Food Chem.*, 2020, **68**, 3806.
114. B. E. Fortier-Mcgill, R. Dutta Majumdar, L. Lam, R. Soong, Y. Liaghati-Mobarhan, A. Sutrisno, R. de Visser, M. J. Simpson, H. L. Wheeler, M. Campbell, A. Gorissen and A. J. Simpson, *J. Agric. Food Chem.*, 2017, **65**, 6779.
115. H. L. Wheeler, R. Soong, D. Courtier-Murias, A. Botana, B. Fortier-Mcgill, W. E. Maas, M. Fey, H. Hutchins, D. Krishnamurthy, R. Kumar,

M. Monette, H. J. Stronks, M. M. Campbell and A. Simpson, *Magn. Reson. Chem.*, 2015, **53**, 735.
116. L. An, J. Ma, H. Wang, F. Li, D. Qin, J. Wu, G. Zhu, J. Zhang, Y. Yuan, L. Zhou and X. Wu, *Food Chem.*, 2018, **269**, 559.
117. D. W. Lachenmeier, H. Eberhard, F. Fang, S. Birk, D. Peter, S. Constanze and S. Manfred, *J. Agric. Food Chem.*, 2009, **57**, 7194.
118. Y. B. Monakhova, T. Kuballa and D. W. Lachenmeier, *Appl. Magn. Reson.*, 2012, **42**, 343.
119. J. Yun, K. J. Shin, J. Choi and C. H. Jo, *Forensic Sci. Int.*, 2018 **286**, 199.
120. K. M. Han, J. H. Lee, H. J. Park, I. Hwang, O. S. Heo and W. S. Kim, *Food Addit. Contam.: Part A*, 2014, **31**, 1.
121. A. Häberli, P. Girard, M. Y. Low and X. Ge, *J. Pharm. Biomed. Anal.*, 2010, **53**, 24.
122. M. Goda, M. Horikawa, T. Iwashita, M. Sawada, D. Tsukazaki, T. Tachibana, K. Sugane and M. Harada, *Shokuhin Eiseigaku Zasshi Food Hyg. Saf. Sci.*, 2013, **54**, 188.
123. Y. Nishiyama, Y. Tamura, A. Ibe, S. Uehara, H. Kamimura, S. Tabata, M. Iida and T. Nishima, *Eisei Kagaku*, 1991, **37**, 89.
124. Y. Kawamura, N. Sugimoto, Y. Takeda and T. Yamada, *Shokuhin Eiseigaku Zasshi Food Hyg. Saf. Sci.*, 1998, **39**, 110.
125. R. Paseiro-Cerrato, S. MacMahon, C. D. Ridge, G. O. Noonan and T. H. Begley, *J. Chromatogr. A*, 2016, **1444**, 106.
126. Q. Li, Z. Yu, D. Zhu, X. Meng, X. Pang, Y. Liu, R. Frew, H. Chen and G. Chen, *J. Sci. Food Agric.*, 2017, **97**, 2875.
127. B. Horn, S. Esslinger, C. Fauhl-Hassek and J. Riedl, *Food Control*, 2021, **128**, 108205.
128. T. A. do Nascimento, T. I. B. Lopes, C. E. D. Nazario, S. L. Oliveira and G. B. Alcantara, *Food Res. Int.*, 2021, **144**, 110362.
129. L. R. Cagliani, G. Maestri and R. Consonni, *Food Control*, 2022, **133**, 108574.
130. C. H. Yong, S. A. Muhammad, F. A. Aziz, F. I. Nasir, M. Z. Mustafa, B. Ibrahim, S. D. Kelly, A. Cannavan and E. K. Seow, *Food Chem.*, 2022, **368**, 130808.
131. M. Spraul, B. Schütz, E. Humpfer, M. Mörtter, H. Schäfer, S. Koswig and P. Rinke, *Magn. Reson. Chem.*, 2009, **47**, S130.
132. I. W. Burton, C. F. Martinez Farina, S. Ragupathy, T. Arunachalam, S. Newmaster and F. Berrué, *J. Agric. Food Chem.*, 2020, **68**, 14643.
133. S. Esslinger, C. Fauhl-Hassek and R. Wittkowski, *Advances in Wine Research*, ACS Symposium Series, 2015, vol. 1203, ch. 6, pp. 85–108.
134. R. Godelmann, F. Fang, E. Humpfer, B. Schütz, M. Bansbach, H. Schäfer and M. Spraul, *J. Agric. Food Chem.*, 2013, **61**, 5610.
135. A. Palmioli, D. Alberici, C. Ciaramelli and C. Airoldi, *Food Chem.*, 2020, **327**, 127025.
136. J. Klare, M. Rurik, E. Rottmann, A. Bollen, O. Kohlbacher, M. Fischer and T. Hackl, *J. Agric. Food Chem.*, 2020, **68**, 14353.

137. M. E. Dimitrakopoulou, K. Matzarapi, S. Chasapi, A. Vantarakis and G. A. Spyroulias, *J. Food Sci.*, 2021, **86**, 4417.
138. S. Beteinakis, A. Papachristodoulou, E. Mikros and M. Halabalaki, *Phytochem. Anal.*, 2022, **33**, 83.
139. F. Longobardi, V. Innamorato, A. Di Gioia, A. Ventrella, V. Lippolis, A. F. Logrieco, L. Catucci and A. Agostiano, *Food Chem.*, 2017, **237**, 743.
140. F. Madrid-Gambin, C. Brunius, M. Garcia-Aloy, S. Estruel-Amades, R. Landberg and C. Andres-Lacueva, *J. Agric. Food Chem.*, 2018, **66**, 6997.
141. H. Köbler, Y. B. Monakhova, T. Kuballa, C. Tschiersch, J. Vancutsem, G. Thielert, A. Mohring and D. W. Lachenmeier, *J. Agric. Food Chem.*, 2011, **59**, 6877.
142. M. Inoue, Y. Sumii and N. Shibata, *ACS Omega*, 2020, **5**, 10633.
143. P. Grandjean and R. Clapp, *Public Health Rep.*, 2014, **129**, 482.
144. J. M. Tront and F. M. Saunders, *Environ. Pollut.*, 2007, **145**, 708.
145. J. C. D'Eon, A. J. Simpson, R. Kumar, A. J. Baer and S. A. Mabury, *Environ. Toxicol. Chem.*, 2010, **29**, 1678.
146. Y. Jin, R. W. Molt and G. M. Blackburn, *Top. Curr. Chem.*, 2017, **375**, 35.
147. V. I. Saloutin, Z. E. Skryabina and Y. V. Burgart, *Bull. Russ. Acad. Sci., Div. Chem. Sci.*, 1992, **41**, 1700.
148. K. Lux and K. Karaghiosoff, *Phosphorus, Sulfur Silicon Relat. Elem.*, 2009, **184**, 1091.
149. F. Manzenrieder, A. O. Frank and H. Kessler, *Angew. Chem., Int. Ed.*, 2008, **47**, 2608.
150. I. S. Alfer'ev and N. V. Mikhalin, *Bull. Russ. Acad. Sci., Div. Chem. Sci.*, 1992, **41**, 1709.
151. H. Pucknat, J. Grobe, D. Le Van, B. Broschk, M. Hegemann, B. Krebs and M. Läge, *Chem. – Eur. J.*, 1996, **2**, 208.
152. P. Singh, R. Tripathi, R. K. Verma, G. C. Nandi, A. Gupta and M. S. Singh, *Phosphorus, Sulfur Silicon Relat. Elem.*, 2010, **185**, 2142.
153. S. Schneider, A. Vij, J. A. Sheehy, F. S. Tham, T. Schroer and K. O. Christe, *J. Inorg. Gen. Chem.*, 2001, **627**, 631.
154. J. Bresien, K. Faust, A. Schulz and A. Villinger, *Angew. Chem., Int. Ed.*, 2015, **54**, 6926.
155. N. K. Gusarova, S. F. Malysheva, N. A. Belogorlova, O. N. Kazheva, A. N. Chekhlov, G. G. Aleksandrov, O. A. D'yachenko, L. M. Sinegovskaya and B. A. Trofimov, *J. Struct. Chem.*, 2010, **51**, 120.
156. J. F. Stebbins, N. Kim, F. Brunet and T. Irifune, *Eur. J. Mineral.*, 2009, **21**, 667.
157. H. El-Rifai, M. Heerboth, T. E. Gedris, S. Newman, W. Orem and W. T. Cooper, *Eur. J. Soil Sci.*, 2008, **59**, 517.
158. W. Negassa, D. Michalik, W. Klysubun and P. Leinweber, *Soil Syst.*, 2020, **4**, 11.
159. R. Shinohara, A. Imai, N. Kawasaki, K. Komatsu, A. Kohzu, S. Miura, T. Sano, T. Satou and N. Tomioka, *Environ. Sci. Technol.*, 2012, **46**, 10572.

160. M. I. Makarov, T. I. Malysheva, L. Haumaier, H. G. Alt and W. Zech, *Geoderma*, 1997, **80**, 61.
161. S. Backnäs, H. Laine-Kaulio and B. Kløve, *Geoderma*, 2012, **189–190**, 50.
162. B. L. Turner, B. J. Cade-Menun and D. T. Westermann, *Soil Sci. Soc. Am. J.*, 2003, **67**, 1168.
163. A. Zhang, Z. Chen, G. Zhang, L. Chen and Z. Wu, *Eur. J. Soil Biol.*, 2012, **52**, 73.
164. M. Tchienkoua and W. Zech, *Agric., Ecosyst. Environ.*, 2003, **100**, 193.
165. O. Koukol, F. Novák, R. Hrabal and M. Vosátka, *Soil Biol. Biochem.*, 2006, **38**, 3372.
166. B. S. Ghotra, T. Vasanthan, M. Wettasinghe and F. Temelli, *Food Hydrocolloids*, 2007, **21**, 1056.
167. C. Joqueviel, R. Martino, V. Gilard, M. Malet-Martino, P. Canal and U. Niemeyer, *Drug Metab. Dispos.*, 1998, **26**, 418.
168. K. Petzold, A. Olofsson, A. Arnqvist, G. Gröbner and J. Schleucher, *J. Am. Chem. Soc.*, 2009, **131**, 14150.
169. T. Schäfer, G. Buntkowsky and T. Gutmann, *ACS Omega*, 2020, **5**, 5412.
170. B. Ringler and C. Von Hänisch, *Z. Anorg. Allg. Chem.*, 2016, **642**, 294.
171. W. Adam, C. van Barneveld and D. Golsch, *Tetrahedron*, 1996, **52**, 2377.
172. L. Schwarz, U. Girreser and B. Clement, *Eur. J. Org. Chem.*, 2014, **2014**, 1961.
173. C. M. VanZomeren, H. Knicker, W. T. Cooper and K. R. Reddy, *Methods in Biogeochemistry of Wetlands*, John Wiley & Sons Inc, Chichester, 2015, pp. 439–463.
174. B. P. Kelleher, M. J. Simpson and A. J. Simpson, *Geochim. Cosmochim. Acta*, 2006, **70**, 4080.
175. M. F. Dignac, S. Derenne, P. Ginestet, A. Bruchet, H. Knicker and C. Largeau, *Environ. Sci. Technol.*, 2000, **34**, 3389.
176. J. Wen, P. Ning, H. Cao, H. Zhao, Z. Sun and Y. Zhang, *J. Hazard. Mater.*, 2019, **364**, 91.
177. X. Q. Lu, W. D. Johnson and J. Hook, *Environ. Sci. Technol.*, 1998, **32**, 2257.
178. S. Rana, B. Pandey, A. Dey, R. Haque, G. Rajaraman and D. Maiti, *ChemCatChem*, 2016, **8**, 3367.
179. C. R. Cornman, K. M. Geiser-Bush and P. Singh, *Inorg. Chem.*, 1994, **33**, 4621.
180. T. Zandvakili, S. J. Fatemi, S. Y. Ebrahimipour, H. Ebrahimnejad, M. Dusek and V. Eigner, *ChemistrySelect*, 2021, **6**, 10405.
181. G. Tricot, O. Mentré, S. Cristol and L. Delevoye, *Inorg. Chem.*, 2012, **51**, 13108.
182. V. Sharma, S. Agarwal, R. Bohra and V. K. Jain, *J. Chem. Res.*, 2004, **2004**, 273.
183. A. Morsali and A. R. Mahjoub, *Chem. Lett.*, 2004, **33**, 64.
184. L. Pu, P. P. Power, I. Boltes and R. Herbst-Irmer, *Organometallics*, 2000, **19**, 352.
185. L. P. Cruz, J. M. Savariault, C. M. Morais, C. Fernandez, J. D. P. De Jesus and J. Rocha, *J. Solid State Chem.*, 2004, **177**, 1549.

186. K. H. Chung and C. H. Moon, *J. Chem. Soc., Dalton Trans.*, 1996, 75.
187. D. J. Kubicki, D. Prochowicz, A. Hofstetter, B. J. Walder and L. Emsley, *ACS Energy Lett.*, 2020, **5**, 2964.
188. C. K. Larive, A. Rogers, M. Morton and W. R. Carper, *Environ. Sci. Technol.*, 1996, **30**, 2828.
189. W. H. Otto, S. D. Burton, W. R. Carper and C. K. Larive, *Environ. Sci. Technol.*, 2001, **35**, 4900.
190. N. Hertkorn, E. M. Perdue and A. Kettrup, *Anal. Chem.*, 2004, **76**, 6327.
191. A. J. Simpson, *Magn. Reson. Chem.*, 2002, **40**, S72.
192. N. Goudarzi, *J. Mol. Struct.*, 2013, **1046**, 57.
193. Y. Kim, *Chemosphere*, 2015, **119**, 803.
194. J. J. Fitzgerald, A. I. Hamza, C. E. Bronnimann and S. F. Dec, *Solid State Ionics*, 1989, **32–33**, 378.
195. M. Schmücker, H. Schneider, K. J. D. MacKenzie, M. E. Smith and D. L. Carroll, *J. Am. Ceram. Soc.*, 2005, **88**, 2935.
196. B. Herreros, S. W. Carr and J. Klinowski, *Science*, 1994, **263**, 1585.
197. J. F. Stebbins and M. Kanzaki, *Science*, 1991, **251**, 294.
198. M. L. Helm, D. G. Vanderveer and G. J. Grant, *J. Chem. Crystallogr.*, 2003, **33**, 625.
199. G. J. Grant, *Struct. Bonding*, 2006, **120**, 107–141.
200. D. Fallaise, H. Balkwill Tweedie, J. Konzuk, C. Cheyne, E. E. Mack and J. G. Longstaffe, *Magn. Reson. Chem.*, 2019, **57**, 93.
201. G. Pagès, A. Gerdova, D. Williamson, V. Gilard, R. Martino and M. Malet-Martino, *Anal. Chem.*, 2014, **86**, 11897.
202. L. H. Antonides, R. M. Brignall, A. Costello, J. Ellison, S. E. Firth, N. Gilbert, B. J. Groom, S. J. Hudson, M. C. Hulme, J. Marron, Z. A. Pullen, T. B. R. Robertson, C. J. Schofield, D. C. Williamson, E. K. Kemsley, O. B. Sutcliffe and R. E. Mewis, *ACS Omega*, 2019, **4**, 7103.
203. G. Assemat, B. Gouilleux, D. Bouillaud, J. Farjon, V. Gilard, P. Giraudeau and M. Malet-Martino, *J. Pharm. Biomed. Anal.*, 2018, **160**, 268.
204. M. Defernez, E. Wren, A. D. Watson, Y. Gunning, I. J. Colquhoun, G. Le Gall, D. Williamson and E. K. Kemsley, *Food Chem.*, 2017, **216**, 106.
205. W. Jakes, A. Gerdova, M. Defernez, A. D. Watson, C. McCallum, E. Limer, I. J. Colquhoun, D. C. Williamson and E. K. Kemsley, *Food Chem.*, 2015, **175**, 1.
206. T. Parker, E. Limer, A. D. Watson, M. Defernez, D. Williamson and E. K. Kemsley, *TrAC, Trends Anal. Chem.*, 2014, **57**, 147.
207. D. McDowell, M. Defernez, E. K. Kemsley, C. T. Elliott and A. Koidis, *LWT*, 2019, **111**, 490.
208. B. Gouilleux, N. V. Christensen, K. G. Malmos and T. Vosegaard, *Anal. Chem.*, 2019, **91**, 3035.
209. A. Krause, Y. Wu, R. Tian and T. A. Van Beek, *Planta Med.*, 2018, **84**, 953.
210. K. Heerah, S. Waclawek, J. Konzuk and J. G. Longstaffe, *Magn. Reson. Chem.*, 2020, **58**, 1160.
211. P. Conte, S. Bubici, E. Palazzolo and G. Alonzo, *Spectrosc. Lett.*, 2009, **42**, 235.

212. A. Maccotta, C. de Pasquale, A. Caruso, C. Cosentino, G. Alonzo and P. Conte, *Environ. Sci. Pollut. Res.*, 2013, **20**, 4847.
213. V. A. Laudicina, C. de Pasquale, P. Conte, L. Badalucco, G. Alonzo and E. Palazzolo, *J. Soils Sediments*, 2012, **12**, 1222.
214. P. Conte, V. Marsala, C. de Pasquale, S. Bubici, M. Valagussa, A. Pozzi and G. Alonzo, *GCB Bioenergy*, 2013, **5**, 116.
215. C. de Pasquale, V. Marsala, A. E. Berns, M. Valagussa, A. Pozzi, G. Alonzo and P. Conte, *J. Soils Sediments*, 2012, **12**, 1211.
216. P. Conte, A. Maccotta, C. de Pasquale, S. Bubici and G. Alonzo, *J. Agric. Food Chem.*, 2009, **57**, 8748.
217. G. Butera, C. de Pasquale, A. Maccotta, G. Alonzo and P. Conte, *Cellulose*, 2011, **18**, 1499.
218. A. E. Berns, S. Bubici, C. de Pasquale, G. Alonzo and P. Conte, *Org. Geochem.*, 2011, **42**, 978.
219. P. Conte, C. Abbate, A. Baglieri, M. Nègre, C. de Pasquale, G. Alonzo and M. Gennari, *Org. Geochem.*, 2011, **42**, 972.
220. P. Conte, V. Loddo, C. de Pasquale, V. Marsala, G. Alonzo and L. Palmisano, *J. Phys. Chem. C*, 2013, **117**, 5269.
221. A. Pukalskas, T. A. van Beek and P. de Waard, *J. Chromatogr. A*, 2005, **1074**, 81.
222. O. Corcoran and M. Spraul, *Drug Discovery*, 2003, **8**, 624.
223. S. H. Hansen, A. G. Jensen, C. Cornett, I. Bjørnsdottir, S. Taylor, B. Wright and I. D. Wilson, *Anal. Chem.*, 1999, **71**, 5235.
224. A. J. Simpson, L. H. Tseng, M. J. Simpson, M. Spraul, U. Braumann, W. L. Kingery, B. P. Kelleher and M. H. B. Hayes, *Analyst*, 2004, **129**, 1216.

CHAPTER 3

Sensitivity Enhancement in Environmental NMR: Current Technologies and Future Potential

W. W. WOLFF, D. H. LYSAK, P. M. COSTA AND A. J. SIMPSON*

Environmental NMR Center, University of Toronto, Canada
*Email: andre.simpson@utoronto.ca

3.1 Introduction

NMR is a highly versatile technique for environmental analysis. It can simultaneously localize, quantify, and identify the structure of molecules through non-destructive and non-invasive methods. In addition, NMR can analyze dynamic systems, changes in materials over time, and non-covalent binding. This makes it a critical technique for environmental research although its Achilles heel is relatively low sensitivity.[1–3] This can limit the development and application of many novel experiments that would otherwise aid in characterizing and understanding the behaviour of environmental samples.

The application of NMR to environmental samples is particularly challenging as the behaviour of the systems that one investigates must be analyzed at environmentally relevant concentrations (*e.g.*, *in vivo* toxicology) and in some cases samples cannot be concentrated, dried, or extracted without losing information from the natural state.[3–5] While there are many methods available for improving the sensitivity of NMR, many of these have strict requirements as to the nature of the sample or technically challenging instrumentation.

New Developments in NMR No. 32
The Environment in a Magnet: Applications of NMR Techniques to Environmental Problems
Edited by Pellegrino Conte, Delia Francesca Chillura Martino and Paolo Lo Meo

Published by the Royal Society of Chemistry, www.rsc.org

In addition, environmental samples are difficult to analyze as they have broad variations of matrices, including air particles, complex aqueous (and sometimes salty) mixtures, slurries, heterogenous soils, sediments,[1] and even live organisms.[3] The lack of control over the sample matrix further challenges the development of NMR spectroscopy for environmental applications, as many of these challenging matrices cannot be optimized (*i.e.*, *in vivo* NMR). Nevertheless, despite these challenges, there are many new and emerging methods in cutting-edge NMR that have the potential to provide substantial sensitivity enhancements moving forward. Many of these approaches have yet to be developed for or applied to environmental samples. In part this may be down to the fact that environmental researchers are not exposed to such approaches, which tend to be presented at the main NMR conferences and NMR niche journals, but never cross the pedagogical boundaries in the environmental field. The purpose of this chapter is to provide a summary of some of the main approaches and hopefully act as a resource that could promote further research in the future in environmental science. Due to the lack of environmental applications to date, this chapter will mainly focus on describing the techniques, the basic theory behind them, example applications in other fields, and future potential for environmental research. We hope that the chapter will represent a stepping-stone and bridge to more comprehensive treatises in the specialized literature.

3.1.1 The Origin of an NMR Signal

It is worth briefly discussing the origin of the NMR signal to understand the intrinsic limit of sensitivity, the basis of the sensitivity enhancement approaches, and the decades of continuing effort which have gone into boosting the sensitivity of NMR. NMR relies on the splitting of degenerate nuclear spin states in a magnetic field (often called Zeeman splitting), which produces a vanishingly small population difference (at room temperature) according to the Boltzmann distribution. This population difference is proportional to the magnetic field strength (B_0) and the gyromagnetic ratio of the nuclei of interest, which can be thought of as the "sensitivity of the nucleus to an external magnetic field" (Figure 3.1).[6]

It is important here to emphasize that this energy gap is small – smaller than thermal energy at room temperature, which ultimately leads to a very small population difference.[6] The signal is ultimately measured from this very small population difference and almost always requires multiple repeated scans to obtain spectra with appropriate signal-to-noise ratios. The signal-to-noise ratio increases only with the square root of the number of scans, placing a significant upper practical limit on how many scans one can collect. The ever-increasing cost of liquid helium (required by NMR magnets) and the high maintenance costs required to operate these systems make efficient use of instrument time a necessity and makes environmental applications particularly challenging, where samples are often complex mixtures of hundreds of compounds with low concentrations.

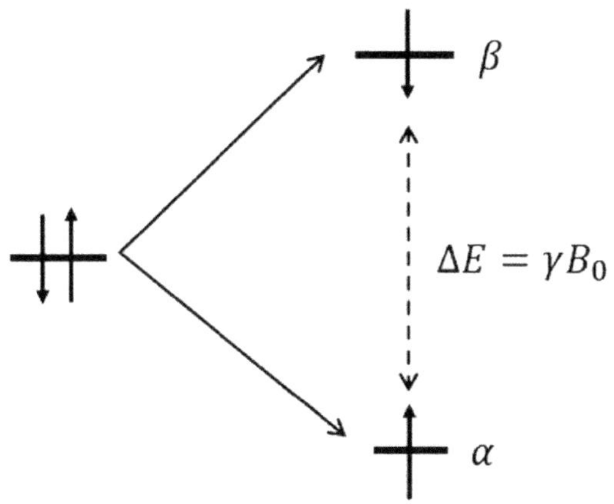

Figure 3.1 Zeeman splitting of the two degenerate spin states from a spin-1/2 nucleus in a strong magnetic field, B_0.

There have been a lot of developments in NMR which have been proven to enhance the sensitivity of NMR which is the subject of this chapter. These techniques can be grouped together and broken down into three broad categories: hardware, experimental design (NMR pulse sequences), and hyperpolarization techniques. Although these categories may be combined in practice, each approach has different strengths, challenges, limitations, and compatibility with other sensitivity enhancing NMR techniques, depending on the sample and instrumentation employed. Sensitivity enhancement is usually achieved by designing improved instrument hardware as well as directly enhancing the signals from the matrix that one studies beyond the Boltzmann distribution (hyperpolarization). As the cost of these methods can be prohibitively expensive, an "indirect" enhancement in sensitivity can instead be afforded by time savings *via* pulse sequences or acquisition schemes, such as by non-uniform sampling techniques[7] and NMR supersequences,[8] among other means of spin manipulation. Finally, it is important to note that changing the sample itself is not considered here, as well documented in previous reviews.[9] However, proper preparation of a sample, be it *via* concentration,[10] chemical treatment,[11] isotopic enrichment,[3,12] sample compression[13] or extraction,[14] is always a key first step towards the best NMR data possible.

While the main purpose of this chapter is to provide information on the progress and potential of sensitivity enhancement techniques in environmental NMR, a brief theoretical explanation of each technique will be provided to illustrate the challenges in its implementation.

3.2 Hyperpolarization Experiments

Hyperpolarization experiments aim to address the root cause of the intrinsic low sensitivity of NMR directly by artificially boosting the population difference in the nuclear spins, resulting in significantly greater sensitivity. Despite being one of the most challenging methods to implement in "realistic" samples, hyperpolarization can equally be the most rewarding, as it is the only technique that can ultimately solve the root cause of the low sensitivity in NMR and, in optimized cases, produce a population difference that approaches 100% of the spins within the sample! For example, signal enhancements producing only a 34% population difference yield enhancements as high as 250 000 times (at 0.4 T), as reported for naphthalene crystals doped with pentacene.[15]

In hyperpolarization, a polarizing agent with a population difference in excess of the targeted nucleus is added to one's sample. This polarization can then be transferred to the desired nuclear spins in order to make measurements with much greater sensitivity. The hyperpolarization schemes described below can often seem highly specific to a set of conditions, targets or solvents, but often the degree of specificity in a hyperpolarization experiment can either be used to explore the behaviour of the sample or combined with conventional NMR pulse sequence techniques in order to expand the scope of the sensitivity enhancement.

In general, hyperpolarization techniques may make a significant difference in the limit of detection for NMR, further enhancing the sample scope for which NMR may be applied, particularly in environmental monitoring or metabolomics, where sensitivity is particularly limiting.[16] Many hyperpolarization techniques could otherwise be used to support qualitative experiments, or those in which quantitation is not required, such as in critical multidimensional NMR spectra used for the assignment or resolution of molecules in environmental samples (*e.g.*, HSQC, COSY, HETCOR, TOCSY, DOSY), or to support highly insensitive NMR techniques which yield additional resolution (in-phase intermolecular single quantum coherence (IP-iSQC)[17] and pure shift NMR)[18] without the usual reduction in sensitivity. Despite the apparent loss of true quantitation, with the appropriate pulse sequence, many of these techniques are amenable to at least quantitation by spiking.[19] Finally, some hyperpolarization methods can be made useful in environmental NMR as semi-selective techniques, exploiting the differences in binding, molecule dynamics, and distributions that drive some of these sensitivity enhancement techniques to support NMR characterizations.[20] Many other hyperpolarization techniques that will not be discussed here have been described in literature reviews on different scopes of hyperpolarization.[21–23] Here some of the more popular approaches with future potential for environmental NMR are briefly considered.

3.2.1 Dynamic Nuclear Polarization (DNP)

DNP uses the 658-fold greater sensitivity of electrons compared to protons to enhance the sensitivity of NMR experiments in both solid and solution states. A polarizing agent with an unpaired electron is integrated into the sample, which can be manipulated by microwave irradiation to hyperpolarize a variety of matrices. Conceptually, it is analogous to the familiar NOE and INEPT techniques,[6] which exploit the relatively large polarization of protons compared to heteronuclei (*e.g.*, ^{13}C and ^{15}N). In such experiments, the polarization is transferred from the protons to the heteronuclei to enhance the sensitivity of measurements made on these heteronuclei. In DNP, the transfer is from the electron spins to the nuclear spins. DNP has a very broad scope relative to some of the other hyperpolarization techniques that will be discussed and can be executed on solids and solutions with additional experimental variations on the core processes. Mechanistically, the applications of this technique are very different and have developed into multiple experimental approaches which will be described here. In order to understand the challenges and possibilities of the practical applications of DNP, it is prudent to briefly summarize the theory behind DNP, and the experimental procedures and transfer mechanisms by which it operates. While this is intended to be introductory, the interested reader is directed to extensive reviews of DNP theory.[24–30]

Electrons have spin-1/2 and thus undergo Zeeman splitting and precess in a magnetic field, with an energy gap corresponding to frequencies in the microwave range.[26,31] Unpaired electrons form a population difference which can be manipulated to perform electron paramagnetic resonance (EPR) experiments or hyperpolarize nuclei.[26] For DNP, the key interaction is the hyperfine interaction, an analog of nuclear *J*-coupling, which contains isotropic and anisotropic (orientation independent and dependent coupling contributions), respectively, similar to dipole–dipole coupling in NMR (Figure 3.2).[32]

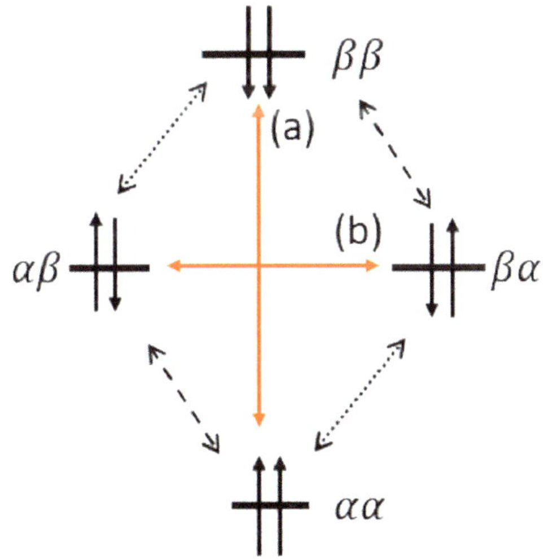

Figure 3.2 An example of an energy level diagram of a nucleus (first letter) and an electron (second letter) hyperfine-coupled together in an NMR spectrometer. The population difference of the nuclear spins (the transitions indicated by the dotted lines) may be amplified by manipulating the population differences of the electrons: directly (*via* transitions indicated by the dashed lines) or indirectly by allowing double quantum (a) and zero quantum (b) transitions between the states $\alpha\alpha$ and $\beta\beta$, and $\alpha\beta$ and $\beta\alpha$, respectively.

By applying microwave frequencies at an appropriate frequency to the sample, the polarization of the electrons may be transferred to the nuclei by four main mechanisms which depend on the conditions of the experiment and properties of the sample: the solid effect (SE),[25] Overhauser effect (OE),[33] thermal mixing (TM),[28] and cross-effect (CE).[25] While each transfer mechanism is unique, DNP in general is limited by the difficulty of generating microwaves at fields greater than 3.4 T, which has traditionally required a gyrotron (a separate superconducting magnet system) to generate microwaves at the necessary frequencies, which adds prohibitive cryogen and instrument costs to the experiment.[34] However, recently introduced solid-state diodes offer future potential to clear this hurdle in the future.[35] In addition, there is also a challenge in generating microwave frequencies in a certain range (the "terahertz gap") which makes such experiments impossible at some magnetic fields.[27,34,36,37] Special emphasis is currently being placed on the development of DNP hardware, and this may drive further proliferation by reducing costs and making it more accessible to a wider range of NMR facilities. Despite these serious complexities at higher field, DNP is a very versatile technique spanning different sample matrices. While the technique and microwave instrumentation are still under active development, it has many potential applications as a tool to non-specifically enhance the sensitivity of environmental NMR.

3.2.1.1 DNP in Solution: Overhauser DNP

For solution-state DNP, Overhauser DNP (ODNP) exploits the only transfer mechanism (the OE) available for one-step hyperpolarization of solution-state experiments, using transient through-space interactions that allow the spins of saturated radical polarizing agents (*e.g.*, nitroxide radicals[38]) to cross-relax to the nuclei, boosting the population difference by up to 658-fold for protons, and even higher for heteronuclei with smaller gyromagnetic ratios.[31]

Unfortunately, the transfer mechanism of ODNP becomes significantly less effective with increasing magnetic field,[27] as the key condition for efficient transfer, where the product of the frequency and correlation time of the polarizing agent less than one is difficult to achieve at higher magnetic fields (and frequencies).[27] These, as well as the difficulties associated with obtaining microwave generators for high magnetic field, have led to many research groups using field cycling[39] or sample shuttling[40] in order to obtain both large ODNP enhancements and high resolution at high field.

While much of the ODNP literature focusses on rapid sample shuttling or transfer from a polarizing field to a high field, an under-exploited opportunity for Overhauser DNP (especially enhanced by triplet DNP; see Section 3.2.1.4) may permit low field NMR systems to play a larger role in environmental NMR, particularly as portable NMR systems, or to aid in their proliferation in environmental monitoring. Despite the lower resolution at low magnetic fields, it is possible to run multidimensional experiments using ODNP, as the repetitive nature of ODNP (contrasted with sample shuttling, and dDNP) allows complex experiments requiring phase cycling and multiple scans to be collected without the hinderance of the much lower sensitivity of low field NMR systems, such as 2D correlation experiments (Figure 3.3).[41,42]

Another particularly interesting example is in the acquisition of a ^{1}H JRES spectrum hyperpolarized 300-fold by ODNP directly at 0.35 T as a proof of concept, as many multidimensional NMR techniques in environmental NMR are exceedingly limited by sensitivity.[41] To put this into perspective, the 300-fold enhancement at 0.35 T (14.90 MHz) results in an effective sensitivity of a 15.69 T (667.73 MHz) spectrometer. Furthermore, collecting directly at a low field circumvents the challenges associated with extremely rapid relaxation during sample shuttling and the complex instrumentation needed for field cycling.[40] ODNP has variable polarization enhancements on different molecules and sites and although it may be an excellent tool for qualitative analysis in complex mixtures,[43] ODNP is particularly potent for multidimensional NMR experiments, where quantitation is typically not a concern. However, prohibitively long acquisition times prevent collection of spectra with sufficient resolution along the indirect dimension or with sufficient sensitivity. For extremely mass limited samples, ODNP has been combined with microfluidics and microcoil detection to further amplify the sensitivity enhancements.[44]

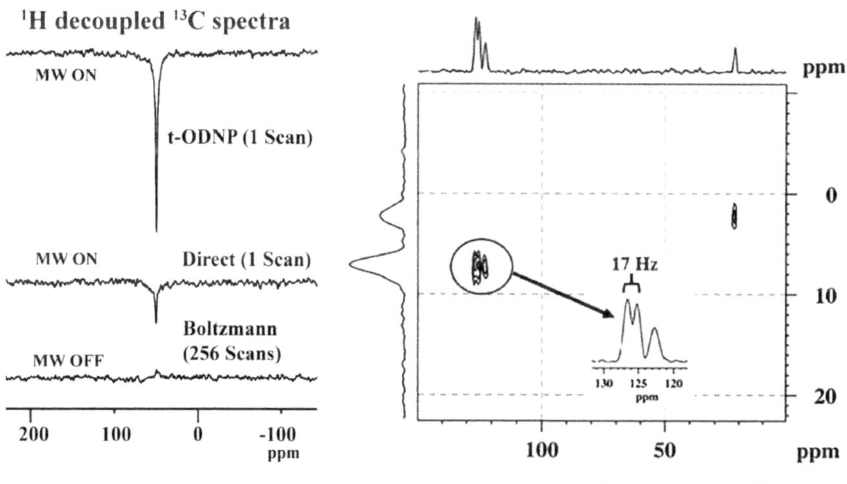

Figure 3.3 Examples of ODNP. Left: Solution-state ODNP ^{1}H-decoupled ^{13}C spectra of ^{13}C-enriched methanol with 40 mM TEMPO as the polarizing agent polarized and acquired at 1.2 T. From top to bottom: proton-hyperpolarized refocussed INEPT experiment (52.3 SNR per scan), direct ODNP of ^{13}C (11.9 SNR per scan), and an unpolarized ^{13}C NMR spectrum (0.2356 SNR per scan). The enhancements for the INEPT transferred DNP enhancement and direct DNP enhancement correspond to 222× and 51×, respectively. Right: HETCOR spectrum of neat toluene (8 µl) with 20 mM TEMPO acquired with transferred ODNP (1.2 h total time). Reproduced from ref. 42 with permission from American Chemical Society, Copyright 2017.

3.2.1.2 Magic Angle Spinning (MAS-DNP): Hyperpolarization of Solids

DNP may also be applied to solid-state NMR samples in magic angle spinning DNP (MAS-DNP) at high field (Figure 3.4).[45]

MAS-DNP experiments primarily exploit a variation of SE, TM, or CE mechanisms,[28,29] which exploit the electron–nuclear interactions in different ways to enhance the population difference of the nuclei. Which mechanism is most efficient for a particular sample is dependent on the experimental conditions (magnetic field strength, polarizing agent concentration, structure, and MAS spinning rate) under which the experiment is performed and the properties (*e.g.*, lineshape) of the radical agent.[28,29]

The SE mechanism directly drives the typically spin forbidden double and zero quantum transitions to increase the population differences of the nuclear spin states (see Figure 3.2(a) and (b)).[30] The cross effect manipulates the differences between two electron spin populations (in a biradical) coupled to a nucleus in order to enhance the differences in the nuclear spin populations.[28] Thermal mixing operates similarly, using transitions between

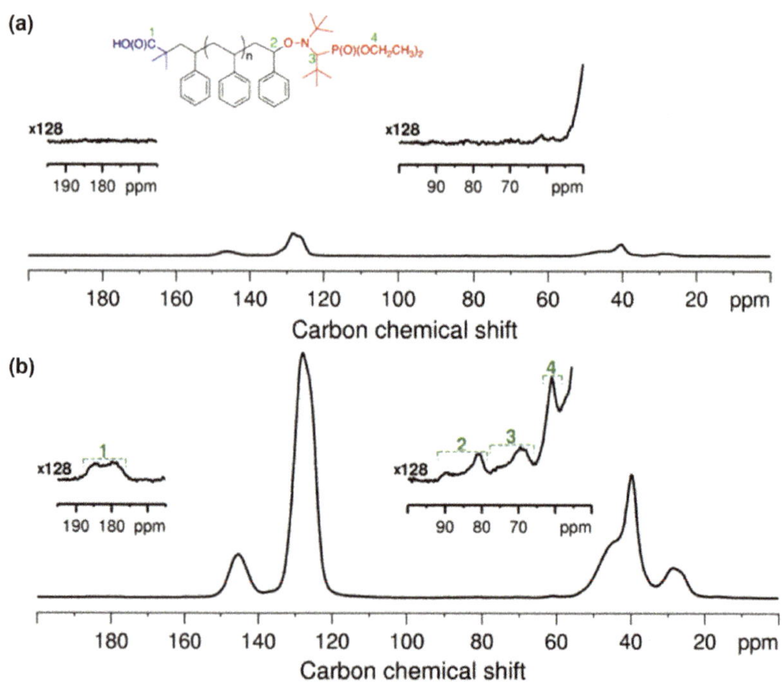

Figure 3.4 ^{13}C cross-polarized (CP)-MAS NMR spectra of an acrylate-terminated poly (ethylene oxide) with a 15 h duration. (a) Conventional CP experiment. (b) Hyperpolarized CP-MAS experiment at an identical intensity scale, corresponding to a 13-fold increase in sensitivity. Reproduced from ref. 45 with permission from Elsevier, Copyright 2019.

the states of coupled electrons to induce a change in a hyperfine-coupled nuclear spin state, resulting in an increase in the population difference.[28]

Many variations of the solid effect exist in order to maximize the efficiency of this polarization transfer scheme;[30,45] however, they all become less efficient at higher magnetic field due to their transfer mechanisms. For example, in the SE, the polarization decreases since the probability of the "forbidden" transition decreases with the square of the magnetic field.[29]

An extra emphasis should be placed on the possible future developments of MAS-DNP, especially given recent experiments at high (21.15 T) fields which yielded 200-fold signal enhancements.[46] Most samples on which MAS-DNP experiments are performed are often crystalline matrices, with the polarizing agent dispersed evenly throughout the sample.[47] To be useful for environmentally relevant solids (*e.g.*, soils, small organisms and seeds), the technique must be compatible with solids that may be either completely amorphous and have nonuniform distributions of the polarizing agent: both of these shortcomings may have been sufficiently addressed for application in environmental NMR, as examples of successful MAS-DNP in microcrystalline powders[47] and amorphous solids[48] have been presented in the literature, overcoming the reliance on crystalline materials by spin diffusion to hyperpolarize nonuniform matrices. Microcrystalline solids are particularly interesting in terms of environmental relevance: while one expects no significant crystallinity in soils, a demonstration that sensitivity enhancements can occur by spin diffusion with a non-uniform distribution of nuclear spins suggests potential usefulness of DNP as applied to soil samples.[48]

In addition, comprehensive multiphase (CMP) NMR may benefit from MAS-DNP (or DNP in general) indirectly in the case where the enhancement is not uniform, as the differences in spin diffusion would yield different polarization enhancements and consequently provide information about the solvent accessibility of portions of the sample to the solvent or polarizing agent by the differences in signal enhancement from a solvent containing a polarizing agent.[48]

3.2.1.3 Single Shot: Chemically and Dissolution Induced Hyperpolarization

An interesting alternative approach to solution-state hyperpolarization is with dissolution DNP (dDNP), which can yield more effective experiments in solution-state DNP, capable of producing maximal population differences that approach 100% under appropriate conditions.[49] Dissolution DNP hyperpolarizes a solid sample, after which it is dissolved in a warm solvent prior to measurement. The capability of hyperpolarizing the sample as a solid allows the experimenter to use extremely low temperatures, at which the largest population difference can be produced, optimize the hyperpolarization efficiency for the polarizing agent, and avoid unwanted destruction of the radical polarizing agent prior to hyperpolarization with an element of the mixture.[49] Like field transfer protocols for ODNP, dDNP requires polarization in an external field prior to dissolution in the sample to be measured.[50] Like all field cycling or shuttling methods, this allows the hyperpolarized material to be used at any field strength independently of the polarizing field.

Single shot experiments carry the apparent disadvantage of being one-time only. This makes it impossible to perform experiments requiring phase cycling or multiple transients. By contrast, this technique has the possibility to perform high throughput hyperpolarization experiments on a broader scope of matrices (*e.g.*, wastewater samples), as one optimized procedure can be applied to hyperpolarize a larger set of samples. In addition, "ultrafast NMR" which generates a 2D NMR spectrum with spatial selection techniques can also be applied with single shot hyperpolarization.[51]

As such, it is possible to obtain important information in a single, shot and in environmental NMR, dDNP may be useful in characterizing the metabolic products of environmental contaminants to aid in understanding the pathways to sublethal and lethal effects on organisms. dDNP has been applied to enzyme catalyzed reactions to obtain a ^{13}C signal enhancement of approximately 1000-fold, permitting acquisition of a ^{13}C NMR spectrum of approximately 33.6 mM of ^{13}C labelled lactate in a single scan, though the authors note that this concentration could be lowered while still acquiring a sufficient signal in one scan, as can be evidenced by the high SNR of the hyperpolarized spectrum.[40]

Another particularly interesting approach for environmental samples afforded by dDNP is dissolution in hyperpolarized water to minimize the changes made to the sample and maximize the scope for dissolution DNP.[38,52] The polarization may be then transferred from the water molecules to any targets with labile protons by exchange, and by employing mixing sequences, relayed across the entire molecule, similarly to the reverse process described later in this chapter.[53] In addition to a broad scope, it is possible to avoid adding organic radicals, instead of generating them by exposure to UV light.[38] Compatibility with particularly sensitive samples that may react with the radical agent could be improved by adding radical scavengers to the solution during the dissolution phase to destroy the radicals after the polarization stage, reducing possibly unwanted reactions with high concentrations of polarizing agents. Similarly, this avoids the potential for low hyperpolarization efficiency of environmental samples containing radicals which causes quenching of the polarizing radicals, improving the performance of DNP in challenging environmental samples.

3.2.1.4 *Triplet DNP: Beyond the Boltzmann Distribution*

One of the most exciting aspects of DNP is the capacity to reach 100% polarization within a sample, which can be achieved by performing experiments at near-zero temperatures or by further modification of the experimental scheme to boost the electron population difference beyond the Boltzmann distribution.[54] DNP can further be combined with laser pulses in order to generate triplet states to boost the population difference of the electron spins significantly above the difference allowed by the Boltzmann distribution, which allows obtaining massive enhancement factors compared to "standard" DNP (Figure 3.5).

Triplet DNP relies on intersystem crossing (ISC) which allows the formation of a triplet excited state of a molecule during relaxation (Figure 3.5).[54] As the electron relaxes, a change in vibrational energy levels with little to no energy difference occurs so that a change in the spin state is preferred over relaxation to the original ground state.[54] This allows the electron to relax to a unique energy level which is unpaired and relaxes much more slowly compared to the original excited state.[54] Usually, triplet DNP is achieved by molecules with polarizing agents with triplet excited states such as dyes and aromatic functional groups.[54] More research was performed using crystal samples[54–56] to ensure uniform distribution and exploit the orientation-dependent electron–nuclear interactions and to yield optimized enhancements. However, these are often not broadly applicable to many molecules or mixtures, which will not crystallize appropriately into a matrix containing the polarizing agent. Despite many of the challenges of triplet DNP, more stable and robust polarizing agents and strategies have been developed which broaden the usefulness of this technique to make it compatible with solution-state experiments.[54,55] One of the more exciting avenues for environmental NMR is the development of nanocrystals which host a pentacene-doped *p*-terphenyl polarizing agent, suspended in an aqueous solution with the assistance of a surfactant.[55] This extends the scope of this technique to aqueous solutions and opens an avenue for employing triplet DNP in solution-state environmental NMR for the analysis of trace contaminants in wastewater.[55] Similarly to Overhauser DNP, this could allow highly selective and multidimensional

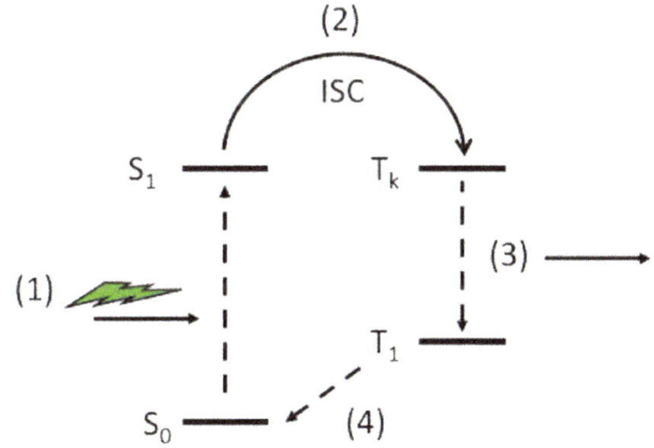

Figure 3.5 Triplet DNP scheme. A singlet ground state of a molecule is excited by a laser pulse of the appropriate frequency (1) to the excited state S_1. Intersystem crossing (ISC) allows the excited singlet state to interconvert to an excited triplet state (2), which relaxes to a ground-state triplet state (3). The huge population difference produced by the triplet state allows the polarization of the nuclei of interest (by the SE, CE, OE, and TM mechanisms) to obtain the sensitivity enhancement, and the system eventually relaxes to the true ground state S_0 (4).

NMR experiments to be vastly accelerated with repetitive polarization of the solution. Avoiding the enormous signals from the polarizing agent can be achieved with the use of T_2 filtering of the comparably large nanocrystals to minimize the impact on the NMR spectrum, as is usually performed with *in vivo* NMR.[57]

3.2.2 Signal Amplification by Reversible Exchange (SABRE)

SABRE uses hyperpolarized hydrogen gas (*para*-hydrogen, pH_2) in order to increase the signal intensity of target substrates in solution that are compatible with the iridium catalyst to facilitate the transfer.[58,59]

SABRE requires a polarized hydrogen generator to generate stable singlet states of H_2 (*para*-H_2, pH_2) (with appropriate subscript for number) at cryogenic temperatures in an oscillating field.[58,59] Once generated, these stable singlet states are unable to relax to thermal equilibrium under appropriate storage conditions. When pumped into a medium containing the spin-transfer catalyst, the catalyst reversibly coordinates with both *para*-hydrogen and the substrate of interest. In doing so, dipolar relaxation from the *para*-hydrogen to the substrate occurs, transferring the polarization to amenable substrates (Figure 3.6).[59]

The hydrogen and substrate bound to the spin-transfer catalyst readily exchange with other pH_2 molecules, allowing the repeated polarization of substrates capable of binding to the spin-transfer catalyst.[60] Although SABRE is more selective compared to DNP, as it requires that the target molecule reversibly exchanges with the iridium catalyst, SABRE has the advantage of having the support of several pulse sequences which permit the polarization transfer of pH_2 to the substrate directly at high field (LIGHT-SABRE),[60–62] which does not result in demanding RF requirements for the MR hardware.[62] At a glance, this

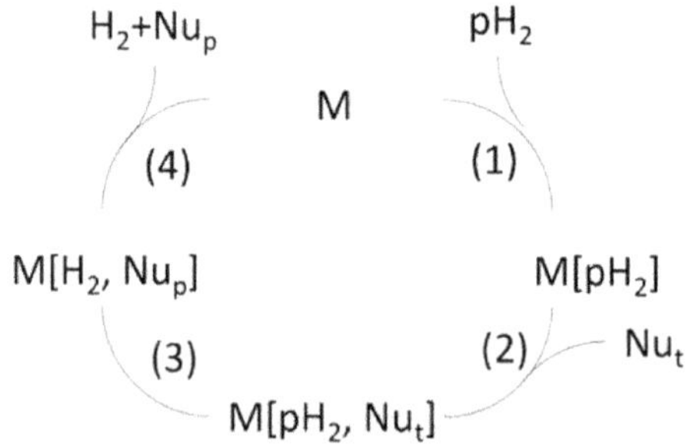

Figure 3.6 Catalytic cycle for SABRE. (1) Polarized parahydrogen gas (pH_2) is bubbled into the solvent containing a symmetry-breaking transition metal complex to which it binds (M, ligands not shown). (2) A reversibly binding ligand binds to another site on the transition metal complex. (3) Under RF application, the polarization is transferred from pH_2 to the reversibly bound ligand. (4) Both the ligand and H_2 dissociate and can exchange with an unpolarized nucleus and pH_2 to repeat the hyperpolarization procedure.

technique appears highly specific, which is useful in some applications requiring targeted analysis, with examples including hyperpolarization of amino acids[63,64] and metabolites in urine.[65] It is possible to extend the scope of a SABRE experiment using SABRE-RELAY, which uses an intermediary carrier (usually the solvent or co-solvent) in order to extend the polarization from molecules that directly bind to the spin-transfer catalyst, to any molecule with exchangeable protons, which significantly expands the scope of SABRE, and further enables the solvent accessibility of contaminants to be probed by this reaction-dependent transfer scheme.[63,66–68]

The recent developments of water-soluble catalysts,[68–70] theoretical evaluations of water as the intermediate polarization transfer catalyst,[71] solid-state catalysts,[72] and improvements in the ability to store the pH_2 for long periods of time[72] yield several interesting possibilities for environmental NMR. While soluble catalysts are inappropriate for *ex vivo* NMR, bead bound spin-transfer catalysts which are stable in water may allow SABRE-RELAY to be applied *ex vivo* (albeit unlikely *in vivo* as the experimental setup requires pure, pressurized hydrogen gas!), allowing significant enhancements of metabolite signals for more rapid acquisition of multidimensional NMR spectra and qualitative NMR experiments. Like ODNP, SABRE has the advantage of repeatable hyperpolarization of liquid-state samples and has potential in supporting low field NMR for environmental monitoring and enabling the acquisition of otherwise time-consuming multidimensional or highly selective experiments.[58] Conversely, SABRE's selectivity could be used in order to probe the interactions between environmental contaminants and other components of the sample, as the catalyst can be tuned in order to promote binding and subsequent hyperpolarization of appropriate substrates. This could then be used in interaction studies with a greater degree of sensitivity to explore how contaminants interact with different matrices.[2] The ability to transport pH_2 without loss of polarization for weeks[72] could further enable this method to be used to support the deployment of portable low field NMR systems, as the hyperpolarized pH_2 may be transported to the field or monitoring stations whilst only requiring one centre capable of generating the pH_2.

3.2.3 Chemically Induced DNP (CIDNP)

Chemically induced DNP similarly uses radical activity to polarize the sample but uses a chemical reaction rather than microwave irradiation to achieve the polarization transfer of the nuclei of interest.[73] CIDNP relies on spin chemistry, an unexpected outcome in radical reactions in which the electronic spin states play a role in the reaction rates, producing very large population differences of electrons with radical reactions in both solids and solutions (Figure 3.7).[74,75]

In photo-CIDNP, the triplet state radical is first formed by photoexcitation, which quickly reacts with the sample to abstract an electron from another molecule.[73] The back reaction may occur between the triplet radical pair and

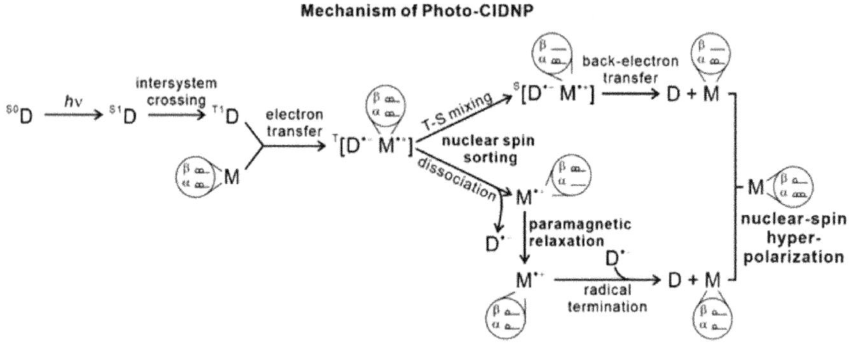

Figure 3.7 CIDNP mechanism. A dye molecule (D) is excited from the ground state (S_0) to the singlet state (S_1). Following ISC to the triplet excited state T_1, the system undergoes a radical reaction with the target molecule (M) which is selective for the spin state, which ultimately results in a hyperpolarized target molecule (M). Reproduced from ref. 77 with permission from American Chemical Society, Copyright 2022.

the new radical, but only upon the conversion of the triplet pair to the singlet state owing to spin selection rules which otherwise forbid the recombination.[73] A large population difference develops in the sample because this propensity for recombination depends on the spin state of the nuclei which are hyperfine-coupled to the electron. Those in the alpha state are more likely to undergo a recombination reaction (reforming the original reactants), while those in the beta state are more likely to undergo the escape reaction (Figure 3.7).[73] Under the appropriate conditions, CIDNP is further a repeatable/replicable hyperpolarization method which allows multi-scan, phase cycled or multidimensional experiments with a significant improvement in sensitivity and the limit of detection (Figure 3.8).[73]

Like DNP, CIDNP can be performed in both solid- and solution-state experiments.[75] One advantage of CIDNP compared to SABRE and microwave-based DNP is the much lower RF requirements of the experiment compared to the other techniques, which reduces power demands on NMR probes. In environmental NMR, there are also lots of viable targets for CIDNP which allow the experimenter to avoid modifying the sample: flavins, photosynthetic systems,[75] aromatic amino acids *in vivo*,[73–77] and environmentally relevant aromatic contaminants and dyes.[73,78–80] Furthermore, it may not just prove useful for sensitivity enhancement of qualitative NMR experiments,[77] but can yield information on the dynamics and solvent accessibility of systems,[81] and could further be used to probe radical reactions at low concentrations seen in the environment or to better understand the toxicological pathways of environmental reactions *in vivo*.[74,76] Another unique option for this experiment is in the imaging of fluorinated contaminants, which could be used to explore the partitioning of these contaminants in soil.[80]

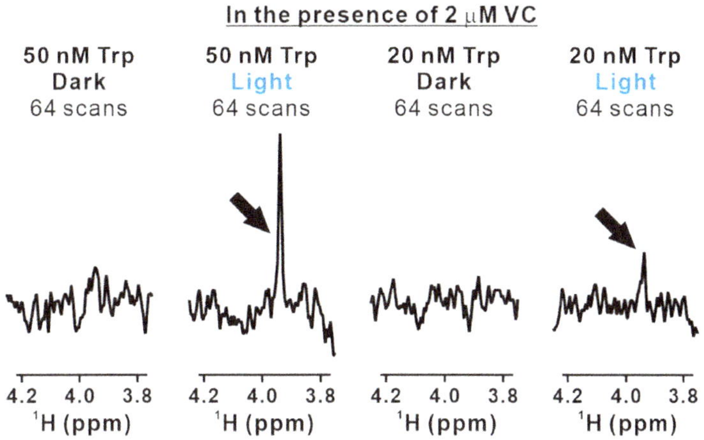

Figure 3.8 CI-DNP of nanomolar tryptophan in the presence of 2 μM ascorbic acid. The hyperpolarization allowed a spectrum of tryptophan to be collected at 20 nM, which corresponds to a limit of detection of 3 ng! Reproduced from ref. 77 with permission from American Chemical Society, Copyright 2022.

3.3 Hardware

Arguably the most well-established and most broadly applicable way to improve sensitivity, hardware improvements have made significant contributions to sensitivity enhancement in all fields of magnetic resonance. These span a wide range of modifications, from designing specialized radiofrequency (RF) microcoils and microcoil arrays[82–84] to several improvements to the strength of the main field of instruments and detection hardware.[85] While significant work has gone into developing sophisticated NMR spectrometers and supporting electronics, optimizing and improving their design can yield significant enhancements in sensitivity which do not depend on alterations made to the sample or the dynamics of the system, in contrast to hyperpolarization experiments. The advent of advanced manufacturing techniques (3D printing,[86] CNC machining,[87] *etc.*) and the progressive miniaturization of electronic components has allowed more elaborate designs to be produced and, ultimately, greater sensitivity.[87] In this section, sensitivity enhancements made to every significant electrical or hardware component of an NMR spectrometer will be discussed.

As NMR has an intrinsically low signal, the noise generated from the sample, as well as electrical noise from the RF receiver coils and the core NMR electronics and cables, must be considered because these can have comparable intensity to the actual signals.[88] Understanding the sources of noise and signal loss in the receiving electronics provides a context for the approaches for sensitivity enhancement at the hardware level.

Some of the sources of noise or signal losses, such as magnetic losses stemming from the sample and skin resistance of the RF coil material, cannot be changed.[88] Coil (and other) ohmic resistance, electrical losses, mass sensitivity, and coupling between the coil and sample can all be adjusted by optimizing each element along the signal path.[88] Maximizing the NMR signal while minimizing these sources of electrical noise is ultimately the goal for hardware sensitivity enhancements. The techniques discussed in this chapter attempt to enhance sensitivity by addressing the source of both signal and noise: by careful design of the receiver hardware, massive enhancements in sensitivity (or reductions in noise) can result in impressive gains in the signal-to-noise ratio.[89] Much of the literature focusses on improving these aspects independently, although most of these techniques can be combined for the ultimate enhancement in sensitivity.

3.3.1 Anatomy of an NMR Spectrometer

The entire path that the signal from the sample takes can be altered or optimized to boost the sensitivity of NMR by either amplifying/maintaining the signal or reducing the noise contributions (Figure 3.9).

NMR is a resonance phenomenon where the weak signal from the precessing nuclei induces a weak current in a receiver, an RF coil forming part of a tuned resonant LC circuit. This RF coil is typically tuned and matched to both be resonant at the frequency of interest (tuned, allowing the signal to "build up" in the receiver circuit) and reflect as little of the signal from the sample as possible (matched) (Figure 3.9).[88] The weak NMR signal is amplified by low noise amplifiers positioned as close as reasonably possible to the probe to minimize the amount of electrical noise prior to amplification before the second round of amplification; this first round of amplification is very crucial for signal intensity. Finally, the signal is down-converted to the audio-frequency range for converting the analog signal into a digital signal, and to achieve digital quadrature detection. Each step before digitization can be optimized to minimize the noise contributions. While it is worth noting the noise figures of the electronics and transmission cables can be

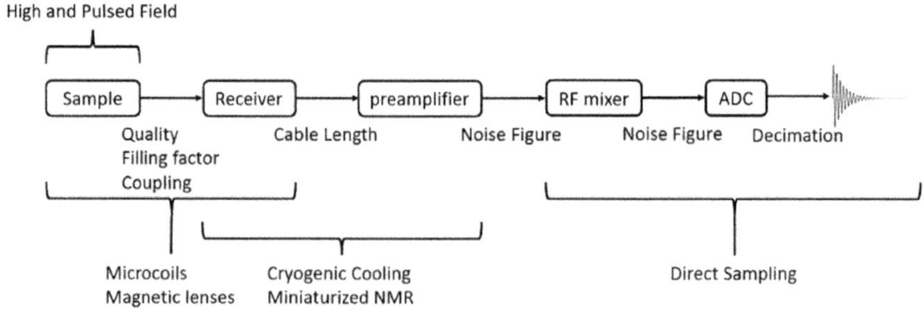

Figure 3.9 Block diagram of an NMR backend. The diagram shows the pathway of the signal through the electronics from the sample through the receiver coil, amplification, and processing and filtering before the final treatment of the raw signal; conversion into digital data by the analog-to-digital computer.

adjusted (especially before the first round of amplification), the focus here will be on approaches that boost sensitivity significantly: the use of microcoils and magnetic lenses, higher magnetic fields, and cryogenic cooling of electronics.

Maximizing the magnetic field is an expensive, but otherwise convenient method to boost the sensitivity of NMR measurements: with the exception of microcoil design, the remaining sensitivity enhancements to be made in NMR all rely on reducing the noise of the NMR system through either optimizing or working around noise-producing components or multiplexing measurements.

3.3.2 Field Strength, Pulsed Field, and Apparatus

No discussion of sensitivity enhancement would be complete without discussion of the two best known methods to enhance the sensitivity of NMR experiments: higher field strengths and cryogenically cooled electronics.[89] The sensitivity of the NMR signal is related to the population difference given by the Boltzmann distribution in a magnetic field. As such, increasing the field strength increases the population difference and the observable signal.[6] Unfortunately, increasing field comes with exponentially increasing cost of both instruments and cryogens to maintain the superconductor's current. At the time of writing, the highest commercially available field was 1.2 GHz.[89] This achievement was made by a breakthrough in manufacturing techniques with high temperature superconductors (HTS)[90] (*e.g.*, REBCO tapes), which can sustain a higher current density and hence generate a much more intense magnetic field compared to conventional low temperature superconductors. While most current ultra-high field systems use HTS inserts,[91] some experimental systems are being produced with all HTS coils, which can both maximize the field strength and, in future, drastically reduce the cost of ownership for high field systems by reducing reliance on liquid helium. HTS tapes only require liquid nitrogen to act as superconductors, although they are often still used with liquid helium.[92] A 400 MHz HTS system which was recently developed[92] has the advantage of being cryogen-free, further reducing the footprint of the system, costs of ownership, and engineering requirements for installing and maintaining the instrument, which could further drive the proliferation of NMR systems. It can be installed in a conventional fume hood in a chemistry laboratory to aid in reaction monitoring and may play a role in onsite environmental monitoring.[2,92]

Similarly, pulsed field NMR can provide a dramatic improvement in the sensitivity of NMR measurements. LTS magnets are limited by their maximum current density limit that can be sustained, while HTS systems are very difficult to manufacture, can sustain much higher current densities, and have been responsible for the development of ultra-high (1.2 GHz) commercial NMR spectrometers.[93] As the cost of liquid helium is often one of the major contributions to the cost of ownership, it is important to mention newer developments in magnetic field strengths, particularly in cryogen-free LTS and HTS magnet systems,[92] which use cryocoolers to maintain the temperature of the superconductors in the magnet. While there are several technical challenges for these systems that remain to be solved to achieve widespread adoption, in the future, these cryogen-free systems may make high fields accessible to more users by reduction of instrument upkeep costs. It is worth mentioning resistive magnets (non-superconducting electromagnets) which, when hybridized with superconducting inserts, can reach up to 45 T;[94] however, they have very high upkeep costs owing to the constant electrical demand and numerous engineering challenges to cool the driven coils.[94]

Pulsed field NMR systems (*not* to be confused with pulsed NMR, or Fourier-transform NMR) use magnetic field pulses with the goal of obtaining much higher sensitivity and resolution. Pulsed magnetic fields of up to 100 T have been produced,[94] although this puts immense strain on magnet hardware through resistive heating of the magnet coils, which requires long delays between pulses in order to avoid damaging the system. These systems are still in the early stages of development owing to the technical difficulties of maintaining a homogenous high magnetic field on the second timescale, and as such, many are only suitable for systems that relax very quickly.[94] Pulsed field systems do not require liquid helium, and despite their many technical challenges which prohibit their use for environmental NMR at this time, pulsed field NMR may yield significant enhancements in sensitivity simply due to the massive field strengths that could be produced, allowing a uniform boost to the population difference of the system.

3.3.3 Microcoils and RF Lenses

Microcoils boast a much greater signal-to-noise ratio for mass limited samples compared to conventional NMR coils.[83,84,87] RF coils otherwise vary significantly in their characteristics, including geometry, frequency response, and size: many of the specifics of these coils are discussed exhaustively in excellent reviews in the literature.[88] Designing microcoils is very complex, as the coils must be of high quality (to maintain field homogeneity) but can only be constructed out of materials that do not significantly affect the magnetic susceptibility of the sample.[95] The coil must also have a high "quality factor", whilst allowing the receiver LC circuit to be matched. In general, the resistance (R) should be small compared to the impedance (χ) (eqn (3.1) and (3.2)).[95]

$$Q = \frac{\chi}{R} = \frac{L\omega_0}{R} \tag{3.1}$$

$$50 = \chi^2 + R^2 \tag{3.2}$$

The coil must be designed such that the resistance (ohmic and other losses) is as small as possible compared to the impedance, while the magnitude of the resistance is matched with the rest of the circuit (usually 50 Ohms). This can alternatively be expressed as maximizing the inductance of the circuit L at the particular resonant frequency ω_0 for the nucleus one is observing.[88,95] The restrictions on probe body materials as well as the requirements that the coil can be tuned to the desired frequency and couple to the sample effectively in addition to the difficulty (and cost) of manufacturing coils on such a small scale still present challenges to the proliferation of microcoil NMR. However, it is possible to customize microcoils to individual samples and produce them at low costs using computer-numerical controlled (CNC) machining, which may allow their use for high throughput environmental monitoring of organisms.[87] The continuing development of microcoils, however, has several exciting implications for environmental NMR, and NMR in general. The requirements for B_1 homogeneity can be somewhat looser if the microcoils are used only to receive the signal from the sample, if the transmitter coils from a conventional NMR probe are used.

For environmental NMR, the additional mass sensitivity allows the huge advantage of allowing the measurement of individual organisms and eggs (*e.g.*, very small (<50 μm) fish eggs,[95] or *Daphnia magna* eggs and organisms[87]). For example, in the case of individual fish eggs, a sensitivity enhancement of 3000-fold was obtained by using a planar microcoil with a 15 pL sample well instead of a conventional 5 mm NMR tube.[95]

Microcoil NMR also allows the unique possibility of constructing microcoil arrays, which can vastly improve the throughput for environmental NMR, directly increasing the sensitivity of the instrument by enhancements in mass sensitivity, and indirectly by time savings afforded by performing the same experiments on multiple samples at the same time.

It is difficult to construct such microcoils and particularly expensive to obtain the remainder of the electronics required for a microcoil probe, often requiring a dedicated probe to be constructed to house and connect the microcoil. An alternative to constructing microcoils is to improve the coupling between samples and NMR receiver coils (greater sensitivity for mass-limited samples) is using Lenz lenses.[96] Lenz lenses exploit Lenz's law to improve the coupling between the sample and receiver, enhancing the sensitivity of the measurement by increasing the effective filling factor, without redesigning the hardware. The precessing nuclei induce a current inside the lens, which in turn generates the corresponding RF frequencies.[96] A wide variety of lens geometries exist, and the interested reader is directed to recent publications that employ Lenz lenses.[96,97]

One further benefit of Lenz lenses is that they may serve to boost the sensitivity of mass limited samples where microcoils may not be usable, for example, to improve the performance of cryogenic probes, which typically have larger coils, a lower filling factor, and thus, geometrically, a lower mass sensitivity, omitting the benefit of lower thermal noise, since extra insulation is required between the RF coils and NMR tube. Lenz lenses also present an alternative approach for tailored sensitivity enhancement for mass limited samples for high-throughput environmental NMR; these may be more cost-effective compared to planar microcoils for onsite analysis, as these can be constructed solely with copper (or other suitable metal) wires.[96]

3.3.4 Detection: Multiplexing, Arrays and Cryogenic Cooling

Arrayed detection is another relatively simple, but underutilized technique to boost the sensitivity of NMR, with several unique approaches to improve instrument throughput and the effective sensitivity per unit time. This is a technically demanding task since two samples acquired simultaneously without any extra consideration would be indistinguishable from one another once transferred to the receiver. The simplest solution is to use separate receivers, but it quickly becomes unreasonably expensive for commercial NMR systems. However, with miniaturized electronics, the pre-amplifiers can be placed as close to the NMR coil as physically possible. This makes it possible to construct arrays of such systems, allowing the same experiment to be performed on multiple samples simultaneously.[98]

As much of the time in NMR experiments is spent waiting for the spins to relax so the population difference can recover, another solution is to stagger acquisition by using gradients to select different samples, wells, or regions of the sample. As a result, the experiments could be performed in parallel, executing an experiment on one sample while another is relaxing to maximize instrument throughput. One major drawback of this technique is that it is very hard on the instrument's hardware where normally the inter-scan delay serves the second function of preventing the probe, amplifiers, and sample from overheating during acquisition. However, specialized probes that permit the power handling capabilities of solid state NMR probes for solution-state experiments may be able to sustain these extended duty cycles more effectively. This staggered acquisition stands to benefit quantitative NMR the most, where the inter-scan delay is often much longer than conventionally performed qualitative spectroscopy.[99]

Cryogenic cooling cools the entirety of the RF circuit up to the pre-amplifiers: coils, transmission lines, and preamplifiers. As most of the noise in an NMR experiment comes from thermal noise from the conductors, this affords an effective increase in sensitivity of 2–4-fold by reducing the amplitude of the noise until the very weak signal is amplified. Liquid nitrogen and helium cooled probes are commercially available but both are very expensive. Liquid helium is much more expensive than liquid nitrogen, though even the nitrogen cooled hardware produces sensitivity gains of 2–4-fold. As previously mentioned, the additional insulation required for cryogenic cooling results in a decreased filling factor, as the samples are kept usually around room temperature (298 K), while the cryogenically cooled hardware is kept at 77 K and 4 K for nitrogen and helium cooled systems, respectively. Cryogenic cooling is well-known, but is exclusively performed at high field, although a possibly useful development may be to develop cryogenically cooled low field systems to strike a balance between the cost of ownership and sensitivity.

Similarly, to boost the sensitivity of the measurement, one could perform experiments at very low temperatures, although this is accompanied by longer relaxation times and is further limited by the sample's tolerance to ultralow temperatures.[100]

3.4 Pulse Sequences, Processing, and Mathematics

3.4.1 Saving Time: Supersequences, Non-uniform Sampling, and Ultrafast Acquisition

The last stage at which the sensitivity of NMR can be enhanced is by improvement of the pulse sequences and experimental and mathematical techniques by which the signals are acquired and transformed. There are many experimental improvements to be made in NMR before and after data acquisition, and in doing so one can maximize the final sensitivity of an experiment. Similarly, to optimize NMR hardware for acquisition, many of the mathematical and experimental techniques have a similarly broader scope, greater versatility and fewer apparent limitations as compared to some of the hyperpolarization techniques and are broadly applicable for many types of samples and matrices. The limitations that some of these methods do have are important, particularly for environmental samples with many broad signals or fast relaxation times, but are largely unique to each sensitivity enhancement approach.

In this section, the impact of unique acquisition approaches, pulse sequences, and sampling techniques developed in NMR will be discussed. Pulse sequences that indirectly enhance the sensitivity of a spectrum by producing significant time savings will also be discussed, as they achieve the same goal as direct sensitivity enhancement: an increase in the number of useful spectra and increased instrument throughput in a reduced timeframe.

The most well-known acquisition method for indirectly enhancing the sensitivity of NMR data is by non-uniform sampling (NUS): 2D NMR data are acquired as a series of 1D NMR spectra with a variable evolution delay. The Fourier transform requires uniformly spaced time increments, which requires very long experiment times for high resolution experiments.[6,7] In short, NUS exploits the usual properties of most NMR spectra, randomly skipping increments in the indirect dimension and reconstructing the missing data with assumptions about the appearance of the spectrum (*e.g.*, lineshape, sparsity, or signal content) to produce the uniformly spaced increments required to produce the NMR spectrum (Figure 3.10).[7]

The implicit assumption that an NMR spectrum is mostly empty space, for example, is particularly potent with an increasing number of dimensions and becomes more useful with increasing dimensionality (3D or higher), where the likelihood that peaks end up clustered together is lower, but also makes NUS more potent for samples with fewer signals or highly selective pulse sequences, allowing fewer samples to meaningfully reconstruct the NMR spectrum. NUS sampling schemes can also affect the sensitivity of an

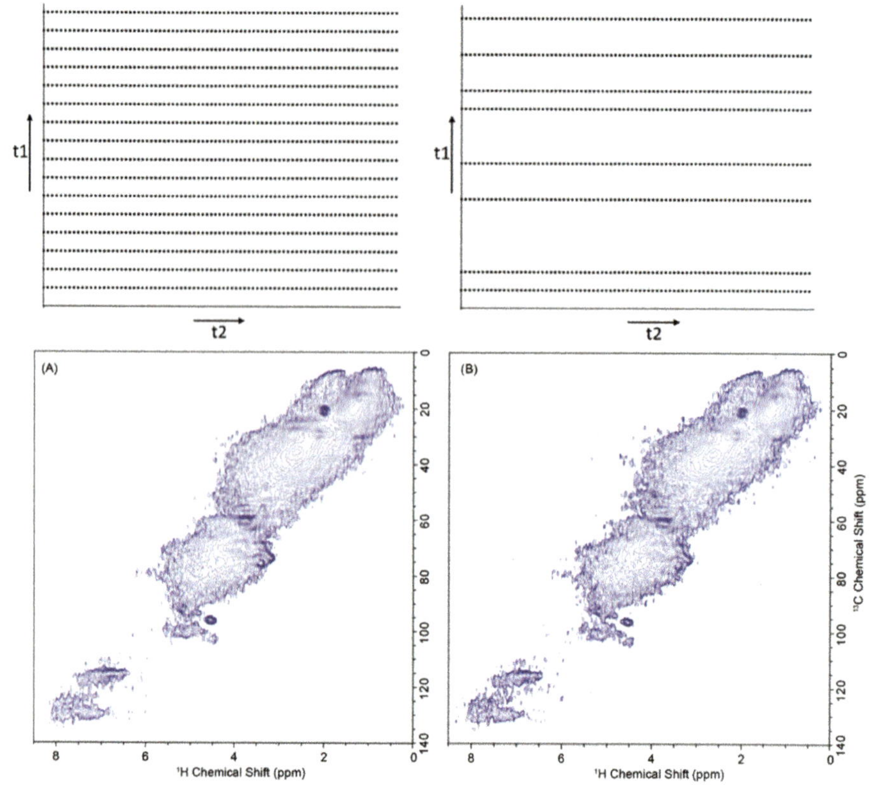

Figure 3.10 NUS sampling scheme. Top left: A series of FIDs for a 2D NMR spectrum uniformly sampled with six samples in the indirect dimension. Top right: A series of FIDs for a 2D NMR spectrum sampled with 50% sampling sparsity 2D ^1H–^{13}C HSQC of the Suwannee River. Dissolved organic matter at 500 MHz acquired using standard direct Fourier transform (A) and NUS mode using 50% of the data points out of 196 in the indirect dimension with T_2 weighting set to 960 μs (B). (B) represents a 50% time savings required for data acquisition. Reproduced from ref. 103 with permission from Bentham Science, Copyright 2013.

experiment similarly to window functions commonly used in NMR data processing: exponential or sine/cosine weighted sampling schedules can increase the sensitivity of experiments or the resolution of experiments depending on the sampling schedule employed.[101,102] This minimizes the number of samples made at the end of the indirect dimension, which is highly advantageous for rapidly relaxing samples, where the slices acquired at the end of the indirect dimension have a low signal.[102] Soils are one such environmental sample in which this exponential T_2 weighting has demonstrated a large sensitivity enhancement in multidimensional NMR spectra in addition to the sensitivity enhancement afforded by the time savings of NUS.[103]

One of the main obstacles to the use of NUS in environmental NMR is the requirement for a minimum number of samples to successfully reconstruct the data based on the signal density in the spectrum, which is particularly challenging when the exact number of signals is unknown. The time savings from NUS are particularly useful in allowing the use of multidimensional NMR in time-resolved NMR experiments, which is useful for *in vivo* NMR experiments exploring toxicological exposure of organisms to environmental contaminants, which often require multidimensional NMR to sufficiently resolve metabolite peaks.[7]

In addition to its direct implementation in NMR data acquisition, NUS also forms the foundation for other NMR methods to resample data and allow more sophisticated acquisition schemes discussed later in this section.

The vast amount of time in NMR experiments is spent waiting for the nuclei to relax to thermal equilibrium following a scan. Several acquisition techniques that yield an effective sensitivity enhancement overcome this shortcoming in two unique ways: either by optimizing measurement times or by making multiple simultaneous measurements of the spins by a pulse sequence in order to obtain more information per scan. These methods use either phase or space encoding to perform multiple measurements simultaneously or prepare the spin system to make measurements one after the other either sequentially or in an interleaved manner.[104] Phase encoding allows two separate sets of data to be acquired at the same time and, by either adding or subtracting the individual scans, allows two independent sets of data to be extracted from the raw data (Figure 3.11).[104,105]

One simple approach is to combine multiple experiments together and acquire time-consuming 2D experiments at the same time, which is crucial for peak assignment in organic chemistry, biochemistry, and environmental chemistry (*e.g.*, HMBC, HSQC, J-HSQC, and TOCSY).[104] These are known as NMR supersequences and are an excellent example of indirect sensitivity enhancement by addressing the time-consuming relaxation delay of an NMR experiment simply by acquiring them at the same time: one after the other, in an interleaved fashion, or by phase encoding.[104] An excellent example of NMR experiments is the set of NMR by ordered ^{1}H acquisition (NOAH) supersequences, in which the unused or recovered magnetization from a previous experiment is used in the following experiment (Figure 3.12).[104]

In the case of the ^{13}C or ^{15}N HSQC, at natural abundance for carbon-13 and nitrogen-15 (approximately 1.1% and 0.37%, respectively), approximately 99% of the magnetization from protons bound to ^{12}C or ^{15}N remains unused and is returned to the *Z*-axis prior to measurement of the HSQC signal and can be used for a homonuclear experiment (*e.g.*, COSY or TOCSY).[8,106] There is some signal loss in this acquisition protocol due to pulse imperfections and slight relaxation during the usual delays in HSQC sequences, which result in a loss of per-scan sensitivity in each of the subsequent pulse sequence blocks.[106] Overall, there is a boost in sensitivity gained from the additional scans that can be collected by faster acquisition, resulting in at least a 25% enhancement in the signal intensity of the

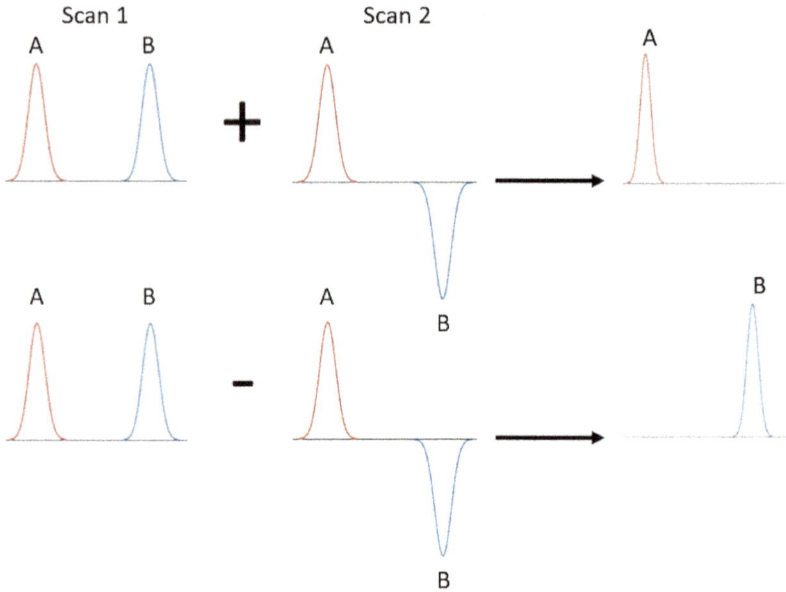

Figure 3.11 IPAP method. Top left: The "in-phase" transient, where signals A and B are acquired in-phase. Bottom left: The "anti-phase" transient in which signals A and B are out of phase relative to one another. Top right and bottom right images show the sum and difference of both transients, respectively, in which one of the two signals is cancelled, separated from one another in each experiment.

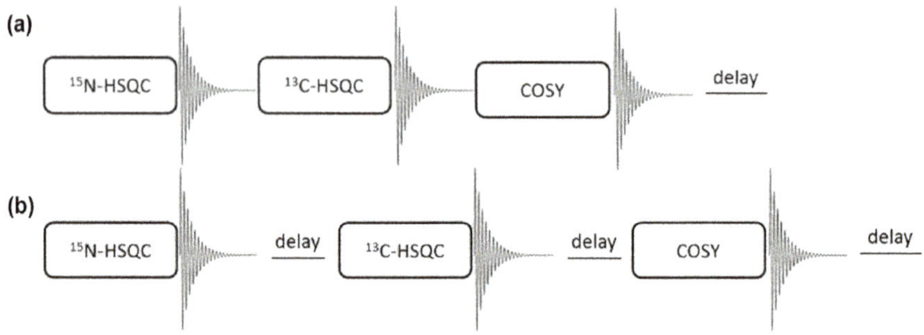

Figure 3.12 NOAH supersequences. Top: An example of a NOAH supersequence which executes three experiments that share the same D1 delay. Bottom: The same three experiments run back-to-back with independent delays. For an experiment with an acquisition time of ~100 ms and an inter-scan delay of 1 s, scheme (b) requires 3.3 s to obtain one scan for each spectrum, while scheme (a) requires only 1.3 s, a 2.5× time saving.

individual experiments in a four-experiment NOAH supersequence.[106] By both using "unused" magnetization (*e.g.*, protons not coupled to ^{13}C in an HSQC) and minimizing the amount of time required for inter-scan delays by allowing experiments to share the same delay, sensitivity is maximized without fundamentally altering the appearance or interpretation of the data. The time savings afforded by combining even only two experiments back-to-back in this fashion can still present a modest gain in sensitivity per unit time despite the fewer number of combined experiments.

This idea of using as much of the magnetization from a sample to boost sensitivity is not new – not all of the magnetization from a sample can be measured by the detector, only a fraction of this signal can be detected. NMR signals evolve as "quantum coherences", and only the (−1)-quantum coherence can be detected.[6] The proportion of magnetization converted to the observable (−1)-quantum coherence is determined by careful design of the phase cycle and pulse sequence to select a subset of signals to observe.[6] Other coherences can contain the same information (the (+1)-quantum coherence), but the information is lost as they are normally not detected.[6] A well-known example to allow two quantum coherences to be observed simultaneously in the sensitivity enhanced (se) HSQC is a well-known example in which both are preserved, affording up to a 2-fold gain in sensitivity by preserving both transfer pathways.[106]

One immediate benefit of NOAH supersequences lies in the use of NMR as a discovery tool.[8] The time savings allow multiple 2D correlation spectra often necessary for the elucidation of complex small molecules to be collected with a reduced overall experiment time compared to independent acquisition, which usually prevents some of these experiments from being performed in the first place.[8] Similarly, reducing the overall measurement time required for each of these sequences may be useful for unstable or rapidly changing samples, and may allow time-resolved acquisition with greater information content.

An equally useful alternative to supersequences for rapid acquisition of natural abundance heteronuclear correlation experiments is by using the nearby unused magnetization from protons not bound to the heteronucleus (*e.g.*, carbon) or bound to inactive isotopes (*e.g.*, ^{12}C) to replenish the magnetization of protons bound to the heteronucleus.[107] Acceleration by sharing adjacent polarization (ASAP) works in this way and allows the relaxation delay between scans to be shortened considerably by using the unused bulk magnetization in the sample.[107] While this allows very rapid pulsing, this puts considerable stress on the NMR hardware owing to the high duty cycle of the resulting pulse sequence, although supporting techniques such as extended acquisition time (EXACT) allow the repeated acquisition with broadband decoupling by acquiring chunks of data and reconstructing the missing gaps, during which the decoupler can be turned off.[107] This removes the limits on the acquisition time and repetition rate for multidimensional NMR experiments.[107] Like other D1-delay reducing methods, this increases the number of scans per unit time that can be performed by the experimenter, and the overall experiment sensitivity per unit time.[108]

Accelerated NMR methods attempt to take a different approach altogether to avoid the D1 delay, working to artificially return or minimally perturb the bulk magnetization along the Z-axis to reduce the inter-scan delay time and the overall time required for an NMR experiment.[109] Driven equilibrium Fourier transform (DEFT) sequences work in this way: by refocussing and returning the signal to the Z-axis after the measurement has been completed, the inter-scan delay can be significantly reduced.[109] An example using slowly relaxing carbon atoms in camphor allows a reduction of the inter-scan delay from 400 seconds to 5 seconds per scan, allowing comparable sensitivity of the 14-hour experiment to be achieved in 10 minutes.[109]

Band-selective optimized-flip-angle short-transient (SOFAST) and alternate SOFAST (ALSOFAST) experiments reduce D1 by employing a significantly reduced flip angle, allowing a much lower D1 and faster signal averaging.[108,110,111] While it sounds like a trivial approach, SOFAST-HMQC can allow interscan delays on the order of hundreds of milliseconds, and although it can result in a sensitivity enhancement of 30–100%, the sequence can be very hard on probes, as the duty cycle for decoupling becomes much higher.[110]

3.4.2 Pulse Sequences: Indirect Measurements

A classic example of indirect measurements of nuclei is in the well-known INEPT and HSQC experiments, which indirectly probe coupled heteronuclei by transferring polarization to and from the less sensitive heteronucleus with a more sensitive probe nucleus (*e.g.*, protons). This carries the combined benefits of faster relaxation rates of protons compared to conventionally used heteronuclei (^{13}C and ^{15}N) as well as a greater sensitivity (gyromagnetic ratio) compared to these heteronuclei.[112] While these are not new developments, this basic approach has been used in more advanced experiments which rely on the transfer of populations between spins to achieve a signal enhancement.[33,42] Many hyperpolarization techniques (*e.g.*, DNP) exploit *J*-coupling, the nuclear Overhauser effect (NOE) or cross-polarization (CP) to efficiently hyperpolarize protons, then transfer the hyperpolarized signals to less sensitive nuclei in solution and solid states, respectively.[33,42,45] Another promising approach for indirect measurement of nuclei is by frequency labelled exchange (FLEX) and chemical exchange saturation (CEST) approaches, which measure molecules by the interaction between molecules with exchangeable protons and a bulk solvent with a labile proton (*e.g.*, water) (Figure 3.13).[53]

Both CEST and FLEX methods rely on the reaction rate between the labile protons and the bulk water signal to produce a signal which depends on the reaction rate between the exchangeable site and water, instead of the concentration of the molecule with the exchangeable site.[53]

FLEX or relayed FLEX (REFLEX) methods exploit Ramsey modulation to measure the frequencies of the molecules of interest encoded on water for detection.[53] In the case of the time-domain method CEST, the labile site is irradiated, which exchanges with and attenuates the water signal, allowing a difference spectrum establishing this signal intensity (*z*-spectrum) to be

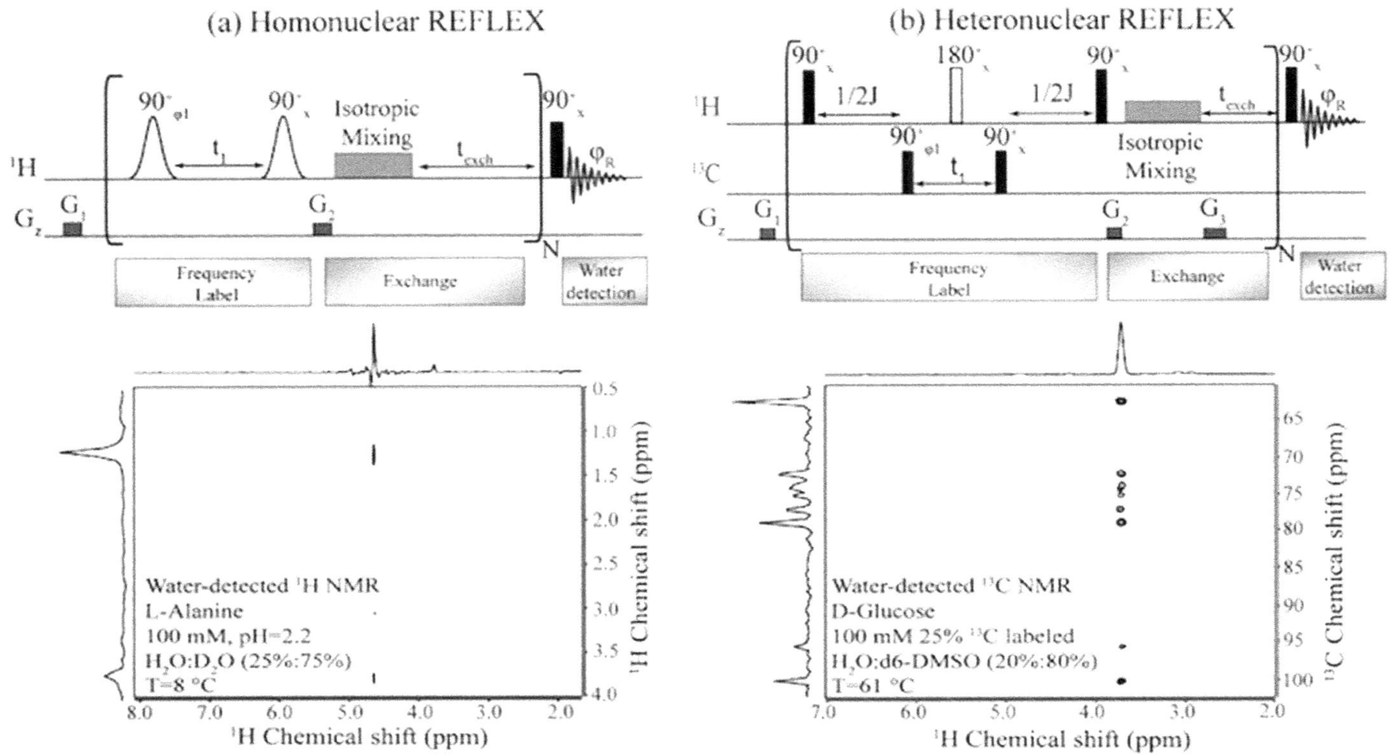

Figure 3.13 Sensitivity enhancement by frequency labelled exchange. A labile proton is excited by an RF pulse, over which the magnetization is allowed to evolve, which encodes the frequencies of the signals of the molecules into the water signal. A period of homo- or heteroisotropic mixing allows the signals for either carbon or protons to be transferred to the labile site to encode the non-labile atoms into the water signal for detection. Reproduced from ref. 53 with permission from the Royal Society of Chemistry.

computed, which can yield enhancements in the signal intensity of labile protons by up to three orders of magnitude, owing to the vast excess of the solvent spins compared to the solute spins.[113] A solid-state analog of this experiment, progressive saturation of the proton reservoir (PROSPR), exploits a similar property in solids, although as the spins are not physically exchanging in the solid, it uses cross-polarization (CP) to achieve a similar goal, allowing the sensitivity of the experiment to become determined by abundant nearby protons, and the rate of cross relaxation between the molecule of interest and these abundant protons.[113] Unlike other sensitivity enhancement procedures that may be ineffective on rapidly relaxing samples, or even result in a lower signal intensity owing to the rapid relaxation of most environmental samples (*e.g.*, SE-HSQC), CEST relies on the T_1 time of the spin pool (such as water) instead of the remainder of the sample.[113]

For environmental NMR, FLEX may be used to establish the solvent accessibility of molecules with exchangeable protons in soils and *in vivo* with greater sensitivity. The introduction of heteronuclear and homonuclear mixing schemes (PROSPR, relayed frequency labelled exchange (REFLEX), and HetFLEX experiments) allows this amplification to be transferred between the solvent and entire metabolites by spin-locks in solution-state experiments. This allows, for example, ^{13}C NMR spectra to be indirectly collected by evaluating the modulation of the water by repeatedly applying heteronuclear mixing to transfer the spins from carbon atoms to protons, amplifying the water signal prior to measurement (Figure 3.13).[53] The sensitivity of the non-labile nuclei in small molecules is thus enhanced by encoding their signals in the water signal.

The immediate apparent disadvantage of these techniques is that they must be acquired as a 2D NMR experiment.[53] Despite this, non-uniform sampling allows significant time savings (as large as an order of magnitude) and results in a two-fold net gain in sensitivity when comparing HetFLEX with the analogous HMBC experiment.[53]

3.4.3 Mathematical Approaches: Addressing the Root Cause

As the aim is to increase the signal-to-noise ratio, an alternative approach to increase the signal-to-noise ratio where the signal cannot be increased any further is to decrease the noise level.[114] Not all noise is the same – some of this noise comes from imperfections in the NMR hardware because of a particular pulse sequence (artefacts) that are sometimes proportional to the signal intensity. Several noise suppression approaches exploit the "randomness" of noise compared to the signal to suppress it. T_1 noise is one such noise source which cannot be removed by highly selective pulse sequences.[114] This noise can obscure signals in spectra with a high dynamic range, since T_1 noise from a high-intensity peak can appear with similar intensity to a real signal with lower intensity.[114] The observation that T_1 noise appeared to increase with the scan number led to acquisition of separate spectra with the minimum number of scans to complete the phase cycle and co-adding the spectra together, leading to spectra with an equivalent signal-to-noise ratio with T_1 noise attenuated by 75% or more.[114] These can lead to gains in the signal-to-noise ratio of the entire spectrum, provided T_1 noise contributes more noise than random thermal noise.[114]

It is worth briefly mentioning that noise suppression is not always a silver bullet for NMR, where methods such as wavelet transforms to suppress noise may unintentionally suppress signals close to the noise level for example.[115] Many of these noise suppression methods work by thresholding, which unfortunately can result in cosmetic changes to the spectra without extensive calculations.[116] Noise and artefact suppression using resampling (NASR) is one such noise suppression approach which increases the effective signal-to-noise ratio by re-sampling data to discriminate between real signals and noise peaks, exploiting the fact that the noise changes randomly with different sub-sampling of the spectra.[116] The authors demonstrate the "real" increase in sensitivity by demonstrating the elimination of "noise peaks" during the denoising of a ^{13}C spectrum of $CDCl_3$.[116] This denoising technique is particularly useful for the assignment of low-concentration contaminants in the environment, as the ability to eliminate noise peaks in spectra of any number of dimensions allows greater confidence in signal assignment for low concentration species in spectra.

Although NMR is at present a comparatively insensitive technique, the techniques presented in this chapter provide current and future opportunities to boost the sensitivity of NMR for environmental research by improvement of almost every element of an NMR spectrometer and experiment. For environmental NMR, these, among other unmentioned techniques, should one day accelerate the proliferation of high and low field NMR in academic and industrial settings, as well as support the development of novel NMR experiments to broaden our understanding of the environment.

Acknowledgements

A. J. S. would like to thank the Natural Sciences and Engineering Research Council of Canada (NSERC) [Alliance (ALLRP 549399), Alliance (ALLRP 555452) and Discovery Programs (RGPIN-2019-04165)], the Canada Foundation for Innovation (CFI), the Ontario Ministry of Research and Innovation (MRI), the Krembil Foundation for providing funding, and the Government of Ontario for an Early Researcher Award.

References

1. M.-C. Chalbot and I. Kavouras, in *Multidimensional Analytical Techniques in Environmental Research*, ed. R. M. B. O. Duarte and A. C. Duarte, Elsevier, 2020, ch. 2, pp. 27–55.
2. A. J. Simpson, M. J. Simpson and R. Soong, *Anal. Chem.*, 2018, **90**, 628.

3. M. Bastawrous, A. Jenne, M. T. Anaraki and A. J. Simpson, *Metabolites*, 2018, **8**, 35.
4. A. J. Simpson, M. J. Simpson and R. Soong, *Environ. Sci. Technol.*, 2012, **46**, 11488.
5. M. T. Anaraki, D. H. Lysak, K. Downey, F. V. C. Kock, X. You, R. D. Majumdar, A. Barison, L. M. Lião, A. G. Ferreira, V. Decker, B. Goerling, M. Spraul, M. Godejohann, P. A. Helm, S. Kleywegt, K. Jobst, R. Soong, M. J. Simpson and A. J. Simpson, *Prog. Nucl. Magn. Reson. Spectrosc.*, 2021, **126–127**, 121.
6. M. Levitt, *Spin Dynamics: Basics of Nuclear Magnetic Resonance*, John Wiley & Sons Ltd., Chichester, 2nd edn, 2008.
7. D. Gołowicz, P. Kasprzak, V. Orekhov and K. Kazimierczuk, *Prog. Nucl. Magn. Reson. Spectrosc.*, 2020, **116**, 40.
8. Ē. Kupče and T. D. W. Claridge, *Angew. Chem., Int. Ed.*, 2017, **56**, 11779.
9. A. J. Simpson, D. J. McNally and M. J. Simpson, *Prog. Nucl. Magn. Reson. Spectrosc.*, 2011, **58**, 97.
10. D. W. Bell, P. Pellechia, L. R. Chambers, A. F. Longo, K. M. McCabe, E. D. Ingall and C. R. Benitez-Nelson, *Limnol. Oceanogr.: Methods*, 2017, **15**, 436.
11. Y. Gélinas, J. A. Baldock and J. I. Hedges, *Org. Geochem.*, 2001, **32**, 677.
12. A. Spence, A. J. Simpson, D. J. Mcnally, B. W. Moran, M. V. McCaul, K. Hart, B. Paull and B. P. Kelleher, *Geochim. Cosmochim. Acta*, 2011, **75**, 2571.
13. I. Fugariu, B. Fortier-McGill, R. Soong, A. Sutrisno and A. J. Simpson, *Magn. Reson. Chem.*, 2015, **53**, 769.
14. D. M. Selegato, A. C. Pilon and F. C. Neto, in *NMR-Based Metabolomics Methods and Protocols*, ed. G. A. N. Gowda and D. Raftery, Plant Metabolomics Using NMR Spectroscopy, Humana, New York, NY, 2019, pp. 345–362.
15. K. Tateishi, M. Negoro, S. Nishida, A. Kagawa, Y. Morita and M. Kitagawa, *Proc. Natl. Acad. Sci. U. S. A.*, 2014, **111**, 7527.
16. A.-H. Emwas, R. Roy, R. T. McKay, L. Tenori, E. Saccenti, G. A. N. Gowda, D. Raftery, F. Alahmari, L. Jaremko, M. Jaremko and D. S. Wishart, *Metabolites*, 2019, **9**, 123.
17. I. Fugariu, W. Bermel, D. Lane, R. Soong and A. J. Simpson, *Angew. Chem., Int. Ed.*, 2017, **56**, 6324.
18. K. Zangger, *Prog. Nucl. Magn. Reson. Spectrosc.*, 2015, **86–87**, 1.
19. P. Giraudeau, *Magn. Reson. Chem.*, 2017, **55**, 61.
20. B. E. Fortier-McGill, R. Dutta Majumdar, L. Lam, R. Soong, Y. Liaghati-Mobarhan, A. Sutrisno, R. de Visser, M. J. Simpson, H. L. Wheeler, M. Campbell, A. Gorissen and A. J. Simpson, *J. Agric. Food Chem.*, 2017, **65**, 6779.
21. M. E. Halse, *TrAC, Trends Anal. Chem.*, 2016, **83**, 76.
22. P. Pham, R. Mandal, C. Qi and C. Hilty, *J. Magn. Reson. Open*, 2022, **10–11**, 100052.

23. D. A. Barskiy, A. M. Coffey, P. Nikolaou, D. M. Mikhaylov, B. M. Goodson, R. T. Branca, G. J. Lu, M. G. Shapiro, V.-V. Telkki, V. V. Zhivonitko, I. V. Koptyug, O. G. Salnikov, K. V. Kovtunov, V. I. Bukhtiyarov, M. S. Rosen, M. J. Barlow, S. Safavi, I. P. Hall, L. Schröder and E. Y. Chekmenev, *Chem. – Eur. J.*, 2017, **23**, 725.
24. A. Equbal, S. K. Jain, Y. Li, K. Tagami, X. Wang and S. Han, *Prog. Nucl. Magn. Reson. Spectrosc.*, 2021, **126–127**, 1.
25. B. Corzilius, *Phys. Chem. Chem. Phys.*, 2016, **18**, 27190.
26. B. Plainchont, P. Berruyer, J.-N. Dumez, S. Jannin and P. Giraudeau, *Anal. Chem.*, 2018, **90**, 3639.
27. T. Maly, G. T. Debelouchina, V. S. Bajaj, K. N. Hu, C. G. Joo, M. L. Mak-Jurkauskas, J. R. Sirigiri, P. C. A. van der Wel, J. Herzfeld, R. J. Temkin and R. G. Griffin, *J. Chem. Phys.*, 2008, **128**, 052211.
28. W. T. Wenckebach, *Appl. Magn. Reson.*, 2021, **52**, 731.
29. K. R. Thurber and R. Tycko, *J. Chem. Phys.*, 2012, **137**, 084508.
30. K. O. Tan, R. T. Weber, T. V. Can and R. G. Griffin, *J. Phys. Chem. Lett.*, 2020, **11**, 3416.
31. E. Ravera, C. Luchinat and G. Parigi, *J. Magn. Reson.*, 2016, **264**, 78.
32. M. Bennati, *eMagRes*, 2017, **6**, 271.
33. J. L. Yoder, P. E. Magnelind, M. A. Espy and M. T. Janicke, *Appl. Magn. Reson.*, 2018, **49**, 707.
34. T. Idehara, M. Glyavin, A. Kuleshov, S. Sabchevski, V. Manuilov, V. Zaslavsky, I. Zotova and A. Sedov, *Rev. Sci. Instrum.*, 2017, **88**, 094708.
35. I. V. Sergeyev, F. Aussenac, A. Purea, C. Reiter, E. Bryerton, S. Retzloff, J. Hesler, L. Tometich and M. Rosay, *Solid State Nucl. Magn. Reson.*, 2019, **100**, 63.
36. C.-H. Du, S. Pan, Z.-C. Gao, F.-H. Li, Z.-W. Zhang and P.-K. Liu, *presented in part at 2020 IEEE International Conference on Plasma Science (ICOPS)*, Singapore, December, 2020.
37. S. Sabchevski, T. Idehara, M. Damyanova, I. Zhelyazkov, E. Balabanova and E. Vasileva, *J. Phys.: Conf. Ser.*, 2018, **992**, 012001.
38. A. C. Pinon, A. Capozzi and J. H. Ardenkjær-Larsen, *Commun. Chem.*, 2020, **3**, 57.
39. B. Gizatullin, C. Mattea and S. Stapf, *J. Magn. Reson.*, 2021, **322**, 106851.
40. R. Balzan, L. Fernandes, A. Comment, L. Pidial, B. Tavitian and P. R. Vasos, *J. Visualized Exp.*, 2016, **108**, e53548.
41. T. J. Keller and T. Maly, *Magn. Reson.*, 2021, **2**, 117.
42. A. Dey, A. Banerjee and N. Chandrakumar, *J. Phys. Chem. B*, 2017, **121**, 7156.
43. G. Pintér and H. Schwalbe, *Angew. Chem., Int. Ed.*, 2017, **56**, 8332.
44. S. Z. Kiss, N. MacKinnon and J. G. Korvink, *Sci. Rep.*, 2021, **11**, 4671.
45. A. G. M. Rankin, J. Trébosc, F. Pourpoint, J.-P. Amoureux and O. Lafon, *Solid State Nucl. Magn. Reson.*, 2019, **101**, 116.
46. P. Berruyer, S. Björgvinsdóttir, A. Bertarello, G. Stevanato, Y. Rao, G. Karthikeyan, G. Casano, O. Ouari, M. Lelli, C. Reiter, F. Engelke and L. Emsley, *J. Phys. Chem. Lett.*, 2020, **11**, 8386.

47. A. J. Rossini, A. Zagdoun, F. Hegner, M. Schwarzwälder, D. Gajan, C. Copéret, A. Lesage and L. Emsley, *J. Am. Chem. Soc.*, 2012, **134**, 16899.
48. T.-C. Ong, M. L. Mak-Jurkauskas, J. J. Walish, V. K. Michaelis, B. Corzilius, A. A. Smith, A. M. Clausen, J. C. Cheetham, T. M. Swager and R. G. Griffin, *J. Phys. Chem. B*, 2013, **117**, 3040.
49. J. H. Ardenkjaer-Larsen, *J. Magn. Reson.*, 2016, **264**, 3.
50. A. S. Kiryutin, B. A. Rodin, A. V. Yurkovskaya, K. L. Ivanov, D. Kurzbach, S. Jannin, D. Guarin, D. Abergel and G. Bodenhausen, *Phys. Chem. Chem. Phys.*, 2019, **21**, 13696.
51. L. Frydman and D. Blazina, *Nat. Phys.*, 2007, **3**, 415.
52. K. W. Lipsø, S. Bowen, O. Rybalko and J. H. Ardenkjær-Larsen, *J. Magn. Reson.*, 2017, **274**, 65.
53. M. Novakovic, R. P. Martinho, G. L. Olsen, M. S. Lustig and L. Frydman, *Phys. Chem. Chem. Phys.*, 2018, **20**, 56.
54. H. Kouno, Y. Kawashima, K. Tateishi, T. Uesaka, N. Kimizuka and N. Yanai, *J. Phys. Chem. Lett.*, 2019, **10**, 2208.
55. K. Nishimura, H. Kouno, Y. Kawashima, K. Orihashi, S. Fujiwara, K. Tateishi, T. Uesaka, N. Kimizuka and N. Yanai, *Chem. Commun.*, 2020, **56**, 7217.
56. T. R. Eichhorn, M. Haag, B. van den Brandt, P. Hautle and W. T. Wenckebach, *Chem. Phys. Lett.*, 2013, **555**, 296.
57. F. Rastrelli, S. Jha and F. Mancin, *J. Am. Chem. Soc.*, 2009, **131**, 14222.
58. P. M. Richardson, A. J. Parrott, O. Semenova, A. Nordon, S. B. Duckett and M. E. Halse, *Analyst*, 2018, **143**, 3442.
59. K. X. Moreno, K. Nasr, M. Milne, A. D. Sherry and W. J. Goux, *J. Magn. Reson.*, 2015, **257**, 15.
60. S. Knecht and K. L. Ivanov, *J. Chem. Phys.*, 2019, **150**, 1.
61. A. N. Pravdivtsev, A. v. Yurkovskaya, H.-M. Vieth and K. L. Ivanov, *J. Phys. Chem. B*, 2015, **119**, 13619.
62. T. Theis, M. Truong, A. M. Coffey, E. Y. Chekmenev and W. S. Warren, *J. Magn. Reson.*, 2014, **248**, 23.
63. W. Iali, P. J. Rayner and S. B. Duckett, *Sci. Adv.*, 2018, **4**, 1.
64. A. N. Pravdivtsev, G. Buntkowsky, S. B. Duckett, I. V. Koptyug and J. Hövener, *Angew. Chem., Int. Ed.*, 2021, **60**, 23496.
65. I. Reile, N. Eshuis, N. K. J. Hermkens, B. J. A. van Weerdenburg, M. C. Feiters, F. P. J. T. Rutjes and M. Tessari, *Analyst*, 2016, **141**, 4001.
66. P. M. Richardson, R. O. John, A. J. Parrott, P. J. Rayner, W. Iali, A. Nordon, M. E. Halse and S. B. Duckett, *Phys. Chem. Chem. Phys.*, 2018, **20**, 26362.
67. S. S. Roy, K. M. Appleby, E. J. Fear and S. B. Duckett, *J. Phys. Chem. Lett.*, 2018, **9**, 1112.
68. F. Shi, P. He, Q. A. Best, K. Groome, M. L. Truong, A. M. Coffey, G. Zimay, R. V. Shchepin, K. W. Waddell, E. Y. Chekmenev and B. M. Goodson, *J. Phys. Chem. C*, 2016, **120**, 12149.
69. P. Spannring, I. Reile, M. Emondts, P. P. M. Schleker, N. K. J. Hermkens, N. G. J. van der Zwaluw, B. J. A. van Weerdenburg,

P. Tinnemans, M. Tessari, B. Blümich, F. P. J. T. Rutjes and M. C. Feiters, *Chem. – Eur. J.*, 2016, **22**, 9277.
70. H. Zeng, J. Xu, M. T. McMahon, J. A. B. Lohman and P. C. M. van Zijl, *J. Magn. Reson.*, 2014, **246**, 119.
71. K. L. Ivanov and G. Bodenhausen, *J. Magn. Reson.*, 2018, **292**, 48.
72. J. Hövener, A. N. Pravdivtsev, B. Kidd, C. R. Bowers, S. Glöggler, K. V. Kovtunov, M. Plaumann, R. Katz-Brull, K. Buckenmaier, A. Jerschow, F. Reineri, T. Theis, R. V. Shchepin, S. Wagner, P. Bhattacharya, N. M. Zacharias and E. Y. Chekmenev, *Angew. Chem., Int. Ed.*, 2018, **57**, 11140.
73. L. T. Kuhn, in *Hyperpolarization Methods in NMR Spectroscopy*, ed. L. T. Kuhn, Photo-CIDNP NMR Spectroscopy of Amino Acids and Proteins, Springer, Berlin, Heidelberg, 2013, pp. 229–300.
74. F. Torres, A. Sobol, J. Greenwald, A. Renn, O. Morozova, A. Yurkovskaya and R. Riek, *Phys. Chem. Chem. Phys.*, 2021, **23**, 6641.
75. J. Matysik, Y. Ding, Y. Kim, P. Kurle, A. Yurkovskaya, K. Ivanov and A. Alia, *Appl. Magn. Reson.*, 2022, **53**, 521.
76. I. V. Zhukov, A. S. Kiryutin, M. S. Panov, N. N. Fishman, O. B. Morozova, N. N. Lukzen, K. L. Ivanov, H.-M. Vieth, R. Z. Sagdeev and A. V. Yurkovskaya, *Magn. Reson.*, 2021, **2**, 139.
77. H. Yang, S. Li, C. A. Mickles, V. Guzman-Luna, K. Sugisaki, C. M. Thompson, H. H. Dang and S. Cavagnero, *J. Am. Chem. Soc.*, 2022, **144**, 11608.
78. S. Shomaji, N. V. R. Masna, D. Ariando, S. Deb Paul, K. Horace-Herron, D. Forte, S. Mandal and S. Bhunia, *Foods*, 2021, **10**, 2232.
79. R. E. D. Facts, *US EPA – Pesticides – Fact Sheet for TFM (3-trifluoro-methyl-4-nitro-phenol)*, https://www3.epa.gov/pesticides/chem_search/reg_actions/reregistration/fs_PC-036201_1-Nov-99.pdf (accessed Sept 1, 2022).
80. J. Bernarding, F. Euchner, C. Bruns, R. Ringleb, D. Müller, T. Trantzschel, J. Bargon, U. Bommerich and M. Plaumann, *Chem. Phys. Chem.*, 2018, **19**, 2453.
81. K. Maeda, C. E. Lyon, J. J. Lopez, M. Cemazar, C. M. Dobson and P. J. Hore, *J. Biomol. NMR*, 2000, **16**, 235.
82. B. Keil and L. L. Wald, *J. Magn. Reson.*, 2013, **229**, 75.
83. D. L. Olson, M. E. Lacey and J. V. Sweedler, *Anal. Chem.*, 1998, **70**, 257A.
84. R. E. Hopson and W. Peti, in *Structural Proteomics High-Throughput Methods*, ed. B. Kobe, M. Guss and T. Huber, Microcoil NMR Spectroscopy: a Novel Tool for Biological High Throughput NMR Spectroscopy, Humana Press, 2008, pp. 447–458.
85. A. Hassan, C. M. Quinn, J. Struppe, I. V. Sergeyev, C. Zhang, C. Guo, B. Runge, T. Theint, H. H. Dao, C. P. Jaroniec, M. Berbon, A. Lends, B. Habenstein, A. Loquet, R. Kuemmerle, B. Perrone, A. M. Gronenborn and T. Polenova, *J. Magn. Reson.*, 2020, **311**, 106680.
86. E. Montinaro, M. Grisi, M. C. Letizia, L. Pethö, M. A. M. Gijs, R. Guidetti, J. Michler, J. Brugger and G. Boero, *PLoS One*, 2018, **13**, e0192780.

87. V. Moxley-Paquette, D. Lane, R. Soong, P. Ning, M. Bastawrous, B. Wu, M. Z. Pedram, M. A. Haque Talukder, E. Ghafar-Zadeh, D. Zverev, R. Martin, B. Macpherson, M. Vargas, D. Schmidig, S. Graf, T. Frei, D. al Adwan-Stojilkovic, P. de Castro, F. Busse, W. Bermel, T. Kuehn, R. Kuemmerle, M. Fey, F. Decker, H. Stronks, R. M. A. Sullan, M. Utz and A. J. Simpson, *Anal. Chem.*, 2020, **92**, 15454.
88. J. Mispelter, M. Lupu and A. Briguet, *NMR Probeheads for Biophysical and Biomedical Experiments*, Imperial College Press, London, 2nd edn, 2015.
89. E. Moser, E. Laistler, F. Schmitt and G. Kontaxis, *Front. Phys.*, 2017, **5** DOI: 10.3389/fphy.2017.00033.
90. H. Schwalbe, *Angew. Chem., Int. Ed.*, 2017, **56**, 10252.
91. S. Hahn, D. K. Park, J. Voccio, J. Bascunan and Y. Iwasa, *IEEE Trans. Appl. Supercond.*, 2012, **22**, 4302405.
92. M. V. Silva Elipe, A. Cherney, R. Krull, N. Donovan, J. Tedrow, D. Pooke and K. L. Colson, *Org. Process Res. Dev.*, 2020, **24**, 1428.
93. A. Molodyk, S. Samoilenkov, A. Markelov, P. Degtyarenko, S. Lee, V. Petrykin, M. Gaifullin, A. Mankevich, A. Vavilov, B. Sorbom, J. Cheng, S. Garberg, L. Kesler, Z. Hartwig, S. Gavrilkin, A. Tsvetkov, T. Okada, S. Awaji, D. Abraimov, A. Francis, G. Bradford, D. Larbalestier, C. Senatore, M. Bonura, A. E. Pantoja, S. C. Wimbush, N. M. Strickland and A. Vasiliev, *Sci. Rep.*, 2021, **11**, 2084.
94. Q. Liu, S. Liu, Y. Luo and X. Han, *Matter Radiat. Extremes*, 2021, **6**, 024201.
95. M. Khelifa, D. Mounier and N. Yaakoubi, *Sensors*, 2020, **21**, 170.
96. N. Spengler, P. T. While, M. V. Meissner, U. Wallrabe and J. G. Korvink, *PLoS One*, 2017, **12**, e0182779.
97. T. Meier, N. Wang, D. Mager, J. G. Korvink, S. Petitgirard and L. Dubrovinsky, *Sci. Adv.*, 2017, **3**, eaao5242.
98. V. Badilita, K. Kratt, N. Baxan, J. Anders, D. Elverfeldt, G. Boero, J. Hennig, J. G. Korvink and U. Wallrabe, *presented in part at 2011 IEEE 24th International Conference on Micro Electro Mechanical Systems*, Cancun, Mexico, January, 2011.
99. Y. Liu, Z. Liu, H. Yang and L. He, *J. Anal. Methods Chem.*, 2016, **2016**, 1.
100. R. Tycko, *Acc. Chem. Res.*, 2013, **46**, 1923.
101. F. Delaglio, G. S. Walker, K. A. Farley, R. Sharma, J. C. Hoch, L. W. Arbogast, R. G. Brinson and J. P. Marino, *Am. Pharm. Rev.*, 2017, **20**, 339681.
102. M. R. Palmer, C. L. Suiter, G. E. Henry, J. Rovnyak, J. C. Hoch, T. Polenova and D. Rovnyak, *J. Phys. Chem. B*, 2015, **119**, 6502.
103. H. Farooq, D. Courtier-Murias, R. Soong, W. Bermel, W. Kingery and A. J. Simpson, *Curr. Org. Chem.*, 2013, **17**, 3013.
104. Ē. Kupče, J. R. J. Yong, G. Widmalm and T. D. W. Claridge, *JACS Au*, 2021, **1**, 1892.
105. T. Kern, P. Schanda and B. Brutscher, *J. Magn. Reson.*, 2008, **190**, 333.

106. J. R. J. Yong, A. L. Hansen, Ē. Kupče and T. D. W. Claridge, *J. Magn. Reson.*, 2021, **329**, 107027.
107. I. E. Ndukwe, A. Shchukina, K. Kazimierczuk and C. P. Butts, *Chem. Commun.*, 2016, **52**, 12769–12772.
108. M. R. M. Koos and B. Luy, *J. Magn. Reson.*, 2019, **300**, 61.
109. M. Piotto, M. Bourdonneau, K. Elbayed, J.-M. Wieruszeski and G. Lippens, *Magn. Reson. Chem.*, 2006, **44**, 943.
110. B. Brutscher, in *Encyclopedia of Biophysics*, ed. G. C. K. Roberts, SOFAST HMQC, Springer, Berlin, Heidelberg, 2013, p. 2407.
111. P. Schanda, Ē. Kupče and B. Brutscher, *J. Biomol. NMR*, 2005, **33**, 199.
112. G. A. Morris, *J. Am. Chem. Soc.*, 1980, **102**, 428.
113. M. J. Jaroszewicz, A. R. Altenhof, R. W. Schurko and L. Frydman, *J. Am. Chem. Soc.*, 2021, **143**, 19778.
114. H. Mo, J. S. Harwood, D. Yang and C. B. Post, *J. Magn. Reson.*, 2017, **276**, 43.
115. F. Dancea and U. Günther, *J. Biomol. NMR*, 2005, **33**, 139.
116. B. Jiang, F. Luo, Y. Ding, P. Sun, X. Zhang, L. Jiang, C. Li, X. Mao, D. Yang, C. Tang and M. Liu, *Anal. Chem.*, 2013, **85**, 2523.

CHAPTER 4

Comprehensive Multiphase NMR: Natural Samples in Their Natural State

R. GHOSH BISWAS, A. JENNE, M. BASTAWROUS, L. LABINE, R. SOONG AND A. SIMPSON*

University of Toronto (Scarborough), 1265 Military Trail, ON M1C 1A4, Canada
*Email: andre.simpson@utoronto.ca

4.1 Introduction

Most, if not all, natural samples are complex mixtures of multiple components interacting together to give rise to their function. These components exist as heterogeneous phases (*e.g.*, solution-like dissolved components, structural gels and solids, or dissolved material trapped in solid matrices) with varying physical and chemical properties. To understand processes and reactivity within such complex systems, it is not sufficient to observe isolated phases on their own. As these components exist synergistically, in order to study them in their natural state, it is necessary to establish a non-destructive technique capable of observing all phases in intact samples. This review will explore a novel technique known as comprehensive multiphase NMR (CMP-NMR) capable of this.

Nuclear magnetic resonance (NMR) spectroscopy is one of the most powerful, non-invasive analytical techniques used to observe complex natural samples. It is used in numerous fields to elucidate complex structures, molecular processes and interactions in heterogeneous samples. Traditional

New Developments in NMR No. 32
The Environment in a Magnet: Applications of NMR Techniques to Environmental Problems
Edited by Pellegrino Conte, Delia Francesca Chillura Martino and Paolo Lo Meo

Published by the Royal Society of Chemistry, www.rsc.org

NMR has evolved into two separate branches of research: solution- and solid-state NMR. Both have their own specialized hardware (probes), experiments and sample preparation techniques [extraction and dissolution (solution), dry packing (solid)]. Unfortunately, such sample preparation can lead to the loss of vital information such as phase changes, interactions at interfaces, or biological function.

Mobile components (freely diffusing), such as metabolites in plants and animals, or dissolved contaminants in terrestrial and aquatic environments are observed using solution-state NMR. Solution-state probes are equipped with a pulsed field gradient, lock, and all components are magnetic-susceptibility-matched for the best lineshape possible. As the intrinsic molecular tumbling of dissolved molecules narrows the NMR lineshape considerably, only low power circuitry is required to observe and decouple these components. Contrastingly, due to their rigid nature, solid-like materials (*i.e.*, structural components – exoskeleton, soil lignin) do not exhibit the same tumbling, resulting in broad resonances that cannot be observed with traditional, low power, solution-state hardware. Instead, solid-state probes are equipped to handle the high-power radiofrequency (RF) to observe rigid constituents (*i.e.*, high power ^{1}H–^{13}C decoupling and cross-polarization). To observe crystalline solid structures, the samples must be spun at the magic angle (MAS at 54.7° – reduces dipolar and anisotropic interactions) and dry-packed into sample holders (rotors), eliminating the addition of a lock and gradient in solid-state probes.

However, most natural samples do not exist in either pure solution or solid-state. In fact, in-between 'gel-like' and 'semi-solid-like' phases often host the bulk of interactions and reactions, giving rise to the primary system functions. In 1996, a technique known as high-resolution magic angle spinning (HR-MAS) was introduced and was an 'intermediate' between the two fields.[1] HR-MAS combines the low-power circuity of solution state, as well as magic angle spinning, typically employed by solid-state probes to provide narrow lineshape of the less mobile gel/semi-solid components. HR-MAS has made great strides in monitoring structural changes and metabolic profiling in diverse fields of research such as intact tissues for medical research,[2–6] food sciences[7,8] and environmental research.[9]

However, to understand processes in the natural environment (*e.g.*, binding/sequestration in soils), a technique that can observe interactions between solids, gels, and solutions must be used. This was achieved in 2012, when a technique known as comprehensive multiphase NMR (CMP-NMR) was introduced.[10] As CMP-NMR combines the hardware from both solution/gel-state (HR-MAS) and solid-state NMR, all molecular-level changes (*i.e.*, structural, interactions at phase interfaces, binding orientation, kinetics across phases) can be observed in complex, intact natural samples. As a result, CMP probes consist of a lock, a magic angle gradient, magnetic susceptibility matched coils, and the ability to also handle the high-power RF required to observe rigid components. With this tool, the observation of all phases (solids, solutions and gels), their structural organization and interactions in unaltered samples is possible. The approach has potential applications in a wide range of research fields, where phase transformations and/or movement of molecules between phases can be observed. For example, kinetic and dynamic studies of the molecular processes involved in environmental toxins transporting from aqueous environments (wastewater) and binding as gel-like structures (soil organic components) or being trapped in a solid-state matrix (soil hydrophobic domains) can be measured. The same logic can be applied in the medical field to follow the molecular process of a drug assimilating within the various matrices of the body, including transporting from blood, tissues or bones.

This review follows the major applications of CMP-NMR in environmental research to date, such as structural organization and interactions in soil, plants and intact organisms (*ex vivo* and *in vivo*). Prior to this, a brief explanation of how CMP-NMR works and isolates individual phases is provided; however, for in-depth details, we encourage readers to refer to the original CMP manuscript.[10] This review will finish with a brief look at the potential for future applications and directions.

4.2 Spectral Editing Techniques

Traditionally, for CMP-NMR experiments to discriminate multi-phases, the swollen constituents (solution and gel) are observed using the ^{1}H nuclei, where solid-state experiments are observed using ^{13}C CP-MAS (cross polarization MAS). ^{1}H is used more often due to its high natural abundance (99.9% *vs.* 1.1% of ^{13}C) and gyromagnetic ratio (4$\times$ ^{13}C). However, in rigid solids, ^{1}H–^{1}H dipolar couplings dominate, leading to large broad resonances (often >100s of kHz wide) which, in large part, removes all chemical shift information in 1D ^{1}H NMR. CP utilizes the strong ^{1}H–^{13}C dipolar coupling in solids, where the magnetization from the proton is transferred to the ^{13}C nuclei for detection. This not only increases the sensitivity of ^{13}C 3–4$\times$, but makes use of the improved spectral dispersion of ^{13}C which has a larger chemical shift range (250 ppm).

To gauge a general overview of the phases and constituents present in a quantitative manner, a stepped decoupling approach can be used.[11] This technique is a 1D ^{13}C NMR experiment which makes use of the different ^{1}H decoupling environments required for the different phases. Due to the strong ^{1}H–^{13}C interactions, protons present in the solid state require a very short, high-power decoupling (often >100 W for no longer than 50 ms) to prevent probe damage. On the other hand, to effectively observe the solution fraction, longer low power decoupling (often <1 W for seconds) is required due to longer translational T_2^* relaxation in liquids (s *vs.* ms in solids). The stepped decoupling approach combines both decoupling regimes by beginning with a short high-power decoupling (SPINAL64) for approximately 20 ms (depending on probe specifications) to observe solid carbons, followed by stepping down to a longer low power decoupling (WALTZ64) scheme to detect the carbons of the solution and gel fractions. Using this step-down approach, a quantitative overview of all phases can be detected using a single NMR experiment.[11] The added benefit of this sequence is that a multiphase probe is not required, as the decoupling schemes can be utilized by a traditional solid-state hardware. However, a CMP probe is required to isolate the phases into different sub-spectra for a detailed analysis of the fractions.[10]

In CMP-NMR, spectral editing is the key to isolating the individual phases. A general classification of the constituents in each fraction (using diffusion-based editing) is shown in Figure 4.1. Figures 4.2–4.4 demonstrate how spectral editing is applied to isolate each fraction using standard samples of nicotinamide (solution), 1-aminohexylamide gel (semi-solid) and cholesterol acetate (solid). It is important to note that spectral editing can be done using either diffusion- or relaxation-based editing; however, the former is more commonly utilized as it provides cleaner separation of the fractions. The following sections will walk readers through diffusion-based spectral editing; however, similar spectra can be achieved *via* relaxation-based (^{1}H T_2 filtered) editing, discussed later in this section.

4.2.1 Isolating the Solution Fraction

The solution fraction consists of small, tumbling molecules that freely diffuse. When considering a living organism, this phase may consist of metabolites such as small carbohydrates and amino acids. In the context of soil, freely diffusing molecules can encompass those not interacting with the soil components/biomass. For example, when considering chemical spills,[12] this phase may provide details about the initial moment of the spill prior to sequestration and/or leaching. To isolate this fraction, CMP-NMR uses a technique known as diffusion editing (DE). DE applies a bipolar pulse pair longitudinal encode–decode (BPPLED) sequence[13] whereby a pulsed field gradient spatially encodes the initial position of all species (*i.e.*, twists their magnetization along the rotor axis). It is important to note that solid-state signals cannot be observed *via* DE as the ^{1}H–^{1}H dipolar interactions broaden the true solids below detection. Following a set of diffusion delays (typically ~180 ms for biological samples), the positions of the molecules are decoded (*i.e.*, magnetization is untwisted) and detected. If any of the molecules move positions during the diffusion delay, their signals would not 'untwist' and, as a result, would be attenuated. Therefore, signals from large or bound molecules which do not move will be detected. Considering an average CMP probe gradient strength of 60 G cm^{-1}, the ^{1}H line broadening associated with every micron of movement is ~25 Hz (11.7 T field strength), and thus molecules moving faster than 10 μm s^{-1} are broadened, attenuated and considered to be in the solution fraction. To recover these solution signals, a reference spectrum (Figure 4.2(A)) where diffusion delays are set but the gradients are turned off must be collected. From this, the DE spectrum (Figure 4.2(B)), where both diffusion delays and gradients are turned on, must be subtracted. The difference [inverse diffusion editing (IDE)] results in

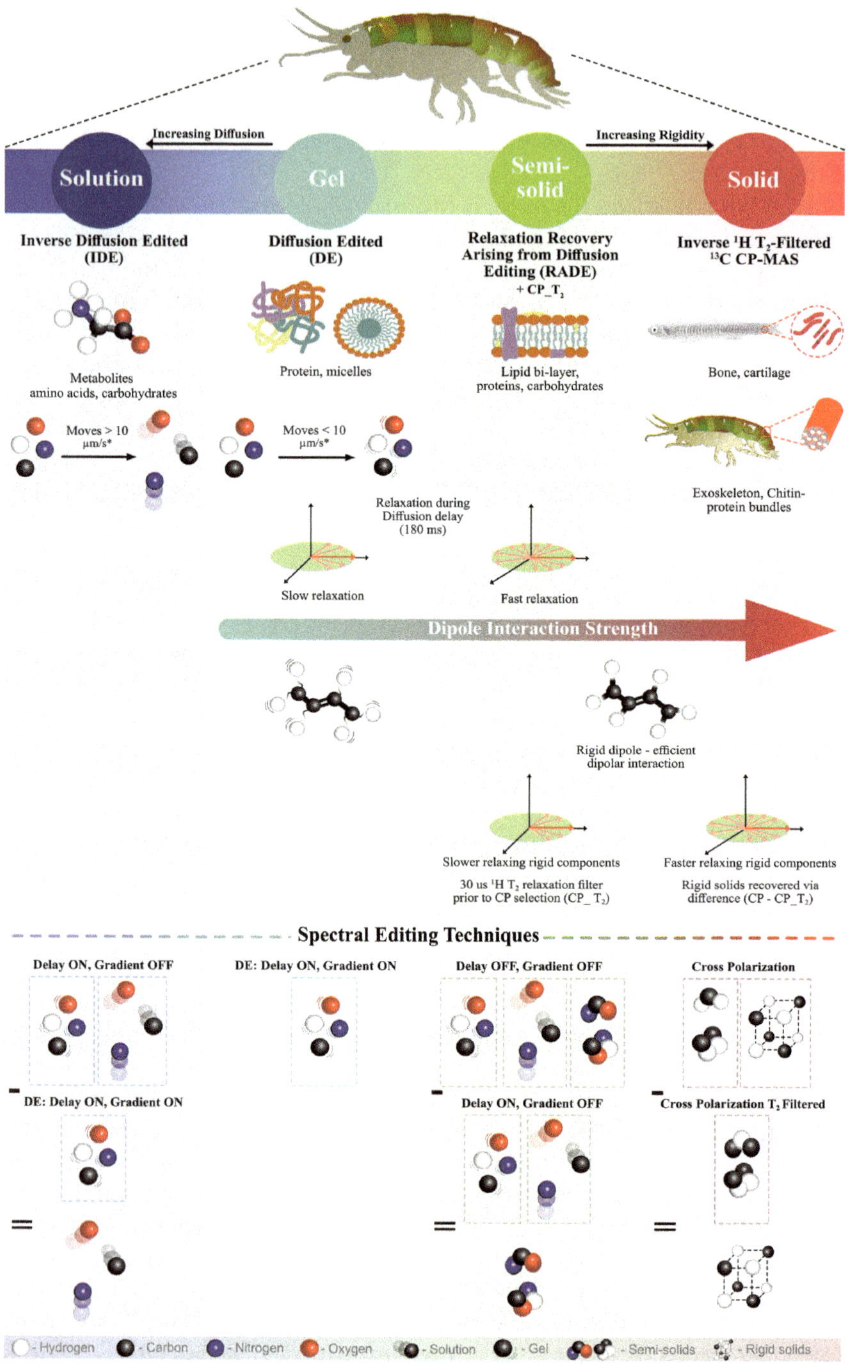

Figure 4.1 Conceptual depiction of multiphase constituents CMP-NMR can isolate using diffusion-based spectral editing. Reproduced from ref. 72 with permission from American Chemical Society, Copyright 2022.

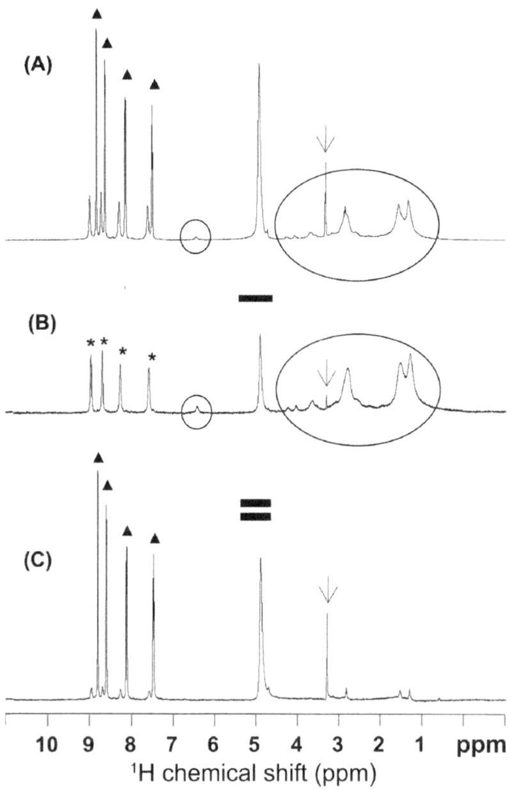

Figure 4.2 Isolating the solution fraction of nicotinamide *via* IDE. (A) Reference spectrum with delays on but gradients off. (B) DE spectrum with both delays and gradients turned on. (C) The difference between (A) and (B) where only mobile components are selected. ▲ Freely diffusing nicotinamide. *Nicotinamide trapped in a gel complex. Arrows depict methanol as a degradation product. Ovals represent signals from the 1-aminohexylamide gel. Reproduced from ref. 10 with permission from Elsevier, Copyright 2012.

any signals that were attenuated when the gradients were used (*i.e.*, freely diffusing solution molecules; Figure 4.2(C)).

4.2.2 Isolating the Gel Fraction (Restricted Diffusion)

DE can also be used to observe gel-like components which have restricted diffusion. As mentioned above, diffusion editing suppresses signals arising from mobile elements, but signals from components that do not move are retained (Figure 4.2(B)). These molecules are anything moving less than 10 $\mu m\, s^{-1}$ using a standard gradient and encompass large macromolecules or bound constituents. In a recent CMP-NMR study observing heterogeneous photocatalytic reactions,[14] it was found that the reaction/catalysis occurs in this gel-like, restricted diffusion phase, where the substrate was loosely interacting with the solid catalyst. In biological systems, these components can be larger protein and carbohydrate structures as well as small lipid vesicles.[15] In other environmental samples, the gel phase may provide information regarding interactions, such as contaminants interacting with soil compartments, or provide information as to which parts of the soil are swellable by water.

4.2.3 Isolating the Semi-solid (Rigid Gel) Fraction

It may seem that all constituents in the swollen phase have been addressed; however, there exists a phase of rigid swollen material known as semi-solids

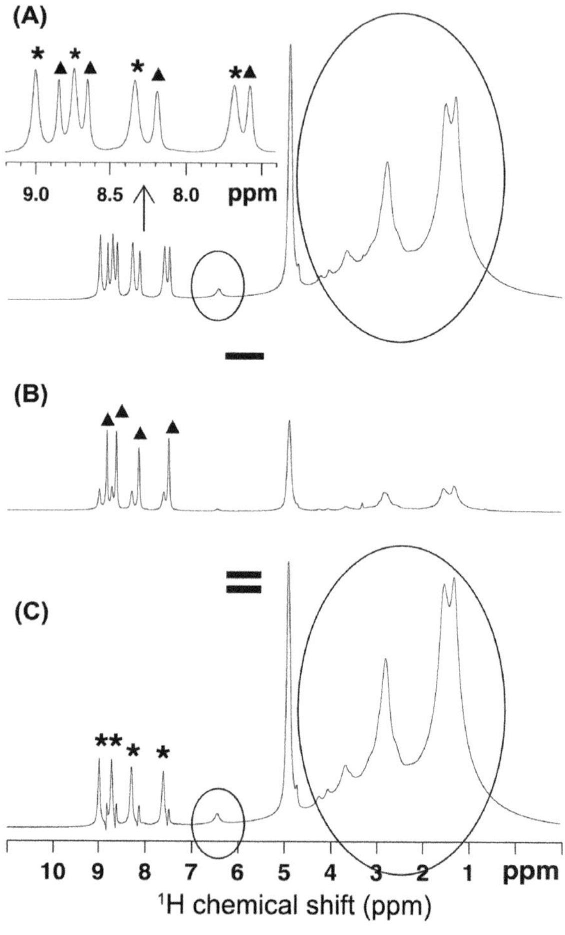

Figure 4.3 Isolating the semi-solid fraction containing 1-aminohexylamide gel *via* RADE. (A) Reference spectrum with both diffusion delays and gradients turned off. (B) DE spectrum with delays turned on but gradients turned off. (C) The difference between (A) and (B) where rigid components relax too quickly during the diffusion delay and may otherwise be missed. ▲ Freely diffusing nicotinamide. *Nicotinamide trapped in a gel complex. Ovals represent signals from the 1-aminohexylamide gel. Reproduced from ref. 10 with permission from Elsevier, Copyright 2012.

or rigid gels. Due to their rigidity, the protons of these components relax very quickly, over the long diffusion periods required for DE. As a result, the semi-solid fraction can be missed if diffusion editing is used alone. There are two techniques that can be used to isolate this fraction. The first technique is known as recovering relaxation losses arising from diffusion editing (RADE), which uses a reference spectrum where both the diffusion delays and gradients are turned off (Figure 4.3(A)), allowing the observation of all swollen (solution, gels and semi-solids) constituents in the sample. From this, another spectrum, where the diffusion delay is turned on, but the gradient remains off, is subtracted (Figure 4.3(B), reference spectrum used to isolate solution components). The difference between these two spectra (Figure 4.3(C)) recovers any signals that may have been attenuated during the diffusion delay period (*i.e.*, semi-solid or rigid gels). These signals arise from larger, rigid macromolecules in the swollen fraction, such as collagen, the lipid bilayer, or proteins and carbohydrates bound to the bilayer.

The second method for isolating the semi-solids utilizes ^{13}C CP, as these components are rigid enough to cross-polarize. To select for this fraction, a short (often 15–30 μs for biological samples) Carr–Purcell–Meiboom–Gill (CPMG) ^{1}H T_2 filter is used prior to CP. This filter attenuates signals with very short T_2 relaxation which are characteristic of crystalline solids. This leaves the less rigid components that can still cross-polarize such as semi-solids (Figure 4.4(B)).

4.2.4 Isolating the Solid Components

To isolate the rigid, crystalline solid components, CP-MAS is used. As mentioned, here the proton magnetization is transferred to ^{13}C *via* the strong ^{1}H–^{13}C dipolar interactions. As solution and dynamic gel-like components exhibit fast molecular/bond motion, the transfer of magnetization from ^{1}H to ^{13}C is inefficient due to modulation of the dipolar interactions. In turn, this results in these components not being detected.[10] After cross polarization, protons are decoupled from ^{13}C during acquisition. The CP spectrum (Figure 4.4(A)) will consist of signals from both crystalline solids as well as semi-solids. Applying a ^{1}H T_2 filter will attenuate signals from the crystalline components resulting only in signals from semi-solids (Figure 4.4(B)). Subtracting the T_2 filtered CP from a standard CP (Figure 4.4(C)) will recover any signals that were attenuated by the filter. This leaves behind signals arising from only rigid, crystalline components, such as chitin exoskeletons, bones or lipid reserves in biological samples, hydrophobic components in soil, or strongly bound compounds in heterogeneous systems (*i.e.*, catalyst surface fouling[14]). Solid-state information provides details as to structural changes, such as the breakdown of large macromolecules (*i.e.*, calcium carbonate dissolution in the exoskeleton of crustaceans and coral reefs due to ocean acidification[16,17]). Note that the rigid solids can be observed *via* the ^{1}H nuclei, using experiments such as

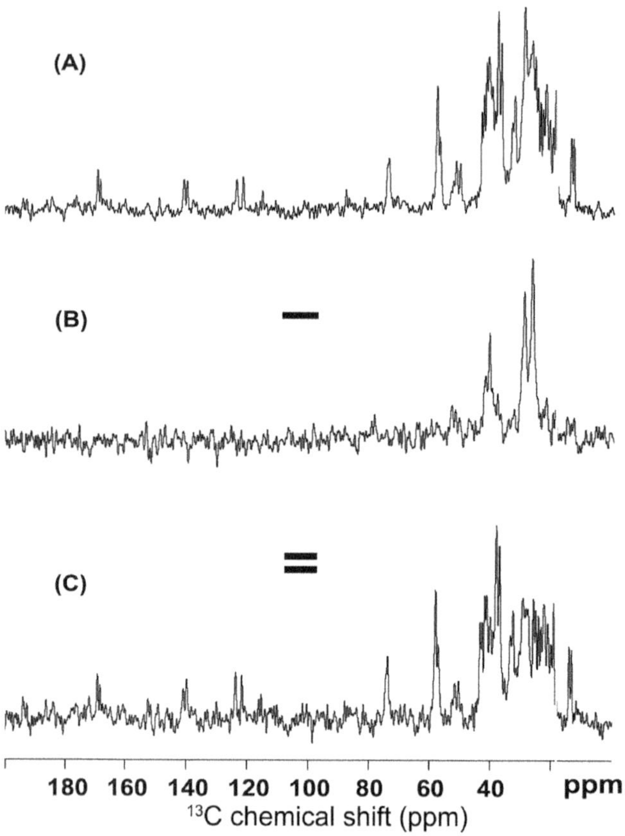

Figure 4.4 Isolating the rigid solid fraction *via* ^{13}C CP-MAS. (A) Reference CP spectrum containing signals from both crystalline solids and semi-solids/rigid gels. (B) ^{1}H T_2 filtered CP, where rigid crystalline structures with short T_2 relaxation are attenuated. This spectrum contains only signals from semi-solids. (C) The difference between (A) and (B) selectively isolates the rigid crystalline solids. Reproduced from ref. 10 with permission from Elsevier, Copyright 2012.

combined rotation and multipulse spectroscopy (CRAMPS);[18] however, anisotropic interactions (chemical shift, dipolar and quadripolar couplings) result in broad resonances which require stringent conditions (challenging NMR setup) with only limited chemical shift information being recovered. Therefore, due to the low resolution of 1D ^{1}H NMR, solid-state resonances are predominantly observed *via* ^{13}C CP-MAS.

4.2.5 Relaxation Based Spectral Editing

Diffusion-based editing is more commonly used for CMP-NMR studies due to its robust nature providing cleaner separation of the phases.[10] A key difference between diffusion and relaxation based spectral editing techniques is that DE isolates the phases by considering the motion of the entire compound, whereas relaxation-based editing can isolate molecular information about mobile functional groups that could be constituents of larger macromolecules. Consider for example a lipoprotein; while definitely a large macromolecular complex, the long mobile lipid chains have considerable

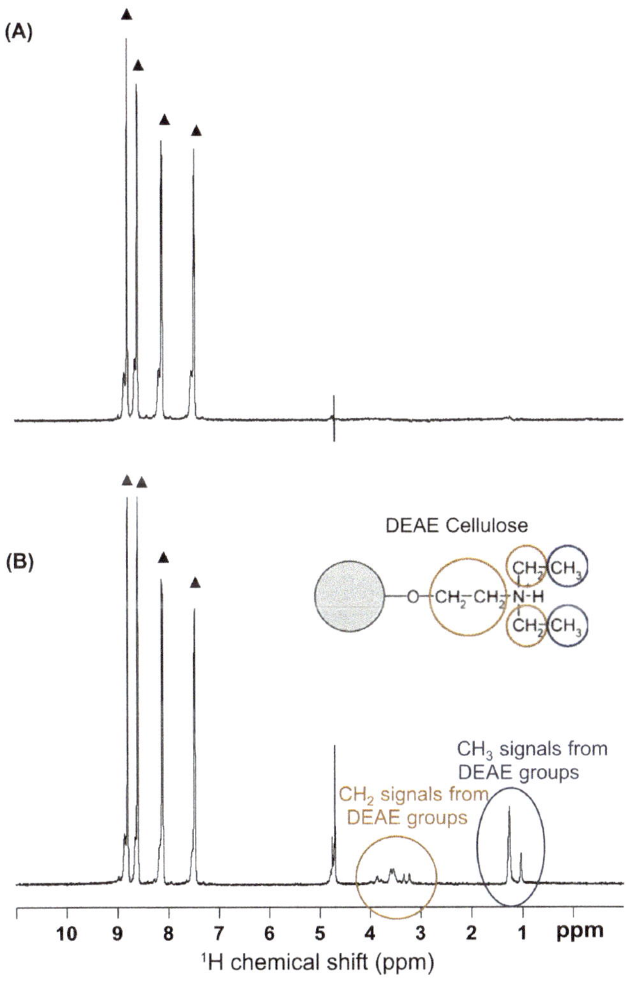

Figure 4.5 Solution phase isolated by ^{1}H IDE and T_2 filtered experiments of gel-like DEAE cellulose and freely diffusing nicotinamide (▲). (A) DE experiment where mobile constituents are isolated by IDE which considers the diffusion of the entire compound, and thus only nicotinamide is detected. (B) Relaxation based experiment using ^{1}H T_2 filters (CPMG) where all constituents exhibiting motion are detected. Here the mobile CH_2 and CH_3 moieties (circled) of DEAE cellulose are also observed though the molecule as a whole exhibits no diffusion. Reproduced from ref. 10 with permission from Elsevier, Copyright 2012.

dynamics that could make them appear "mobile" in relaxation-based editing. For mobile components, CPMG spin-echo pulses can be used to selectively isolate the mobile groups by attenuating signals from faster relaxing (T_2 relaxation) rigid components (gels, semi-solids, and solids). Courtier *et al.* demonstrate the complementary nature of relaxation-based editing to DE using DEAE cellulose which behaves like a gel but has several, highly mobile ethyl groups.[10] Figure 4.5 compares the solution phase obtained *via* IDE (Figure 4.5(A)) and CPMG filters (^{1}H T_2 filtered; Figure 4.5(B)). The spectral profile of the dissolved nicotinamide remains similar between the two; however, the mobile aliphatic groups of the DEAE cellulose can only be isolated using relaxation-based filters. As the entire molecule of cellulose does not diffuse during the diffusion period, its signals are suppressed in the IDE. However, as the ethyl groups are quite mobile, their protons will have a larger T_2 relaxation compared to the rest of the molecules allowing detection. In this manner, both IDE and relaxation-based editing provide similar yet complementary information to each other, where the information obtained from both can help understand local motions and flexibility of macromolecules such as mobile moieties of large organic acids in soil.

In this manner, a single probe can be used to isolate and observe all fractions (solution, gels, semi-solids and solids) in complex heterogeneous samples, where minimal sample preparation is required. All applications of CMP-NMR to elucidate environmental processes to date will be discussed in the following sections. Readers interested in applying CMP-NMR should refer to the tutorial by Liaghati Mobarhan *et al.*[19] One aspect to note is that CMP-NMR is best run with a spectrometer lock. This can be added to the sample (for example D_2O spiked into a wet soil) or be added in a separate capillary at the bottom of the rotor.[20] The advantage of the latter is that the sample remains completely unchanged.

4.3 Multiphase Analysis of Molecular Structure in Natural Samples

As mentioned, historically NMR research relied on altering the samples' composition to meet instrument requirements (*i.e.*, drying and packing into rotors for solid-state NMR or extraction/dissolution for solution-state NMR). However, natural samples, such as soil and plants, arise from the synergy between solid crystalline structures, dissolved components, and everything in between. As such, CMP-NMR is the only technique that can decipher molecular structure, interactions and processes closely resembling those occurring in the environment naturally. This section will review the capabilities of CMP-NMR to resolve molecular structures in complex heterogeneous systems such as soil, plants, edible food, and biofuel production as well as non-living systems (automobile rubber and photocatalysts).

4.3.1 Elucidating the Complex Structure of Soil Organic Matter and Oil Contaminated Soils

One of the early applications of CMP-NMR was to investigate the organization of organic matter within whole soil.[21] Soil is often difficult to study in its native state as it is a complex mixture of organic and inorganic materials. Since its isolation, soil organic matter (SOM) has been the subject of many studies.[22] The composition, structure, and reactivity with the surrounding environment of SOM are just some of the mysteries of this complex structure. For a long time, it was widely believed that SOM was composed of humic substances cross-linked together; however, further studies suggest that it is better defined as a complex mixture.[22] The organization of SOM in its native state must be well characterized in order to determine the fate of environmental contaminants, nutrient/water retention, microbial biodegradation, and consequences of anthropogenic activities (*e.g.*, leaching of compounds following soil treatment or acid rain). CMP-NMR can provide information about the molecular structure of the various soil layers, and thus the potential interactions taking place in the soil. By isolating the fractions *via* the CMP-NMR spectral editing techniques mentioned above (Section 4.2), a simplified schematic of the components of soil was created (Figure 4.6).[21]

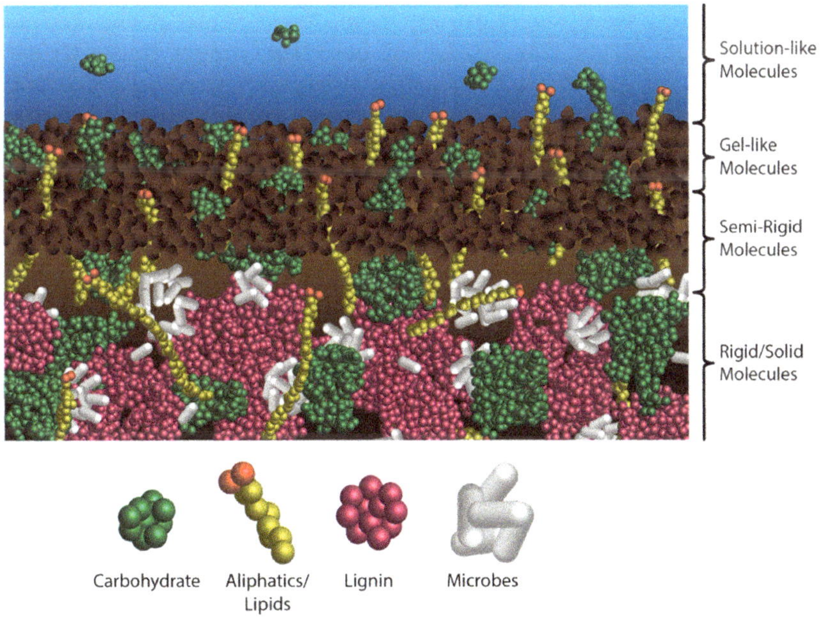

Figure 4.6 Conceptual diagram representing the organization of soil organic matter using CMP-NMR. The mobile components (carbohydrates and polar heads of lipids) make up the solution fraction at the soil–water interface. Moving deeper into soil layers (down), the accessibility to water decreases and the constituents (lignin and microbes) become more rigid. Reproduced from ref. 20 with permission from American Chemical Society, Copyright 2016.

The study found that major biopolymers such as carbohydrates, lignin, and proteins were found to be present in roughly equal portions, while lipids made up a smaller fraction (~10%) of whole soil.[21] Both diffusion- and relaxation-based CMP-NMR experiments were used to understand the mobile constituents at the soil–water interface (top layer). Carbohydrates were found to dominate this solution fraction. Using relaxation filters, the mobile COOH groups of lipids were also found in this phase. These components, though mobile, are only subsections of larger macromolecules that cannot diffuse, suggesting the polar heads are oriented towards the aqueous phase but the bulk of the molecule (CH_2 and CH_3 chains) is secured within the soil fraction with restricted molecular motion.[21]

The next layer, with less access to water, consists of species with restricted motion (gel and semi-solid). Lipids dominated these fractions, though some carbohydrates also remained swollen, indicating they were interacting with components at the interface and demonstrated internal mobility. Finally, rigid components, such as lignin and microbial matter, are believed to make up the bottom layers, where physical barriers to water accessibility can also play a role in maintaining soil structure and health.[21] Interestingly, the aromatic domains of lignin and protein remained impenetrable by water, indicating they may be integrated within microbial matter and are less accessible to water at a neutral pH.[21]

Understanding solvent accessibility in soil can be further explored using CP. CP becomes less efficient as the soil becomes swollen, thus attenuating the solid-state signals. This becomes particularly important when looking at contaminated soil, such as during oil spills, which hinders water accessibility and transport throughout the soil layers.

Oil spills are a major source of environmental pollution and have a detrimental impact on marine and terrestrial organisms as well as entire ecosystems.[23] As many components of petroleum are toxic and carcinogenic, it is important to understand where these contaminants may sequester to determine their ramifications and devise effective remediation techniques.[24] Oil spills may also impose a physical barrier to soil swelling, which can lead to erosion, leaching of nutrients/pesticides and thus subsequent deterioration of plant and microbial health. Farooq *et al.* applied CMP-NMR to aged (1973) oil contaminated soil to understand the mechanisms by which oil coats the soil and elicits structural changes, thereby impacting interactions between different soil compartments.[25] It was found that oil creates a hydrophobic layer within the top layer of soil and reduces accessibility to water, inhibiting most components from swelling. This has numerous consequences on soil fertility and plant growth in that environment.[25]

Solution/gel-state capabilities of CMP-NMR (IDE and DE, respectively) confirm that interactions occur between oil and water (forming a layer). This was observed by an increase in aromatic CH_3, typically from the aromatic domains of asphaltene and BTEX (benzene, toluene, ethylbenzene and xylene) as they become more mobile (Figure 4.7).[25] Their presence at the soil–water interface suggests they can leach into groundwater from affected soils.

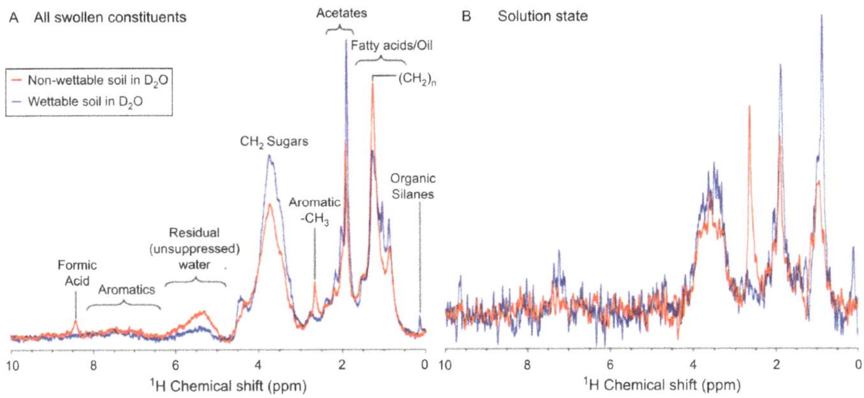

Figure 4.7 ^{1}H NMR depicting control (wettable) and oil contaminated (non-wettable) soils. (A) All swollen constituents labelled. (B) IDE depicting all dissolved components in the solution phase. Reproduced from ref. 25 with permission from CISRO Publishing, Copyright 2015.

Reduced carbohydrate resonances in oil contaminated soils also emphasize the inability of water to penetrate the oil coating (Figure 4.7(A)).[25] Moreover, microbial activity was reduced in the contaminated soil as indicated by the reduced acetate signal (Figure 4.7(A)).[25] It is also interesting to note that polar compounds, such as formic acid, in the top layer would normally leach out of non-contaminated soils (as seen by the absence of its resonance in the wettable soil spectrum). However, in the oil contaminated soil, a broad peak of formic acid was seen, implying that it was trapped in the water matrices and unable to pass the non-polar oil at the oil-water interface. Overall, CMP-NMR can provide an overview of the organization of the soil matrix which can be used to understand how and where xenobiotics move and interact with the soil, which can be advantageous in determining which layer to target for soil remediation.

4.3.2 CMP-NMR to Study ^{13}C Enriched Seeds and Germination Process

Plants are also mixtures of both organic and inorganic components and are important for environmental research, as they play a major role for food, shelter, nutrients and carbon/nitrogen cycling in all ecosystems. Previously, plant and seed analysis often required invasive sample preparation techniques (grinding, dissolution into buffers, centrifuging) that can destroy natural interactions and structural integrity. As minimal sample preparation is required for CMP-NMR, it can be used to gain understanding of both structural and molecular compositions (*i.e.*, oil, protein, moisture content, binding, associations, *etc.*) and can be used to understand metabolic variations resulting from genetic mutations and/or nutrient deficiencies. CMP-NMR is also the only technique capable of observing volatile metabolites such as methanol[26,27] and ethylene gas (dissolved in the solution fraction)[26] which are required for plant cell wall metabolism and germination, respectively. These metabolites would otherwise be lost during sample preparation. Dynamic changes such as movement of nutrients across various phases or growth and development of roots and shoots during germination can also be observed.

Seeds in particular are an essential source of fibre, nutrients, and antioxidants. Lam *et al.* used CMP-NMR to study the structural composition of ^{13}C labelled broccoli, wheat and corn seeds to represent globally consumed plants and legumes.[26] It was found that the gel and semi-solid phases of the seeds were dominated by triacylglycerides (TAG) consistent with long-term energy stores. Semi-solid-like carbohydrates were believed to be from larger starch polymers in the endosperm, cell walls and seed coat. Furthermore, the solid-state spectra consisted largely of starch and crystalline cellulose/hemicellulose.[26]

Figure 4.8 shows the use of CMP-NMR to identify the phase distribution of key biomolecules across phases in different types of seeds (wheat, broccoli, and corn). For example, compared to the other seeds, broccoli has the least

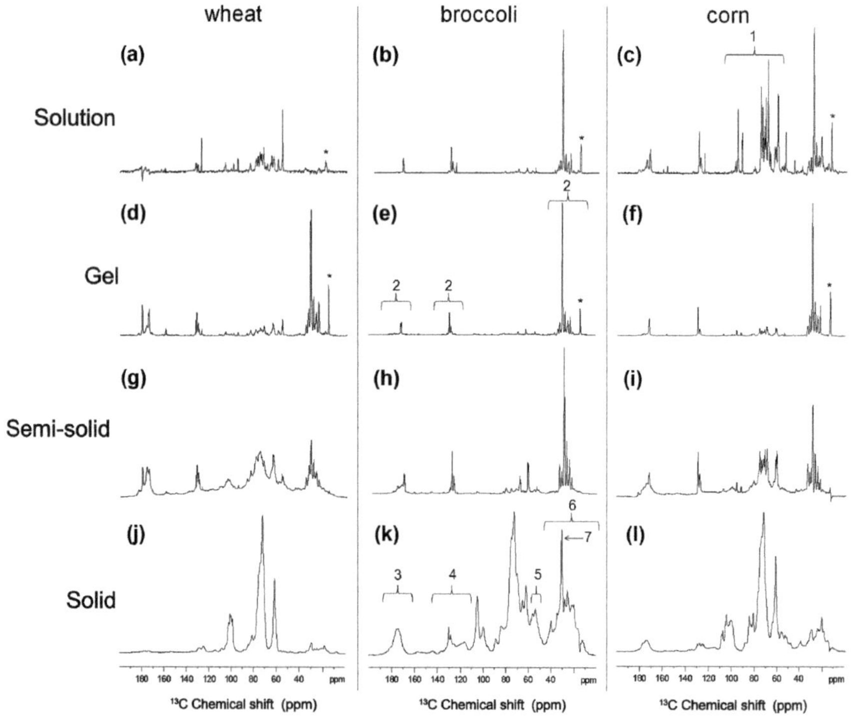

Figure 4.8 Multiphase NMR of the contents of (columns) wheat, broccoli and corn seeds isolated using the spectral editing techniques of CMP-NMR. The individual fractions are found across the rows. Reproduced from ref. 26 with permission from American Chemical Society, Copyright 2014.

amount of carbohydrates in the solution state, while the majority of its biopolymers (proteins and long chain fatty acids) exist as true solids and consist of the highest lipid content relative to the other seeds.[26] Interestingly, Lam *et al.* found that the broccoli seeds still germinated even after spinning.[26] This holds the potential to monitor both structural and molecular flux during seed germination and was the next step in CMP plant research.

The study by Fortier-McGill *et al.*[28] follows the metabolic and structural changes within a growing ^{13}C labelled wheat seed (*Triticum aestivum*) every 24 hours for 4 days (Figure 4.9).[28] Prior to germination (while the seed was still dry), most biopolymers (lipids and carbohydrates) exhibited restricted diffusion in the gel and semi-solid phases, compared to after 4 days of germination where the carbohydrates and amino acids (aliphatic and aromatic side chains) had swollen considerably, and entered a more mobile gel and solution-like state (Figure 4.9). IDE which highlights the dissolved phase showed a reduction in TAG signals as lipids are metabolised more efficiently for energy intensive processes like germination.[28]

Following 4 days of germination, there was an increase in dissolved metabolites as the seed became swollen in the presence of water. Using RADE, it was observed that the semi-solid fraction consisted of largely TAG resonances, which are immobilized within subcellular droplets.[28] Although very little changes in crystalline solid-state NMR were seen over 4 days of germination, T_2-filtered solid-state experiments (dynamic solids) enabled the differentiation of the C_6 of dynamic/less rigid starch from that of crystalline cellulose (Figure 4.10(D)).[28]

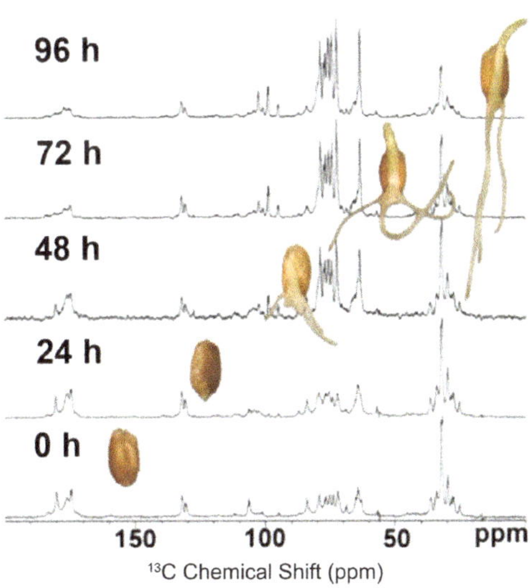

Figure 4.9 Depiction of germinating ^{13}C enriched *Triticum aestivum* (wheat seed) over 4 days. Changes in the semi-solid fraction are showcased. Reproduced from ref. 28 with permission from American Chemical Society, Copyright 2017.

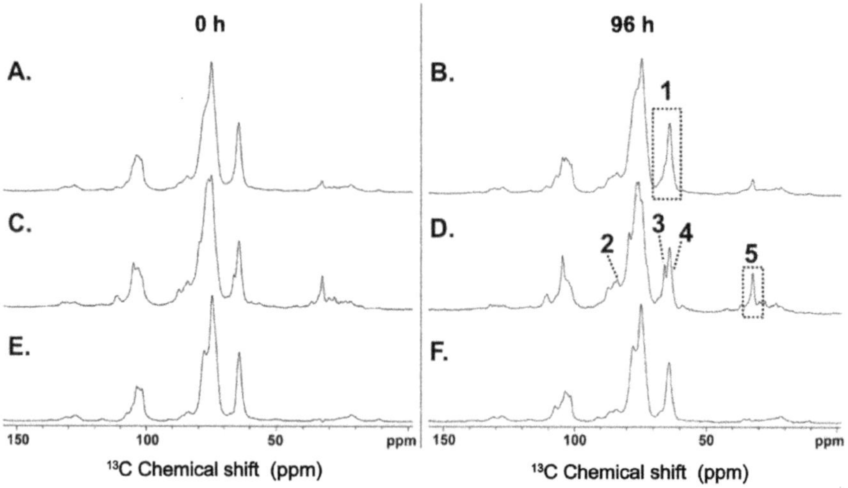

Figure 4.10 ^{13}C CP-MAS of germinating seeds up to 4 days (96 h). (A and B) CP-MAS detecting all solid components. (C and D) Dynamic solids obtained *via* ^{1}H T_2 relaxation filtration. (E and F) Crystalline solid fraction obtained by the inverse ^{1}H T_2 filtration from subtracting (C) and (A) as well as (D) and (B). (1) C6 of starch and cellulose, (2) dynamic C4 of starch, (3) dynamic C6 of starch, (4) dynamic C6 of cellulose, and (5) dynamic CH_2 of TAG. Reproduced from ref. 28 with permission from American Chemical Society, Copyright 2017.

An additional application of CMP-NMR is the ability to understand changes in processes and interactions resulting from genetic deficiencies. A study by Wheeler *et al.* used CMP-NMR to look at *Arabidopsis thaliana* and the genetic mutation (labelled as *ectopic lignification1*, *eli1*) which impacts the protein subunit for cellulose production, resulting in an overall change in cell wall composition.[27] Relative to the wild type, the *eli1* mutant was found to have higher levels of starch and lignin to compensate for the cellulose deficiency.[27] The increase in rigid starch production correlated to a decrease in production of soluble, free carbohydrates in the solution phase. The *eli1* mutant was also found to have higher levels of long chain fatty acids (change in energy store metabolization or production of more seed derived lipids), nucleic acids, methanol (demethylation of pectin in cell wall metabolism), phenylalanine (precursor to lignin monomer), glutamine (deamination of phenylamine reduces glutamate to avoid the presence of toxic NH_2) and pentaglycine.[27]

The solid phase spectra of both the mutant and wild type in the *A. thaliana* seedlings contained signals from similar biopolymers (cellulose, starch and lignin). Interestingly, both wild type and mutant *A. thaliana* had similar levels of lignin content. Using relaxation filters (T_2 or CPMG filter prior to CP to attenuate signals from crystalline solids), the *eli1* mutant was seen to have higher levels of amorphous cellulose. This suggests that the mutation does not render the targeted enzyme functionless, but may have produced the typically crystalline cellulose[29] in a more mobile form. This may have further consequences on the structural integrity of these genetically mutated plants.[27]

4.3.3 Applications of CMP-NMR to Monitor Structural Changes During Cooking

Changes in starch mobility also have considerations with respect to food for human consumption. The swelling, solubility, and change in viscosity as carbohydrates interact with lipids in most starch-based food will differ during hydrothermal treatment (cooking) and can be measured by CMP-NMR.[30] Traditionally, in the food sciences, HR-MAS NMR is used to observe the changes in the swollen phase;[7,8] however as many starch-based foods are often dry-packaged, changes in the crystalline solids must also be considered. Barison *et al.* used CMP-NMR to observe changes in molecular flux across various phases of starch-based foods such as spaghetti, black turtle beans and long-grain white rice.[30]

In their uncooked, dry states, the swollen fractions of all three foods were dominated by TAG (beans > spaghetti > rice) and water (may correlate to shelf-life longevity) trapped in the structural matrix, where the carbohydrates are present in the solid phase as starch networks (confirmed by CP). CMP-NMR was used to differentiate between mobile and trapped TAG, where the latter can be found as micelles or interacting as starch–lipid complexes in the semi-solid fraction (confirmed by their presence in ^{13}C experiments with low power decoupling).[30] Upon cooking, the trapped structural water (solution) and starch (gel/semi-solid) become dynamic and variations in amylopectin and amylose polymorphs[31,32] were observed, providing insight into the branching arrangement. Constituents in the solution fraction, such as metabolites, stay within the swollen gelatinized structure rather than leach into the surrounding water. Overall, understanding the molecular structural changes during the gelatinization process of food can provide insight into the nutritional changes (*e.g.*, from growth conditions, preservatives, spoilage) and improved food processing/quality control measures can be implemented. CMP-NMR may also provide complementary information to understand the impact of other cooking techniques (grilling, frying, baking, *etc.*) on the structural changes of biopolymers.

4.3.4 Extraction of Biofuel from Algae

As the global demand for fossil fuel increases, the search for alternative, renewable energy sources continues. The extraction of biofuel from organic matter is another example of a complex process involving changes in structural biopolymers which can be monitored *via* CMP-NMR. Algae are a popular feedstock source for biofuel production as they are photosynthetic, and rapidly reproduce, and a diverse range of strains with different compositions and properties exist.[33] However, to optimize the extraction efficiency, their composition must be understood, and depending on the material being extracted (lipids, carbohydrates, or proteins), targeted extraction techniques may be required. Ideally, for biofuel, the lipids should be in the solution state or be easily solubilized for facile conversion, whereas the waste should be more solid-like for easy separation and disposal/re-purposing.

Ning *et al.* used CMP-NMR to observe changes in the biochemical composition of algal biomass under different extraction conditions (temperature, presence of catalyst) and with different algal feedstocks.[33] Here, dissolved metabolites (amino acids, carbohydrates) made up the solution fraction, whereas their less mobile counterparts were in a gel-like state. Unsaturated fatty acids dominated the semi-solid fraction and were targeted for extraction. Solids such as long chain lipids, proteins (glycoproteins) and structural carbohydrates (cellulose cell wall) required harsher conditions to improve extraction efficiency. The authors of this study used the quantitative stepped decoupling approach (Section 4.2.1) to estimate the contributions of each macromolecule in all phases.

In this study, three main extraction conditions were explored to determine optimal practice for biofuel extraction. Four temperatures commonly used in industrial practices (125, 150, 200, and 250 °C; Figure 4.11) were examined and 200 °C was determined to be the optimal temperature to mobilize most energy-rich lipids while avoiding unwanted side reactions.[33] Interestingly, at 250 °C the rigid solids increased slightly, which was attributed to side reactions (*i.e.*, condensation reaction), encouraging the repolymerization of the carbohydrates into a solid structure (Figure 4.11(d) and e). At this temperature (250 °C), both TAG and carbohydrates were believed to be hydrolyzed by the appearance of their monomers in the solution phase, which may aid the production of bioethanol.

Different feedstock sources (*i.e.*, species of algae) vary in their molecular composition which can impact the intended biofuel use. Ning *et al.* compared the biomass composition of 2 green algal and 2 cyanobacterial (high CH_2 content in long chain lipids) species at 200 °C. The presence of a cellulose-rich (green algae) or peptidoglycan-rich (gram-negative bacteria) cell wall hinders the accessibility of solvent during extraction and reduces the amount of lipid that can be mobilized, thus requiring further parameter optimization. Interestingly, the presence of polyhydroxybutyrate (PHB), a suitable replacement for plastics, was found in the solid-like fractions of the *Synechocystis* cyanobacterial strain.[34,35] Under optimal extraction conditions, crystalline solid PHB can be easily separated and used commercially.[33]

Lastly, catalysts can be used to aid in the optimization of extraction efficiency, without altering the metabolic profile of the constituents and preventing side reactions. The presence of a sodium carbonate catalyst helped mobilize macromolecules (solution phase doubled) from the solid fractions. However, a large portion of the biomass remained rigid, suggesting more careful optimization is required for large-scale biofuel production.

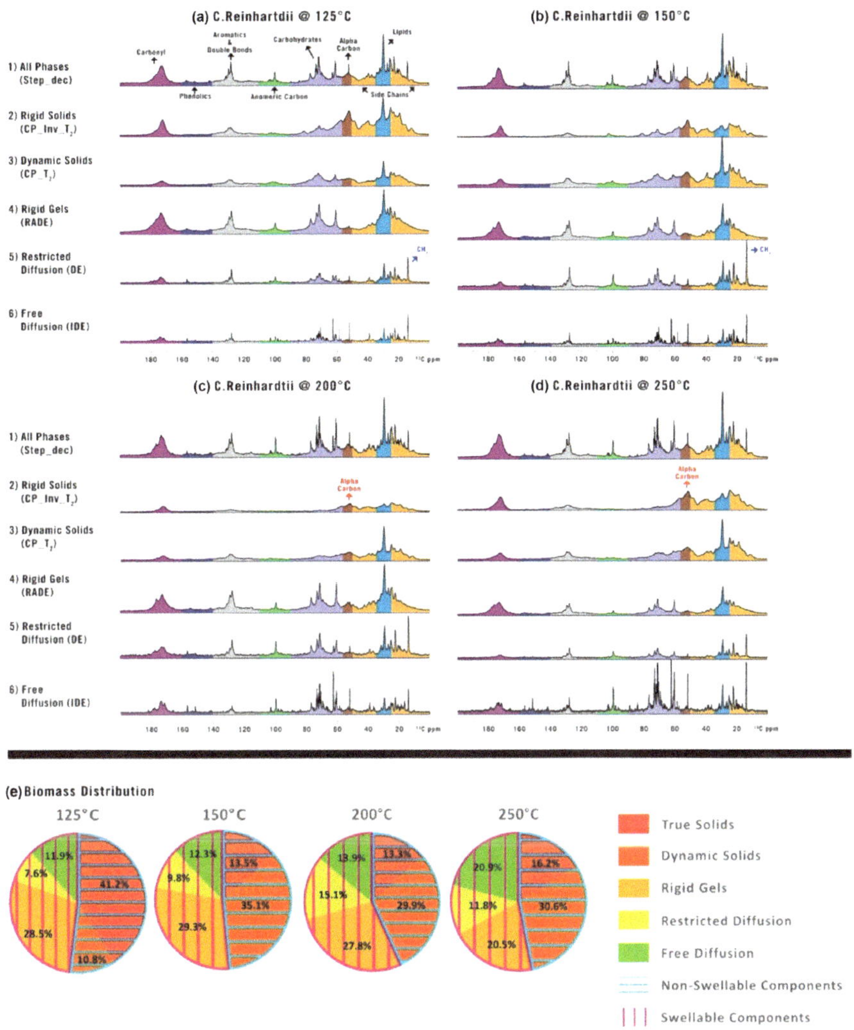

Figure 4.11 ^{13}C CMP-NMR of biomass extracted from *Chlamydomonas reinhardtii* at different extraction temperatures (a–d). (e) Graphical representation of % multiphase distribution of CR biomass using the stepped decoupling sequence. Reproduced from ref. 33 with permission from American Chemical Society, Copyright 2020.

4.3.5 Degradation of Car Engine Rubber by Biofuel

CMP-NMR not only provides an avenue to observe the systematic optimization of biofuel extraction but can also observe its potential interactions with automobile components (*i.e.*, rubber). Biodiesel is a less toxic, biodegradable (vegetable and animal fats derived) alternative to petroleum products and has been used to supplement petrodiesel in many countries.[36] However, as biodiesel is more polar than petrodiesel and contains more unsaturated fatty acids (double bonds), the swelling and breakdown of rubber *via* oxidative degradation increases. Silva *et al.* applied CMP-NMR to observe the structural changes associated with rubber swelling by soybean-derived biodiesel in automobile engines.[37] Both natural rubber ((NR), more mobile domains) and commercially used (highly rigid) synthetic rubbers (styrene-butadiene/SBR, nitrile/NBR, ethylene–propylene–diene monomer/EPDM) were used to understand the aging (long-term contact, 90 days) and degradation process.[37]

As the solid, intact rubber swells and breaks down, the rubber takes on gel-like characteristics (restricted diffusion), whereas the degradation products and fuel comprise the solution phase. It was found that the biodiesel was able to penetrate and become trapped within the structures of all rubbers, and an increase in rubber mobile domains (*i.e.*, local motion) suggested that the biodiesel had also induced structural changes (excluding EPDM).[37] The loss of all rigid characteristics for the other 3 rubbers was confirmed by RADE, leading to degradation especially in NBR. The ease of rubber swelling correlated with its rigidity. EPDM was the most rigid rubber tested, and no degradation was seen, which may attest to its long-term performance.[37] Using CP, the mobility of the aliphatic and aromatic regions of EPDM was seen to increase, indicating that CMP-NMR would be ideal to monitor longer term interactions for highly rigid rubbers. In this study, the authors ranked the compatibility of the different rubbers with biodiesel as EPDM > SBR > NR > NBR. However, this can vary with biodiesel feedstock sources (differing degrees of saturation).[37] The impact of dissolved/solution-state rubber degradation products on the surrounding environment (*i.e.*, soil sequestration, toxic impacts) may also be worth considering. For example, recent findings have demonstrated that additives in tire rubber can be transformed into a highly toxic compound responsible for the death of coho salmon.[38]

Typically, structural changes are elucidated using solid-state NMR. However, within a CMP system, subsequent molecular fluxes in swollen fractions can be observed in intact natural samples. Thus far, the various structural changes monitored using CMP-NMR in multiple biotic and abiotic systems have been reviewed. However, the next section will discuss how CMP-NMR can also provide in-depth information regarding *molecular interactions* during complex processes.

4.4 Multiphase Analysis of Molecular Interactions and Processes

To gain a better understanding of a system, all interactions within the sample and its surrounding environment must be observed. Here, observing both structural changes (phase transformation) in addition to molecular interactions (binding, movement between phases) becomes vital to understand broad-scale processes. This section explores the application of CMP-NMR to observe interactions within natural and inorganic systems such as biomass–clay complexes[39] and monitor entire processes such as following from spill to sequestration of contaminants in soil,[12] and the photodegradation of phenol *via* catalyst activity.[14]

4.4.1 Biomass and Clay Interactions

CMP-NMR was used to elucidate the organization of the complex organic soil matrix.[21] It is the synergy between these compartments that gives rise to soil structure, reactivity, carbon cycling/storage, mineral formation as well as contaminant binding. As such, it is crucial to understand the interactions between microbes, SOM and clay minerals to gauge overall soil health. SOM–mineral complexes are a result of microbial contributions. The molecular mechanism by which microbial cells would interact with mineral constituents to gain protection and access to nutrients can only be understood *via* a non-invasive technique such as CMP-NMR. Olivelli *et al.* used CMP-NMR to understand structural and molecular interactions (*e.g.*, binding preferences) of fungal (*Aphanocladium* sp./*Apha.* and *Acremonium* sp./*Acre.*) and bacterial (*Actinobacteria*) biomass with montmorillonite clay (bentonite).[39] The biomass–clay (BMMT) mechanism of attachment differed between fungi and bacteria. Here fungi demonstrated interactions with clay using high molecular weight carbohydrates and lipid aliphatic chains by *Acre.* and low molecular weight lipids and proteins by *Apha*. On the other hand, bacteria demonstrated non-specific interactions/preferences with clay, possibly through extracellular polymeric substances (EPS).[39] The presence of large molecular weight macromolecules in the gel and solid states suggested whole cells may attach to the clay surfaces. In addition, solid-state NMR alludes to biomass protection from lysis when attached to clay surfaces, as well as access to nutrients and stress.[39] By understanding the structure and interactions within a natural system (*i.e.*, soil), the behaviour of the changes can be explained. For example, in the event of a chemical spill, how does the contaminant interact and sequester within the various soil compartments?

4.4.2 CMP-NMR to Monitor Perfluorinated Pollutant Sequestration in Soil

Perfluorinated carbons (PFCs) are stable, labelled as 'forever chemicals' with lifetimes ranging from weeks to years, depending on the structure.[40,41] They are typically used as surfactants for numerous consumer products,[42] anti-foaming and agrochemicals,[40] and can easily escape into the surrounding environment (terrestrial to atmospheric). Due to their persistence, biomagnification, and ability for long-range transportation,[43] these compounds pose a threat to human/animal health, soil microbe populations and ecosystems in general. As traditional soil partitioning metrics (K_{OC}) do not consider interactions (microbial population, soil complexity),[12] a technique that examines systems in their entirety is required to determine the true molecular fate of these species.

Masoom *et al.* used ^{19}F CMP-NMR and multiphase kinetics (sorption mechanisms, binding orientation) to follow perfluorooctanoic acid (PFOA) and pentafluorophenol (PFP) through natural intact soil from the moment they were spilled (solution-state) to their sequestration to the true solid state.[12,21] The study found that PFOA is predominantly dissolved in the solution state, where less than half (~41%) is sequestered in the solid state (Figure 4.12I-d). Conversely, PFP demonstrates solvent-like behaviour (increased swelling and permeation) where 88% of the ^{19}F signal was found in the semi-solid phase (RADE) before sequestration (Figure 4.12I-d).[12] This behaviour could not be predicted from the traditional K_{OC} of the compound and may vary with pH, which can be a significant factor impacting contaminant retention and release. Using solid-state ^{19}F–^{13}C CP (magnetization pool from ^{19}F), 3× more PFP was found to sequester in soil relative to PFOA.

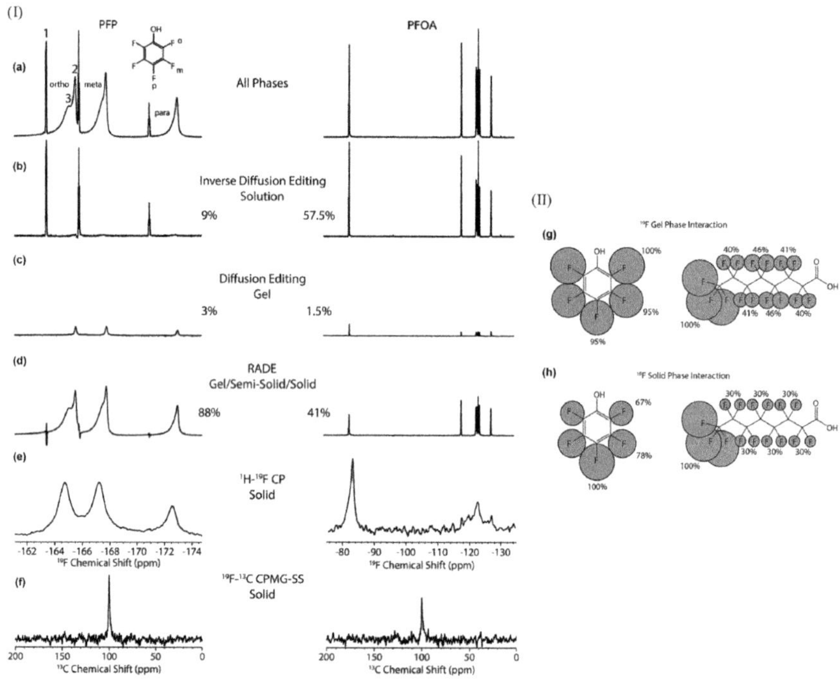

Figure 4.12 (I) Overview of all phases isolated by spectral editing: (a) 1D ^{19}F spectra of all phases. (b) Solution phase containing mobile constituents isolated by inverse diffusion editing (IDE). (c) Diffusion edited spectra depicting the gel fraction highlighting the contaminants at the soil–water interface. (d) Semi-solid fraction obtained by RADE highlighting larger constituents which relax quickly over the diffusion delay. (e) Solid fraction obtained by ^{1}H–^{19}F cross polarization emphasizing the rigid dipoles. (f) Single spike experiment obtained by CP-CPMG-SS to focus the ^{19}F–^{13}C signals to improve sensitivity. (II) Epitope maps of PFP and PFOA in the (g) gel and (h) solid phases using HSTD and CP. Reproduced from ref. 12 with permission from American Chemical Society, Copyright 2015.

Kinetic studies of the solution and gel phases supported this, where both compounds immediately adsorbed to the soil–water interface (gel, DE), and PFOA rapidly saturated and achieved equilibrium in solution, whereas the solvent-like behaviour of PFP was attributed to swelling of soil prior to sequestration. Both compounds spent limited time in this gel fraction, before sequestration (solid, CP), occurring within the first 10 min.[12]

To investigate the binding mechanism, heteronuclear saturation transfer difference (HSTD) in the gel and solid phases was used.[44] Briefly, this works by measuring the transfer of saturation from ^{1}H in soil to ^{19}F in the contaminants based on proximity. A variation of this technique is known as reverse HSTD (RHSTD) which transfers saturation from ^{19}F to ^{1}H to determine where/which macromolecule in soil the contaminants interact with.[45] From the HSTD results, it was found that PFP demonstrated non-specific binding (due to its small size), whereas the CF_3 groups of the long chain in PFOA showed the strongest interactions with the hydrophobic groups in soil (Figure 4.12II). At the soil–water interface, stronger interactions (relative to the rest of the compound) of the ^{19}F atom closest to the OH in PFP suggested that the polar OH group had a role in driving interactions between soil–aqueous interfaces and SOM (Figure 4.12II). The opposite was seen in the solid phase, where the hydrophobic groups drove sequestration.[12,46,47] Using RHSTD, it was determined that PFP interacts with an array of soil macromolecules (lipids, carbohydrates, lignin and protein), where sequestration occurred *via* the aromatic resonances in lignin in SOM (in line with the hydrophobicity of PFP and soil organization).[12,21] In contrast, PFOA interacted very little with the soil, though in extracted SOM it was shown to have a high affinity for proteins (lysed microbial cells). This suggests that PFOA is quite specific in its binding, and in natural intact soil, where microbial cells are found deeper in soil layers,[21] PFOA will remain in solution at the soil–aqueous interface, free to leach into the rest of the environment.[12] This emphasizes the importance of CMP-NMR's ability to observe interactions in natural, unaltered complex samples. In addition, the molecular behaviours of these PFCs can be used to understand their persistence in organisms. For example, the interactive aliphatic chains of PFOA along with its ability to bind to microbial proteins correspond to the binding of PFOA in albumin proteins interfering with the metabolism of fatty acids in living organisms.[48]

In this manner, molecular insight into the movement and binding of pollutants (*i.e.*, chemicals, microplastics, *etc.*) in soil can be advantageous when considering remediation techniques and their toxic-modes-of-action in other living systems in their natural state. For example, a recent study observing the binding behaviour of polystyrene nanoparticles with amino acids used CMP-NMR to demonstrate that phenylalanine and tryptophan bind to the outer shell of the nanoparticles in a gel-like phase.[49] This poses harmful impacts when considering the widespread presence of plastics in soil and aquatic environments, where small molecules (*e.g.*, pharmaceuticals, pesticides) may be sequestered or taken up by organisms.[49] This study showcases the potential applicability of CMP-NMR in other fields of research (*e.g.*, drug binding to components in blood or tissues[50]).

4.4.3 CMP-NMR to Monitor Photocatalytic Reactions

Many remediation techniques (wastewater treatment, soil treatment)[51–53] integrate the use of biological and inorganic catalysts to break down contaminants and remove heavy metals (*e.g.*, mercury)[54,55] from soil and water bodies. These reactions are traditionally measured by isolating and observing changes in either the solution (substrate breakdown, by-product formation) or solid state (catalyst surface changes) using specialized techniques.[56–60] Unfortunately, the methods involved restrict analysis to only solution or solid phases, overlooking interactions or reactions occurring in the swollen (*i.e.*, gel) phases or the system as an intact whole. Ghosh Biswas *et al.* explored the potential of CMP-NMR to understand these molecular level processes, in the context of the photodegradation of phenol by titanium dioxide (Degussa, P25–TiO_2)[61]-based photocatalysts under increasing periods of illumination.[14]

Three phases were observed in this system: (1) solution – freely diffusing phenol, water, and degradation by-products (*e.g.*, hydroquinone); (2) gel – molecules with limited motion near the catalyst surface and (3) solid – molecules strongly bound to the catalyst surface.[14] In the dark, phenol was dispersed amongst all 3 phases (Figure 4.13(A)). Upon illumination, charge carriers move to the surface of the catalyst to generate reactive oxidative species (ROS) (Figure 4.13(B)). These species migrate away from the catalyst surface to the gel phase to initiate the breakdown of phenol. Freely diffusing phenol from the solution fraction would replenish those in the gel phase over the course of the reaction. During prolonged illumination (72–200 h, Figure 4.13(C) and (D)), a broadening of solid-state phenol resonances indicated the fouling of the catalyst surface which further reduced the photocatalytic active sites and slowed down the reaction overall. This was supported by the buildup of phenol signals in the gel phase due to the lack of efficient degradation.[14]

Overall, this study demonstrates the complementary nature of CMP-NMR to observe multiphase changes owing to the function of the entire heterogeneous system, currently impossible using traditional methods of analysis.

4.5 CMP-NMR as a Tool to Observe Metabolic Profiles of Intact Natural Samples

Biological function arises from the synergism between the multiple phases in living systems. Understanding the constituents of each phase helps determine how organisms function and is vital for environmental and medical research. Magnetic resonance imaging (MRI) is one of the main tools used to study living organisms;[62,63] however, it has its limitations (low sensitivity, can only detect dissolved compounds, requires a large sample volume, and lacks the capability to lock causing signal drifting leading to broad lines) which may be alleviated by CMP-NMR. Due to its non-invasive nature, CMP-NMR is ideal for observing all interactions in intact organisms,

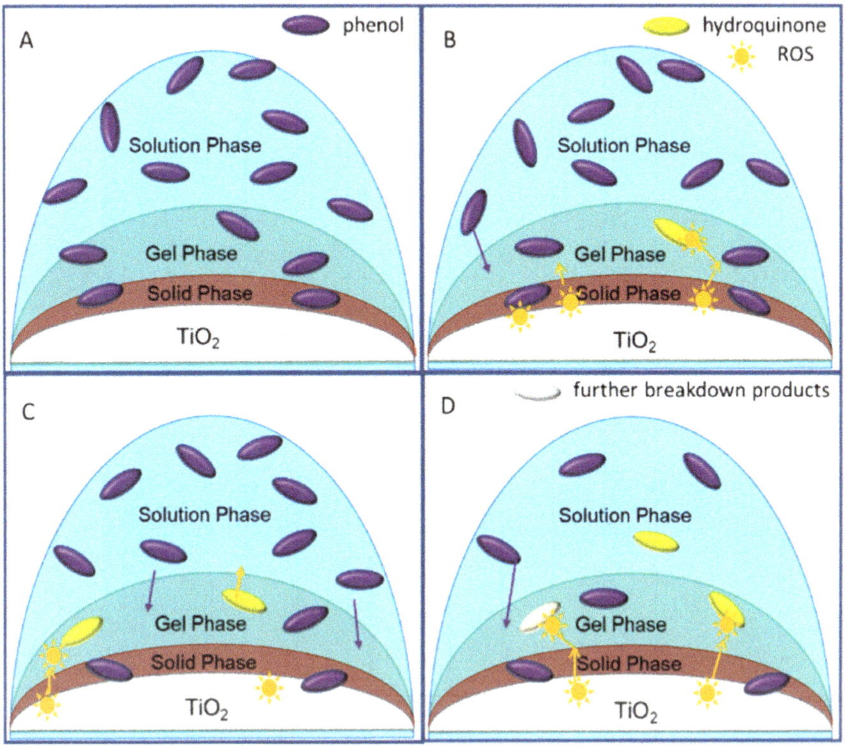

Figure 4.13 Schematic depiction of the photocatalysis of phenol using P25–TiO_2. (A) Before illumination, phenol exhibits adsorption and desorption in all phases. (B) Upon illumination, reactive oxygenated species are generated on the catalyst surface and react with phenol in the gel phase. (C) Upon further illumination (48–72 h), organic molecules diffuse between the phases. (D) During long-term illumination (200 h), membrane fouling occurs as phenol becomes tightly bound to the catalyst surface and further degradation products are observed as the reaction continues. Reproduced from ref. 14 with permission from the Royal Society of Chemistry.

both *ex vivo* and *in vivo*. Here, structural changes are monitored using CMP NMR's solid capabilities, while biochemical and metabolic fluxes are observed using its solution-state capabilities.

Thus far, CMP-NMR has been applied to observe intact organisms such as freshwater fleas (*Daphnia magna*) and freshwater shrimp (*Hyalella azteca*). These are both model organisms in environmental toxicity studies and are keystone species, playing a key role in the transfer of nutrients and toxins in the freshwater food web.[64–66] Both are highly susceptible to environmental stress (*e.g.*, temperature, pH, nutrient, salinity, oxygen, toxin/pollutants),[67–69] and population decline may lead to larger-scale consequences such as an ecosystem collapse. As such, it is vital to understand the biochemical pathways impacted during stress responses to gain insight into the

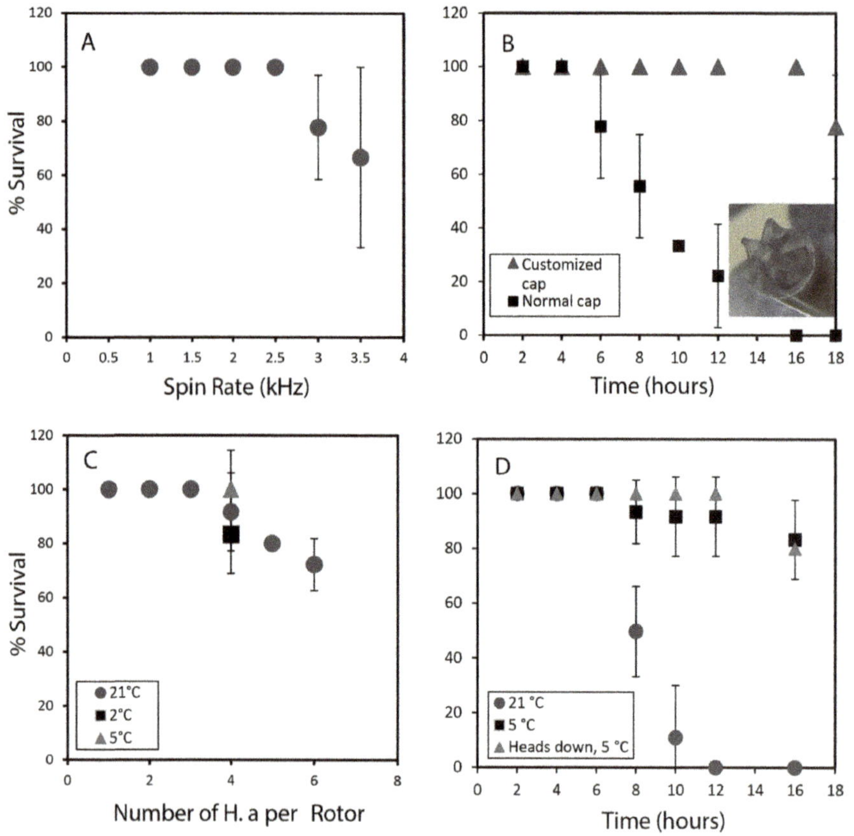

Figure 4.14 Percentage of *Hyalella azteca* survival rate under different conditions. (A) Impact of the spin rate (1 h of spinning). (B) Organism survival increasing with the use of a customized rotor cap drilled with a hole to permit oxygen exchange. (C) Impact of increasing the number of organisms in the rotor at varying temperatures. (D) Duration of survival at differing temperatures (21° and 5 °C). When the organism is loaded headfirst, such that the gills are submerged, the survival rate is 100% up to 12 h. Reproduced from ref. 15 with permission from the Royal Society of Chemistry.

toxic-mode-of-action of chemicals and potential recovery processes. Although *in vivo* responses are ideal to understanding real-time molecular changes, there are many considerations[15] (Figure 4.14) for such studies in CMP-NMR, for example:

- *Magic angle spinning*: spinning is required to narrow broad resonances and anisotropic interactions from rigid materials.[1] However, this is a major stressor for living organisms (Figure 4.14(A)), as biological tissue tends to rupture above 3000–4000 Hz in a 4 mm rotor.[4,70,71] As such, organisms must be spun more slowly, which introduces spinning artifacts (sidebands) often masking vital metabolites, making their assignment difficult.[72] Traditionally, 2500 Hz is used in CMP-NMR to move spinning artifacts from the ^{1}H resonances out of the spectral window on a 500 MHz magnet. At this speed, *H. azteca* can survive up to 14 hours, which can be increased with slower spinning (*i.e.*, 500 Hz for >48 hours);[19] however, artifacts become more problematic at such speeds.

- *Presence of water*: to reduce magnetic susceptibility inhomogeneity and ensure survival, organisms must be placed in a rotor full of water (gills facing down). This not only reduces the detectable biomass, spectral resolution and signal-to-noise ratio (SNR), but also introduces broad water resonances, and subsequent spinning artifacts from the water render traditional ^{1}H experiments unusable.[72,73]
- *Oxygen requirement*: for survival, organisms must be provided with oxygen (Figure 4.14(B)). For this, a 560 μm hole can be drilled into the rotor cap to permit air exchange as the sample spins. This works by creating a 'vortex' where the centrifugal forces keep the water spinning against the rotor walls, preventing leaks.[15,19] Unfortunately, some evaporation of water is seen which reduces the survival and thereby the ability to perform temporal studies.
- *Number of organisms*: the presence of >1 organism in a rotor, while increasing the detectable biomass, greatly limits survival and temporal studies (Figure 4.14(C)).[19] The oxygen exchange from the hole in the rotor cap is not efficient enough to support the presence of multiple organisms for long periods of time and is indicated by a build-up of lactic acid and alanine (anaerobic stress metabolites).[15]
- *Temperature control*: to slow down metabolism and reduce oxygen consumption, the temperature of the studies must be reduced (Figure 4.14(D)). Typical *in vivo* CMP-NMR studies were conducted at ~5 °C which may not be feasible in all labs.[15,19]
- *Lipid resonances*: *in vivo* NMR spectra acquired using a stationary system (solution-state flow system without spinning) were found to be dominated by broad lipid resonances that mask signals from other metabolites (Section 4.5.3.5). Although CMP-NMR alleviates this somewhat by isolating (spectral editing, diffusion and relaxation filters)[10] the gel/semi-solid fractions, smaller lipid metabolites can still be present. During *ex vivo* CMP-NMR, when using freeze-dried organisms, it was seen that spinning caused the mobile lipids to rise to the top of the rotor and out of the coil region, uncovering vital metabolites that may be key to understanding biological function.[73] Other 1D and 2D lipid suppression NMR techniques may also be employed.[74–77]
- *^{13}C enrichment*: ^{1}H NMR is commonly used due to its high natural abundance (99.9%) and resulting sensitivity. However, the spectral window of 1D ^{1}H NMR is quite limited (10 ppm) resulting in decreased resolution and overlapping resonances. ^{13}C NMR offers improved resolution owing to its wide spectral window (250 ppm); however, it is underutilized because of its low natural abundance and gyromagnetic

ratio (1.1%, $\frac{1}{4}$ ^{1}H). This makes solid-state NMR of organisms at natural abundance difficult.[72] To make ^{13}C NMR more feasible, the organisms can be isotopically enriched by feeding them ^{13}C labelled food sources (algae) for the duration of their lifecycle; however, this can be cost prohibitive over time. Previous studies have shown that it takes at least 2 generations for metabolites to be fully labelled.[73] Prior to experimentation, the guts of the organisms must be cleared *via* feeding the organisms ^{12}C food sources or microbeads which clear the digestive tract.[78]

4.5.1 *Ex Vivo* CMP-NMR as a Complementary Tool to Understand *In Vivo* Processes

During *in vivo* CMP-NMR, it can sometimes be difficult to observe living organisms in their natural state. *Ex vivo* CMP-NMR can be used when real-time stress responses are not a priority, and can provide complementary information to understand *in vivo* processes.[73] The advantages of *ex vivo* CMP-NMR include the use of deuterium oxide (D_2O) to swell samples which reduces broad water resonances and their associated sidebands. Additionally, all conditions used to ensure organism survival are not required, organisms can be stored and transported between labs with no constraints, and more biomass can be used, along with faster spinning to move artifacts out of the spectral window.[73] Ghosh Biswas *et al.* applied *ex vivo* CMP-NMR to observe structural and metabolic changes within *Daphnia magna* occurring during growth and reproduction on a generational scale.[73]

Within *D. magna* the phases encompass dissolved metabolites (solution), lipids and carbohydrates for energy metabolism (gel and semi-solid), alongside solid lipid reserves, and proteins and carbohydrates (solid) which make up the chitin exoskeleton. To understand molecular flux during aging, the different life stages (pregnant daphnids, juveniles (1 week old), and adults (>3 weeks old)) of *D. magna* were observed (Figure 4.15).[73]

The storage and mobilization of lipids within pregnant and adult daphnids, observed using CMP-NMR, was interesting to note. *D. magna* reproduce asexually (under non-stressed conditions)[79] and do not have the ability to produce all of their own lipids, some of which must be obtained from their diet.[80,81] When not required for energy intensive processes (*i.e.*, reproduction), *Daphnia* store their lipids in a solid-like phase. This is mobilized into a semi-solid structure to produce lipid-rich eggs, which manifests itself as the pregnant *Daphnia* possessing the highest levels of unsaturated fatty acid signals in the RADE (Figure 4.15(e)), which can be correlated to the health of the neonates.[73,82] With age, the lipids are stored in a solid form once again until they are required for reproduction and aging (nearing end of lifespan). 2D NMR (Section 4.5.4) can provide additional information about metabolic changes occurring within the solution fraction.[15,78] For example, increased levels of precursor nucleic acid metabolites (xanthosine, hypoxanthine, inosine and uracil) suggest that during pregnancy the

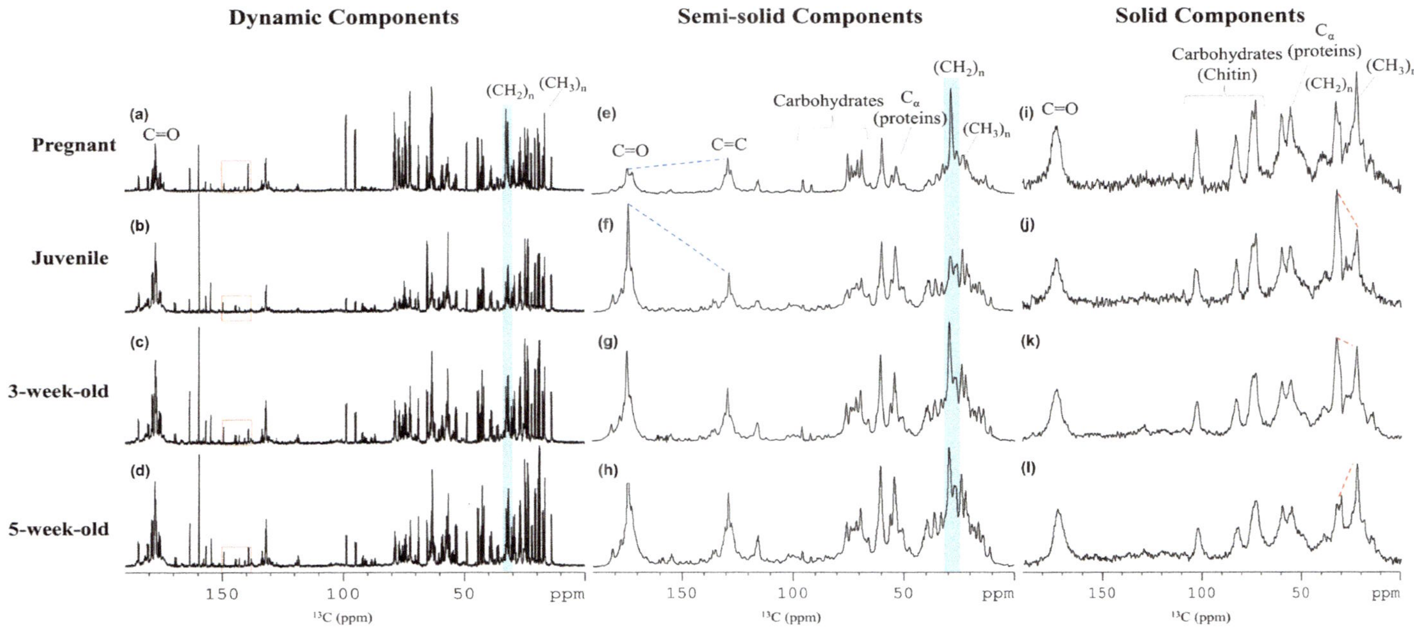

Figure 4.15 1D ^{13}C NMR spectra of *Daphnia magna* at various life stages and ages: pregnant, juvenile (1-week-old), 3-week-old adult and 5-week-old adults (nearing the end of their lifecycle). (a–d) Solution phase containing mobile components such as metabolites. (e–h) Semi-solid fraction obtained from RADE, dominated by signals from mobilized lipids as can be seen by the ratio (blue dashed lines) of C=C and C=O bonds where C=C depicts the unsaturated lipids. (i–l) Rigid solid components obtained from inverse ^{1}H T_2-filtered CP-MAS. The ratio between CH_3 and CH_2 (red dashed lines) correlates to the protein in the chitin structure (CH_3) and stored lipid reserves (CH_2). Reproduced from ref. 73 with permission from Elsevier, Copyright 2020.

mothers store the building blocks of nucleic acids within the egg sacs for their young to build their own DNA and RNA.[73]

4.5.2 *In Vivo* CMP-NMR

Changes in an organism's biochemical pathway can be an indicator of stress and disease. Using NMR, real-time, sub-lethal stress responses alongside recovery of living systems can be measured.[83] As mentioned, *in vivo* studies often utilize solution-state NMR;[78,84,85] however, many biological responses are often complex processes involving multiple phases. These include the transport of drugs and metabolites between different mediums such as tissues and bones,[50] crystallization/aggregation of cells and protein,[86] degradation/breakdown of solids such as shell thinning or bone breakdown.[87] Thus, a technique capable of monitoring the transformation of one phase to another, or the movement of compounds between phases is required.

There have been a limited number of studies using CMP-NMR to living organisms (*D. magna* and *H. azteca*).[15,19,71,72] The first application of using CMP-NMR to monitor a living organism was conducted by Liaghati Mobarhan *et al.* in 2016.[15] Here *H. azteca* was used as its rigid exoskeleton provides protection under MAS allowing for longer experimentation. Due to the complexity of living systems, spectral overlap from broad resonances such as rigid lipids (cell wall, micelles) can mask important metabolites. Diffusion-based spectral editing can be used to isolate individual phases to elucidate metabolites in overlapping regions.[10] The solution phase (Figure 4.16, IDE) comprised signals from a variety of metabolites that, without spectral editing, are buried under the strong lipid signals. Changes in these metabolites can occur quickly in response to stress, whereas the return to their 'normal' levels can be indicative of recovery. The compounds with restricted diffusion (Figure 4.16, restricted diffusion DE) consist largely of lipids, consistent with cell walls, stored lipids and bound species. Rigid lipids of cell membranes as well as aromatic amino acids, and proteins of muscles were found in a semi-solid state (Figure 4.16, RADE). Solid-state CP (Figure 4.16, rigid solids) was dominated by crystalized carbohydrates and CH_3 groups from amino acid groups (*i.e.*, valine, leucine) in the chitin-exoskeleton structure and lipid stores.[15]

2D NMR spectroscopy (Section 4.5.4) was used to identify ~40 metabolites. Assignments from 2D NMR can be used to label peaks in 1D spectra more often used in other fields. Many of the amino acids observed were components of muscles, neurotransmitters and biomarkers to understand cardiometabolic processes.[15] Detailed NMR assignments and the function of major of metabolites found in freshwater crustaceans have been documented.[73,88] By observing all phases of the organism *in situ*, one may be able to follow the metabolic processes occurring.

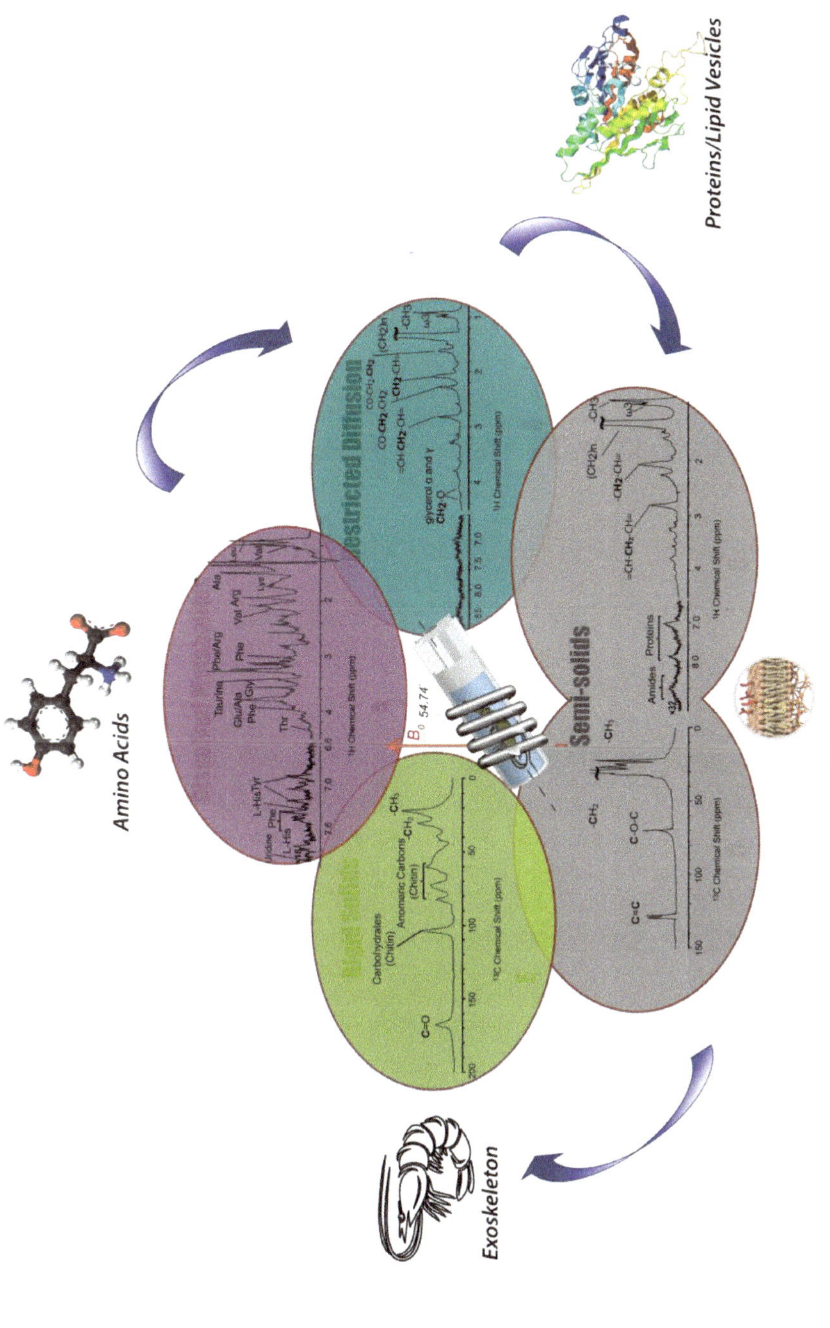

Figure 4.16 ^{1}H and ^{13}C spectral editing techniques applied to a living *Hyalella azteca*. Example constituents of each phase are depicted: Solution phase – dissolved metabolites (amino acids), restricted diffusion (gel phase) – macromolecules (lipid vesicles, proteins), semi-solid phase – large macromolecules (cell membrane), and solid phase – (exoskeleton). Reproduced from ref. 19 with permission from John Wiley & Sons, Copyright 2020.

4.5.3 Considerations and *In Vivo* CMP-NMR Progress

As spinning is one of the largest stressors for *in vivo* CMP studies, techniques to lower the spin rate to significantly increase organisms' survival while reducing the rate of water evaporation have been considered.

4.5.3.1 *Slow Spinning Techniques to Reduce Stress*

In the context of measuring real-time stress responses, it is important to eliminate the impacts of spinning as this can induce biochemical changes (*i.e.*, increased alanine production)[15] and making it harder to discern responses from the stressor of interest. Slow spinning is required to increase organism survival under MAS; however, this can introduce spinning artifacts (spinning sidebands) masking resonances of interest. A technique called localized magic angle turning (LOCMAT) demonstrated improved resolution on biological samples at slow spinning (~4 Hz); however, due to the reduced sensitivity, a large sample volume is required.[89,90] Moreover, contributions from lipid resonances dominated the spectral profile, making identifying and assigning underlying metabolites difficult. Here lipid suppression techniques may offer some improvements. Another technique known as phase-corrected magic angle turning (PHORMAT) allows the slow spinning of living organisms with high resolution. PHORMAT has been applied in HR-MAS studies of large animals, such as mice, at 1 Hz with success.[91] Using this technique in conjunction with CMP-NMR may help to keep organisms alive during *in vivo* analyses.

4.5.3.2 *Ultraslow Spinning*

Liaghati Mobarhan *et al.* have explored ultraslow spinning of *D. magna* and *H. azteca* (down to 50 Hz) using ^{2}H/^{13}C enrichment to improve sensitivity.[71] At such slow speeds (at 25 °C), the survival of *D. magna* (4 h) and *H. azteca* (6 h) increases to 24 h and 48 h, respectively.[71] In the solid state, ^{2}H shows a very broad spectral profile that is kHz wide. However, chemical shift anisotropy and homo/heteronuclear dipolar couplings narrow quickly with molecular motion, which helps selectively emphasize the solution fraction. Without ^{2}H added as a lock there is no HOD signal, which eliminates the impact from water and its subsequent sidebands during slow spinning *in vivo* experiments (Figure 4.17), where changes in the organism's metabolome could be assigned and monitored.

4.5.3.3 *Other Techniques to Attenuate Spinning Artifacts Arising from Slow Spinning*

Spinning sidebands are also a major concern for *in vivo* spectra, specifically from rigid/solid-like materials which have broadened resonances resulting from chemical shift anisotropy. There are a number of techniques available for suppressing these artifacts. Total suppression of spinning sidebands (TOSS)[92] is a commonly used technique; however, this can also reduce the SNR of the spectra. A variation of this technique known as TOSS.243 (uses a phase cycle of 243) showed further success. Finally, phase-adjusted spinning sideband (PASS) simplifies spinning artifacts in 2D NMR by separating the isotropic and anisotropic chemical shifts, thus completely isolating the spinning sidebands with minimal loss of signal intensity.[93,94] Another major factor to consider in slow spinning *in vivo* studies is the broad water resonance and its associated sidebands. This requires accurate water suppression and sideband suppression techniques to reduce their contributions.

4.5.3.4 Water Suppression

Due to the presence of high volumes of water (organism and surrounding water in the rotor), its detected signal will also be large, masking signals from key metabolites. Traditional water suppression techniques (*e.g.*, W5-SPR-WATERGATE, presaturation) are only effective at high spinning rates (>1000 Hz) and may also remove underlying metabolite peaks near the water peak and lead to spectral distortion.[95] This is emphasized at slow spinning, where both the water resonance and its associated sidebands become problematic. Typically, PASS is combined with commonly used water suppression techniques, such as presaturation and DANTE (delays alternating for nutation for tailored excitation); however, it has found limited success at suppressing large water resonances or suppressing resonances from metabolites near the solvent resonance.[95] To overcome this, Liaghati Mobarhan *et al.* found success in using a sequence known as presaturation utilizing relaxation gradients and echoes (PURGE) water suppression and coupled it to existing sideband suppression techniques (PASS, TOSS, and TOSS.243) to effectively attenuate the water sidebands without impacting real signals from the organism (Figure 4.18).[95] Using this technique, Liaghati Mobarhan *et al.* found PURGE-PASS and PURGE-TOSS.243 to be most effective at reduced spinning speeds down to ~100 Hz.[95]

4.5.3.5 Lipid Suppression

As mentioned (Section 4.5), resonances from rigid lipids can mask underlying solution-state metabolites (Figure 4.19). This makes it difficult to obtain real-time changes of metabolites in the solution phase. To overcome this, lipid suppression techniques[74–77] may be used to diminish the dominating peak without relying on DE. Figure 4.19 compares small earthworms (left) and freshwater shrimp (right) with and without lipid suppression applied. Many metabolites previously masked by the lipids are recovered after lipid suppression, making assignment and monitoring more feasible.

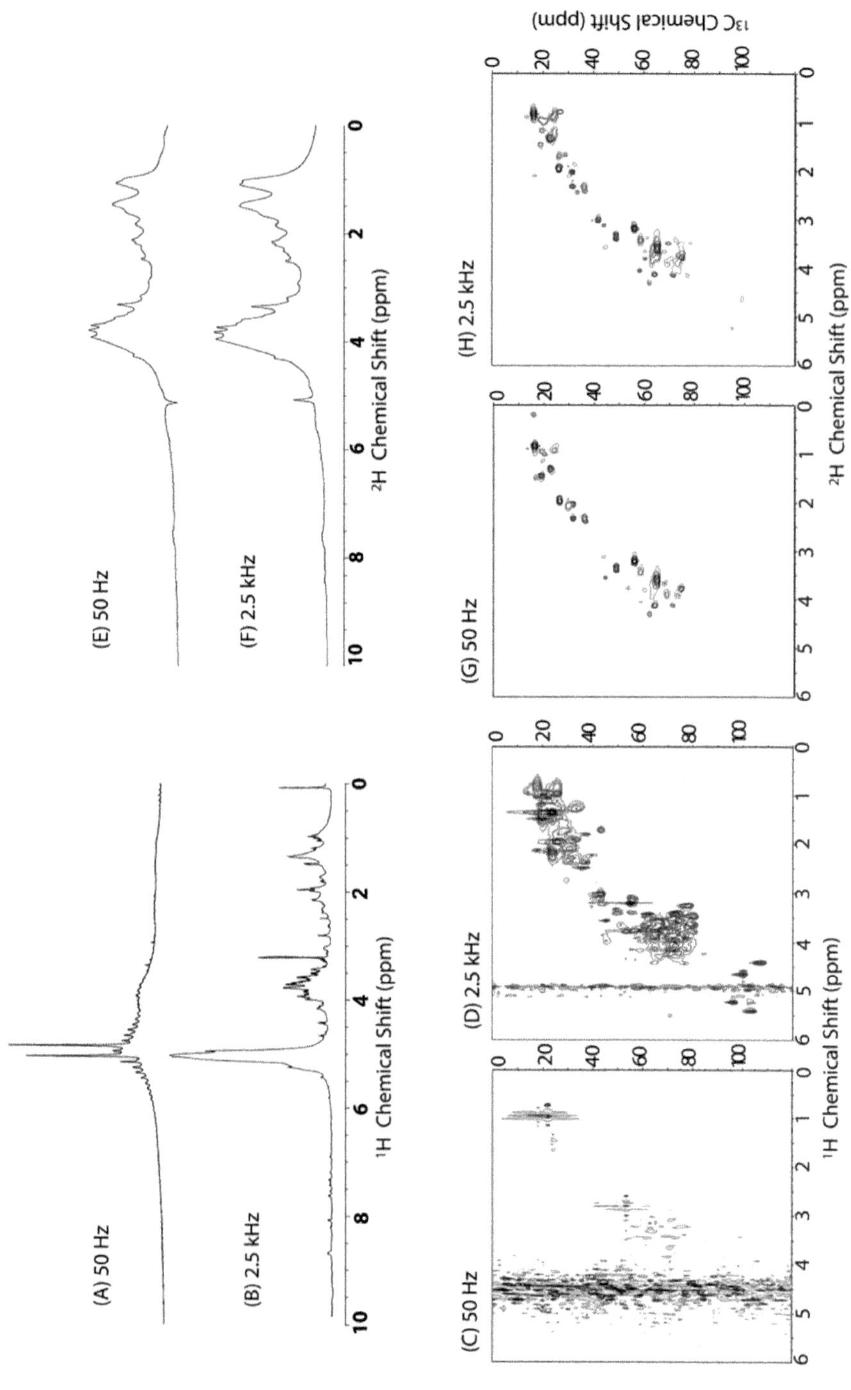

(A) 50 Hz
(B) 2.5 kHz
¹H Chemical Shift (ppm)
(E) 50 Hz
(F) 2.5 kHz
²H Chemical Shift (ppm)
(C) 50 Hz
(D) 2.5 kHz
¹H Chemical Shift (ppm)
(G) 50 Hz
(H) 2.5 kHz
²H Chemical Shift (ppm)
¹³C Chemical Shift (ppm)

4.5.3.6 Sample Heating During Solid-state NMR

The ability to handle the high power required for solid-state CP and decoupling is CMP-NMR's most distinguishing feature. However, this can generate radio-frequency heating which potentially could stress and denature proteins within the organism during *in vivo* CMP-NMR. The advent of a new E-Free CMP probe helps combat this issue.[96] This probe uses a loop gap resonator (LGR) design,[97,98] placing the ^{1}H coil further from the sample to reduce heating by 3X compared to a standard CMP coil, thereby increasing organism survival significantly (Figure 4.20), allowing for temporal observation of molecular structural changes.[96] The placement of the inner carbon detection coil also increases the filling factor to improve the ^{13}C sensitivity by a factor of 2.[72]

Overall, CMP-NMR has proven to be an effective technique to monitor molecular-level processes occurring in living organisms.

4.5.4 Multidimensional Metabolomics

4.5.4.1 2D NMR for Metabolomic Assignments

In contrast to traditional solid-state NMR, an advantage of CMP probes is the installation of a magic angle gradient which is important for multidimensional NMR experiment that performs gradient based coherence selection. Both homonuclear and heteronuclear 2-dimensional (2D) experiments can be used to provide increased spectral dispersion, making identification and assignment of metabolites ideal. However, one caveat of multidimensional NMR is that it biases the mobile constituents of the swollen fraction. Hertkorn *et al.* reported the peak capacity of 1D ^{1}H NMR to be ~3000 peaks, whereas that of ^{13}C increased to 30 000 peaks.[99] However, 2D NMR such as heteronuclear single quantum correlation (HSQC) spectroscopy improves the resolution to 2 000 000.[99] HSQC correlates ^{1}H–^{13}C, where a cross peak occurs only if a carbon and a proton are directly bonded. A database of metabolites, such as the Bruker Bioreference NMR database

---

Figure 4.17 Comparing ^{13}C enriched *C. vulgaris* at 2.5 kHz and 50 Hz spinning. (A) At 50 Hz the 1D spectrum is dominated by spinning sidebands making identification of metabolites impossible. (B) Spinning faster to 2.5 kHz moves spinning sidebands out of the ^{1}H spectral window. 2D ^{1}H–^{13}C HSQC at (C) 50 Hz and (D) 2.5 kHz. Comparing ^{2}H/^{13}C enriched *C. vulgaris* using 1D ^{2}H NMR at (E) 50 Hz and (F) 2.5 kHz, as well as 2D ^{2}H/^{13}C HMQC at (G) 50 Hz and (H) 2.5 kHz. Both 1D and 2D spectra appear similar at each spinning speed, indicating that the ^{2}H removes broad resonances from water signals typically observed in ^{1}H NMR. Note: due to the smaller bandwidth of ^{2}H (~76 Hz at 11.7 T), the 1D spectra are less resolved relative to 1D ^{1}H (~500 Hz at 11.7 T) NMR. Reproduced from ref. 71, https://pubs.acs.org/doi/10.1021/acsomega.8b02882, under the terms of the CC BY-NC-ND 4.0 license, https://creativecommons.org/licenses/by-nc-nd/4.0/.

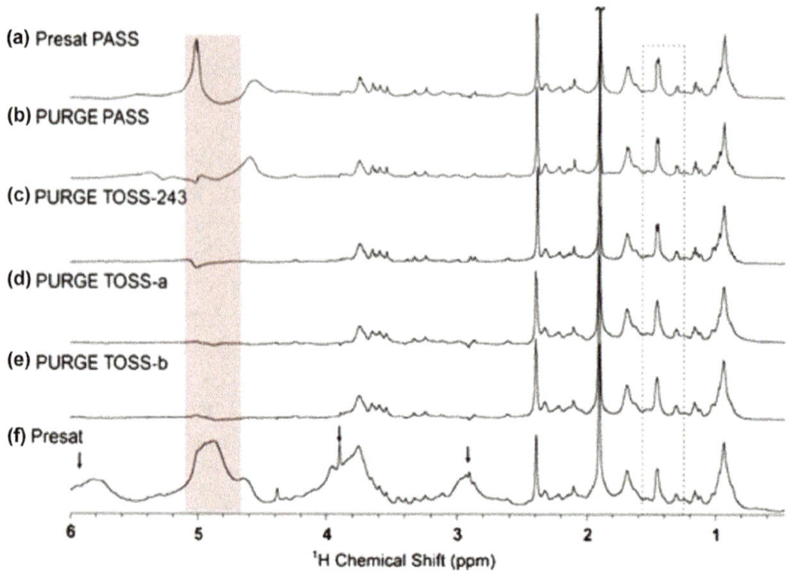

Figure 4.18 Comparing water suppression and sideband suppression techniques to an *E. fetida* (earthworm) spun at 500 Hz. The various techniques include (a) presaturation and PASS, (b) PURGE and PASS, (c) PURGE and TOSS-243, (d) PURGE and TOSS-a, (e) PURGE and TOSS-b and (f) presaturation. The dotted rectangle indicates the improvement in resolution in (b and c). Arrows depict spinning sideband artifacts. Pink highlights emphasize suppression of the spinning sidebands using each technique. Reproduced from ref. 95 with permission from Springer Nature, Copyright 2017.

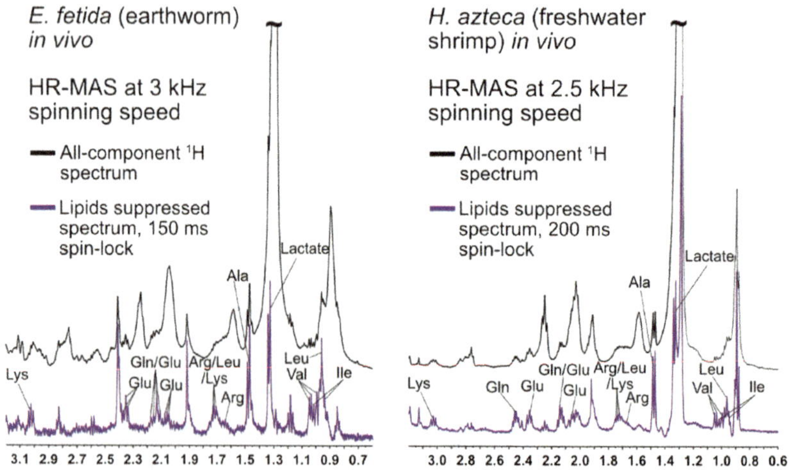

Figure 4.19 1D *in vivo* ^{1}H HR-MAS NMR spectra of (left) earthworm (*Eisenia fetida*) and (right) freshwater shrimp (*Hyalella azteca*) with (purple) and without (black) lipid suppression. Residual lipid resonances in *H. azteca* arise from slow spinning (2.5 kHz) to keep the organism alive. Within *E. fetida* the lipid resonances are almost completely suppressed. Reproduced from ref. 77 with permission from John Wiley & Sons, Copyright 2019.

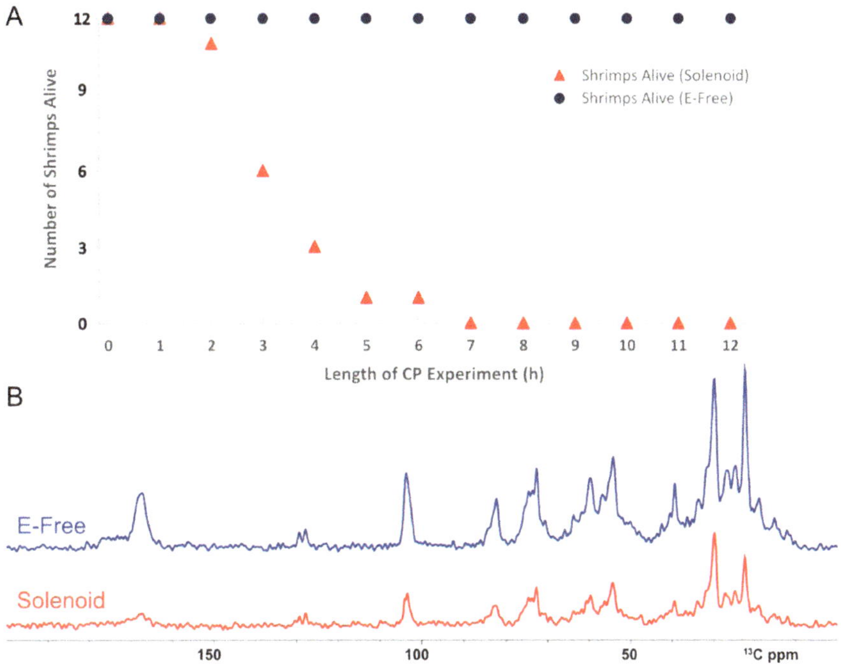

Figure 4.20 Comparing the novel E-Free MAS CMP coil and a standard solenoid CMP coil. (A) Survival of 12 (24 total) *Hyalella azteca* under high-power decoupling (100 kHz) over 12 hours. Using the E-Free coil design, 100% survival was observed, whereas only 25% of the shrimp survived after 4 hours. After 7 hours of experiment, all of the shrimps died in the solenoid coil. (B) *In vivo* CP-MAS at 4 hours on both probes. The spectral profile remains the same; however, due to the inner placement of the carbon coil in the E-Free design, a higher SNR was observed. Reproduced from ref. 96 with permission from American Chemical Society, Copyright 2021.

(~650 compounds), can be used to assign these peaks to their corresponding metabolites. Figure 4.21 shows the assignment of *in vivo* $^{1}H–^{13}C$ HSQC of ^{13}C enriched *Hyalella azteca*, which would be extremely important for monitoring the impact of a stressor on the metabolome. Another example is the mutations seen in *eli1* in seedlings (Section 4.3.2) which contained higher levels of metabolites corresponding to lignin biosynthesis relative to the wild type, which can be seen in the aromatic regions of the $^{1}H–^{13}C$ HSQC (Figure 4.22).[27] For example, it can be seen that the seedlings with the *eli1* mutation consist of higher levels of tryptophan, phenylalanine, lignin double bond units and hydroxyphenol units (Figure 4.22(b)).[27] Furthermore, during seed germination, 2D NMR demonstrated that dissolved sugars (raffinose family of oligosaccharides) were also used as an energy source for growing seeds and were consumed.[26] Finally, it is worth mentioning that recently developed selective NMR sequences such as DREAMTIME NMR can also be used (in both 1D and 2D NMR) to isolate user-desired metabolites from complex

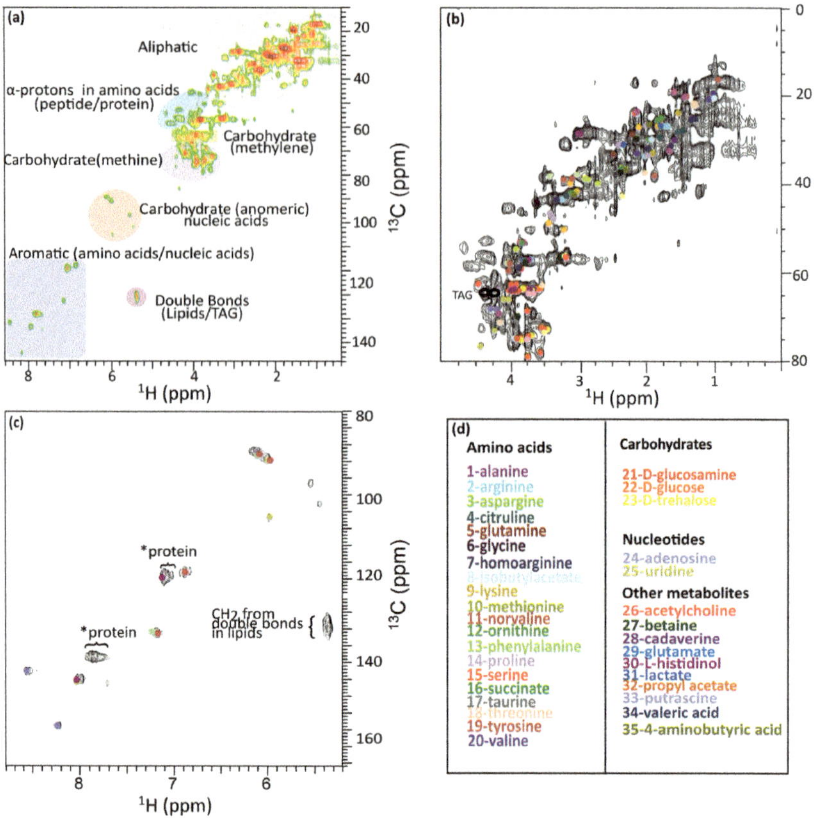

Figure 4.21 *In vivo* ^{1}H–^{13}C HSQC of ^{13}C enriched *Hyalella azteca*. (a) Major macromolecules are categorized. Assignment of ∼40 metabolites in the (b) aliphatic region and (c) aromatic region. (d) List of assigned metabolites colour-coded with the assignments in (b). Reproduced from ref. 15 with permission from the Royal Society of Chemistry.

samples.[100] DREAMTIME allows users to simply obtain spectra by selecting only suites of molecules that they want to see. We very recently published DREAMTIME for HR-MAS systems.

4.5.4.2 *Slow Spinning Multidimensional NMR*

Ghosh Biswas *et al.* applied a novel ^{13}C optimized CMP-NMR probe to perform heteronuclear correlation spectroscopy (HETCOR) which observes ^{13}C–^{1}H bond correlations *via* ^{13}C NMR, in contrast to HSQC which observes these bonds along the ^{1}H axis.[72] This eliminates the detection of broad water resonances during *in vivo* NMR, allowing slower spinning to significantly increase the survival of the organisms. Figure 4.23 compares the HSQC and HETCOR spectra of a single freshwater shrimp, *in vivo*, where peak

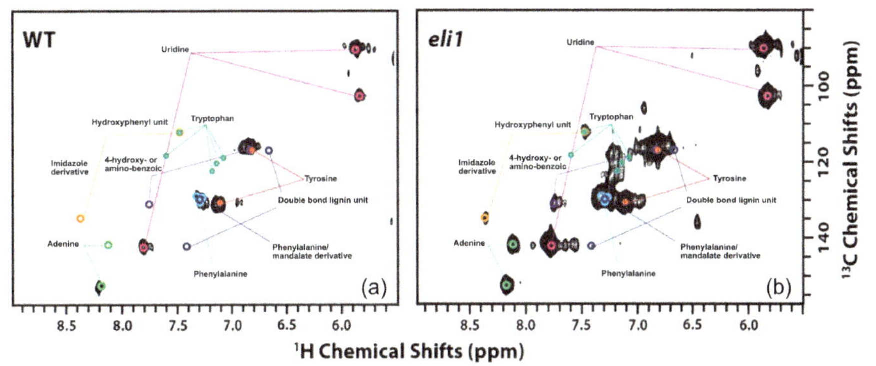

Figure 4.22 2D ^{1}H–^{13}C HQSC of (a) wildtype and (b) *eli1* (*ectopic lignification1*) *Arabidopsis thaliana* seedlings. Improved resolution from 2D NMR depicts the differences in resonances within the aromatic region arising from the *eli1* mutation. Reproduced from ref. 27 with permission from John Wiley & Sons, Copyright 2015.

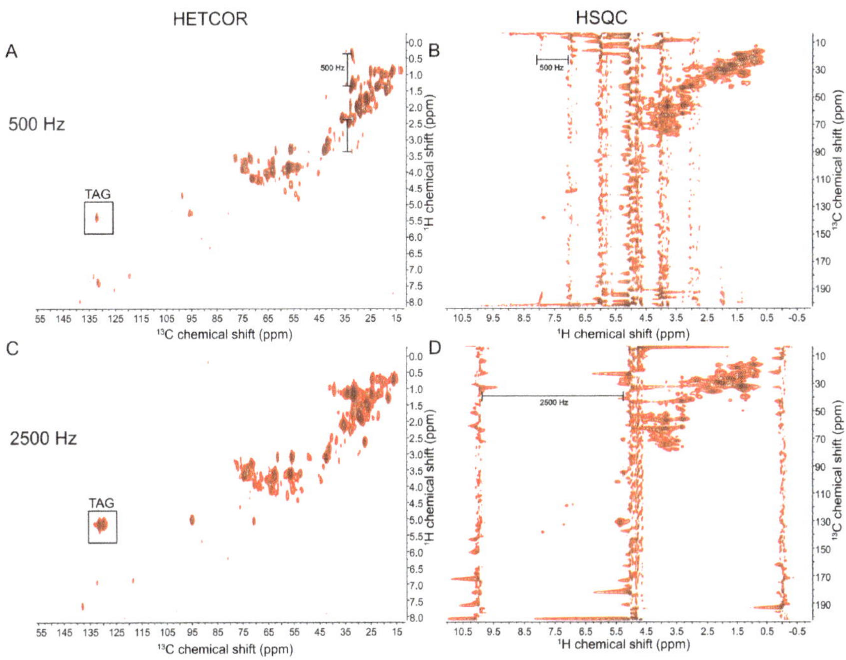

Figure 4.23 2D NMR of a single ^{13}C enriched *Hyalella azteca* spinning under MAS, *in vivo*, at (A and B) 500 Hz and (C and D) 2500 Hz. ^{1}H–^{13}C HSQC (B and D) highlights the spinning sidebands from water resonances which mask metabolite information. This is particularly emphasized at slower spinning speeds (500 Hz). Faster spinning (2.5 kHz) can be used to move artifacts outside the spectral window. Comparatively, ^{13}C–^{1}H HETCOR (A and C) depicts the lack of spinning sidebands arising from water when ^{13}C is the nucleus under direct observation. Note: some sidebands appear in the ^{1}H dimension at 500 Hz. Reproduced from ref. 72 with permission from American Chemistry Society, Copyright 2022.

identification in slow spinning HSQC is impossible due to spinning sidebands from water, but can be performed using HETCOR.[72] A series of temporal metabolic changes can be observed with the increased time window over which the organisms survive under slow spinning.

4.6 Conclusions and Future Directions

This review recapitulates the applications of CMP-NMR reported to date. We have explored the advantages of non-invasive multiphase analysis on a number of heterogeneous samples, both abiotic and biotic. This review showcases the use of CMP-NMR to monitor structural changes, interactions, and entire natural processes demonstrating its applicability in living organisms to study complex systems with minimal sample preparation. This review truly highlights the versatility of CMP-NMR which can be applied to diverse fields of research. There are, however, considerations with respect to the widespread use of CMP-NMR in diverse study areas. The largest consideration is the inherently low sensitivity of NMR. This is further compounded by the tuning in CMP-NMR probes which must be balanced between ^{13}C (for solid detection) and ^{1}H (for liquid/gel detection). Conversely, in dedicated HR-MAS probes, the sensitivity is optimized only for ^{1}H, whereas in dedicated CP-MAS probes the sensitivity is optimized for ^{13}C (or X). The result of balancing the tuning in CMP-NMR probes can lead to up to 40% SNR loss compared to other dedicated probes. That stated, recent ^{13}C optimized CMP probes that use two coils have largely overcome this drawback.

4.6.1 Improving ^{13}C Sensitivity

Carbon is the backbone of most natural organic compounds and thus important to monitor when studying metabolic and structural changes (^{13}C CP-MAS). Moreover, due to the larger carbon spectral width (~250 ppm), ^{13}C NMR offers improved spectral dispersion. Thus, metabolites masked in ^{1}H NMR (only ~10 ppm spectral window) due to spectral overlap can be identified and assigned in ^{13}C NMR. However, carbon detection becomes difficult due to the low gyromagnetic ratio of ^{13}C, its low natural abundance and the cost of isotopic enrichment. To improve its sensitivity, an alternative detection coil design can be used. For example, the novel CMP E-Free MAS probe uses a 2-coil design, placing the ^{13}C solenoid detection coil on the inside of a ^{1}H loop gap resonator (for high power decoupling, cross-polarization).[72] The goal of this design is to reduce the heating of lossy samples, with the added benefit of a 110–113% increase in carbon sensitivity when compared to a traditional multi-tuned solenoid CMP coil.[72] Unfortunately, due to the placement of the ^{1}H detection coil, a 24% loss in ^{1}H sensitivity is observed.[72] However, this compromise may be necessary when considering the importance of ^{13}C NMR in monitoring temperature sensitive or complex natural samples.

4.6.2 Larger Diameter Probes to Increase Biomass

It is important to note that when observing complex multiphase samples, the total observable NMR signal is split between the components in each phase. This is further compounded by the fact that samples must either fit into standard 4 mm sample rotors or be dissected to fit. In the latter case, susceptibility to solvent penetration could skew results in the individual phases (*e.g.*, dissolving metabolites which may otherwise exist as a gel in natural samples).[26] The development of the 7 mm (5.8 mm inner diameter) CMP-NMR probe helps alleviate this by allowing larger samples to be monitored intact (*e.g.*, drug capsules, large seeds), as well as packing more biomass into the rotor, reducing experiment time.[101] From the increased sample volume, the larger diameter probe demonstrates a 2.4× increase in signal, relative to the standard 4 mm (3.2 mm inner diameter) design.[101] This poses significant advantages for observing samples in their native state (*e.g.*, organisms sampled from their environment) at natural abundance.

4.6.3 Cryogenically Cooled Hardware

NMR sensitivity is largely limited by the thermal noise (from the receiver coil, preamplifier, sample) of the NMR system.[102] Thus to improve the SNR the thermal noise must be reduced. The development of cryogenically cooled probes addressed this by increasing the sensitivity by a factor of 3–4.[102] Since its introduction, cryogenically cooled probes have been vital for solution-state NMR. Typically, low temperature solid-state experiments rely on using cryogenically cooled gas to spin samples in order to improve the sensitivity (along with dynamic nuclear polarization), monitor low-temperature phenomena and observe kinetic reactions with unstable intermediates.[103] Recently a cryogenically cooled solid-state MAS probe, where the probe hardware and electronics are cooled while the sample temperature can be controlled *via* a variable temperature unit, was introduced.[104] As CMP-NMR probes have similar designs to solid-state probes, cryo-CMP-NMR may be on the horizon and should find important applications when monitoring natural samples and small volume samples, especially at natural abundance.

4.6.4 Micro-coils

As the sample volume increases so does the sensitivity. Large diameter probes may be used to increase the detectable sample volume (Section 4.6.2).[101] However, when considering the impacts of environmental stressors and toxins, it is often the young (neonates) and eggs that are most susceptible.[105,106] For example, *Daphnia magna* reproduce asexually; however, in times of stress, they will produce dormant eggs which will hatch when conditions are optimal again. If these eggs were to be impacted by environmental conditions, an ecosystem collapse could ensue. As such, it is important to understand the metabolic changes of early development. Unfortunately, due to their limited size, many organisms/eggs would be required to fit in standard NMR hardware. To address this, NMR accessories that meet the constraints of volume-limited samples can be created. An example of this is micro-coils which use a technique known as magic angle coil spinning (MACS), for HR-MAS and solid-state observation. MACS is a method, whereby a tuned solenoid micro-coil surrounds the sample, and both are rotating in the NMR together, thus inductively coupling the system to the standard outer coil.[107] By decreasing the size of the 'detection coil', the filling factor is significantly improved and offers enhancements up to a factor of 14.[107] This technique offers the opportunity to study samples at the nanolitre level[108] where biological tissue samples can be monitored while intact.[109,110] When combined with CMP-NMR, MACS coils offer significant potential to monitor volume-limited, individual phases to examine the system as a whole.

All in all, NMR is a powerful tool to monitor molecular-level interactions and processes, and it has yet to be applied to its full potential in environmental research. CMP-NMR bridges the two existing fields of NMR (solution/gel-state and solid-state) to allow the study of all phases (solution, gel, and solid) in intact natural samples down to the bond level. Due to the differences in these two fields of research (hardware, sample preparation, experimental), information regarding the whole system can be missed. With the ever-increasing presence of xenobiotics in the environment, their consequences in the ecosystem (organism, soil, vegetation, *etc.*) must be considered together, in their native state. CMP-NMR enables the study of anthropogenic toxins and their modes-of-action, binding behaviour and ability to bioaccumulate and transform within biotic systems. To date, no other analytical technique is capable of providing such molecular-level detail, non-invasively. Moreover, heteronuclear CMP-NMR broadens the potential for process monitoring across diverse areas of research. Thus far, CMP research has heavily focused on understanding interactions from an environmental perspective but should be applicable in other areas of research including mutation and disease monitoring, understanding pharmacokinetics, and manufacturing processes. Overall, CMP-NMR is a highly versatile technique and poses significant potential for a holistic understanding of complex systems.

Acknowledgements

We would like to thank the Natural Sciences and Engineering Research Council of Canada (NSERC) [Alliance (ALLRP 549399), Alliance (ALLRP 555452) and Discovery Programs (RGPIN-2019-04165)], the Canada Foundation for Innovation (CFI), the Ontario Ministry of Research and Innovation (MRI), and the Krembil Foundation for providing funding and the Government of Ontario for an Early Researcher Award.

References

1. W. E. Maas, F. H. Laukien and D. G. Cory, *J. Am. Chem. Soc.*, 1996, **118**, 13085–13086.
2. O. Beckonert, M. Coen, H. C. Keun, Y. Wang, T. M. D. Ebbels, E. Holmes, J. C. Lindon and J. K. Nicholson, *Nat. Protoc.*, 2010, **5**(6), 1019–1032.
3. E. C. Y. Chan, P. K. Koh, M. Mal, P. Y. Cheah, K. W. Eu, A. Backshall, R. Cavill, J. K. Nicholson and H. C. Keun, *J. Proteome Res.*, 2009, **8**, 352–361.
4. J. L. Taylor, C. L. Wu, D. Cory, R. G. Gonzalez, A. Bielecki and L. L. Cheng, *Magn. Reson. Med.*, 2003, **50**, 627–632.
5. K. Shet, S. M. Siddiqui, H. Yoshihara, J. Kurhanewicz, M. Ries and X. Li, *NMR Biomed.*, 2012, **25**, 538–544.
6. C. M. Rocha, A. S. Barros, A. M. Gil, B. J. Goodfellow, E. Humpfer, M. Spraul, I. M. Carreira, J. B. Melo, J. Bernardo, A. Gomes, V. Sousa, L. Carvalho and I. F. Duarte, *J. Proteome Res.*, 2010, **9**, 319–332.
7. A. D. C. Santos, F. A. Fonseca, L. M. Lião, G. B. Alcantara and A. Barison, *TrAC, Trends Anal. Chem.*, 2015, **73**, 10–18.
8. M. Vermathen, M. Marzorati, D. Baumgartner, C. Good and P. Vermathen, *J. Agric. Food Chem.*, 2011, **59**, 12784–12793.
9. A. J. Simpson, M. J. Simpson, W. L. Kingery, B. A. Lefebvre, A. Moser, A. J. Williams, M. Kvasha and B. P. Kelleher, *Langmuir*, 2006, **22**, 4498–4503.
10. D. Courtier-Murias, H. Farooq, H. Masoom, A. Botana, R. Soong, J. G. Longstaffe, M. J. Simpson, W. E. Maas, M. Fey, B. Andrew, J. Struppe, H. Hutchins, S. Krishnamurthy, R. Kumar, M. Monette, H. J. Stronks, A. Hume and A. J. Simpson, *J. Magn. Reson.*, 2012, **217**, 61–76.
11. P. Ning, R. Soong, W. Bermel, D. Lane, M. J. Simpson and A. J. Simpson, *Anal. Bioanal. Chem.*, 2018, **410**, 7055–7065.
12. H. Masoom, D. Courtier-Murias, R. Soong, W. E. Maas, M. Fey, R. Kumar, M. Monette, H. J. Stronks, M. J. Simpson and A. J. Simpson, *Environ. Sci. Technol.*, 2015, **49**, 13983–13991.
13. D. H. Wu, A. Chen and C. S. Johnson, *J. Magn. Reson., Ser. A*, 1995, **115**, 260–264.
14. R. Ghosh Biswas, M. P. Croxall, R. T. Lawrence, R. Soong, M. C. Goh and A. J. Simpson, *Nanoscale*, 2022, **14**, 9869–9876.
15. Y. L. Mobarhan, B. Fortier-McGill, R. Soong, W. E. Maas, M. Fey, M. Monette, H. J. Stronks, S. Schmidt, H. Heumann, W. Norwood and A. J. Simpson, *Chem. Sci.*, 2016, 7, 4856–4866.
16. O. Hoegh-Guldberg, P. J. Mumby, A. J. Hooten, R. S. Steneck, P. Greenfield, E. Gomez, C. D. Harvell, P. F. Sale, A. J. Edwards, K. Caldeira, N. Knowlton, C. M. Eakin, R. Iglesias-Prieto, N. Muthiga, R. H. Bradbury, A. Dubi and M. E. Hatziolos, *Science*, 2007, **318**, 1737–1742.
17. N. M. Whiteley, *Mar. Ecol.: Prog. Ser.*, 2011, **430**, 257–271.

18. D. P. Burum, *Concepts Magn. Reson.*, 1990, **2**, 213–227.
19. Y. L. Mobarhan, R. Soong, D. Lane and A. J. Simpson, *Magn. Reson. Chem.*, 2020, **58**, 427–444.
20. H. Masoom, A. Adamo, A. J. Simpson, H. Masoom, A. Adamo and A. J. Simpson, *Environ. Chem.*, 2016, **13**, 767–775.
21. H. Masoom, D. Courtier-Murias, H. Farooq, R. Soong, B. P. Kelleher, C. Zhang, W. E. Maas, M. Fey, R. Kumar, M. Monette, H. J. Stronks, M. J. Simpson and A. J. Simpson, *Environ. Sci. Technol.*, 2016, **50**, 1670–1680.
22. J. Lehmann and M. Kleber, *Nature*, 2015, **528**, 60.
23. J. M. Teal and R. W. Howarth, *Environ. Manage.*, 1984, **8**, 27–43.
24. A. Jernelöv, *Ambio*, 2010, **39**, 353–66.
25. H. Farooq, D. Courtier-Murias, M. J. Simspon, W. E. Maas, M. Fey, B. Andrew, J. Struppe, H. Hutchins, S. Krishnamurthy, R. Kumar, M. Monette, H. J. Stronks and A. J. Simpson, *Environ. Chem.*, 2015, **12**, 227.
26. L. Lam, R. Soong, A. Sutrisno, R. de Visser, M. J. Simpson, H. L. Wheeler, M. Campbell, W. E. Maas, M. Fey, A. Gorissen, H. Hutchins, B. Andrew, J. Struppe, S. Krishnamurthy, R. Kumar, M. Monette, H. J. Stronks, A. Hume and A. J. Simpson, *J. Agric. Food Chem.*, 2014, **62**, 107–115.
27. H. L. Wheeler, R. Soong, D. Courtier-Murias, A. Botana, B. Fortier-Mcgill, W. E. Maas, M. Fey, H. Hutchins, S. Krishnamurthy, R. Kumar, M. Monette, H. J. Stronks, M. M. Campbell and A. Simpson, *Magn. Reson. Chem.*, 2015, **53**, 735–744.
28. B. E. Fortier-McGill, R. Dutta Majumdar, L. Lam, R. Soong, Y. Liaghati-Mobarhan, A. Sutrisno, R. de Visser, M. J. Simpson, H. L. Wheeler, M. Campbell, A. Gorissen and A. J. Simpson, *J. Agric. Food Chem.*, 2017, **65**, 6779–6788.
29. R. H. Newman, L. M. Davies and P. J. Harris, *Plant Physiol.*, 1996, **111**, 475–485.
30. A. Barison, R. G. Biswas, P. Ning, F. V. C. Kock, R. Soong, M. C. B. Di Medeiros, A. Simpson and L. M. Lião, *Food Chem.*, 2022, **397**, 133800.
31. A. S. Sivam, G. I. N. Waterhouse, Z. D. Zujovic, C. O. Perera and D. Sun-Waterhouse, *Food Bioprocess Technol.*, 2011, **6**, 110–123.
32. X. Hu, B. Wei, B. Zhang, H. Li, X. Xu, Z. Jin and Y. Tian, *Int. J. Biol. Macromol.*, 2013, **61**, 329–332.
33. P. Ning, D. Lane, R. Soong, D. Schmidig, T. Frei, P. De Castro, I. Kovacevic, S. Graf, S. Wegner, F. Busse, J. Struppe, M. Fey, H. Stronks, M. Monette, M. J. Simpson and A. J. Simpson, *ACS Sustainable Chem. Eng.*, 2020, **8**, 17551–17564.
34. J. I. Choi and S. Y. Lee, *Bioprocess Eng.*, 1997, **17**, 335–342.
35. R. Carpine, F. Raganati, G. Olivieri, K. J. Hellingwerf, A. Pollio, P. Salatino and A. Marzocchella, *Algal Res.*, 2018, **29**, 49–60.
36. J. Janaun and N. Ellis, *Renewable Sustainable Energy Rev.*, 2010, **14**, 1312–1320.

37. L. M. A. Silva, E. G. A. Filho, A. J. Simpson, M. R. Monteiro and T. Venâncio, *Fuel*, 2016, **166**, 436–445.
38. Z. Tian, H. Zhao, K. T. Peter, M. Gonzalez, J. Wetzel, C. Wu, X. Hu, J. Prat, E. Mudrock, R. Hettinger, A. E. Cortina, R. G. Biswas, F. V. C. Kock, R. Soong, A. Jenne, B. Du, F. Hou, H. He, R. Lundeen, A. Gilbreath, R. Sutton, N. L. Scholz, J. W. Davis, M. C. Dodd, A. Simpson, J. K. McIntyre and E. P. Kolodziej, *Science*, 2021, **371**, 185–189.
39. M. S. Olivelli, I. Fugariu, R. M. Torres Sánchez, G. Curutchet, A. J. Simpson and M. J. Simpson, *ACS Earth Space Chem.*, 2020, **4**, 2061–2072.
40. W.-T. Tsai, H.-P. Chen and W.-Y. Hsien, *J. Loss Prev. Process Ind.*, 2002, **15**, 65–75.
41. Long-chain Perfluorinated Chemicals (PFCs) Action Plan, in Assessing and Managing Chemicals under TSCA, U.S. Environmental Protection Agency, 2009.
42. Agency for Toxic Substances and Disease Registry, *Toxicological Profile for Perfluoroalkyls*, 2020.
43. B. Liu, H. Zhang, Y. Yu, L. Xie, J. Li, X. Wang and W. Dong, *Arch. Environ. Contam. Toxicol.*, 2020, **78**, 182–189.
44. J. G. Longstaffe and A. J. Simpson, *Environ. Toxicol. Chem.*, 2011, **30**, 1745–1753.
45. D. H. Lysak, M. J. Simpson and A. J. Simpson, Applications of nuclear magnetic resonance for the study of soils in *Encyclopedia of Soils in the Environment*, Academic Press, 2nd edn, 2023, pp. 605–624.
46. A. Shirzadi, M. J. Simpson, Y. Xu and A. J. Simpson, *Environ. Sci. Technol.*, 2008, **42**, 1084–1090.
47. J. G. Longstaffe, D. Courtier-Murias and A. J. Simpson, *Chemosphere*, 2013, **90**, 270–275.
48. M. Forsthuber, A. M. Kaiser, S. Granitzer, I. Hassl, M. Hengstschläger, H. Stangl and C. Gundacker, *Environ. Int.*, 2020, **137**, 105324.
49. R. Ghosh Biswas, R. Soong, D. Schmidig, P. De Castro, S. Graf, F. Busse, A. J. Simpson and L. B. Casabianca, *ACS Appl. Nano Mater.*, **5**, 16519–16527.
50. P. K. Deb, O. Al-Attraqchi, M. R. Prasad and R. K. Tekade, in *Dosage Form Design Considerations*, Academic Press, 2018, vol. 1, pp. 371–399.
51. G. Ren, H. Han, Y. Wang, S. Liu, J. Zhao, X. Meng and Z. Li, *Nanomaterials*, 2021, **11**, 1804.
52. S. Yadav, M. Kumari and A. K. Saroha, *J. Environ. Manage.*, 2018, **228**, 169–188.
53. Y. Cao, X. Yuan, Y. Zhao and H. Wang, *Chem. Eng. J.*, 2022, **431**, 133833.
54. V. M. Alamillo-López, V. Sánchez-Mendieta, O. F. Olea-Mejía, M. G. González-Pedroza and R. A. Morales-Luckie, *Catalysts*, 2020, **10**, 727.
55. X. Gao and X. Meng, *Processes*, 2021, **9**, 1729.
56. Y. Zhang, G. Zhao, Y. Xuan, L. Gan and M. Pan, *Cellulose*, 2021, **28**, 991–1009.

57. M. P. Croxall, R. T. Lawrence and M. C. Goh, *J. Photochem. Photobiol., A*, 2019, **382**, 111965.
58. H. Sheng, H. Zhang, W. Song, H. Ji, W. Ma, C. Chen and J. Zhao, *Angew. Chem., Int. Ed.*, 2015, **54**, 5905–5909.
59. Z. Wang, J. Huang, R. Amal and Y. Jiang, *Appl. Catal., B*, 2018, **223**, 16–21.
60. W. Xu and D. Raftery, *J. Phys. Chem. B*, 2001, **105**, 4343–4349.
61. T. Ohno, K. Sarukawa, K. Tokieda and M. Matsumura, *J. Catal.*, 2001, **203**, 82–86.
62. T. R. Neu, B. Manz, F. Volke, J. J. Dynes, A. P. Hitchcock and J. R. Lawrence, *FEMS Microbiol. Ecol.*, 2010, **72**, 1–21.
63. P. Jezzard, P. M. Matthews and S. M. Smith, *J. Magn. Reson. Imaging*, 2003, **17**, 383–383.
64. R. Guan and W.-X. Wang, *Environ. Toxicol. Chem.*, 2004, **23**, 2689.
65. X.-J. Liu, I.-H. Ni and W.-X. Wang, *Water Res.*, 2002, **36**, 4563–4569.
66. H. C. Poynton, S. Hasenbein, J. B. Benoit, M. S. Sepulveda, M. F. Poelchau, D. S. T. Hughes, S. C. Murali, S. Chen, K. M. Glastad, M. A. D. Goodisman, J. H. Werren, J. H. Vineis, J. L. Bowen, M. Friedrich, J. Jones, H. M. Robertson, R. Feyereisen, A. Mechler-Hickson, N. Mathers, C. E. Lee, J. K. Colbourne, A. Biales, J. S. Johnston, G. A. Wellborn, A. J. Rosendale, A. G. Cridge, M. C. Munoz-Torres, P. A. Bain, A. R. Manny, K. M. Major, F. N. Lambert, C. D. Vulpe, P. Tuck, B. J. Blalock, Y. Y. Lin, M. E. Smith, H. Ochoa-Acuña, M. J. M. Chen, C. P. Childers, J. Qu, S. Dugan, S. L. Lee, H. Chao, H. Dinh, Y. Han, H. Doddapaneni, K. C. Worley, D. M. Muzny, R. A. Gibbs and S. Richards, *Environ. Sci. Technol.*, 2018, **52**, 6009–6022.
67. B. Adam, S. Zofia and S. Tadeusz, *J. Comp. Physiol., B*, 2014, **184**, 976.
68. M. Bastawrous, D. Lane, R. Soong, M. T. Anaraki, D. Schmidig, T. Frei, P. De Castro, S. Graf, T. Kuehn, R. Kümmerle, F. Busse, H. Heumann, H. Boenisch, M. Gundy and A. J. Simpson, *J. Magn. Reson. Open*, 2022, **12–13**, 100079.
69. Y. Duan, S. I. Guttman, J. T. Oris and A. J. Bailer, *Environ. Toxicol. Chem.*, 2000, **19**, 1414–1421.
70. S. W. Provencher, *Magn. Reson. Med.*, 1993, **30**, 672–679.
71. Y. Liaghati Mobarhan, R. Soong, W. Bermel, M. J. Simpson, J. Struppe, H. Heumann, S. Schmidt, H. Boenisch, D. Lane and A. J. Simpson, *ACS Omega*, 2018, **3**, 17023–17035.
72. R. G. Biswas, R. Soong, P. Ning, D. Lane, M. Bastawrous, A. Jenne, D. Schmidig, P. de Castro, S. Graf, T. Kuehn, R. Kümmerle, W. Bermel, F. Busse, J. Struppe, M. J. Simpson and A. J. Simpson, *Anal. Chem.*, 2022, **94**, 8756–8765.
73. R. Ghosh Biswas, B. Fortier-McGill, M. Akhter, R. Soong, P. Ning, M. Bastawrous, A. Jenne, D. Schmidig, P. De Castro, S. Graf, T. Kuehn, F. Busse, J. Struppe, M. Fey, H. Heumann, H. Boenisch, M. Gundy, M. J. Simpson and A. J. Simpson, *Anal. Chim. Acta: X*, 2020, **6**, 100051.
74. W. Block, J. Pauly, A. Kerr and D. Nishimura, *Magn. Reson. Med.*, 1997, **38**, 198–206.

75. M. Liu, J. K. Nicholson and J. C. Lindon, *Anal. Chem.*, 1996, **68**, 3370–3376.
76. G. Diserens, M. Vermathen, C. Precht, N. T. Broskey, C. Boesch, F. Amati, J. F. Dufour and P. Vermathen, *Analyst*, 2014, **140**, 272–279.
77. Q. Hassan, R. Dutta Majumdar, B. Wu, D. Lane, M. Tabatabaei-Anraki, R. Soong, M. J. Simpson and A. J. Simpson, *Magn. Reson. Chem.*, 2019, **57**, 69–81.
78. R. D. Majumdar, M. Akhter, B. Fortier-McGill, R. Soong, Y. Liaghati-Mobarhan, A. J. Simpson, M. Spraul, S. Schmidt and H. Heumann, *eMagRes*, 2017, **6**, 133–148.
79. D. Ebert, *Ecology, Epidemiology, and Evolution of Parasitism in Daphnia*, National Library of Medicine US National Center for Biotechnology Information, Bethesda, Maryland, 2005.
80. A. Wacker and D. Martin-Creuzburg, *Funct. Ecol.*, 2007, **21**, 738–747.
81. C. Becker and M. Boersma, *Limnol. Oceanogr.*, 2005, **50**, 388–397.
82. A. Putman, D. Martin-Creuzburg, B. Panis and L. D. Meester, *J. Plankton Res.*, 2015, **37**, 752–763.
83. M. Tabatabaei Anaraki, R. Dutta Majumdar, N. Wagner, R. Soong, V. Kovacevic, E. J. Reiner, S. P. Bhavsar, X. Ortiz Almirall, D. Lane, M. J. Simpson, H. Heumann, S. Schmidt and A. J. Simpson, *Anal. Chem.*, 2018, **90**, 7912–7921.
84. R. Soong, Y. Liaghati Mobarhan, M. Tabatabaei, M. Bastawrous, R. G. Biswas, M. Simpson and A. Simpson, *Magn. Reson. Chem.*, 2019, **58**, 411–426.
85. R. Soong, E. Nagato, A. Sutrisno, B. Fortier-McGill, M. Akhter, S. Schmidt, H. Heumann and A. J. Simpson, *Magn. Reson. Chem.*, 2015, **53**, 774–779.
86. P. C. A. van der Wel, *Solid State Nucl. Magn. Reson.*, 2017, **88**, 14.
87. O. Nikel, D. Laurencin, S. A. McCallum, C. M. Gundberg and D. Vashishth, *Langmuir*, 2013, **29**, 13882.
88. M. Bastawrous, M. Tabatabaei-Anaraki, R. Soong, W. Bermel, M. Gundy, H. Boenisch, H. Heumann and A. J. Simpson, *Anal. Chim. Acta*, 2020, **1138**, 168–180.
89. R. A. Wind, J. Z. Hu and P. D. Majors, *Magn. Reson. Med.*, 2006, **55**, 41–49.
90. R. A. Wind, J. Z. Hu and P. D. Majors, *Drug Discovery Today: Technol.*, 2005, **2**, 291–294.
91. R. A. Wind, J. Z. Hu and D. N. Rommereim, *Magn. Reson. Med.*, 2003, **50**, 1113–1119.
92. W. Dixon, J. Schaefer, M. Sefcik, E. Stejskal and R. McKay, *J. Magn. Reson.*, 1982, **49**, 341–345.
93. W. T. Dixon, *J. Chem. Phys.*, 1982, 77, 1800–1809.
94. O. N. Antzutkin, S. C. Shekar and M. H. Levitt, *J. Magn. Reson., Ser. A*, 1995, **115**, 7–19.
95. Y. L. Mobarhan, J. Struppe, B. Fortier-McGill and A. J. Simpson, *Anal. Bioanal. Chem.*, 2017, **409**, 5043–5055.

96. P. Ning, D. Lane, R. G. Biswas, R. Soong, D. Schmidig, T. Frei, P. De Castro, I. Kovacevic, S. Graf, S. Wegner, F. Busse, T. Kuehn, J. Struppe, M. Fey, H. J. Stronks, M. Monette, M. J. Simpson and A. J. Simpson, *Anal. Chem.*, 2021, **93**, 10326–10333.
97. S. A. McNeill, P. L. Gor'kov, K. Shetty, W. W. Brey and J. R. Long, *J. Magn. Reson.*, 2009, **197**, 135–144.
98. P. L. Gor'kov, E. Y. Chekmenev, C. Li, M. Cotten, J. J. Buffy, N. J. Traaseth, G. Veglia and W. W. Brey, *J. Magn. Reson.*, 2007, **185**, 77–93.
99. N. Hertkorn, C. Ruecker, M. Meringer, R. Gugisch, M. Frommberger, E. M. Perdue, M. Witt and P. Schmitt-Kopplin, *Anal. Bioanal. Chem.*, 2007, **389**, 1311–27.
100. A. Jenne, W. Bermel, C. A. Michal, O. Gruschke, R. Soong, R. G. Biswas, M. Bastawrous and A. J. Simpson, *Angew. Chem., Int. Ed.*, 2022, **61**, e202110044.
101. P. Ning, D. Lane, R. Ghosh Biswas, A. Jenne, M. Bastawrous, R. Soong, D. Schmidig, T. Frei, P. De Castro, I. Kovacevic, S. Graf, S. Wegner, W. Bermel, F. Busse, T. Kuehn, R. Kuemmerle, J. Struppe, M. Fey, H. J. Stronks, M. Monette, M. J. Simpson and A. J. Simpson, *Analyst*, 2021, **146**, 4461–4472.
102. H. Kovacs, D. Moskau and M. Spraul, *Prog. Nucl. Magn. Reson. Spectrosc.*, 2005, **46**, 131–155.
103. M. Concistrè, O. G. Johannessen, E. Carignani, M. Geppi and M. H. Levitt, *Acc. Chem. Res.*, 2013, **46**, 1914–1922.
104. A. Hassan, C. M. Quinn, J. Struppe, I. V. Sergeyev, C. Zhang, C. Guo, B. Runge, T. Theint, H. H. Dao, C. P. Jaroniec, M. Berbon, A. Lends, B. Habenstein, A. Loquet, R. Kuemmerle, B. Perrone, A. M. Gronenborn and T. Polenova, *J. Magn. Reson.*, 2020, **311**, 106680.
105. M. Möst, A. C. Chiaia-Hernandez, M. P. Frey, J. Hollender and P. Spaak, *Environ. Toxicol. Chem.*, 2015, **34**, 338–345.
106. S. Navis, A. Waterkeyn, T. Voet, L. D. Meester and L. Brendonck, *Ecotoxicology*, 2013, **22**, 803–814.
107. D. Sakellariou, G. L. Goff and J.-F. Jacquinot, *Nature*, 2007, **447**, 694–697.
108. H. Janssen, A. Brinkmann, E. R. H. van Eck, P. J. M. A. van Bentum and A. P. M. Kentgens, *J. Am. Chem. Soc.*, 2006, **128**, 8722–8723.
109. E. V. Vonhof, M. Piotto, E. Holmes, J. C. Lindon, J. K. Nicholson and J. V. Li, *Anal. Chem.*, 2020, **92**, 11516–11519.
110. S. S. Adhikari, L. Zhao, T. Dickmeis, J. G. Korvink and V. Badilita, *Analyst*, 2019, **144**, 7192–7199.

CHAPTER 5

Environmental In Vivo *NMR: Explaining Toxicity and Processes at the Biochemical Level*

D. H. LYSAK, W. W. WOLFF, P. M. COSTA AND A. J. SIMPSON*

Environmental NMR Centre, Scarborough Campus, University of Toronto, 1265 Military Trail, ON M1C 1A4, Canada
*Email: andre.simpson@utoronto.ca

5.1 Introduction

Since its inception and first uses in the mid-20th century,[1,2] nuclear magnetic resonance (NMR) has become a powerful tool for the analysis of a wide range of samples. Whether used to identify an unknown organic compound,[3] elucidate the composition of environmental mixtures such as wastewater effluents[4] or examine the interactions between small molecules and supramolecular humic complexes,[5] NMR can provide valuable information for nearly any research goal. Going further, the non-destructive nature and analytical characteristics of NMR make it uniquely poised for studying living organisms. By studying the metabolic profile of organisms exposed to environmentally relevant pollutants, sublethal toxicity information can be extracted in real time, which can then be used to help inform disposal techniques and environmental regulations.[6] In this chapter, the advantages, approaches, challenges and potential of *in vivo* NMR will be

New Developments in NMR No. 32
The Environment in a Magnet: Applications of NMR Techniques to Environmental Problems
Edited by Pellegrino Conte, Delia Francesca Chillura Martino and Paolo Lo Meo

Published by the Royal Society of Chemistry, www.rsc.org

discussed, in order to demonstrate the substantial utility of this technique in understanding environmental toxicity at a biochemical level.

For the purpose of this chapter, the term "*in vivo* NMR" refers to the study of small living organisms sustained within a vertical bore high resolution NMR spectrometer, in contrast to larger horizontal bore magnetic resonance imaging (MRI) systems. As an example, a flow system may be employed to provide oxygen and nutrients to the organisms contained within a standard 5 mm NMR tube.[7] This can be contrasted with MRI systems with horizontal bores, in which the study of living organisms of various sizes is in fact the main goal. While MRI is without doubt *in vivo* analysis, the aim is typically to develop an image for diagnostic purposes. On the other hand, this chapter focuses on the application of high-resolution NMR techniques to explore the metabolic profiles of small organisms *in vivo* and examine sublethal responses to environmentally relevant stressors such as pollutants or lack of oxygen/nutrients. Here, the focus will be on millimeter to centimeter scale organisms (*e.g.*, *Daphnia magna*, *Hyalella azteca*, and *Eisenia fetida*), as opposed to bacteria/microorganisms, though examples of the latter will also be considered.

5.2 Pollution, Toxicity and the Environment

It is without doubt that human activity has had a substantial impact on the environment; recent estimates indicate that over 350 000 chemicals and chemical mixtures have been registered for use worldwide,[8] and over 400 million tons of hazardous waste are produced and disposed of each year.[9] In just the United States alone, approximately 1500 new substances are produced each year,[9] including proprietary compounds with publicly unavailable chemical structures.[10] In addition to the sheer number of compounds entering the environment, the released chemical contaminants can become transformed into a diverse set of breakdown products, through either anthropogenic treatment processes[11] or natural degradation.[12] These products can have strikingly different chemical and toxicological profiles. As an example, irbesartan, a common blood pressure medication, has been shown to have a disinfection by-product (resulting from wastewater treatment) that shows 12-fold larger ecotoxicity than irbesartan itself![13] As such, an understanding of the behaviour and toxicity of compounds in the environment is critical to maintaining ecosystem health.

While the acute toxicity of a single compound can be tested quite readily, the environment represents an extremely complex matrix of chemicals and pollutants.[14] The traditional toxicity tests that most environmental regulations are based on focus on apical endpoints such as loss of movement, reproduction or death in the test organisms.[15] Arguably the most commonly utilized metric, the lethal concentration 50(LC_{50}) denotes the concentration at which 50% of the test organisms have died.[15] While this method is, of course, useful in comparing the acute toxicity of different chemicals, the resulting information provided is limited. Such tests do not offer any information on sublethal toxicity, toxic-modes-of-action, biotransformation, recovery or contaminant interactions. Further, organisms in the environment are exposed to not only a single compound, or even its group of degradation products, but a veritable "cocktail" of environmental stressors and pollutants of varying chemical properties, concentrations and toxicities. Interactions between these pollutants can also change their toxicity compared to individual exposure.[14] With the vast number of both emerging and legacy pollutants, it is impossible to individually screen the parent compound, break down products and then evaluate the toxicity of each, especially when mixtures of potentially interacting compounds are considered.

A recent report from the US environmental protection agency (EPA) has stated that a paradigm shift in toxicity testing is needed.[16] As opposed to focusing on apical endpoints resulting from a single compound, there is a need to understand why chemicals are toxic (*i.e.*, understand the biochemical toxicity pathways) and to examine concurrent, cumulative exposure to multiple chemicals at once[16] – goals to which *in vivo* metabolomics is well-suited. Defined as the study of fluxes in the low molecular weight components (typically of molecular weight <1500) of a living system,[17] metabolomics can help identify toxicological changes long before organismal death or loss of movement occurs. Furthermore, the metabolic response of an organism can be used to extract information about the toxic-mode-of-action and progression of disease. For example, a decrease in the tricarboxylic acid cycle intermediates in response to polychlorinated biphenyl exposure was linked to the inhibition of oxidative phosphorylation in rat liver cells.[18]

NMR has numerous characteristics that make it ideal for metabolomics studies.[19,20] First, although selective experiments exist,[21–24] NMR is by nature non-targeted and is thus able to examine changes in a wide variety of metabolites concurrently. Non-targeted metabolomics has the potential to take a "high level" look at the changes in an organism's metabolism, which, when paired with statistical techniques such as principal component analysis (PCA), can reveal changes between samples that would simply be missed in individual metabolite monitoring experiments.[20] While the potential for non-targeted analysis is a great benefit, the unique advantage of NMR is its ability for non-targeted studies of intact samples.

Due to its non-destructive nature, NMR can be used to analyze the metabolome of living organisms in real-time. To consider an example, living water fleas (*D. magna*) can be sustained using a flow system under ideal conditions (*i.e.*, without stressors) in an NMR spectrometer in order to gather a control data set. They can then be exposed to a low concentration of an environmentally relevant pollutant, such as bisphenol A (an endocrine disruptor), and the metabolic changes in the exact same organisms could be tracked.[25] Since the control groups and the exposed groups can be the exact same organisms,[26] there is high confidence that any changes in the metabolic profile are due to exposure to the pollutant, since there is no additional natural variability due to a different cohort of organisms. Further, since the organisms do not die, if the flow system is then flushed with "clean" water, it becomes possible to gauge the organism's recovery from stress. This can help distinguish between a temporary metabolic flux that later returns to homeostasis and a permanent metabolic change, which can be indicative of disease. Such information can be used by regulatory agencies to set effective, science-based environmental policies.

5.3 Types of *In Vivo* NMR

5.3.1 Solution State NMR

In vivo NMR in a solution state probe usually involves a flow system that provides the organisms being studied with food as well as oxygen. The most commonly used organisms in solution state NMR are water fleas[27] (*D. magna*), though earthworms[28,29] and other larger aquatic organisms such as bivalves[30] have also been studied. An example of a model flow system can be found in Figure 5.1.

In brief, thin tubing along with a set of pumps is used to both slowly add additional, oxygen-rich, fresh media to the NMR tube from a reservoir, and remove used media. The suction pump runs at a faster rate, maintaining the same sample height in the flow cell (required for optimal shimming of cryoprobes) and providing a zero-pressure environment for the organisms. This has been shown to be an effective way to keep *D. magna* alive under minimal stress, which is an excellent control prior to chemical exposure.[7] To further enhance spectral quality, a capillary filled with a deuterated solvent can be inserted to act as a lock to reduce the effects of field drift over long experiments. Further, both the top and bottom of the NMR tube are plugged in order to restrict the organisms to the active coil region and maximize the signal. Despite the simple concept, this flow system provides both a powerful way to sustain organisms, as well as a means to introduce pollutants and stressors for metabolomics studies. Solution state NMR is highly versatile due to the vast array of pulse sequences that can be applied (see Section 5.5). While the most commonly studied nucleus is ^{1}H due to its high natural abundance and gyromagnetic ratio, there have been studies employing heteronuclei including ^{13}C, ^{31}P and ^{15}N due to the different information and resolving power afforded,[6] and there is potential for the study of more "exotic" nuclei (see Section 5.5.1.5).

5.3.2 Magic Angle Spinning (MAS) NMR

Despite the considerable versatility of solution state NMR for metabolomics, it is unable to provide information on the gel-like or solid components of an organism.[31] Gel-like and solid components typically show very broad lineshapes, due to relatively strong anisotropic interactions, and this effect is amplified in true solids.[32] However, by spinning a sample at the "magic angle" (~54.7°), the effects of this broadening can be mitigated, allowing the analysis of gels and solids with lineshape that approaches that of solution

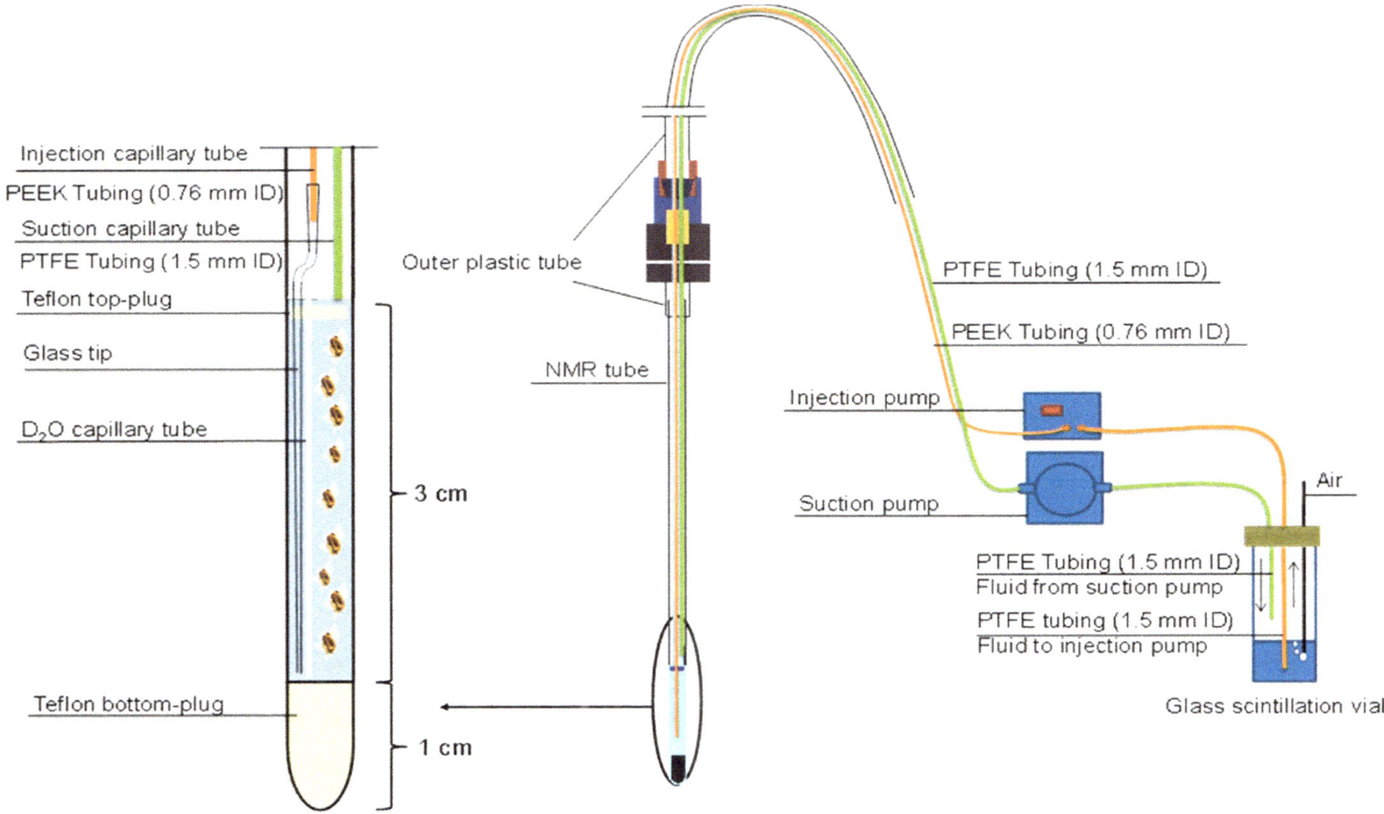

Figure 5.1 An example of a low volume flow built in-house to sustain *D. magna* for *in vivo* NMR. Reproduced from ref. 7 with permission from American Chemical Society, Copyright 2018.

state spectra.[33] MAS NMR techniques can be divided into three main groups: solid state NMR, high resolution magic angle spinning (HR-MAS) NMR and comprehensive multiphase (CMP) NMR. For optimal signal to noise, true solid-state NMR requires dried, well-packed solids and since arguably all living organisms are hydrated to some degree, solid-state NMR is typically not used for *in vivo* studies. Further, many solid-state probes are not susceptibility matched, leading to broad lineshapes and lack a lock and a gradient, limiting the types of experiments that can be performed. However, there are certain important, true solid components such as chitin in the exoskeletons of aquatic organisms that can be investigated using CMP NMR.[31]

HR-MAS can be a powerful technique to analyze organisms *in vivo* due to the additional information provided on gel-like components. In contrast to solid state probes, HR-MAS probes are typically susceptibility matched and equipped with both a gradient and a lock.[34] This allows long duration, high-resolution experiments, along with advanced techniques such as water suppression, diffusion editing, and gradient selected multidimensional NMR. Magic angle spinning acts to reduce the effects of the anisotropic interactions, which significantly improves lineshape for gel-like components, though the stronger interactions in true solids cannot be overcome.[33] However, compared to solution state NMR, this technique can be an excellent analysis tool for "swollen" or "gel-like" samples and has previously been applied to whole organisms including fruit flies (*Drosophila melanogaster*),[35] nematodes (*Caenorhabditis elegans*),[36] earthworms (*Eisenia fetida*)[37] as well as small aquatic species including *Calanus finmarchicus*,[38] *Daphnia magna*[39,40] and *Hyalella azteca*.[41] The additional information provided on gel-like domains can be used to examine processes such as metabolite transfer, mobility changes and interactions between contaminants and membranes.

To push the envelope further, CMP NMR combines the advantages of HR-MAS with electronic components that can tolerate the high power required for true solid NMR experiments.[42] Therefore, CMP NMR can give an overview of every bond in every phase of a living organism: ranging from the soluble metabolites, to the gel-like membranes/swollen components, and finally the true solids such as bones or exoskeletons.[31] Although each component could be investigated on its own with a dedicated probe, the major advantage of CMP is that it allows the study of interactions between phases, which cannot be done on a single-phase probe. As an example, CMP NMR has been used to study the kinetics of fluorinated contaminant binding and sequestration to soil.[43] Since all phases can be detected, the disappearance of signal from the solution state can be directly corroborated by an increase in the signal in the solid state, which confirms sequestration into the solid phase. This has also been applied to study the sequestration and bioconversion of pollutants in a living shrimp.[31] For example, the shells of aquatic shrimp have been shown to change both chemical composition and thickness when exposed to polluted water,[44] and this process could be studied directly *via* CMP NMR. Despite the potential of these techniques, both HR-MAS and CMP NMR have considerable experimental challenges that are unique to *in vivo* analysis, which are discussed in the next section.

5.4 *In Vivo* NMR – Challenges and Approaches

Along with the considerable analytical power afforded by *in vivo* NMR, there are considerable challenges associated with the study of living organisms, both in solution and under magic angle spinning. In this section, the key technical and experimental challenges of *in vivo* NMR and strategies that have been developed to mitigate them will be discussed.

5.4.1 Broad Lineshape

A living organism is a highly complex sample that comprises thousands of metabolites in a constant state of flux. However, just the chemical complexity is not the end of the story. A living organism is made up of a series of "compartments" filled with different chemical species. Due to the magnetic susceptibility distortions caused by the differences between these compartments, a metabolite found in one region can resonate at a slightly different Larmor frequency from the same metabolite found in a different region, which leads to a considerably broadened lineshape.[45] This broadened lineshape can result in both peak overlap, which greatly increases the difficulty of peak assignment/quantification, and decreased splitting information, which further makes peak pattern matching more challenging.

There are a number of approaches that can help mitigate the effects of line broadening, including the conventional, but crucial need to ensure that the sample is well shimmed. Although this may seem trivial for a standard solution state sample, it can be quite challenging with complex, *in vivo* organisms, especially considering the additional devices in/near the coil region required to sustain the organisms. The use of automated gradient shimming algorithms (*e.g.*, TopShim) is highly recommended, with an additional round of adjustments by hand afterwards. For reference, Figure 5.2 shows a comparison of a sample of *D. magna in vivo*, compared to an *ex vivo* sample.[46]

Clearly, the *in vivo* spectrum shows a much poorer lineshape, and this is exacerbated by the organism's movement. The *ex vivo* sample, which consists of crushed *D. magna*, shows a markedly better lineshape since the organism has died (and is thus not moving), and the metabolites have leaked into the surrounding solvent. However, the magnetic inhomogeneities caused by the organisms' bodies are still present, which is why the lineshape is still poorer than that of typical small molecules or metabolite extracts. In addition, in *ex vivo* experiments the lipids that typically dominate an *in vivo*

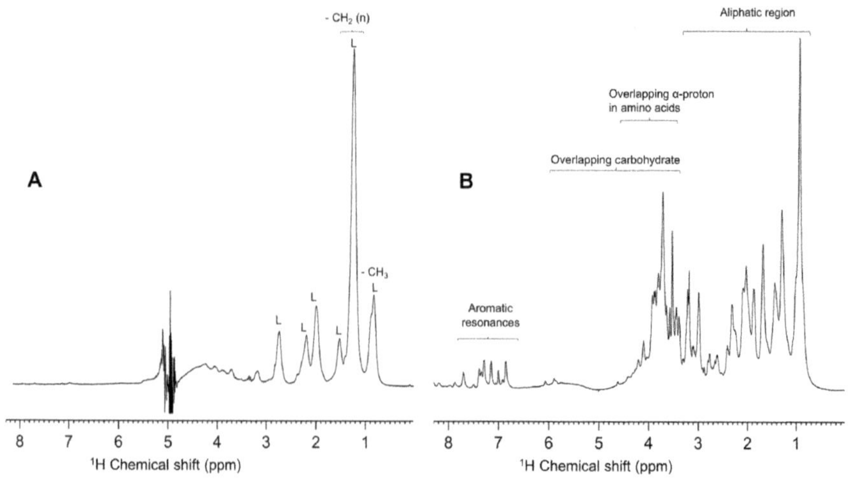

Figure 5.2 A lineshape comparison of (A) *in vivo D. magna* and (B) *ex vivo D. magna* (*i.e.*, organisms crushed after *in vivo* analysis). Lipid peaks dominate the *in vivo* spectrum and are marked with an "L", while some peak regions are labelled in the *ex vivo* spectrum. Reproduced from ref. 46 with permission from the Royal Society of Chemistry.

spectrum float to the top of the water (outside the coil region), making the 0.5–1.7 ppm region of the spectrum more informative. While *ex vivo* experiments can be excellent in helping to assign peaks,[47] these experiments lose a number of the benefits of *in vivo* NMR (such as the ability to study metabolite fluxes in real time, monitor recovery and study the same organisms as control and exposed) and represent a "snapshot" of an organism's metabolome at a single time point. As such, *ex vivo* experiments can be very useful to inform the analysis of *in vivo* spectra, but overall provide less complete information.

The effects of line-broadening can also be reduced through the use of pulse sequences designed to overcome the effects of field inhomogeneity. The in-phase intermolecular single quantum coherence (IP-iSQC) sequence, for example, forms long range dipolar couplings in the solution state, in order to extract chemical shift information free from broadening effects,[28,48] leading to a substantial improvement in linewidth (Figure 5.3).

The comparison to the standard ^{1}H spectrum is a night and day difference, and the IP-iSQC spectrum shows lineshape that even rivals that of a buffer extract.[28] As such, this is a powerful technique that can help extract metabolite information from otherwise "hopelessly broad" spectra. However, this ability comes at a cost: IP-iSQC is around 10 times less sensitive than a standard ^{1}H proton spectrum and can thus require long acquisition times, though for sufficiently broad spectra this is nonetheless a very powerful approach. Furthermore, since it relies only on ^{1}H, it does not require isotopic enrichment and could potentially be applied to organisms

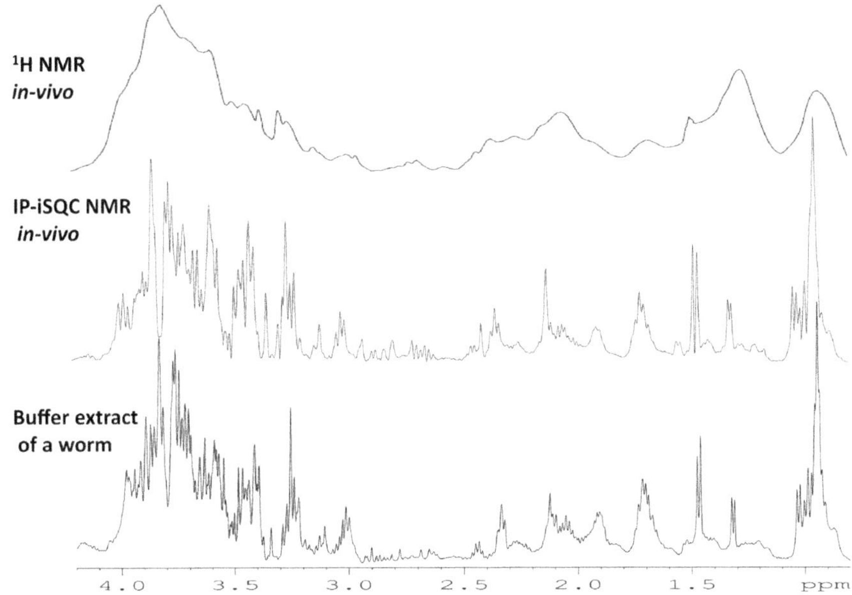

Figure 5.3 A comparison of a standard *in vivo* ^{1}H spectrum, the *in vivo* IP-iSQC sequence of a living *E. fetida*, along with an *E. fetida* buffer extract. Reproduced from ref. 28 with permission from John Wiley & Sons, Copyright 2017.

sampled directly from the environment. It is also important to note that this technique works only for truly dissolved compounds and thus does not provide information on gel-like domains.[28]

5.4.2 Spectral Overlap

The very complex metabolite profile of a living organism, combined with broad lineshapes, can cause spectral overlap to be a serious issue. While techniques such as IP-iSQC can help reduce linewidths, they may not necessarily address the core issue of simply having overlapping metabolite signals. While a standard ^{1}H NMR spectrum is expected to be able to resolve ~3000 peaks under ideal conditions,[49] this number is much lower in practice, since groups of metabolites tend to resonate in the same regions (*e.g.*, carbohydrates or aromatic moieties) and there are "unused" areas in a standard spectrum. On the other hand, techniques such as heteronuclear NMR and multidimensional NMR can have resolving powers that are orders of magnitude higher.[49] Employing these techniques can allow the discrimination of peaks that would otherwise overlap in 1D ^{1}H NMR, by "spreading" peaks across another dimension. More information on the specific 2D and heteronuclear approaches used *in vivo* can be found in Section 5.5.2. In addition to techniques that provide additional spectral dispersion, selective experiments exist which can reduce the overlap by targeting specific metabolites (*e.g.*, DREAMTIME[23]) or metabolite classes (amino acid only experiments[50]).

5.4.3 Water Suppression

Since deuterated solvents are typically toxic to organisms, *in vivo* studies are typically carried out in water or a buffered solution, making the large water signal a considerable challenge. Further, most aquatic organisms are mainly composed of water. For example, *H. azteca* is almost 95% water by mass[51] and the different water environments in and outside of the organism further lead to very broad water peaks. Since the water peak is orders of magnitude more intense than the metabolites of interest, if unsuppressed, it can cause difficulties with analog-to-digital conversion (*i.e.*, receiver overflows) or require a much lower receiver gain, which compromises the signal from lower intensity peaks. The most commonly used solvent suppression techniques such as simple presaturation[52] or PURGE[53] are typically insufficient for *in vivo* experiments. It has been reported that the most effective water suppression technique for *in vivo* NMR is SPR-W5-WATERGATE.[54] This method employs a combination of shaped presaturation, and a pair of W5 binomial-like pulses flanked by gradients.[55] The shaped presaturation block helps decrease the intensity and width of the water resonance, while the W5 pulses selectively invert everything but the water peak. Since the water peak is not inverted, the magnetization is destroyed by the dephasing effects of the gradient pulses, while the rest of the spectrum is recovered (for additional details on water suppression, readers are directed to the review by Zheng and Price[56]). Typically, this results in water signals near the noise level, though there is a possibility of baseline artefacts.

5.4.4 Sensitivity

While sensitivity remains the Achilles' heel of NMR, there are additional concerns specific to *in vivo* analysis. First, the generally broad lineshapes *in vivo* result in lower signal-to-noise ratios (SNRs) simply because the peak intensity is "spread out" over a larger region. Second, living organisms are mainly made up of water, and thus there is a very low biomass present for analysis. As an example, the dry weight of a single adult *D. magna* is ~100 μg.[57] When combined with the 100s to 1000s of compounds present in the organism, the mass of an individual metabolite present can easily be in the nanogram range, which is especially problematic for heteronuclear or multidimensional experiments. Further, since the organisms are alive, there is no potential for concentrating the sample: the biomass is limited by the number of organisms that can fit into the active coil region. For *D. magna* in a standard 5 mm tube this is typically between 20 and 40 organisms,[7] which is equal to a total mass <5 mg. In comparison, when following an extraction protocol, hundreds of *D. magna* could be extracted into a buffer for a single experiment, leading to a much more concentrated solution for analysis.

Numerous approaches have been employed to help mitigate the difficulty of low sensitivity.[58] To address the issue of limited biomass, there are two major approaches: large diameter probes and microcoils. Some pioneering *in vivo* studies on aquatic bivalves and fish embryos were performed in 10 mm probes.[59–61] While larger probe sizes are not common for many laboratories, they provide the key advantage of simply allowing a greatly increased biomass to be studied and thus result in much higher signal intensities. This is, however, a double-edged sword, as a larger number of organisms required may increase costs, especially for isotopically enriched organisms. Although large diameter probes have the disadvantages of long pulse widths and typically lower B_1 field homogeneities,[62] the increase in sample size may open the door to the study of non-isotopically enriched organisms.

At the other extreme, the use of microcoils has potential for the study of tiny individual organisms, neonates or eggs. In general, smaller diameter probes, especially micro-coils, show improved mass sensitivity (that is, the amount of signal per unit mass of sample) compared to larger probes.[63] As an example, a single *D. magna* egg on a 50 μm planar microcoil showed a mass sensitivity 3000 times higher than a standard 5 mm probe, which is equal to a 9-million-fold time saving.[64] While this is of little benefit for routine chemical analysis where the mass of the sample is relatively high, it is of great value in studying mass limited samples such as individual *D. magna* eggs (~200–500 μm in diameter). To fill a standard 5 mm probe with *D. magna* eggs would require thousands of individual eggs,[65] which must be painstakingly extracted by hand, making this approach prohibitively difficult. On the other hand, a microcoil can acquire a ^{1}H NMR spectrum on a single egg, and even 2D ^{1}H–^{13}C HMQC has been reported on a single ^{13}C labelled *D. magna* adult.[66] Despite their potential, microcoils are not commonly found in commercial probes and tend to have low B_1 field homogeneity, making experiments with multiple pulses more challenging.[66] However, the use of microcoils remains an excellent way to study highly environmentally relevant, mass limited samples.

Another way to improve sensitivity is the use of a cryoprobe. By cooling the coil and electrical components of a probe with liquid nitrogen or helium, the thermal noise (which is the main source of noise) in the receiver circuits can be reduced.[67] This can result in up to a 3-to-4-fold improvement in the signal-to-noise ratio compared to a room temperature probe and thus decrease acquisition time by around an order of magnitude.[68] This approach has been quite popular for solution state NMR, with a large number of *in vivo* studies employing cryoprobes.[25,46,69–71] However, cryoprobes are much more costly to purchase and have considerable upkeep costs. Further, they are typically especially sensitive to salt concentrations, and thus if the organism being studied has a high metabolic salt concentration or is sustained in a high-salt buffer, the cryoprobe advantage may be reduced to some degree,[72] but is still expected to outperform a room temperature probe in most scenarios.

While 1D ^{1}H NMR spectra are typically easy to acquire with sufficient sensitivity *in vivo*, heteronuclear or two-dimensional experiments are much more challenging. To address this, isotopic enrichment has been employed to enhance the signal.[41,73] Using ^{13}C as an example, by enriching an organism to >95% ^{13}C, the sensitivity of a carbon detection experiment is expected to improve by nearly two orders of magnitude. This opens the potential for otherwise unfeasibly long experiments such as heteronuclear correlation (HETCOR), or 1D ^{13}C NMR, and significantly improves the sensitivity of ^{1}H–^{13}C experiments such as heteronuclear single quantum coherence (HSQC). Uniform ^{13}C labelling of organisms such as *D. magna*,[74] *H. azteca*[47] and *C. elegans*[75] has been reported in the literature and used to obtain high quality spectra *in vivo*, even on single organisms. For *D. magna* and *H. azteca*, this is done by feeding the organisms with ^{13}C labelled algae, which can either be purchased commercially or be grown in-house using ^{13}C enriched CO_2. To achieve high labelling efficiency, typically multiple generations must be cultured,[74] generally making this a costly approach and limiting studies to organisms cultured in the laboratory as opposed to samples from the environment.

However, there is also a possibility of introducing a single isotopically enriched compound (or a set of compounds) *in vivo* for use as a metabolic tracer (commonly applied in MRI[34]). Since the signal arising from the enriched compound is much higher than the rest of the organism, this can be used to selectively study metabolic transformations of a compound. That said, the number of isotopically enriched compounds available is limited and they are typically very expensive. Interestingly, if ^{13}C enriched organisms are used, non-enriched compounds can be used as tracers, which allows a much broader range of tracer molecules to be employed. By using a pulse sequence with a ^{13}C filter, any ^{1}H magnetization arising from protons bound to ^{13}C (*i.e.*, the organism itself) can be suppressed, leaving only the signals from the ^{12}C bound protons in the tracer molecule.[76] This not only simplifies the spectrum, but also allows the exact fate of the tracer molecule to be directly studied, providing information on metabolism *in vivo*.

5.4.5 MAS Techniques – Organism Survival

The major consideration for *in vivo* MAS NMR is the health and survival of the organism, which can be challenging under spinning conditions. First, while MAS rotors are typically sealed to prevent any sample from leaking out, in order to sustain living organisms, the rotor caps must be modified by drilling a hole in order to allow oxygen exchange.[31] In addition, while typical spinning speeds for solid state NMR can easily exceed 5 kHz, such rates generate very strong centripetal forces which are completely unsuitable for living organisms. While some stress due to spinning is unavoidable, much slower spinning speeds are required *in vivo*. Previous studies[31,40] have indicated that *D. magna* can survive spinning at 50 Hz for 24 hours, but survival drops off to only 4 hours while spinning at 2.5 kHz. While other organisms with a more rigid exoskeleton such as *H. azteca* are more resistant to spinning speeds and can survive up to 6 hours at 2.5 kHz, these are still relatively slow spinning speeds, which can lead to both broader lineshapes in more solid components, as well as spinning sidebands. Methods to deal with spinning sidebands are discussed in Section 5.4.6. Since survival under spinning is unique to each type of organism studied, survivability should be assessed for new organisms prior to conducting longer experiments. In brief, it is crucial to strike a balance between spinning fast enough to obtain a high-quality spectrum and slow enough that the organism is not under significant stress throughout the experiment.

5.4.6 MAS Techniques – Spinning Sidebands

Due to the relatively slow spinning speeds required for organism survival, spinning sidebands, which are typically shifted outside of the spectral region in fast spinning, can overlap with real peaks. Spinning sidebands are spurious signals that result from the incomplete averaging of anisotropic interactions such as dipolar couplings or chemical shift anisotropy.[77] MAS "pushes" spinning sidebands away from the isotropic peak by the frequency of spinning, and thus higher frequency magnets require increased MAS spinning rates, making it especially challenging at higher fields. For one-dimensional work, a combined water suppression and sideband suppression technique known as PURGE-TOSS.243 has been reported *in vivo*.[51] TOSS.243 employs a series of 5 π pulses separated by delays, which results in a FID free of rotational echoes, and thus containing only information on the isotropic resonance,[78] while the 234-step phase cycle helps reduce the effects of pulse imperfections.[79] Despite the requirement of a 234-step phase cycle, this technique was shown to provide excellent performance in living *H. azteca*, down to spinning speeds as low as 100 Hz.[51]

No effective approach has yet been reported[80] for slow spinning, *in vivo* sideband suppression for two-dimensional NMR. However, since sidebands are more likely to occur for intense, sharp peaks, and given the relatively broad and low intensity signals of metabolites *in vivo*, usually the major source of sidebands is simply water. Recent work has indicated that ^{13}C detection experiments are less susceptible to sidebands, even at slower spinning speeds because there is no signal (and thus sidebands) from the large water peak.[80] Figure 5.4 shows a comparison of a HETCOR and an HSQC spectrum of a single *H. azteca*, at spinning speeds of 500 and 2500 Hz. Clearly, sidebands pose little problem in the ^{13}C detected HETCOR, but create considerable overlap in the HSQC, especially at 500 Hz. Given that HETCOR experiments have been shown to provide similar, though slightly less,[70] metabolic information content to HSQC, this could be a viable approach for *in vivo* analysis of ^{13}C labelled organisms, especially for carbon optimized probes.

Another possibility for sideband suppression is variable speed magic angle spinning (VS-MAS).[81] While typical MAS experiments spin the sample

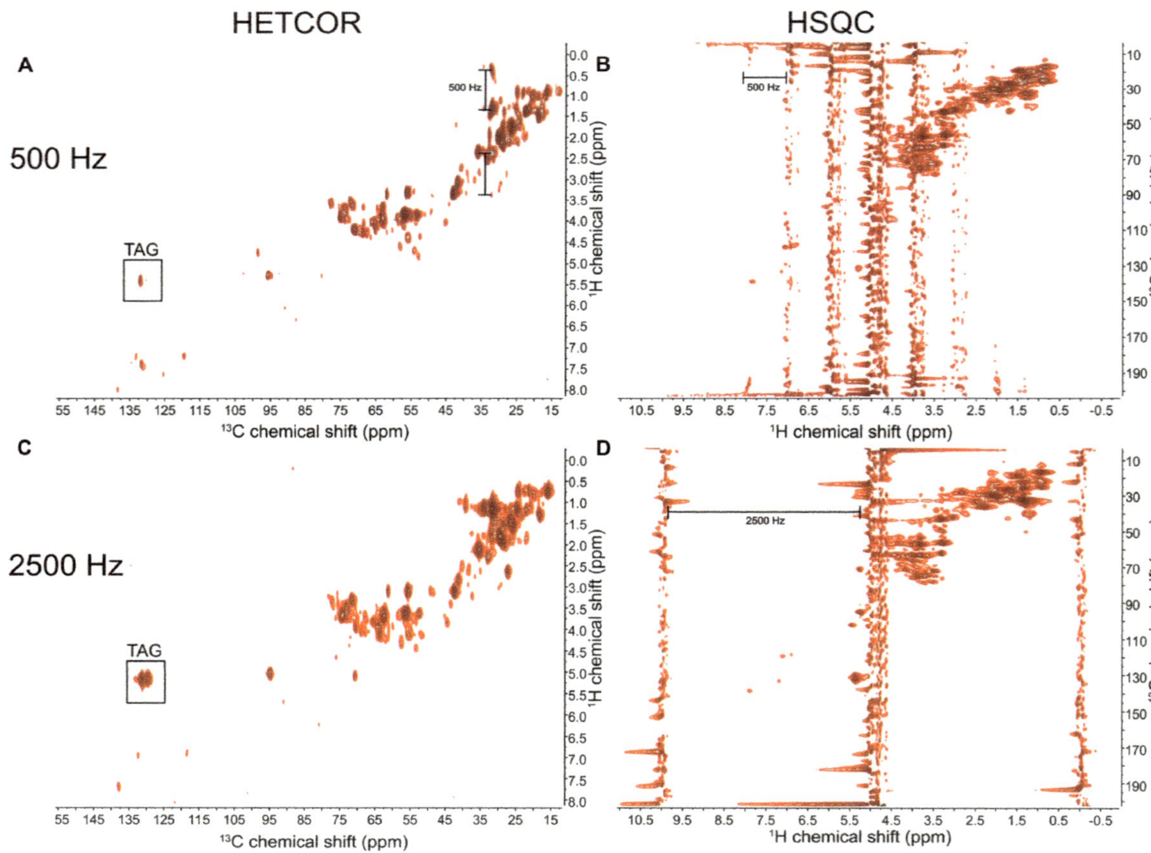

Figure 5.4 A comparison of sidebands in a single ^{13}C-enriched *H. azteca* under MAS. (A) A HETCOR at 500 Hz, (B) an HSQC at 500 Hz, (C) a HETCOR at 2500 Hz, and (D) an HSQC at 2500 Hz. Reproduced from ref. 80 with permission from American Chemical Society, Copyright 2022.

at a single spin rate, as the name implies, VS-MAS uses several different spinning speeds. Thus, each experiment is divided into blocks which are carried out at different spinning speeds and then combined to form the final dataset. By spinning at different speeds, sidebands are generated at different frequencies for each block, while the actual signal itself remains at the same frequency. The sidebands do not increase in intensity (in the same region) with additional scans and are thus attenuated. This technique does not require any additional RF pulses and can be applied to 2D NMR,[82] though it may be more challenging to set up. Although to the authors' knowledge this technique has not yet been applied *in vivo*, it has substantial potential in this area and has been shown to be effective for quantitative CMP NMR of dried *Arabidopsis* tissue.[83]

5.5 *In Vivo* NMR Techniques: Applications and Examples

Having considered the different types of *in vivo* NMR and the associated challenges, it is instructive to turn the discussion towards practical examples and techniques that are either commonly used or have potential to be used *in vivo*. While this is not meant to be an exhaustive review, it is meant to give a "top 10" style overview of the techniques available, along with their advantages and disadvantages. While the majority of techniques discussed here are solution state approaches, CMP NMR (and by extension HR-MAS) will also be considered.

5.5.1 One-dimensional NMR

5.5.1.1 ^{1}H NMR

Although 1D ^{1}H NMR is the most commonly applied form of NMR spectroscopy in the world,[34] a standard ^{1}H spectrum has significant limitations *in vivo*. In addition to challenges with water suppression, the broad lineshapes characteristic of *in vivo* samples greatly reduce the information that can be extracted from a ^{1}H spectrum, making peak assignments very challenging. Furthermore, ^{1}H *in vivo* spectra are typically dominated by large lipid peaks, which overlap and obscure important metabolites such as hydrophobic amino acids. The lipid resonances can be suppressed to some extent through the use of spectral editing techniques (see Section 5.4), based on diffusion or T_2 relaxation, which can allow metabolites to be identified.[37,84] The use of the IP-iSQC sequence is also a powerful approach, although typically the additional spectral dispersion of multidimensional techniques is preferred for the assignment of unknown peaks. However, once the peaks have been assigned from 2D data, the higher sensitivity of 1D ^{1}H NMR makes it amenable for metabolite monitoring over time and can be especially effective when combined with the enhanced lineshape resulting from IP-iSQC.

5.5.1.2 ^{13}C NMR

Compared to ^{1}H NMR, the spectral window of ^{13}C NMR is much wider, improving dispersion. While most metabolites range from 0 to 12 ppm in ^{1}H, for ^{13}C this range is more than an order of magnitude larger, with a shift range of ~250 ppm.[85] However, the difference in sensitivity should not be understated: the relative receptivity (at natural abundance) of ^{13}C is ~5700 times less than that of ^{1}H.[86] The sensitivity of ^{13}C acquisition can be improved by taking advantage of the nuclear Overhauser effect (NOE) enhancement from attached protons[87] or polarization transfer techniques such as distortionless enhancement by polarization transfer (DEPT), which can provide a 3–4-fold improvement in the SNR for proton coupled carbons.[88] Despite its relatively low sensitivity, the omnipresence of carbon in metabolites makes this technique interesting. However, the low biomass present in a typical *in vivo* experiment means that acquiring a ^{13}C spectrum almost invariably requires isotopic enrichment, making this process expensive and limited to lab-grown organisms, or the study of ^{13}C enriched tracers. As an example, *in vivo* ^{13}C NMR was used to track the metabolism of ^{13}C labelled propionate in an activated wastewater sludge.[89] Wastewater sludge is a complex mixture of aerobic and anaerobic bacteria which are used to degrade organic compounds in the secondary phase of wastewater treatment. The sludge was later extracted with chloroform and further ^{13}C experiments were carried out to help identify the exact PHAs formed.

5.5.1.3 ^{14}N/^{15}N NMR

Although nitrogen is less commonly found in metabolites compared to ^{1}H or ^{13}C, it is still present in a number of important metabolites. *In vivo*, nitrogen is found in macromolecules including proteins, nucleic acids, as well as neurotransmitters such as dopamine. In terms of environmental chemistry, nitrate and nitrite conversion (a major part of the nitrogen cycle) can be studied directly by nitrogen NMR,[90] but not by ^{1}H or ^{13}C. Of the two isotopes, ^{14}N is vastly more naturally abundant but the large quadrupole moment makes high resolution NMR much more challenging, especially given the additional broadening *in vivo*. Thus, although ^{15}N NMR is only ~0.37% abundant, it is spin $\frac{1}{2}$ and tends to be the more popular choice for high resolution nitrogen NMR.[86] ^{15}N NMR takes both the advantages and disadvantages of ^{13}C NMR to an extreme: while the spectral window is around ~900 ppm,[91] the receptivity at natural abundance is only 2% of that of ^{13}C. Due to the low abundance, ^{15}N isotopic enrichment provides very substantial signal improvements and nearly all *in vivo* work utilizes enriched organisms or tracers. Certain uniformly ^{15}N labelled organisms such as *D. magna* are commercially available and have been studied,[31] while protocols have been reported for producing ^{15}N enriched plants such as potato tubers and wheat.[92,93] Espen and colleagues used ^{15}N NMR to study the uptake and metabolism of $^{15}NO_3^-$ in living maize roots in the presence and absence of tungstate.[94] While in control experiments, the nitrate was bio-transformed into amino acids including glutamine and alanine, in the presence of tungstate this conversion did not occur. Although a simple example, this study demonstrates the potential of examining biotransformation *via* ^{15}N NMR and could easily be applied to environmentally relevant pollutants such as nitrogen containing pesticides.

5.5.1.4 ^{31}P NMR

Phosphorus is a very important nucleus for the study of energy metabolism. The changing concentrations of energy intermediates such as adenosine triphosphate (ATP), adenosine diphosphate (ADP) as well as inorganic phosphate and polyphosphates can be used to probe an organism's energy balance and are key indicators of metabolic stress.[95] Further, ^{31}P is a high sensitivity nucleus that is 100% naturally abundant, giving a relative receptivity that is ~6% of ^{1}H, which makes ^{31}P an excellent choice for the analysis of unlabelled organisms. In fact, the first reported study examining the toxic effects of organic chemicals within intact aquatic organisms employed ^{31}P NMR, to study the effect of pentachlorophenol on living red abalone (*Haliotis rufescens*).[96] In this study, a large diameter imaging coil in a horizontal magnet was equipped with a flow system and the organisms were exposed to pentachlorophenol (PCP). Pentachlorophenol is a pesticide that is known to uncouple mitochondrial oxidative phosphorylation, making it an excellent choice to study. A ^{31}P NMR timelapse can be found in Figure 5.5.

In this experiment, the abalone were exposed for five hours to 1.2 ppm of PCP, following which the inflowing water was replaced with fresh water and the abalone were then allowed to recover for 13 hours. The most significant changes during exposure were in phosphoarginine, which decreased by 60–80% of pre-exposure levels, and inorganic phosphate which increased to 300–800% of pre-exposure levels. This study is another excellent example of NMR metabolomics: not only does it allow the tracking of the relative change in multiple metabolites in living organisms, but it also permits an examination of metabolic recovery, which is not possible on the same organisms using other, destructive techniques. Here, inorganic phosphate was found to recover to pre-exposure levels after 13 hours, but phosphoarginine did not fully recover in the timeframe of the experiment. Further studies using ^{31}P have investigated the effects of sublethal copper toxicity in abalone,[97] the metabolic response of medaka embryos to hypoxia,[61] exposure to dinoseb,[59] and metabolism of glyphosate by freshwater shrimp.[31]

5.5.1.5 Other Nuclei

While the previously discussed isotopes make up the bulk of *in vivo* studies, NMR can be applied to a much wider range of nuclei, which may have potential for *in vivo* application in the future. ^{2}H, for example, can be added to any protonated chemical without changing its chemistry, making it an

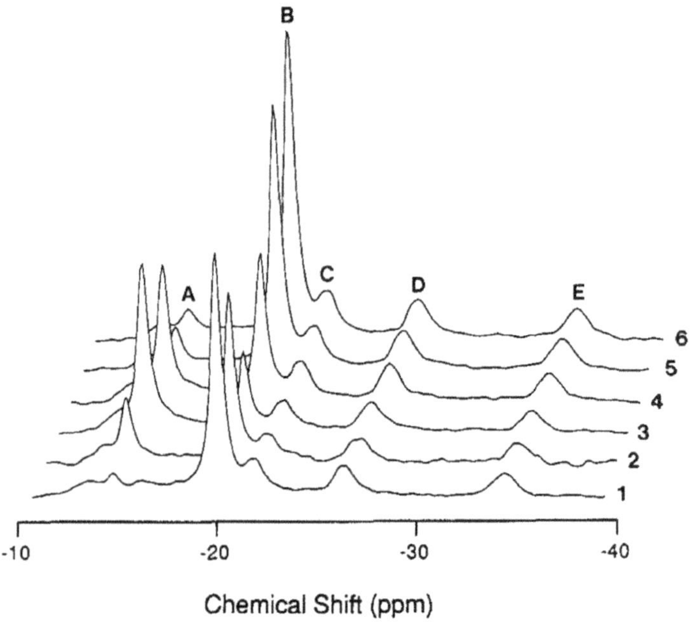

Figure 5.5 A ^{31}P NMR time course of the exposure and recovery of red abalone to pentachlorophenol. Here the letter labels indicate the following: (A) inorganic phosphate, (B) phosphoarginine, (C) overlapping γ-phosphate of ATP and β-phosphate of ADP, (D) overlapping α-phosphate of ATP and ADP as well as the phosphate of nicotinamide adenine dinucleotide (NAD) and (E) β-phosphate of ATP. Reproduced from ref. 96 with permission from John Wiley & Sons, Copyright 1991.

excellent tracer *in vivo*. Further, ^{2}H retains the same chemical shift information as ^{1}H NMR, facilitating the use of databases for peak assignments.[41] Despite relatively low sensitivity and quadrupolar broadening, one study has even shown the uniform ^{2}H/^{13}C labelling of both *H. azteca* and *D. magna*, and demonstrated ^{2}H detected 2D experiments.[41] It was also noted that ^{2}H detection reduced problems associated with the large water peak, by avoiding signals from ^{1}H. In another example, deuterated phenylalanine metabolism in freshwater shrimp was monitored.[31]

^{19}F is another example of a nucleus with great potential *in vivo*. First, ^{19}F has 100% natural abundance and a receptivity that is ~80% of that of ^{1}H, making it one of the most sensitive nuclei that can be studied by NMR.[86] Second, fluorinated compounds are an increasingly important class of pollutants.[98] The non-stick nature of polyfluorinated compounds is exploited in a variety of products ranging from non-stick cooking wares[99] to fire retardants.[100] However, these compounds have been reported to be carcinogenic in rats[98] and have been found everywhere from surface waters[101] to even human blood.[102] Further, fluorine is common in pesticides and is present in ~25% of pharmaceuticals,[103] making this an environmentally important molecular probe that typically does not appear in natural metabolic products. While this may seem like a disadvantage, it means that there is very little expected background from the organism, in stark contrast to ^{1}H, for example, where lipid peaks mask many metabolites. Although to the authors' knowledge ^{19}F based *in vivo* NMR toxicity studies have not yet been reported, ^{19}F NMR has been used to study labelled proteins in living yeast cells,[104] and a CMP ^{19}F spectrum of *H. azteca* in the presence of perfluorooctanesulfonate has also been published.[31] Overall, ^{19}F NMR has great potential to track the metabolism and accumulation of fluorinated compounds *in vivo*.

Metal NMR is another potential avenue for *in vivo* studies. Heavy metals such as lead and mercury have substantial environmental impacts and even very small, sublethal concentrations have been associated with increased risks of human diseases.[105] Lead is particularly toxic to the nervous system, being associated with negative cognitive effects at even very low concentrations,[105] and is a serious environmental contaminant due to its use as a fuel additive as recently as the beginning of the 21st century.[106] Mercury is among the most toxic elements in the world and can bioaccumulate in aquatic organisms.[107] Both lead and mercury have NMR active, spin $\frac{1}{2}$ isotopes. ^{207}Pb, though insensitive, has an enormous chemical shift range of ~17 000 ppm[108] and ^{199}Hg has a chemical shift range of ~4300 ppm.[109] Historically, the NMR of heavy metals was hampered by the extremely toxic chemical shift standards, but with modern external standards and electronic referencing this is less of a concern.[46] *In vivo* studies involving these metals have not yet been reported, but due to their importance as environmental pollutants, this is definitely an area that could be examined. Finally, these are not the only metals that are NMR active, and truly a wide range of metals (*e.g.*, cadmium, sodium, lithium) and "exotic" nuclei could be used to study organisms *in vivo*, depending on the research goal.

5.5.2 Multidimensional Experiments

The major benefit of multidimensional experiments is the greatly improved spectral dispersion, which helps overcome overlap common in 1D. Being able to discriminate peaks that would otherwise overlap affords much more effective compound identification. However, the improved dispersion comes at a cost: multidimensional experiments require much longer experiment times and are generally much less sensitive than 1D detection.

5.5.2.1 Correlation Spectroscopy (COSY)

COSY is an example of a 2D homonuclear technique that provides information on neighboring protons in a chemical structure. Being a ^{1}H–^{1}H correlation, this is among the most sensitive and simple 2D techniques available and has the potential to be applied to unlabelled organisms. The coupling information gained from COSY can map out coupled spin systems in a molecule, helping precisely identify metabolites. Unfortunately, COSY is typically quite uninformative *in vivo*, due to broad lineshapes, as well as challenges with water suppression.[46] SPR-W5-WATERGATE is normally added before the COSY experiment to help reduce the water peak *in vivo* but often the only COSY cross peaks that can be identified *in vivo* result from lipids. However, this technique was shown to be a powerful approach for *ex vivo* analysis and is useful as a complement to other proton detection experiments.[46]

5.5.2.2 *Total Correlation Spectroscopy (TOCSY)*

TOCSY is a technique that allows magnetization transfer through any uninterrupted *J*-coupled homonuclear spin system. ^{1}H–^{1}H TOCSY is most commonly used, but ^{13}C–^{13}C TOCSY is also a possibility though it is much less sensitive. While COSY provides information on immediate neighbours, TOCSY provides additional information about protons that are still coupled, but further away in the spin system. *In vivo*, adiabatic TOCSY is preferred as it helps mitigate the effects of fast relaxation caused by magnetic susceptibility distortions,[28] and it is often possible to see signals beyond the lipid peaks. One disadvantage of TOCSY is that the number of correlations is highly dependent on the mixing times employed.[110] While this may be trivial to optimize for a single compound, an average must be taken for mixtures. This makes library matching more challenging as some expected correlations may be missing (or *vice versa*). However, TOCSY information is quite complementary to COSY and can thus be used to help cross-assign peaks between these spectra, especially *ex vivo*, albeit some correlations can also be seen *in vivo*.[46]

5.5.2.3 *Heteronuclear Single Quantum Coherence (HSQC) and Heteronuclear Multiple Quantum Coherence (HMQC)*

The HSQC experiment is a single bond correlation and is typically used to identify protons directly bonded to ^{13}C. However, HSQC is also possible with any other heteronucleus that is bonded to a proton: ^{1}H–^{15}N HSQC is common in protein research,[111] and ^{1}H–^{31}P HSQC has been applied to study soil extracts.[112] Figure 5.6 shows the greatly improved dispersion afforded by a 2D HSQC, compared to a 1D ^{1}H spectrum. Clearly, peaks that are completely overlapped in 1D become resolvable in 2D, which is the main power of 2D techniques. Further, HSQC is highly reproducible and the only operator set experimental parameter is the ^{1}H–^{13}C one-bond *J*-coupling value, which is nearly always set to an "average" of 140–145 Hz.[46] This greatly increases confidence in library matches, and ^{1}H–^{13}C HSQC spectra for many common metabolites can be found in databases such as the Human Metabolome Database (HMDB)[113] (though environmentally specific databases are greatly needed), which is very valuable when assigning unknown peaks. Overall, ^{1}H–^{13}C HSQC has become a "workhorse" in metabolite assignment *in vivo*.

It is also important to note that there are different types of HSQC experiments, some of which are more amenable to *in vivo* analysis. Since HSQC

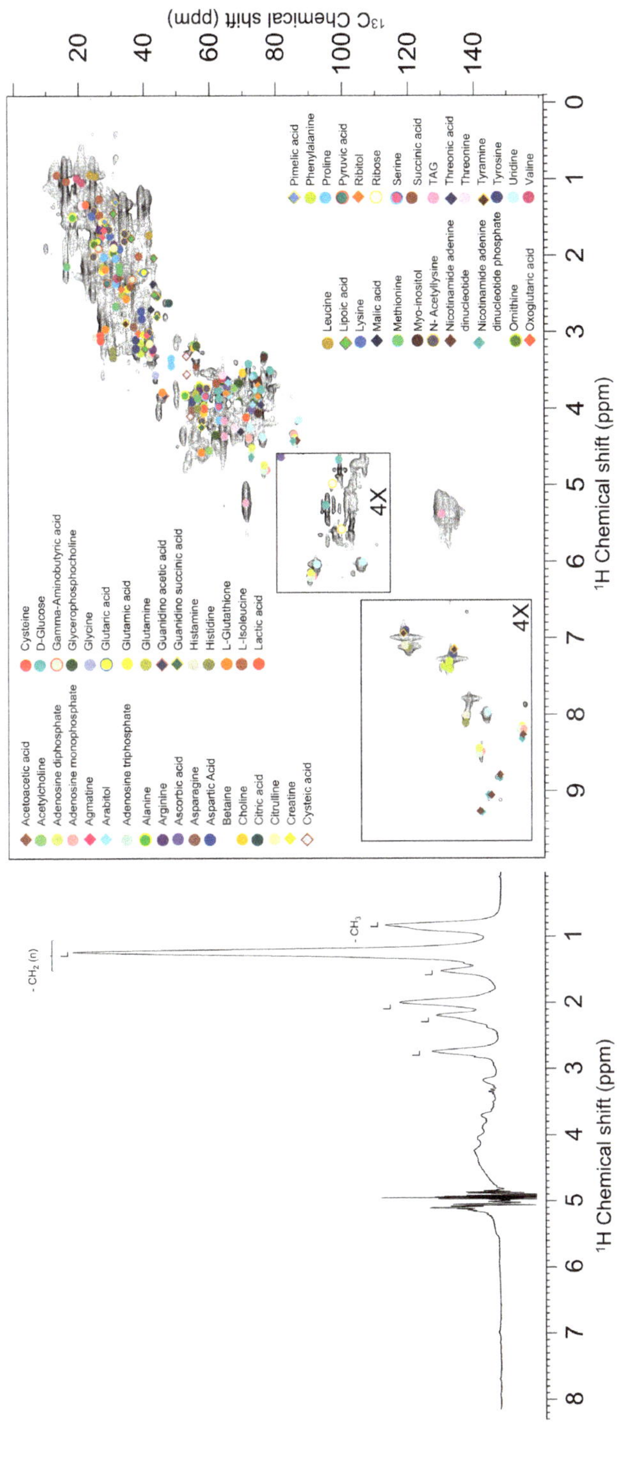

Figure 5.6 A comparison of a ^{1}H 1D NMR spectrum (left) and a 2D ^{1}H–^{13}C HSQC (right) of ^{13}C enriched *D. magna in vivo*. Some metabolite assignments are provided in the HSQC. Reproduced from ref. 46 with permission from the Royal Society of Chemistry.

is a ^{1}H detection experiment, it has the potential to show artefacts from the strong water peak. To help mitigate this effect, modern gradient selected HSQC sequences are preferred *in vivo*, since this leads to cleaner spectra, often without the need for additional water suppression.[114] Although sensitivity enhanced (also known as preservation of equivalent pathways) HSQC sequences offer up to a factor of a square root of 2 improvement in the SNR, they require an additional three pulses and a delay of half of the carbon proton *J*-coupling.[115] Due to fast relaxation *in vivo*, the additional length added to the pulse sequence typically nullifies the SNR advantage and thus "standard" HSQC sequences are preferred.[34] For uniformly ^{13}C labelled organisms, ^{13}C–^{13}C couplings result in splitting in the HSQC spectrum, which can decrease the signal-to-noise ratio. While constant time HSQC has the potential to remove these coupling interactions,[116] it is equivalent to acquiring each 2D increment at the length of the longest increment. This is highly unfavorable due to relaxation, as the later increments of an *in vivo* spectrum typically contain little signal. As such, the general recommendation for optimal HSQC is to use a standard, gradient selected experiment.

Of additional note, the ^{1}H–^{13}C coupling information provided by HSQC can also be obtained through heteronuclear multiple quantum coherence (HMQC).[117] Although they operate on fundamentally different principles, the practical difference between these techniques is that in HMQC ^{1}H magnetization evolves during t_1, while this is not the case in HSQC. This can lead to broadening in the ^{13}C dimension, and thus typically HSQC is preferred for improved resolution.[117] However, HMQC requires fewer inversion pulses and has been shown to be more "rugged" with respect to pulse imperfections and B_1 inhomogeneity.[66] Both HMQC and HSQC spectra have been reported *in vivo* and in general it is recommended that HSQC is employed on standard probes, but care must be taken that the sample is well shimmed and all pulses are carefully calibrated. However, for microcoils (especially planar coils) which show poor B_1 homogeneity, HMQC is the preferred approach.[66]

5.5.2.4 Heteronuclear Correlation (HETCOR)

HETCOR can be thought of as the inverse of an HSQC. In HSQC, magnetization evolves on the heteronucleus during t_1 and is then transferred to ^{1}H for detection, whereas in HETCOR the opposite is true,[118] which results in a number of key differences. First, since the gyromagnetic ratio of ^{13}C is substantially lower than that of ^{1}H, generally speaking ^{13}C–^{1}H HETCOR is less sensitive than ^{1}H–^{13}C HSQC. However, detection on ^{13}C provides greatly improved resolution in the direct dimension and eliminates the need for solvent suppression. A recent publication[70] comparing HETCOR to HSQC *in vivo* noted that, due to enhanced sensitivity, HSQC typically allows identification of slightly more metabolites. However, peaks near the water resonance which would typically be obscured in HSQC (such as phosphoenolpyruvate, a metabolite involved in glycolysis and gluconeogenesis[70]) can be resolved in

HETCOR. Furthermore, if HSQC and HETCOR are used in combination, it can act to confirm metabolite assignments.

5.5.2.5 *Higher Dimensional Techniques*

While the techniques described so far have been 2D, higher dimensional experiments have the potential to further improve dispersion. However, the challenges associated with 2D experiments are further amplified as the number of dimensions increases: not only do these experiments take even longer, fast relaxation *in vivo* can lead to very little signal in later increments. While 3D experiments are quite common in protein research,[119,120] the first 3D spectrum *in vivo*, a $^{1}H–^{13}C$ HSQC-TOCSY of ^{13}C labelled living *D. magna*, was reported in 2020[46] and can be found in Figure 5.7.

The additional dispersion provided in the 3D experiment can be visualized through the 3D cube, but for metabolite assignments the 2D cross sections are very useful. By "slicing" along the F2–F3 plane, the result is essentially a greatly simplified HSQC spectrum that is "filtered" by the selected chemical shift in F1. However, it is important to note that relaxation limits the number of peaks that can be identified *in vivo*. As such, the images shown in Figure 5.7(D)–(F) are from *ex vivo* samples, but demonstrate the principle of 3D NMR well. Although experiments past 3 dimensions are possible (for example, a 4D experiment was used to study soil humic substances[121]), such experiments are yet to be applied *in vivo* and are likely to struggle significantly with relaxation and extremely long experiment times.

5.5.3 Selective Experiments

As has been thoroughly discussed, peak overlap is a key challenge *in vivo*. It is thus sometimes useful to reduce spectral complexity by selecting a series or class of metabolites. A 2019 study[50] reported an HSQC experiment adapted from protein NMR[122] that is selective for amino acids. Known as $(H)C_bC_a(COC_a)H_a$, this technique is a spectral filter that selects for compounds that have the $H_βC_β–H_αC_α–C{=}O$ moiety in which the α, β and carbonyl carbons must be ^{13}C. Due to the number of magnetization transfer steps, this technique has low sensitivity and is only viable for ^{13}C enriched organisms. However, the authors noted that it works remarkably well *in vivo*, showing well-resolved resonances from the 20 standard amino acids, along with lactic acid and glutathione, which contain similar $H_βC_β–H_αC_α–C{=}O$ moieties, and are important biomarkers of oxidative stress.[50]

Although the amino acid selective sequence is powerful *in vivo*, it requires ^{13}C labelling and long experiment times, and can only be used for amino acids. Amino acids are very important stress indicators *in vivo* but it would be ideal to be able to change the selection target to suit the research goal. Selective TOCSY is a simple approach in which a shaped pulse is used to selectively excite the resonance of interest prior to a TOCSY mixing time.[22]

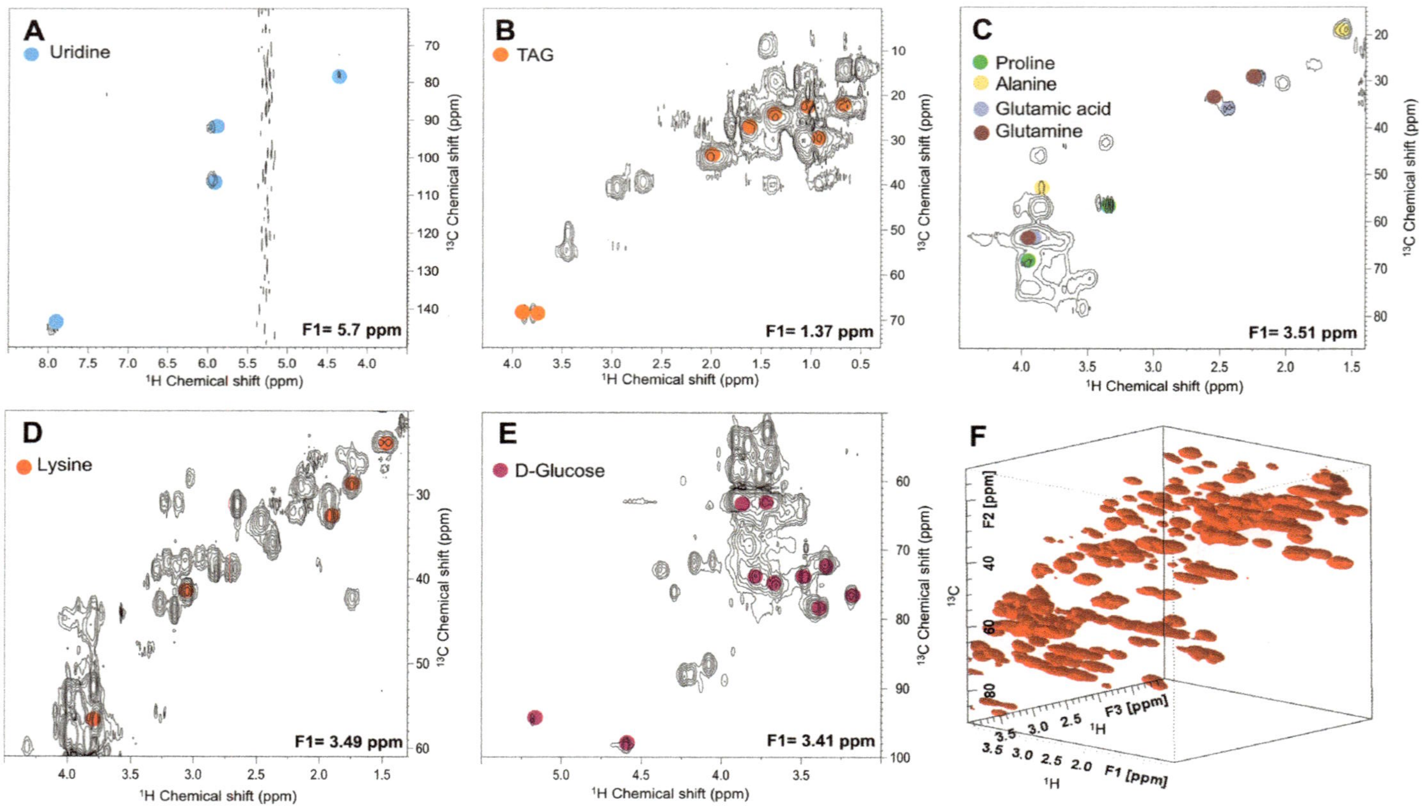

Figure 5.7 (A–C) *In vivo* 2D cross sections along F2–F3 of a 3D ^{1}H–^{13}C HSQC-TOCSY experiment on *D. magna*, with some metabolites labelled. (D–E) *Ex vivo* 2D cross sections, showing additional metabolites and (F) the *ex vivo* 3D cube. Reproduced from ref. 46 with permission from the Royal Society of Chemistry.

Thus, this shows only peaks coupled to the initially excited resonance. The resulting spectrum can be further simplified by the recently reported combination with a pure shift experiment, which collapses multiplet structures into singlets, improving resolution.[24] However, when significant overlap is present in the initial spectrum (as is often the case *in vivo*), the selective excitation pulse can undesirably select multiple components.[21] A recent publication has described a pulse sequence element known as gradient-enhanced multiplet-selective targeted-observation NMR experiment (GEMSTONE) which allows a single signal to be extracted from overlapping resonances, in a single scan.[123] Although this technique has not been applied *in vivo*, the ability to select individual signals from overlapped spectral regions is definitely of interest.

Designed refocused excitation and optional mixing for targets *in vivo* and mixture elucidation (DREAMTIME) NMR is one of the most exciting recent developments in selective NMR.[23] This technique essentially allows the spectroscopist to see only the compounds desired, which is excellent for tracking specific metabolic processes. This technique works by filtering a specific *J*-coupled spin pair in a target molecule, resulting in only 2 peaks per compound selected, and an optional TOCSY mixing block can be added to help with initial identification. The DREAMTIME filter is very effective and will only select a spin pair if both signals resonate at the selected frequencies, the signals are *J*-coupled (*i.e.*, neighbours in the chemical structure) and the *J*-coupling constant is similar to that of the targeted molecule. Figure 5.8 shows an example of DREAMTIME applied to *D. magna* undergoing oxidative stress. Figure 5.8 demonstrates the potential of DREAMTIME for both multiple selection of compounds *in vivo*, which would otherwise not be clearly identifiable due to overlap, and also the ability to track a physiological change.

Furthermore, DREAMTIME is a 1D ^{1}H experiment and retains up to 50% of the signal from a standard ^{1}H spectrum. However, when concatenated with a focusing approach, it can actually provide improved SNR compared to a standard ^{1}H spectrum. This is similar in principle to sensitive, homogeneous, and resolved peaks in real time (SHARPER),[124] but has the additional advantage of being able to focus multiple compounds that are off resonance. Extremely impressive improvements between 700% and 1200% over standard ^{1}H NMR were reported in human urine and blood, and the authors noted that the focusing effect provides even larger advantages for samples with broad lines.[23] The main challenge of DREAMTIME is that it is relatively difficult to set up, though an extensive guide is provided by the authors.[23] While single compound selection is quite simple and single focusing is only slightly more challenging, multiple compound focusing can be quite difficult, requiring significant optimization of experimental parameters. Also, since the result of the DREAMTIME filter is in-phase magnetization, it can be appended onto any standard experiment such as COSY, HSQC or even MRI.[23]

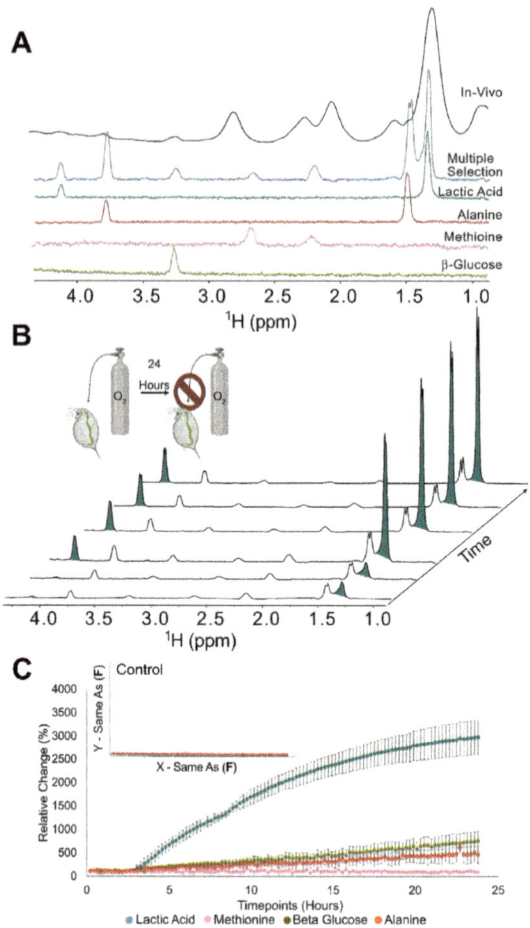

Figure 5.8 An illustration of multiple selection using DREAMTIME NMR to follow the process of anoxic stress in living *D. magna*. (A) The selection of 4 compounds, both individually and as a group, along with a reference ^{1}H spectrum, (B) a time course evolution of the multiply selected spectrum and (C) a graph of the relative intensity changes of the peaks. Reproduced from ref. 23 with permission from John Wiley & Sons, Copyright 2022.

5.5.4 CMP NMR – Spectral Editing

While the major benefit of CMP NMR is that it can be used to study all phases in a sample, it is often useful to separate out the information arising from different phases. Here, the general methods of spectral editing will be discussed, but for additional details readers are directed to the original publications.[31,40,41]

Diffusion editing (DE) is a common method for distinguishing truly dissolved components (*i.e.*, that have unrestricted diffusion) from swollen gels with restricted diffusion.[125] While this can also be done using relaxation filters, it has been reported that DE results in more complete spectral editing.[31] A DE experiment encodes the spatial position of a molecule through the use of a gradient pulse at the beginning of an experiment and decodes the position at the end, with an opposite gradient pulse.[31] Due to the dephasing effect of the gradients, the signal from molecules that have diffused away from the starting point will be attenuated. By correctly setting the diffusion delay, truly dissolved components can be eliminated from the spectrum leaving only the components with restricted diffusion such as bound molecules, large macromolecules, biopolymers and lipid vesicles. By subtracting a DE experiment from a control ^{1}H spectrum, the truly dissolved fraction can be isolated. This is known as inverse diffusion editing (IDE) and is particularly useful as it reduces the intensity of lipid peaks, bringing out the metabolite signals buried underneath.[40]

Moving on to more solid components, there is a phase known as semi-solid (or rigid), which, although restricted in diffusion, relaxes too quickly to be detected in a DE experiment. To recover this signal, a technique known as relaxation recovery arising from diffusion editing (RADE) can be applied.[126] Here, the DE experiment is acquired first with zero diffusion delay, then with the correct diffusion delay, but with zero gradient strength. By subtracting these spectra, the signals lost to relaxation during the diffusion delay are recovered. RADE spectra are mainly dominated by more rigid lipids, but there are also often contributions from proteins.[126] It is also important to note that this category generally represents the limit of what can be detected by standard "low power" ^{1}H NMR.

While the previous techniques are typically ^{1}H detected, they could also be observed on ^{13}C, especially if a direct observe probe is utilized.[80] Due to the very strong dipolar couplings in ^{1}H spectra, solids are best detected using ^{13}C. This is where CMP probes distinguish themselves from HR-MAS probes: CMP is designed to handle the high pulse and decoupling power required for cross polarization (CP-MAS) experiments.[83] CP-MAS is only efficient in components that do not exhibit tumbling or motion, and thus also acts to automatically filter out soluble metabolites.[31] To distinguish between rigid, true crystalline solids and dynamic/semi-solids, a ^{1}H T_2 filter can be applied prior to the CP experiment.[127] Due to the extremely fast ^{1}H relaxation in true solids, this filter removes the signal from rigid solids. The rigid solid signal can then be found by subtracting the T_2 filtered CP from a standard CP experiment. By combining all of these experiments, the individual phases can be distinguished in CMP, and this has been reported in the literature for living *H. azteca*.[40]

5.5.5 Studying Interactions

One of the major benefits of NMR over other analytical techniques is the ability to study non-covalent interactions. By using techniques such as saturation transfer difference (STD), the exact "epitope map" of a compound[128] bound to a macromolecule can be constructed, while reverse heteronuclear saturation transfer difference (RH-STD) can determine which macromolecules a compound binds to.[129] Although STD has been used to study the binding of pesticides to soil,[43] this technique has yet to be applied *in vivo*. This is likely due to the relatively low sensitivity of this technique, along with the fact that typically contaminants are present at very low concentrations in the environment. However, the information that could be gained from this kind of study is very powerful and warrants future investigation. Simply discovering that a pollutant binds a specific enzyme can help explain the biochemical nature of its toxicity, and STD can help determine which specific parts of a molecule are responsible for binding.

5.6 Conclusion

NMR has long been recognized as a tool for chemical analysis of small molecules. However, as the sensitivity and wealth of available techniques have expanded, NMR has evolved into much more than that, becoming a powerful tool for the study of macromolecules and complex mixtures, including living organisms. This chapter has discussed both the benefits and challenges of *in vivo* NMR, while providing examples from the literature and best practices for *in vivo* applications. By studying living organisms, real-time metabolic information can be gathered and used to understand sublethal toxicity, precise toxicity pathways, biotransformation and bioaccumulation. *In vivo* NMR can extract a treasure trove of information and help answer complex, environmentally relevant questions such as "why is this chemical toxic?" or "is the metabolic effect of a pollutant permanent, or transitory?". Despite the power of this technique, it remains underutilized in the study of environmental toxicity, likely due to the practical experimental challenges. The authors hope that the information presented in this chapter will help make NMR of living organisms more accessible to the field of environmental chemistry. In summary, *in vivo* NMR is a powerful tool that gives unparalleled insights into environmental stress and in exploring toxicity in the future.

Acknowledgements

A. J. S would like to thank the Natural Sciences and Engineering Research Council of Canada (NSERC) [Alliance (ALLRP 549399), Alliance (ALLRP 555452) and Discovery Programs (RGPIN-2019-04165)], the Canada Foundation for Innovation (CFI), the Ontario Ministry of Research and Innovation (MRI), the Krembil Foundation for providing funding, and the Government of Ontario for an Early Researcher Award.

References

1. F. Bloch, *Phys. Rev.*, 1946, **70**, 460.
2. P. M. Bloch, W. W. Hanse and M. Packard, *Phys. Rev.*, 1946, **70**, 474.
3. R. W. Silverstein and G. Bassler, *J. Chem. Educ.*, 1962, **39**, 546.
4. E. G. Alves Filho, L. Sartori, L. M. A. Silva, B. F. Silva, P. S. Fadini, R. Soong, A. Simpson and A. G. Ferreira, *Magn. Reson. Chem.*, 2015, **53**, 704.
5. W. H. Otto, D. J. Britten and C. K. Larive, *J. Colloid Interface Sci.*, 2003, **261**, 508.
6. M. Bastawrous, A. Jenne, M. Tabatabaei Anaraki and A. Simpson, *Metabolites*, 2018, **8**, 35.
7. M. Tabatabaei Anaraki, R. Dutta Majumdar, N. Wagner, R. Soong, V. Kovacevic, E. J. Reiner, S. P. Bhavsar, X. Ortiz Almirall, D. Lane, M. J. Simpson, H. Heumann, S. Schmidt and A. J. Simpson, *Anal. Chem.*, 2018, **90**, 7912.
8. Z. Wang, G. W. Walker, D. C. G. Muir and K. Nagatani-Yoshida, *Environ. Sci. Technol.*, 2020, **54**, 2575.
9. R. Naidu, B. Biswas, I. R. Willett, J. Cribb, B. K. Singh, C. P. Nathanail, F. Coulon, K. T. Semple, K. C. Jones, A. Barclay and R. J. Aitken, *Environ. Int.*, 2021, **156**, 106616.
10. Z. Wang, S. Adu-Kumi, M. L. Diamond, R. Guardans, T. Harner, A. Harte, N. Kajiwara, J. Klánová, J. Liu, E. G. Moreira, D. C. G. Muir, N. Suzuki, V. Pinas, T. Seppälä, R. Weber and B. Yuan, *Environ. Sci. Technol.*, 2022, **56**, 2936.
11. K. Watson, G. Shaw, F. D. L. Leusch and N. L. Knight, *Water Res.*, 2012, **46**, 6069.
12. T. Paszko and P. Muszyński, *Int. Agrophys.*, 2017, **31**, 401.
13. V. Romanucci, A. Siciliano, M. Guida, G. Libralato, L. Saviano, G. Luongo, L. Previtera, G. di Fabio and A. Zarrelli, *Sci. Total Environ.*, 2020, **712**, 135625.
14. M. C. Celander, *Aquat. Toxicol.*, 2011, **105**, 72.
15. G. Persoone, R. Baudo, M. Cotman, C. Blaise, K. C. Thompson, M. Moreira-Santos, B. Vollat, A. Törökne and T. Han, *Knowl. Manag. Aquat. Ecosyst.*, 2009, **393**, 1.
16. D. Krewski, D. Acosta, M. Andersen, H. Anderson, J. C. Bailar, K. Boekelheide, R. Brent, G. Charnley, V. G. Cheung, S. Green, K. T. Kelsey, N. I. Kerkvliet, A. A. Li, L. McCray, O. Meyer, R. D. Patterson, W. Pennie, R. A. Scala, G. M. Solomon, M. Stephens, J. Yager, L. Zeise and Staff of Committee on Toxicity Test, *J. Toxicol. Environ. Health, Part B*, 2010, **13**, 51.
17. J. G. Bundy, M. P. Davey and M. R. Viant, *Metabolomics*, 2009, **5**, 3.
18. C. Lu, Y. Wang, Z. Sheng, G. Liu, Z. Fu, J. Zhao, J. Zhao, X. Yan, B. Zhu and S. Peng, *Toxicol. Appl. Pharmacol.*, 2010, **248**, 178.
19. A. H. Emwas, R. Roy, R. T. McKay, L. Tenori, E. Saccenti, G. A. N. Gowda, D. Raftery, F. Alahmari, L. Jaremko, M. Jaremko and D. S. Wishart, *Metabolites*, 2019, **9**, 1.

20. N. V. Reo, *Drug Chem. Toxicol.*, 2002, **25**, 375.
21. P. Sandusky and D. Raftery, *Anal. Chem.*, 2005, **77**, 2455.
22. N. MacKinnon, P. T. While and J. G. Korvink, *J. Magn. Reson.*, 2016, **272**, 147.
23. A. Jenne, W. Bermel, C. A. Michal, O. Gruschke, R. Soong, R. Ghosh Biswas, M. Bastawrous and A. J. Simpson, *Angew. Chem., Int. Ed.*, 2022, **61**, e202110044 (*Angew. Chem.*, 2022, **134**, e202110044).
24. G. D. Poggetto, L. Castañar, G. A. Morris and M. Nilsson, *RSC Adv.*, 2016, **6**, 100063.
25. D. Lane, W. Bermel, P. Ning, T. Y. Jeong, R. Martin, R. Soong, B. Wu, M. Tabatabaei-Anaraki, H. Heumann, M. Gundy, H. Boenisch, A. Adamo, G. Arhonditsis and A. J. Simpson, *Anal. Chem.*, 2020, **92**, 9856.
26. M. Tabatabaei Anaraki, M. J. Simpson and A. J. Simpson, *Magn. Reson. Chem.*, 2018, **56**, 1117.
27. R. Soong, Y. Liaghati Mobarhan, M. Tabatabaei, M. Bastawrous, R. G. Biswas, M. Simpson and A. Simpson, *Magn. Reson. Chem.*, 2020, **58**, 411.
28. I. Fugariu, W. Bermel, D. Lane, R. Soong and A. J. Simpson, *Angew. Chem., Int. Ed.*, 2017, **56**, 6324.
29. D. Bon, V. Gilard, S. Massou, G. Pérès, M. Malet-Martino, R. Martino and F. Desmoulin, *Biol. Fertil. Soils*, 2006, **43**, 191.
30. R. S. Tjeerdema, R. J. Kauten and D. G. Crosby, *Aquat. Toxicol.*, 1991, **21**, 279.
31. Y. Liaghati Mobarhan, R. Soong, D. Lane and A. J. Simpson, *Magn. Reson. Chem.*, 2020, **58**, 427.
32. D. Courtier-Murias, H. Farooq, H. Masoom, A. Botana, R. Soong, J. G. Longstaffe, M. J. Simpson, W. E. Maas, M. Fey, B. Andrew, J. Struppe, H. Hutchins, S. Krishnamurthy, R. Kumar, M. Monette, H. J. Stronks, A. Hume and A. J. Simpson, *J. Magn. Reson.*, 2012, **217**, 61.
33. H. Farooq, D. Courtier-Murias, R. Soong, W. Bermel, W. Kingery and A. Simpson, *Curr. Org. Chem.*, 2013, **17**, 3013.
34. A. J. Simpson, D. J. McNally and M. J. Simpson, *Prog. Nucl. Magn. Reson. Spectrosc.*, 2011, **58**, 97.
35. V. Sarou-Kanian, N. Joudiou, F. Louat, M. Yon, F. Szeremeta, S. Même, D. Massiot, M. Decoville, F. Fayon and J. C. Beloeil, *Sci. Rep.*, 2015, **5**, 1.
36. V. Righi, A. A. Soukas, G. Ruvkun and A. A. Tzika, in *International Society for Magnetic Resonance in Medicine Proceedings*, International Society for Magnetic Resonance in Medicine, Concord, 2010.
37. Q. Hassan, R. Dutta Majumdar, B. Wu, D. Lane, M. Tabatabaei-Anraki, R. Soong, M. J. Simpson and A. J. Simpson, *Magn. Reson. Chem.*, 2019, **57**, 69.
38. B. H. Hansen, D. Altin, A. Booth, S. H. Vang, M. Frenzel, K. R. Sørheim, O. G. Brakstad and T. R. Størseth, *Aquat. Toxicol.*, 2010, **99**, 212.
39. A. Bunescu, J. Garric, B. Vollat, E. Canet-Soulas, D. Graveron-Demilly and F. Fauvelle, *Mol. BioSyst.*, 2009, **6**, 121.
40. Y. L. Mobarhan, B. Fortier-McGill, R. Soong, W. E. Maas, M. Fey, M. Monette, H. J. Stronks, S. Schmidt, H. Heumann, W. Norwood and A. J. Simpson, *Chem. Sci.*, 2016, **7**, 4856.

41. Y. Liaghati Mobarhan, R. Soong, W. Bermel, M. J. Simpson, J. Struppe, H. Heumann, S. Schmidt, H. Boenisch, D. Lane and A. J. Simpson, *ACS Omega*, 2018, **3**, 17023.
42. A. J. Simpson, D. Courtier-Murias, J. G. Longstaffe, H. Masoom, R. Soong, L. Lam, A. Sutrisno, H. Farooq, M. J. Simpson, W. E. Maas, M. Fey, B. Andrew, J. Struppe, H. Hutchins, S. Krishnamurthy, R. Kumar, M. Monette and H. J. Stronks, *eMagRes*, 2013, **2**, 399.
43. H. Masoom, D. Courtier-Murias, R. Soong, W. E. Maas, M. Fey, R. Kumar, M. Monette, H. J. Stronks, M. J. Simpson and A. J. Simpson, *Environ. Sci. Technol.*, 2015, **49**, 13983.
44. F. Gagné, C. Blaise and J. Pellerin, *Environ. Res.*, 2005, **98**, 89.
45. P. Kinchesh, E. W. Randall and K. Zick, *Magn. Reson. Imaging*, 1994, **12**, 305.
46. M. T. Anaraki, D. H. Lysak, R. Soong, M. J. Simpson, M. Spraul, W. Bermel, H. Heumann, M. Gundy, H. Boenisch and A. J. Simpson, *Analyst*, 2020, **145**, 5787.
47. R. Ghosh Biswas, B. Fortier-McGill, M. Akhter, R. Soong, P. Ning, M. Bastawrous, A. Jenne, D. Schmidig, P. de Castro, S. Graf, T. Kuehn, F. Busse, J. Struppe, M. Fey, H. Heumann, H. Boenisch, M. Gundy, M. J. Simpson and A. J. Simpson, *Anal. Chim. Acta: X*, 2020, **100051**, 1.
48. S. Vathyam, S. Lee and W. S. Warren, *Science*, 1996, **272**, 92.
49. N. Hertkorn, C. Ruecker, M. Meringer, R. Gugisch, M. Frommberger, E. M. Perdue, M. Witt and P. Schmitt-Kopplin, *Anal. Bioanal. Chem.*, 2007, **389**, 1311.
50. D. Lane, R. Soong, W. Bermel, P. Ning, R. Dutta Majumdar, M. Tabatabaei-Anaraki, H. Heumann, M. Gundy, H. Bönisch, Y. Liaghati Mobarhan, M. J. Simpson and A. J. Simpson, *ACS Omega*, 2019, **4**, 9017.
51. Y. L. Mobarhan, J. Struppe, B. Fortier-McGill and A. J. Simpson, *Anal. Bioanal. Chem.*, 2017, **409**, 5043.
52. R. T. McKay, *Annu. Rep. NMR Spectrosc.*, 2009, **66**, 33.
53. A. J. Simpson and S. A. Brown, *J. Magn. Reson.*, 2005, **175**, 340.
54. R. Soong, E. Nagato, A. Sutrisno, B. Fortier-Mcgill, M. Akhter, S. Schmidt, H. Heumann and A. J. Simpson, *Magn. Reson. Chem.*, 2015, **53**, 774.
55. B. Lam and A. J. Simpson, *Analyst*, 2008, **133**, 263.
56. G. Zheng and W. S. Price, *Prog. Nucl. Magn. Reson. Spectrosc.*, 2010, **56**, 267.
57. V. Moxley-Paquette, D. Lane, R. Soong, P. Ning, M. Bastawrous, B. Wu, M. Z. Pedram, M. A. Haque Talukder, E. Ghafar-Zadeh, D. Zverev, R. Martin, B. Macpherson, M. Vargas, D. Schmidig, S. Graf, T. Frei, D. A. Adwan-Stojilkovic, P. de Castro, F. Busse, W. Bermel, T. Kuehn, R. Kuemmerle, M. Fey, F. Decker, H. Stronks, R. M. A. Sullan, M. Utz and A. J. Simpson, *Anal. Chem.*, 2020, **92**, 15454.
58. J. H. Lee, Y. Okuno and S. Cavagnero, *J. Magn. Reson.*, 2014, **241**, 18.

59. M. R. Viant, C. A. Pincetich, D. E. Hinton and R. S. Tjeerdema, *Aquat. Toxicol.*, 2006, **76**, 329.
60. M. R. Viant, C. A. Pincetich, J. H. Walton, R. S. Tjeerdema and D. E. Hinton, *Mar. Environ. Res.*, 2002, **54**, 553.
61. C. A. Pincetich, M. R. Viant, D. E. Hinton and R. S. Tjeerdema, *Comp. Biochem. Physiol., Part C: Toxicol. Pharmacol.*, 2005, **140**, 103.
62. M. T. Anaraki, D. H. Lysak, K. Downey, F. V. C. Kock, X. You, R. D. Majumdar, A. Barison, L. M. Lião, A. G. Ferreira, V. Decker, B. Goerling, M. Spraul, M. Godejohann, P. A. Helm, S. Kleywegt, K. Jobst, R. Soong, M. J. Simpson and A. J. Simpson, *Prog. Nucl. Magn. Reson. Spectrosc.*, 2021, **126–127**, 121.
63. G. E. Martin, *Annu. Rep. NMR Spectrosc.*, 2005, **56**, 1.
64. I. Fugariu, R. Soong, D. Lane, M. Fey, W. Maas, F. Vincent, A. Beck, D. Schmidig, B. Treanor and A. J. Simpson, *Analyst*, 2017, **142**, 4812.
65. M. Bastawrous, O. Gruschke, R. Soong, A. Jenne, D. Gross, F. Busse, B. Nashman, A. Lacerda and A. J. Simpson, *Anal. Chem.*, 2022, **94**, 8523.
66. V. Moxley-Paquette, B. Wu, D. Lane, M. Bastawrous, P. Ning, R. Soong, P. de Castro, I. Kovacevic, T. Frei, J. Stuessi, D. A. Adwan-Stojilkovic, S. Graf, F. Vincent, D. Schmidig, T. Kuehn, R. Kuemmerle, A. Beck, M. Fey, W. Bermel, F. Busse, M. Gundy, H. Boenisch, H. Heumann, B. Nashman, R. Dutta Majumdar, A. Lacerda and A. J. Simpson, *Magn. Reson. Chem.*, 2022, **60**, 386.
67. P. Styles, N. F. Soffe and C. A. Scott, *J. Magn. Reson.*, 1989, **84**, 376.
68. H. Kovacs, D. Moskau and M. Spraul, *Prog. Nucl. Magn. Reson. Spectrosc.*, 2005, **46**, 131.
69. M. T. Anaraki, D. Lane, M. Bastawrous, A. Jenne and A. J. Simpson, in *Methods in Molecular Biology*, ed. G. A. Nagana Gowda and D. Raftery, New York, Springer, 1st edn, 2019, ch. 9, vol. 2037, pp. 395.
70. M. Bastawrous, M. Tabatabaei-Anaraki, R. Soong, W. Bermel, M. Gundy, H. Boenisch, H. Heumann and A. J. Simpson, *Anal. Chim. Acta*, 2020, **1138**, 168.
71. D. Lane, T. E. Skinner, N. I. Gershenzon, W. Bermel, R. Soong, R. Dutta Majumdar, Y. Liaghati Mobarhan, S. Schmidt, H. Heumann, M. Monette, M. J. Simpson and A. J. Simpson, *J. Biomol. NMR*, 2019, **73**, 31.
72. M. W. Voehler, G. Collier, J. K. Young, M. P. Stone and M. W. Germann, *J. Magn. Reson.*, 2006, **183**, 102.
73. M. Akhter, R. Dutta Majumdar, B. Fortier-McGill, R. Soong, Y. Liaghati-Mobarhan, M. Simpson, G. Arhonditsis, S. Schmidt, H. Heumann and A. J. Simpson, *Anal. Bioanal. Chem.*, 2016, **408**, 4357.
74. R. D. Majumdar, M. Akhter, B. Fortier-McGill, R. Soong, Y. Liaghati-Mobarhan, A. J. Simpson, M. Spraul, S. Schmidt and H. Heumann, *eMagRes*, 2017, **6**, 133.
75. Y. J. An, W. J. Xu, X. Jin, H. Wen, H. Kim, J. Lee and S. Park, *ACS Chem. Biol.*, 2012, 7, 2012.
76. D. Lane, Y. Liaghati Mobarhan, R. Soong, P. Ning, W. Bermel, M. Tabatabaei Anaraki, B. Wu, H. Heumann, M. Gundy, H. Boenisch,

T. Y. Jeong, V. Kovacevic, M. J. Simpson and A. J. Simpson, *Anal. Chem.*, 2019, **91**, 15000.
77. W. T. Dixon, *J. Chem. Phys.*, 1982, 77, 1800.
78. O. N. Antzutkin, *Prog. Nucl. Magn. Reson. Spectrosc.*, 1999, **35**, 203.
79. O. N. Antzutkin and M. H. Levitt, *J. Magn. Reson., Ser. A*, 1996, **118**, 295.
80. R. Ghosh Biswas, R. Soong, P. Ning, D. Lane, M. Bastawrous, A. Jenne, D. Schmidig, P. de Castro, S. Graf, T. Kuehn, R. Kümmerle, W. Bermel, F. Busse, J. Struppe, M. J. Simpson and A. J. Simpson, *Anal. Chem.*, 2022, **94**, 8756.
81. P. K. Madhu, R. Pratima and A. Kumar, *Chem. Phys. Lett.*, 1996, **256**, 87.
82. T. S. Mahesh, T. G. Ajithkumar, C. S. Nagaraja, G. Bodenhausen and A. Kumar, *Chem. Phys. Lett.*, 2000, **319**, 278.
83. P. Ning, R. Soong, W. Bermel, D. Lane, M. J. Simpson and A. J. Simpson, *Anal. Bioanal. Chem.*, 2018, **410**, 7055.
84. M. Liu, J. K. Nicholson, J. A. Parkinson and J. C. Lindon, *Anal. Chem.*, 1997, **69**, 1504.
85. A. J. Simpson, M. J. Simpson and R. Soong, *Anal. Chem.*, 2018, **90**, 628.
86. R. K. Harris, E. D. Becker, S. M. Cabral de Menezes, R. Goodfellow and P. Granger, *Pure Appl. Chem.*, 2001, **73**, 1795.
87. H. C. Keun, O. Beckonert, J. L. Griffin, C. Richter, D. Moskau, J. C. Lindon and J. K. Nicholson, *Anal. Chem.*, 2002, **74**, 4588.
88. T. J. Henderson, *J. Am. Chem. Soc.*, 2004, **126**, 3682.
89. P. C. Lemos, L. S. Serafim, M. M. Santos, M. A. M. Reis and H. Santos, *Appl. Environ. Microbiol.*, 2003, **69**, 241.
90. A. M. Scharff, H. Egsgaard, P. E. Hansen and L. Rosendahl, *Plant Physiol.*, 2003, **131**, 367.
91. K. A. Thorn and L. G. Cox, *J. Environ. Qual.*, 2012, **41**, 865.
92. A. Ćeranić, M. Doppler, C. Büschl, A. Parich, K. Xu, A. Koutnik, H. Bürstmayr, M. Lemmens and R. Schuhmacher, *Plant Methods*, 2020, **16**, 1.
93. J. H. Ippel, L. Pouvreau, T. Kroef, H. Gruppen, G. Versteeg, P. van den Putten, P. C. Struik and C. P. M. van Mierlo, *Proteomics*, 2004, **4**, 226.
94. L. Espen, F. F. Nocito and M. Cocucci, *J. Exp. Bot.*, 2004, **55**, 2053.
95. M. Florentz and P. Granger, *Environ. Technol. Lett.*, 1983, **4**, 9.
96. R. S. Tjeerdema, T. W. M Fan, R. M. Higashi and D. G. Crosby, *J. Biochem. Toxicol.*, 1991, **6**, 45.
97. M. R. Viant, J. H. Walton, P. L. TenBrook and R. S. Tjeerdema, *Aquat. Toxicol.*, 2002, **57**, 139.
98. A. B. Lindstrom, M. J. Strynar and E. L. Libelo, *Environ. Sci. Technol.*, 2011, **45**, 7954.
99. L. Richter, A. Cordner and P. Brown, *Soc. Stud. Sci.*, 2018, **48**, 691.
100. B. Weiner, L. W. Y. Yeung, E. B. Marchington, L. A. D'Agostino and S. A. Mabury, *Environ. Chem.*, 2013, **10**, 486.
101. N. Saito, K. Harada, K. Inoue, K. Sasaki, T. Yoshinaga and A. Koizumi, *J. Occup. Health*, 2004, **46**, 49.

102. K. Kannan, S. Corsolini, J. Falandysz, G. Fillmann, K. S. Kumar, B. G. Loganathan, M. A. Mohd, J. Olivero, N. van Wouwe, J. H. Yang and K. M. Aldous, *Environ. Sci. Technol.*, 2004, **38**, 4489.
103. C. D. Murphy, *Appl. Microbiol. Biotechnol.*, 2016, **100**, 2617.
104. S. P. Williams, P. M. Haggie and K. M. Brindle, *Biophys. J.*, 1997, 72, 490.
105. R. A. Goyer, *Environ. Health Perspect.*, 1993, **100**, 177.
106. A. L. Wani, A. Ara and J. A. Usmani, *Interdiscip. Toxicol.*, 2015, **8**, 55.
107. T. W. Clarkson, *Crit. Rev. Clin. Lab. Sci.*, 1997, **34**, 369.
108. B. Wrackmeyer and K. Horchler, *Annu. Rep. NMR Spectrosc.*, 1990, **22**, 249.
109. B. Wrackmeyer and R. Contreras, *Annu. Rep. NMR Spectrosc.*, 1992, **24**, 267.
110. B. Celda and G. T. Montelione, *J. Magn. Reson.*, 1993, **101**, 189.
111. P. Kiraly, R. W. Adams, L. Paudel, M. Foroozandeh, J. A. Aguilar, I. Timári, M. J. Cliff, M. Nilsson, P. Sándor, G. Batta, J. P. Waltho, K. E. Kövér and G. A. Morris, *J. Biomol. NMR*, 2015, **62**, 43.
112. J. Vestergren, A. G. Vincent, M. Jansson, P. Persson, U. Ilstedt, G. Gröbner, R. Giesler and J. Schleucher, *Environ. Sci. Technol.*, 2012, **46**, 3950.
113. The Human Metabolome Database, https://hmdb.ca/ (accessed November 2022).
114. W. Willker, D. Leibfritz, R. Kerssebaum and W. Bermel, *Magn. Reson. Chem.*, 1993, **31**, 287.
115. C. J. Turner, P. J. Connolly and A. S. Stern, *J. Magn. Reson.*, 1999, **284**, 281.
116. G. W. Vuister and A. Bax, *J. Magn. Reson.*, 1992, **98**, 428.
117. P. K. Mandal and A. Majumdar, *Concepts Magn. Reson., Part A*, 2004, **20**, 1.
118. J. Keeler, *Understanding NMR Spectroscopy*, Wiley, Hoboken, 2002.
119. S. Grzesiek and A. Bax, *J. Magn. Reson.*, 1992, **96**, 432.
120. L. E. Kay, D. Marion and A. Bax, *J. Magn. Reson.*, 1989, **84**, 72.
121. N. G. A. Bell, A. A. L. Michalchuk, J. W. T. Blackburn, M. C. Graham and D. Uhrín, *Angew. Chem., Int. Ed.*, 2015, **54**, 8382.
122. L. E. Kay, *J. Am. Chem. Soc.*, 1993, **115**, 2055.
123. P. Kiraly, N. Kern, M. P. Plesniak, M. Nilsson, D. J. Procter, G. A. Morris and R. W. Adams, *Angew. Chem., Int. Ed.*, 2021, **60**, 666.
124. A. B. Jones, G. C. Lloyd-Jones and D. Uhrín, *Anal. Chem.*, 2017, **89**, 10013.
125. H. L. Wheeler, R. Soong, D. Courtier-Murias, A. Botana, B. Fortier-Mcgill, W. E. Maas, M. Fey, H. Hutchins, D. Krishnamurthy, R. Kumar, M. Monette, H. J. Stronks, M. M. Campbell and A. Simpson, *Magn. Reson. Chem.*, 2015, **53**, 735.
126. L. Lam, R. Soong, A. Sutrisno, R. de Visser, M. J. Simpson, H. L. Wheeler, M. Campbell, W. E. Maas, M. Fey, A. Gorissen, H. Hutchins, B. Andrew, J. Struppe, S. Krishnamurthy, R. Kumar, M. Monette, H. J. Stronks, A. Hume and A. J. Simpson, *J. Agric. Food Chem.*, 2014, **62**, 107.

127. P. Ning, D. Lane, R. Ghosh Biswas, A. Jenne, M. Bastawrous, R. Soong, D. Schmidig, T. Frei, P. de Castro, I. Kovacevic, S. Graf, S. Wegner, W. Bermel, F. Busse, T. Kuehn, R. Kuemmerle, J. Struppe, M. Fey, H. J. Stronks, M. Monette, M. J. Simpson and A. J. Simpson, *Analyst*, 2021, **146**, 4461.
128. M. Mayer and B. Meyer, *Angew. Chem., Int. Ed.*, 1999, **38**, 1784.
129. J. G. Longstaffe, M. J. Simpson, W. Maas and A. J. Simpson, *Environ. Sci. Technol.*, 2010, **44**, 5476.

CHAPTER 6

Self-diffusion NMR as a Powerful Tool for the Evaluation of Environmental Contamination

R. VALIULLIN

Leipzig University, Germany
Email: valiullin@uni-leipzig.de

6.1 Fundamentals of Diffusion Processes

Diffusion is one of the most fundamental processes occurring in nature.[1,2] In particular, it is responsible for matter redistribution in the presence of concentration gradients. Notably, even if the concentration gradients have already been equilibrated by the process of so-called transport diffusion, translational motion does not freeze out. Because of thermal energy, the molecules continue to perform chaotic Brownian motion or random walk.[3] This process is referred to as self-diffusion. Understanding and quantifying diffusion processes helps not only to predict how fast matter will be transported from one place to another by means of elementary random steps, but also to get deeper insight into various interactions taking place in the system. The latter may include both energetic interactions, *e.g.*, how molecules stick together or accommodate at surfaces, as well as steric ones, if hard matter occurring in the sample excludes some part of space from diffusion. The pulsed field gradient technique of NMR (PFG NMR) is one of the most powerful experimental approaches designed to study diffusion processes.[4–6] In particular, it

New Developments in NMR No. 32
The Environment in a Magnet: Applications of NMR Techniques to Environmental Problems
Edited by Pellegrino Conte, Delia Francesca Chillura Martino and Paolo Lo Meo

Published by the Royal Society of Chemistry, www.rsc.org

probes in the most direct way the diffusion propagator, *i.e.*, the function fully encoding the statistics of molecular motion.[7] In this chapter, we will cover the basics of this technique and show a spectrum of options it provides, which might be helpful in solving environmental problems in various contexts.[8]

It turns out that, under certain conditions, two transport processes, namely equilibration of the concentration gradient and random motion under equilibrium conditions, are described by one and the same set of equations and parameters. Thus, by applying the equilibrium measurement technique, such as NMR, one obtains parameters that are equally valuable under non-equilibrium conditions too.

Figure 6.1 exemplifies different situations just discussed and may serve as an instructive visualization to quantify the diffusion phenomena. Thus, Figure 6.1(a)

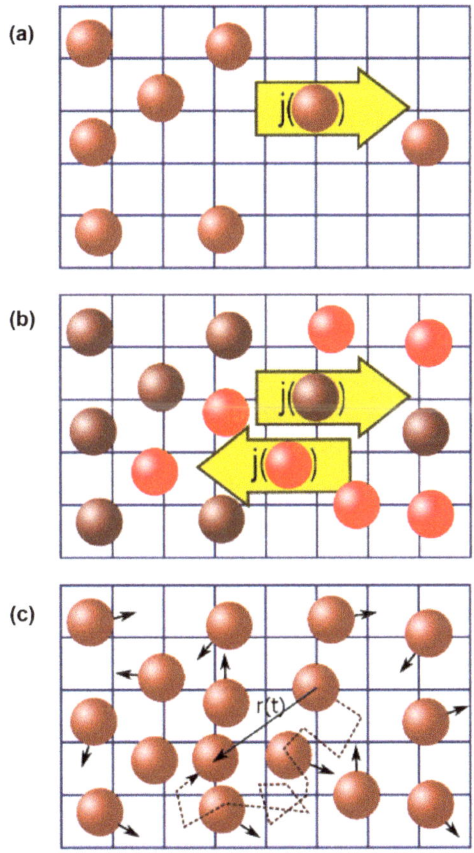

Figure 6.1 Different facets of diffusion. (a) The formation of diffusion fluxes in the presence of the concentration gradient. (b) Diffusive fluxes of the color-labelled, identical molecules under equilibrium conditions. (c) Translational motion of the tracer particles. Reproduced from ref. 9, https://doi.org/10.1039/C5CS00715A, under the terms of the CC BY 3.0 license, https://creativecommons.org/licenses/by/3.0/.

shows a non-equilibrium system where a gradient in the particle concentration is established. With such particle ensembles of spatially-varying densities, one may imply molecules in the gas phase or Brownian particles in aqueous phases. Even if the inter-particle interactions are neglected, the mere of the fact of the existence of the concentration gradient combined with the random motion of each particle results in the emergence of the particle flux J towards the regions with lower concentrations. Note that the inter-particle interactions are indeed negligible for low particle concentrations – the situations typically encountered in the context of environmental sciences, where propagations of only tiny amounts of species under interest (pollutant molecules, radioactive traces, salts) are considered. In this contribution, we confine ourselves to consider only this limit of high particle dilutions.

The first Fick's law relates the flux J to the concentration gradient $\mathrm{d}c/\mathrm{d}x$ (for the sake of reducing complexity, we consider only gradients in one spatial direction) as

$$J = -D_\mathrm{T}\frac{\mathrm{d}c}{\mathrm{d}x} \tag{6.1}$$

where the coefficient of proportionality D_T is called the coefficient of transport diffusion of simply transport diffusivity.[9] In combination with the mass conservation law,

$$\frac{\partial c}{\mathrm{d}t} = -\frac{\partial J}{\partial x} \tag{6.2}$$

stating that the time variation of the number of particles in a selected volume can only proceed *via* particle diffusion into/out of this volume, eqn (6.1) yields Fick's second law of diffusion:

$$\frac{\mathrm{d}c}{\mathrm{d}t} = \frac{\mathrm{d}}{\mathrm{d}x}\left(D_\mathrm{T}\frac{\mathrm{d}c}{\mathrm{d}x}\right) \tag{6.3}$$

or, for constant D_T,

$$\frac{\partial c}{\partial t} = D_\mathrm{T}\frac{\partial^2 c}{\partial x^2} \tag{6.4}$$

The solution of eqn (6.4) for bulk phases with the initial condition of non-zero particle concentration $c(x, t=0) = c_0$ only at $x = x_0$,

$$c(x,t) = \frac{c_0}{\sqrt{4\pi D_\mathrm{T} t}}\exp\left\{-\frac{(x-x_0)^2}{4D_\mathrm{T} t}\right\} \tag{6.5}$$

plays a central role in the diffusion theory. In particular, eqn (6.5) predicts that the width of the particle concentration distribution, as characterized by the full width at half height, will grow only as a square root of time, making mass redistribution by diffusion progressively slower with increasing time.

Let us consider now self-diffusion under equilibrium conditions. This situation is depicted in Figure 6.1(b), where the concentration gradient of the brown particles is eliminated by adding otherwise identical, red particles. With the inter-particle interactions neglected, eqn (6.1) can directly be applied to either of two color-labelled particle ensembles. Two consequences are noteworthy. First, under these conditions the coefficient of self-diffusion D is found to be equal to the transport diffusivity D_T. In this regard, we will use in what follows D and D_T interchangeably. Second, eqn (6.5) describes also mass transfer under equilibrium conditions, where the concentration c may refer to any sub-ensembles of the (labelled) particles.

Because mass redistribution under equilibrium conditions, as described by eqn (6.4), results as a cumulative effect over the random motion of the individual particles, it is intuitively clear that the individual particle trajectories should also bear information about the self-diffusion process of the entire system (Figure 6.1(c)). Indeed, it has been shown by Einstein[10] that, for purely random diffusion processes, the mean-square-displacements (MSDs) $\langle x^2(t)\rangle$ are found to grow linearly with time:

$$\langle x^2(t)\rangle = 2Dt \tag{6.6}$$

Most importantly, the coefficient of proportionality between $\langle x^2(t)\rangle$ and $2t$ is the same diffusivity D as in eqn (6.5). This has an important consequence for experimental sciences by providing an alternative route for assessing the diffusion constant by probing the molecular trajectories, namely MSD. The most prominent examples of such techniques include PFG NMR and single-particle tracking.[11–13]

To complete this section and for the purpose of subsequent presentation, it is worth mentioning that the quantity $p(x,t) \equiv c(x,t)/c_0$ in eqn (6.5) can be considered as a probability density that a particle starting its journey at time $t=0$ at the position $x_0=0$ will be found at the position x given time t to diffuse. This brings about a direct connection between eqn (6.5) and (6.6). Indeed, by computing

$$\langle x^2(t)\rangle = \int_{-\infty}^{\infty} x^2 p(x,t)\mathrm{d}x \tag{6.7}$$

with $p(x,t)=c(x,t)/c_0$, one finds, as expected, eqn (6.6).

6.2 Basics of Diffusion NMR

6.2.1 Fundamentals of Pulsed Field Gradient NMR

Placed in the external magnetic field B_0 nuclear spins perform precession with the Larmor frequency $\omega_0=\gamma B_0$, where γ is the gyromagnetic ratio of the nuclei under study. Among many nuclei with the non-zero spin, such as ^{13}C, ^{19}F or ^{23}Na and many others, hydrogen is the most abundant and predominantly addressed in diffusion NMR. A relatively weak interaction of the nuclear spin with the magnetic field, with the energies of $\hbar\omega_0$ several orders of magnitude weaker than the typical chemical bond energies, renders NMR a powerful non-invasive and non-perturbative tool to study different microscopic processes occurring in nature. The basics of this method are presented in numerous textbooks and reviews, and here we only briefly recapitulate some points essential for the chapter context.

The central quantity in NMR is the function

$$m(t) = \langle e^{i\varphi(t)} \rangle \tag{6.8}$$

representing the (normalized) signal intensity measured in the experiments. Here, $\varphi(t)$ is the time-dependent phase acquired by a spin and $\langle \cdots \rangle$ stands for the ensemble average over the entire spin system. The phase change can be caused by a number of different mechanisms due to interaction of the spins with the surrounding environment.[6,14] This leads, in particular, to an irreversible dephasing referred to as the nuclear magnetic relaxation:

$$m_r(t) = \exp\left(-\frac{t}{T_2}\right)\exp\left(-\frac{t}{T_1}\right) \tag{6.9}$$

where T_2 and T_1 are the transversal and longitudinal relaxation times, respectively.

In diffusion NMR, the conventional pulse sequences are purposefully superimposed with the gradients of the magnetic field causing additional dephasing $\varphi_g(t)$ of the transversal magnetization.[15] The latter process can be measured independently and is found to be directly controlled by diffusion of the spin-bearing molecules as explained in more detail in what follows.

The phases $\varphi_g(t)$ acquired by the spins in a linearly varying magnetic field, $B_0 + gz$, where $g = \partial B/\partial z$ is the magnetic field gradient intensity, become spatially dependent:

$$\varphi_g(t) = \int_0^t \gamma(B_0 + gz)\mathrm{d}\tau = \gamma B_0 t + \int_0^t \gamma g z \mathrm{d}\tau \tag{6.10}$$

Here we assume that the magnetic field gradient points along the main field z-axis, following the convention of the NMR literature. The gradients can, however, be generated along any spatial direction, which is routinely done in magnetic resonance imaging.[6,16] With the application of a single gradient pulse, the magnetization forms a helix with the wavelength λ decreasing with increasing time as shown in Figure 6.2(a). If, however, a pair of the gradient pulses with opposite polarities is applied, dephasing caused by the first gradient pulse is compensated by the second one and the gradient echo is formed as illustrated in Figure 6.2(b).

In typical diffusion NMR experiments, the pulse sequences generating the spin echo signals are applied. The simplest version of such pulse sequences, the Hahn echo pulse sequence, is shown in Figure 6.3. The purpose of the π radio-frequency pulse is to invert the magnetization evolution, so that, as in the case of the gradient echo, dephasing caused by the first gradient pulse is

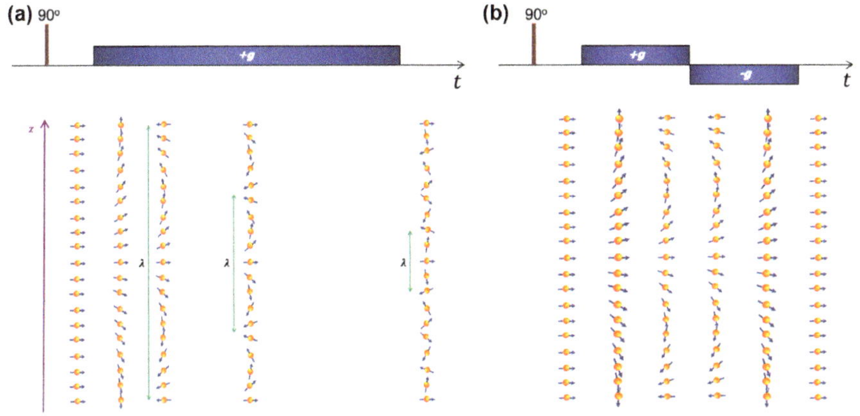

Figure 6.2 Evolutions of the magnetization profiles in the transversal plane after the application (a) of a single gradient pulse and (b) of a pair of identical gradient pulses with opposite polarities. In (a) the formation of the polarization helix with decreasing wavelength λ is noted, while in (b) the polarization helix is rewound back to the end of the gradient pulse pair for immobile spins.

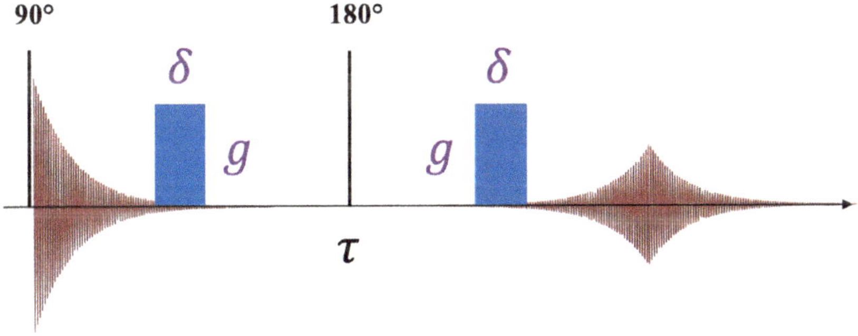

Figure 6.3 The Hahn echo pulse sequence with two magnetic field gradient pulses of intensity g and duration δ.

completely re-phased back by the second one. With eqn (6.8) and (6.10), the echo signal intensity at time 2τ results as

$$m_g(2\tau) = \left\langle \exp\left\{ i\gamma g \left(\int_0^{\delta} z(t)\mathrm{d}t - \int_{\tau}^{\tau+\delta} z(t)\mathrm{d}t \right) \right\} \right\rangle \tag{6.11}$$

Eqn (6.11) reveals that re-phasing will be complete in the case of immobile spins only and demonstrates how NMR becomes sensitive to the molecular motion.

In the so-called short gradient pulse approximation, *i.e.*, when molecular displacements during the gradient pulses are neglected, eqn (6.11) reduces to a simple function:

$$m_g(t_d) = \langle \exp\{ i\gamma g \delta \Delta z(t_d) \} \rangle \tag{6.12}$$

where t_d is the time separation between two magnetic field gradient pulses and $\Delta z(t_d)$ is the z-component of the displacement during t_d. In the literature, t_d is often referred to as the diffusion or observation time. For more complex pulse sequences incorporating several gradient pulses or with the effect of diffusion occurring during the inclusion of gradient pulses, eqn (6.12) still holds, but with a respectively redefined diffusion time, t_d. For example, for normal, unrestricted diffusion occurring during the gradient pulses, the diffusion time t_d is given by $t_d = \Delta - \frac{1}{3}\delta$, where Δ is the time interval between the leading edges of two magnetic field gradient pulses.

The ensemble average in eqn (6.12) can be represented as the integral over different populations of molecules having different displacements, $\Delta z(t_d)$,

$$m_g(t_d) = \int_{\infty}^{\infty} p(\Delta z, t_d) e^{iq\Delta z} d\Delta z \tag{6.13}$$

where $q = \gamma g \delta$ defines the experimentally-controlled wave number (generally vector, keeping in mind that the gradient g is a vector quantity) and $p(\Delta z, t_d)$ is the probability density for having displacement Δz during time interval t_d. This function, encoding all statistical features of molecular motion, is the central quantity in the diffusion theory. In the literature, $p(\Delta z, t_d)$ is also referred to as the diffusion propagator. If the displacement directions over short time intervals $t \ll t_d$ are uncorrelated, the propagator $p(\Delta z, t_d)$ assumes the Gaussian form:

$$p(\Delta z, t_d) = \left(\frac{1}{2\pi\langle\Delta z^2(t_d)\rangle}\right)^{1/2} e^{-\frac{\Delta z^2(t_d)}{2\langle\Delta z^2(t_d)\rangle}} \tag{6.14}$$

In this case, $p(\Delta z, t_d)$ becomes identical with the function $p(x, t)$ as given by eqn (6.5) and discussed in Section 6.1. By combining eqn (6.13) and (6.14), the spin echo diffusion attenuation finally becomes

$$m_g(t_d) = e^{-\frac{1}{2}q^2\langle\Delta z^2(t_d)\rangle} \tag{6.15}$$

In deriving eqn (6.15), the Gaussian form of the diffusion propagator has been assumed. Quite generally, however, the latter may not apply. It turns out that, irrespective of the displacement statistics, eqn (6.15) remains valid as soon as the data at low q-values are considered. Indeed, by expanding $e^{iq\Delta z}$ in eqn (6.13) in a series for $q \to 0$ and evaluating $m_g(t_d)$ with only the first non-vanishing term, eqn (6.15) is obtained.

For normal diffusion, eqn (6.15) can more conventionally be represented as

$$m_g(t_d) = e^{-q^2 D t_d} \tag{6.16}$$

Note that the signal attenuations due to nuclear magnetic relaxation and diffusion in the magnetic field gradient decouple just because they occur independently from each other. Therefore, the echo signal intensity measured, $m(2\tau)$, is given by the product $m(2\tau) = m_r(2\tau)m_g(t_d)$. To separate the contribution to the echo attenuation caused by only diffusion, a simple normalization of the echo signal intensities by that obtained at $q = 0$, *i.e.*, caused by sole nuclear magnetic relaxation, can be used. In order to be consistent with the literature, in what follows we will refer to the diffusion attenuations of the spin echo signals, $m_g(t_d)$, with $S(q, t_d)$. The latter notion expresses a very useful fact that the signal intensity measured is a function of both q and t_d. This is in contrast to the b-factor often used in the literature, which combines q and t_d into one parameter, $b = q^2 t_d$. Its use, for example, by the magnetic resonance community, is justified as soon as the diffusion propagator is Gaussian or a low q-region is considered. Indeed, it has been discussed earlier that in the latter case $S(q, t_d)$ is not sensitive to the propagator shape in the region of low q-values. However, if the propagator is not Gaussian, then variation of q or t_d by keeping the product $b = q^2 t_d$ constant may become inequivalent.

6.2.2 Pulse Sequences

Our preceding discussion was based solely on the Hahn echo pulse sequence, typically referred to as PGSE, for it contains all essential elements of PFG NMR. There is, however, a spectrum of different pulse sequences which can be very helpful for specific applications. In what follows, we briefly consider some generic pulse sequences serving different purposes, including avoiding artefacts in diffusion measurements.[17]

The double PGSE pulse sequence (Figure 6.4(a)) combines two PGSEs to suppress the effects arising from the coherent motion, such as flow or convection.[18] This is often the case when the diffusion measurements are performed under elevated temperatures and the emerging convective flows complicate assessments of the genuine diffusivities. The pulse sequence is designed in a way that the additional phase $\gamma g \delta v t_d$ acquired during the first PGSE due to the velocity term is compensated during the second part. Reverting polarities of the last two gradient pulses still gives rise to the echo formation, but with the uncompensated velocity term. Comparison of the results of two pulse sequences can, thus, be used to check the presence of convective flows.

The stimulated echo pulse sequence (PGSTE, Figure 6.4(b)) is useful when the system under study is characterized by the short transverse nuclear magnetic relaxation time T_2. In many such cases, the longitudinal relaxation time T_1 can still be sufficiently long. This, for example, is often found in polymer melts.[19] Splitting the π-pulse in PGSE into two $\pi/2$-pulses with some time interval between them allows increasing the separation between the gradient pulses and hence the diffusion time, without losing the nuclear magnetization due to the transversal relaxation. In this way, a notably broader range of the diffusion times may be covered.

The Cotts 13-interval pulse sequence (Figure 6.4(c)) has originally been designed to eliminate unwanted effects arising from the occurrences of

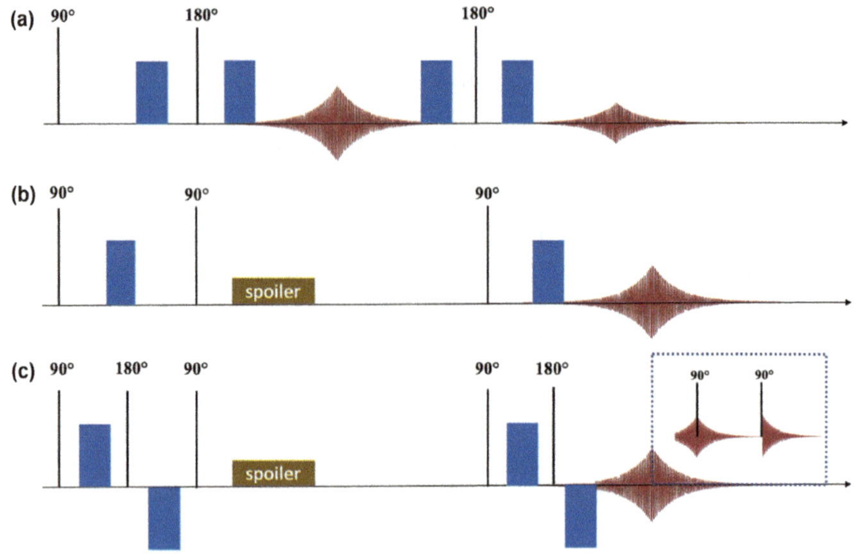

Figure 6.4 Selected PFG NMR pulse sequences: (a) the double spin-echo pulse sequence for velocity compensation; (b) the stimulated echo pulse sequence; (c) the 13-interval Cotts pulse sequence for measurements in inhomogeneous magnetic fields. The inset in (c) shows a longitudinal echo delay block modifying the Cotts pulse sequence.

uncontrolled, internal magnetic field inhomogeneities in the samples.[20] The latter may originate, for example, due to magnetic susceptibility differences in composite or porous materials. The two π-pulses incorporated into PGSTE serve to cancel dephasing due to diffusion in the internal field gradients. In addition, making the gradient pulse amplitudes asymmetric allows for the efficient elimination of the cross-terms between internal and applied magnetic field gradients.[21–23] Another advantage of applying two gradient pulses of opposite polarities with the π-pulse in between is an effective suppression of the induced eddy currents which may interfere with the signal being measured.[24] Alternating the gradient pulse polarities can also be implemented in other pulse sequences, such as those designed for convection suppression.[25]

The 13-interval LED pulse sequence is an alternative version of the Cotts 11-interval pulse sequence adding the option of spectrally-resolved diffusion measurements. Instead of recording only the echo amplitude, a pair of π/2-pulses with a time interval between them and optionally containing a spoiler gradient pulse is used to produce the undistorted FID signal. The latter can be Fourier-transformed to yield the spectrum. In this way, in the samples containing several components giving signals at different resonance frequencies, the respective diffusion rates can be obtained independently. Notably, such blocks of the longitudinal echo delay (LED) can complement any pulse sequences discussed earlier.

Finally, it is worth mentioning a spectrum of 2D correlation NMR experiments in which the pulse sequences are designed to correlate the diffusion rates with other NMR quantities such as the relaxation times T_2 and T_1 or with the diffusions rates by producing, respectively, two-dimensional D–T_2, D–T_1, and D–D maps.[26]

6.2.3 Multi-phase Systems

Often the measured spin echo diffusion attenuations $S(q, t_d)$ do not follow simple exponential form predicted by eqn (6.16). This may be caused by a number of reasons. One of the most frequent ones is related to the occurrence of a spectrum of diffusion rates in the sample. This is typical for mixtures, where all mixture components having different diffusivities D_i bear the nuclei under study and hence contribute to the measured NMR signal. Single-component fluids may also exhibit a spectrum of the diffusion rates if diffusion takes place in an inhomogeneous medium giving rise to spatial regions with different diffusivities. Because the measured NMR signal is cumulative over all resonant nuclei in the sample, quite generally $S(q, t_d)$ can be represented as the sum:

$$S(q, t_d) = \sum_i f_i \, e^{-q^2 D_i t_d} \tag{6.17}$$

where f_i is the relative fraction with which the phase i, characterized by the diffusivity D_i, contributes to the spin-echo diffusion attenuation function. Note that, in the context of diffusion NMR, the word "phase" is conventionally used not in the thermodynamic sense, but rather indicates the occurrences of molecular sub-ensembles with different transport properties. The relative fractions f_i are determined by the relative fractions p_i of the resonant nuclei in each phase and their respective relaxation times T_2 and T_1:

$$f_i = \frac{p_i \exp\left\{-\frac{\tau_{tr}}{T_{2i}} - \frac{\tau_l}{T_{1i}}\right\}}{\sum_i p_i \exp\left\{-\frac{\tau_{tr}}{T_{2i}} - \frac{\tau_l}{T_{1i}}\right\}} \tag{6.18}$$

In eqn (6.18) τ_{tr} and τ_l are the total time intervals in the NMR pulse sequence where the magnetization is stored in the transversal plane and longitudinal direction, respectively. Only when the relaxation effects are negligible, for example, when $\tau_{tr} \ll T_2$ and $\tau_l \ll T_1$ or they are identical in all phases, the relative fractions f_i reduce to the spin fractions p_i in each phase:

$$S(q, t_d) = \sum_i p_i \, e^{-q^2 D_i t_d} \tag{6.19}$$

For the sake of simplicity, we assume in what follows that the relaxation weighting is negligible. If this is not the case, eqn (6.18) must be used to obtain the real spin fractions by independent measurements of the relaxation rates in each phase.

The distributions of p_i are generally unknown. The exceptions to mention are when the sample composition is exactly known or when the distribution shape can be inferred from the nature of the sample. Hence, some model functions, such as normal or log-normal distribution functions, can be used for fitting to the experimental data. Whether or not the distribution function selected truly represents the one in the sample is not trivial to answer and, hence, their interpretations must be done with care. Alternatively, the Laplace transformation can be used to obtain the diffusivity spectrum. The latter is, however, a highly challenging problem demanding very accurate measurements of $S(q, t_d)$ over a sufficiently large range of q-values.[27] In many cases, however, assessing the average diffusion rate $\bar{D}$ is sufficient. To estimate accurately $\bar{D}$, any suitable distribution function approaching $S(q, t_d)$ with a reasonable accuracy can be used to compute $\bar{D}$. Another useful, model-free approach is finding $\bar{D}$ from the initial part of $S(q, t_d)$ by approaching eqn (6.16) to the experimental data at $q \to 0$. To see this, let us evaluate the derivative of eqn (6.19) at low q-values:

$$-\left.\frac{\partial S(q, t_d)}{\partial(q^2 t_d)}\right|_{q\to 0} = \sum_i p_i D_i = \bar{D} \tag{6.20}$$

If the relaxation weighting is not negligible, the average diffusivity obtained in either approach will be given by $\sum_i f_i D_i$ different from $\bar{D}$. Eqn (6.20) also reveals a useful feature of the tangential lines to the experimental data constructed at any $q \neq 0$. These tangents will have the functional form $p \times e^{-q^2 \bar{D}_p t_d}$, yielding that the fraction p of the most slowly diffusing spins in the entire ensemble is characterized by the average diffusivity $\bar{D}_p$.

The concept of $\bar{D}$, which is the average diffusivity over the spin ensembles, is most useful for one-component fluids. In this case, p_i simply coincides with the fractions of the molecules in each phase, and hence $\bar{D}$ represents the particle ensemble average. In the mixtures, the number of the resonant spins in each mixture component may differ from one another and the averages over spin and particle ensembles become unequal. If the relaxation weighting is not negligible, the average diffusivity obtained in either approach will be given by $\sum_i f_i D_i$ different from $\bar{D}$. It is important to appreciate this fact; otherwise, this may become a source of misinterpretations.[28]

As a representative example, let us consider a multi-phase system composed of the high molecular mass polymer dextran with an average molecular mass of 500 kDa dissolved in a deuterated solvent.[29] The dilution is high such that the intermolecular interactions between dextran molecules can be neglected. ^{1}H PFG NMR experiments reveal a notable nonexponential behavior for $S(q, t_d)$ as demonstrated in Figure 6.5. This is caused by the naturally occurring distribution of the molecular masses of the dextran molecules. Indeed,

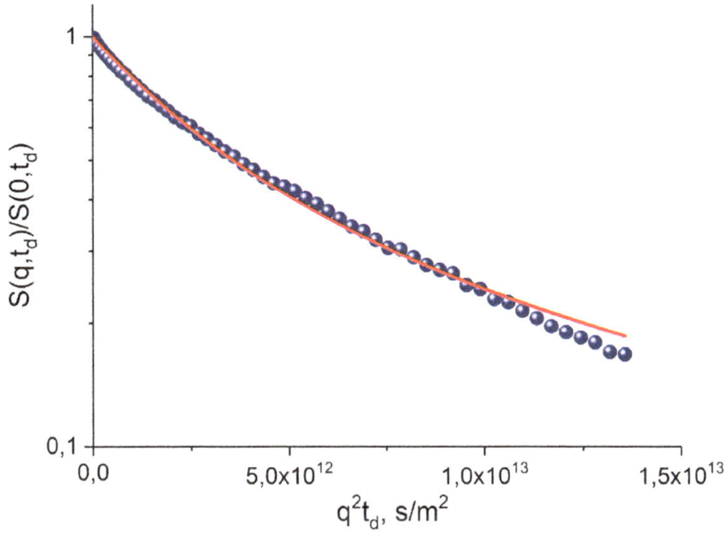

Figure 6.5 ^{1}H spin-echo diffusion attenuation function for 500 kDa dextran molecules dissolved in a deuterated solvent. The solid line shows the best fit of eqn 6.22 to the experimental δ.

the polymer mass M defines its hydrodynamic radius $R_h \propto M^{1/3}$. In turn, the latter results in the mass-dependent diffusivities D_M in agreement with the Einstein–Stokes relation:[30]

$$D_M = \frac{kT}{6\pi\eta R_h} \tag{6.21}$$

where η is the dynamic viscosity of the solvent. Because the chemical structure of all dextran molecules is identical, (i) they all have similar relaxation properties and (ii) the mass and spin fractions coincide. Thus, $S(q, t_d)$ is determined *via* eqn (6.19) with p_i representing the mass fractions of the dextran molecules with different molecular masses. In Figure 6.5, the experimental data were fitted to eqn (6.19) with the log-normal distribution for the diffusivities[31,32]

$$p(D_M) = \left(\sqrt{2\pi}D_M\sigma\right)^{-1/2} \exp\left\{-\frac{(\ln D_M - \ln \bar{D})^2}{2\sigma^2}\right\} \tag{6.22}$$

where σ is the distribution width. This example demonstrates one of the applications of diffusion NMR for relatively easy and accurate measurements of the molecular mass distributions.[33–35] In the same spirit, this approach can be used for the droplet size analysis in, *e.g.*, emulsions or oil–water mixtures.[36,37] In addition, this example illustrates well the concept of diffusion-ordered spectroscopy (DOSY), in which molecular species, which cannot be resolved spectroscopically, can still be distinguished based on their transport properties.[38–40]

In the polymer system just considered, the spins never change their parent molecules (intermolecular magnetization exchange by flip-flop transitions is excluded by high dilution[41]). Thus, for the displacement exceeding the globule sizes, *i.e.*, when segmental motion can be neglected, all spins in each polymer chain will acquire similar MSDs, and hence will be characterized by one diffusion rate.[42] Very generally, such conditions are classified as no-exchange or slow-exchange scenarios. These conditions are not only found in mixtures, but can also be observed for single-component fluids in inhomogeneous media. In the latter case, the spatial extensions of the regions with different diffusivities must be notably larger than the molecular displacements during the diffusion time t_d. Such situations can be found, *e.g.*, with large porous beads with bead sizes of hundreds of μm. With typical displacements probed with PFG NMR of the order of several μm, there will be negligible molecular exchange between the two pools of molecules within the beads (hindered diffusion) and between the beads (bulk-like diffusion).

There are some exotic examples falling into the slow-exchange category even though all molecules have identical translation properties and one would expect a single-component diffusion attenuation given by eqn (6.16). They might originate from a specific feature of the PFG NMR method in which only the projections of the molecular displacements onto the gradient axis are probed. Diffusion is generally a tensor quantity and its principal components may be different:[43]

$$D = \begin{pmatrix} D_{xx} & 0 & 0 \\ 0 & D_{yy} & 0 \\ 0 & 0 & D_{zz} \end{pmatrix} \tag{6.23}$$

Let us consider a limiting case of one-dimensional diffusion, where random translational motion occurs only in one direction so that $D_{xx} \neq 0$ and $D_{yy} = D_{zz} = 0$. This type of motion may occur in, *e.g.*, zeolite crystals or membranes with one-dimensional pore structures, where diffusion takes place only along pore channels. If these channels are orientated under the angle θ with respect to the gradient axis, then any displacement Δx along the channel will contribute only with $\Delta x \cos(\theta)$ to the magnetization dephasing as illustrated in Figure 6.6. As long as the molecules do not leave their parent crystals, this situation resembles the slow exchange scenario with the phases being associated with the molecules differing by the projections of their displacements onto the gradient axis.

With eqn (6.15) and (6.16), the spin echo diffusion attenuation is then found as

$$S(q, t_d) = e^{-q^2 D_{xx} \cos^2(\theta) t_d} \tag{6.24}$$

When the sample under study is not a single crystal, but contains a powder with the individual crystals orientated isotropically in space, the measured $S(q, t_d)$ results as the powder average

$$S(q, t_d) = \int_{-\pi/2}^{\pi/2} e^{-q^2 D_{xx} \cos^2(\theta) t_d} \sin(\theta) d\theta \tag{6.25}$$

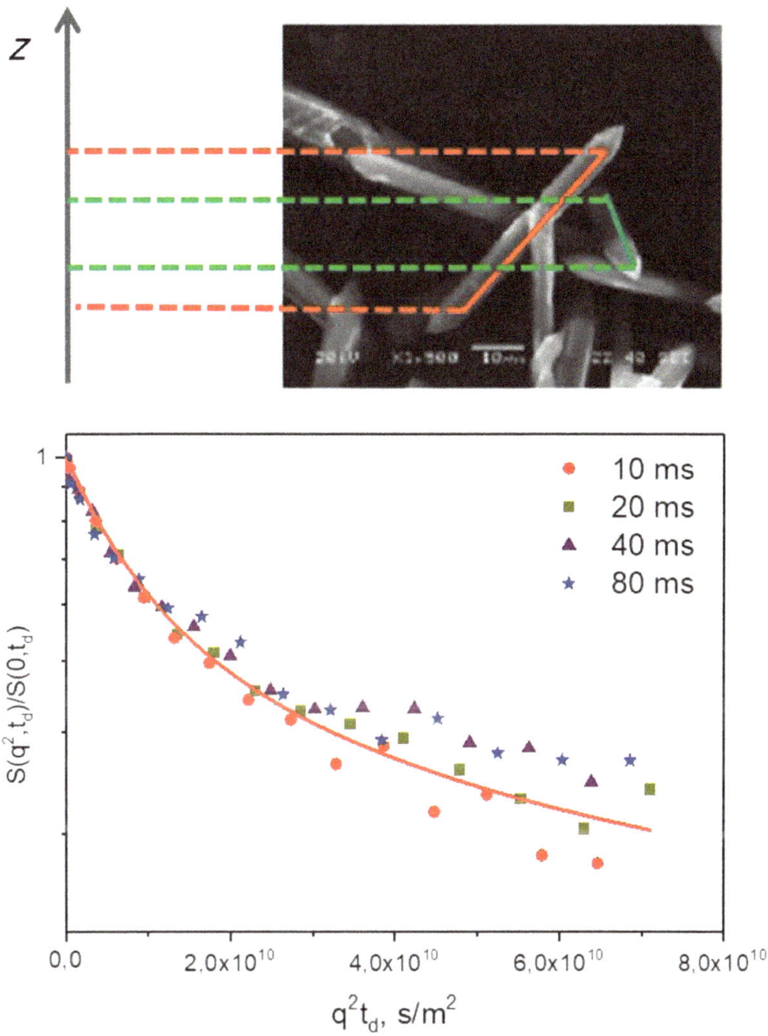

Figure 6.6 Spin-echo diffusion attenuation functions for ethane in the metal–organic-framework $Sc_2(NH_2\text{-}BDC)_3$ with one-dimensional pore structure measured with different diffusion times from 10 ms to 80 ms. The solid line shows eqn (6.25) with the diffusion rate $D_{xx} = 2.2 \times 10^{-10}\ m^2\ s^{-1}$. The upper panel illustrates different projections of the identical displacements in the individual crystal with different orientations.

Figure 6.6 shows a relevant example, where diffusion of ethane in a sample containing powder of a metal–organic framework with a one-dimensional pore structure was studied. It demonstrates that eqn (6.25) accurately reproduces the shape of the spin-echo diffusion attenuation functions measured. Notably, the latter equation has been derived under the assumption of the isotropic distribution of crystal orientations. This might, however, be not the case, *e.g.*, for the plate-like crystals showing a preferential orientation. In this case, the analysis of $S(q, t_d)$ may be used to assess the distribution of the crystal orientations.[44]

Such a system, especially with channels of nanoscopic dimensions and with molecules of anisotropic shapes, needs to be treated with care. As a matter of fact, the transversal relaxation in them can as well be a function of the crystal orientation due to residual dipolar coupling.[45,46] Indeed, the structural anisotropy may give rise to the fact that the intramolecular dipolar interactions, which are effectively averaged out by the isotropic molecular rotations in bulk fluids, may survive the averaging due to the induced anisotropy in molecular orientations.[47] In this case, eqn (6.25) needs to be complemented by the additional relaxation-weighting function given by eqn (6.18) with the orientation-dependent T_2.

What if, however, the molecules have a chance to alternate between the phases with different transport properties? If the crystal sizes in the preceding example were comparable with the molecular displacements, more and more molecules would escape their parent crystals with increasing observation time and enter other crystals with different orientations. Clearly, if during the experimental time the molecules change the crystals multiple times, the memory about orientation of the original crystal will be completely lost. In this case, all molecules in the system become indistinguishable and one may expect intuitively behavior consistent with eqn (6.16).

This indeed can most instructively be shown by considering the MSD $\langle \Delta z^2(t_d) \rangle$ in eqn (6.15).[48] The displacement $\Delta z(t_d)$ is cumulative over displacements during much shorter time intervals t, such that $t \ll t_d$. Let us consider a two-phase system, which we denote with the subscripts a and b. All individual displacements $\Delta z(t)$ in the sum $\Delta z(t_d) = \sum \Delta z(t)$ can be combined as $\Delta z(t_d) = \sum \Delta z_a(t) + \sum \Delta z_b(t)$ so that the displacements in the two phases become separated as illustrated in Figure 6.7. The total MSD thus becomes

$$\left\langle \Delta z^2(t_d) \right\rangle = \left\langle \Delta z_a{}^2(t_a) \right\rangle + \left\langle \Delta z_b{}^2(t_b) \right\rangle + 2\langle \Delta z_a(t_a) \Delta z_b(t_b) \rangle \tag{6.26}$$

where $t_d = t_a + t_b$. With multiple transitions between different phases and if there are no correlations between $\Delta z_a(t)$ and $\Delta z_b(t)$, that may emerge due to specific boundary conditions at the phase interfaces, the last term on the right-hand side of eqn (6.26) averages out to zero. Dividing all remaining MSDs in eqn (6.26) by the respective diffusion times, it is easily seen that

$$\frac{\langle \Delta z^2(t_d) \rangle}{2t_d} = \frac{t_a}{t_d} \frac{\langle \Delta z_a{}^2(t_a) \rangle}{2t_a} + \frac{t_b}{t_d} \frac{\langle \Delta z_b{}^2(t_b) \rangle}{2t_b} \tag{6.27}$$

Under equilibrium conditions, the time fractions t_i/t_d spent by the molecules in each phase are simply equal to the fractions p_i of molecules in the respective phases. By applying in addition eqn (6.6), it is thus found that in the

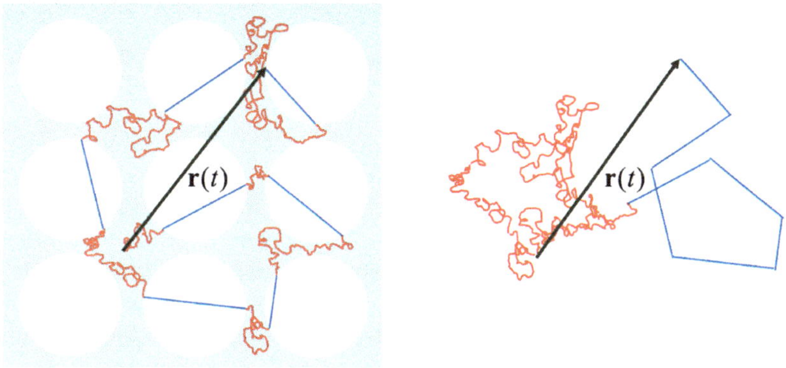

Figure 6.7 Molecular trajectory $\mathbf{r}(t)$ composed of random excursions in the capillary-condensed phase (red) intermittent with the Knudsen flights in the pore space (blue). The right panel demonstrates the recombined trajectory leaving the displacement $\mathbf{r}(t)$ unchanged.

fast exchange limit the entire molecular ensemble is characterized by the single diffusivity

$$\bar{D} = p_a D_a + p_b D_b \tag{6.28}$$

which is identical to the average diffusivity as given by eqn (6.20). Often, these equations are referred to as the fast-exchange diffusion equations.

If the genuine diffusivities in the phases a and b are known or measured independently, the average diffusivity can easily be predicted on the basis of eqn (6.28). There are, however, situations when this approach fails. Indeed, in deriving eqn (6.28) it has been assumed that the combined sub-trajectories in phases a and b represent fully-uncorrelated random trajectories. This might not always be the case. As a particular example, Figure 6.7 demonstrates a situation when one of the sub-trajectories is constructed from single ballistic or so-called Knudsen flights in the gas phase confined in the mesopores of a hierarchical micro–mesoporous solid.[48] These flights are intermittent with diffusive excursions in the adsorbed liquid phase in the micropores. The trajectories composed of the Knudsen flights exhibit the Gaussian statistics given by eqn (6.14) after a sufficiently large number of flights. Hence, they can be characterized by an effective diffusivity, D_{eff}. However, specific boundary conditions at the gas–liquid interface give rise to the anti-correlations between the directions of two consequent Knudsen flights.[48] As a result, these anti-correlations lead to the diffusivities D_{eff} notably smaller than the genuine Knudsen diffusivity:[49,50]

$$D_K = \frac{1}{a} d\bar{v} \tag{6.29}$$

one would measure in this system. In eqn (6.29) d is the pore size, $\bar{v}$ is the average thermal velocity and a is a numerical constant determined by the pore space geometry (*e.g.*, 3 for cylinders or 8 for spheres). Consequently, eqn (6.28) with the Knudsen diffusivity D_K will overpredict $\bar{D}$. Instead, the effective diffusivities in the phases accounting for the possible correlations in the displacements need to be used in the fast-exchange diffusion equation.

In the regime of intermediate molecular exchange, there will be a complex transition from the slow to fast exchange limits with increasing observation or exchange time. Under these conditions, different molecules will exhibit different statistics of their displacements: some molecules will still remain in their original phases, while others will experience different numbers of transitions between different phases. For the two-phase systems, this may respectively be seen as the evolution of the diffusivity spectrum from bimodal (D_a, D_b) at short exchange times to unimodal ($p_aD_a+p_bD_b$) at long times as illustrated in Figure 6.8. In contrast to the two limiting cases of fast and slow exchange, the intermediate exchange regime bears information about the spatial characteristics of the involved phases for the exchange features depend on the spatial extensions of the phases and their geometries.[51] The latter precludes the general solution of the problem, namely the prediction of the shape of $S(q, t_d)$ and its variation with varying diffusion times. Therefore, often some assumptions allowing the exact solutions are made. These include, in particular, the assumption of the constant exchange rate, which is done for the case of chemical exchange for predicting the relaxation spectra.[52,53] How good these assumptions are in considering spatially-extended multi-phase systems is a matter of debate.[54–56]

Even though no general solutions of this problem are available, some properties can be assessed relatively easily. This is illustrated with the help of Figure 6.9 showing a typical exchange pattern as seen by PFG NMR. The measurements have been performed with porous silicon having tubular pores with lengths L between 10 μm and 50 μm.[57] The porous solid was exposed to a cyclohexane atmosphere at a 60 mbar vapor pressure. As revealed by Figure 6.9, the spin-echo diffusion attenuations $S(q, t_d)$ evolve from that having a nearly bi-exponential form at the shortest diffusion time $t_d=5$ ms to that having a highly multi-exponential form at $t_d=300$ ms. Notably, the slopes to $S(q, t_d)$ at large q-values, corresponding to the slowest

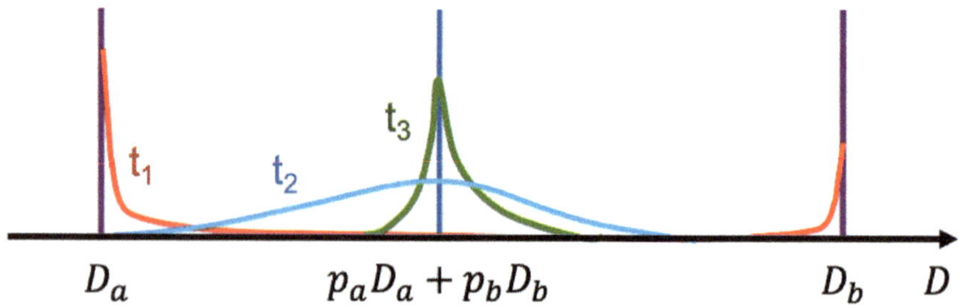

Figure 6.8 Schematics of the evolution of the diffusivity spectrum from the regimes of slow (D_a, D_b) to fast ($p_aD_a+p_bD_b$) exchange. The spectra are shown in the arbitrary scale for better visualization. t_i indicate different exchange times such that $t_1<t_2<t_3$.

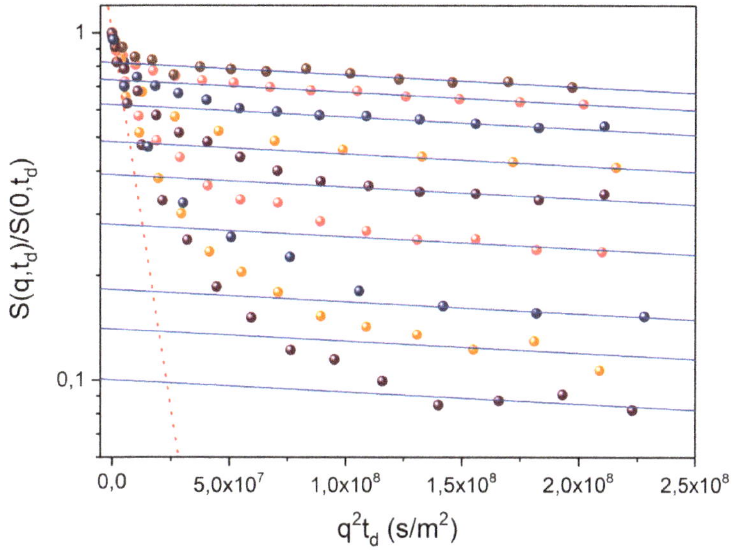

Figure 6.9 Spin-echo diffusion attenuations for cyclohexane in porous silicon for diffusion times t_d increasing from 5 ms to 300 ms (from top to bottom). The broken line shows the slope corresponding to the average diffusivity given by eqn (6.28). The solid lines represent eqn (6.16) with the intrapore diffusivity $D_{ip} = 8.0 \times 10^{-10}$ m^2 s^{-1} multiplied by the time-dependent fractions $p_{ip}(t_d)$.

diffusivities in the system, are identical for all diffusion times. This diffusivity D_{ip} can be associated with the intrapore diffusivity, *i.e.*, the diffusivity of cyclohexane in the porous silicon channels. It is seen that the fraction of such molecules, p_{ip}, decreases with increasing t_d because more and more molecules escape the channels to the external gas phase with the notably higher diffusivity. Indeed, with the diffusivity $D_{ip} = 8.0 \times 10^{-10}$ m^2 s^{-1}, the molecular displacements during $t_d = 5$ ms are of the order of 3 μm, which is notably shorter than the shortest channel length of 10 μm. On the other hand, for $t_d = 300$ ms the displacements are about 22 μm so that many molecules may have a chance to escape the channels. It is instructive to note that, at long diffusion times, $S(q, t_d)$ shows a tendency to converge to a mono-exponential form:

$$S(q, t_d) = e^{-q^2 \bar{D} t_d} \tag{6.30}$$

with the average diffusivity $\bar{D}$ given by eqn (6.28) and determined in line with eqn (6.20). It is seen from Figure 6.9 that eqn (6.30) is a tangent at $q = 0$ to any experimental curve obtained with different diffusion times. This confirms a useful feature of $\bar{D}$ that it remains constant irrespective of the exchange time. This may not hold if, for example, the diffusion rate in one of the phases involved is time-dependent, due to, *e.g.*, the occurrence of restricted diffusion as discussed in Section 6.2.4.

With the diffusivity D_{ip} independently known, the obtained function $p_{ip}(t_d)$ can be analysed to estimate the sizes of the spatially-extended phases.

The solutions for these functions for different geometries may be found elsewhere.[51,58] An unfortunate message is that the geometry shall be known in advance, and this information, however, often is not available. Some useful properties can still be assessed. For example, it can be shown that, for short exchange times, the function $p_{\mathrm{ip}}(t_{\mathrm{d}})$ can be accurately approximated by[57]

$$p_{\mathrm{ip}}(t_{\mathrm{d}}) = 1 - 4\sqrt{\frac{D_{\mathrm{ip}} t_{\mathrm{d}}}{\pi}} \langle L^{-1} \rangle \tag{6.31}$$

where $\langle L^{-1} \rangle$ is the average reciprocal channel length. As shown in Figure 6.10, fit of eqn (6.31) to the experimental data at short t_{d} yields an average channel length of about 22 μm, which is in excellent agreement with the independently measured value of 23 μm.

Another useful parameter is the average residence time $\bar{t}$ of the molecules in the phase, which can be related, *e.g.*, to the release time of pharmaceuticals from a porous host. It is defined as

$$\bar{t} = \int_0^\infty t\psi(t)\mathrm{d}t \tag{6.32}$$

where $\psi(t)$ is the normalized molecular life-time distribution in the phase.[59] On the other hand,

$$1 - p_{\mathrm{ip}}(t_{\mathrm{d}}) = \int_0^{t_{\mathrm{d}}} \psi(t)\mathrm{d}t \tag{6.33}$$

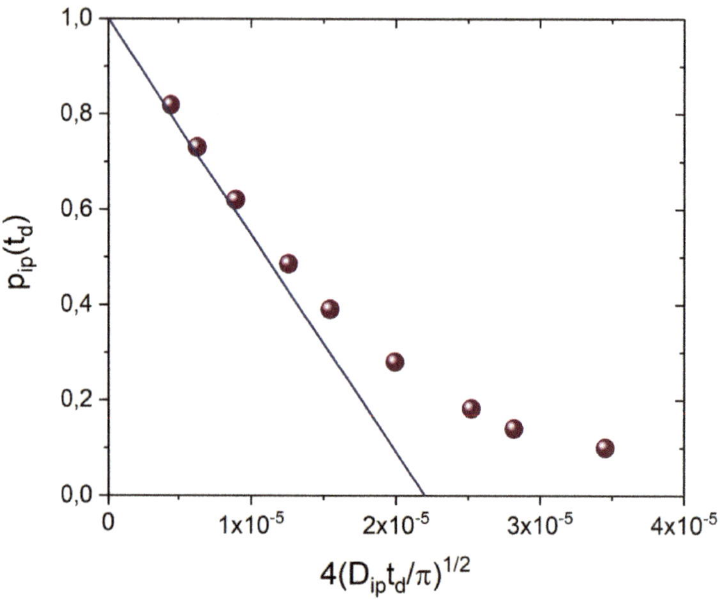

Figure 6.10 Short-time behavior of $p_{\mathrm{ip}}(t_{\mathrm{d}})$ from Figure 6.9. The solid lines represent eqn (6.31) with all parameters independently known.

Combination of eqn (6.30) and (6.31) allows a model-free assessment of the average residence time. The latter can be used for estimating the average size of the phase or permeability at the interface between two phases.[60]

6.2.4 Confined Diffusion

Molecular transport often takes place under conditions of steric restrictions or spatial confinements, or, generally, in heterogeneous environments.[61] It can be diffusion in the pore spaces of porous solids, intracellular spaces, voids and cages formed between solid particles, *etc.* Probing hindered or restricted diffusion is important not only to quantify typically slow transport in these systems, but also to obtain structural details of the medium where diffusion takes place. The latter can be effectively done, thanks to a unique option of PFG NMR allowing tracking of molecular displacements over broad ranges of diffusion times from milliseconds to seconds or, respectively, molecular displacements from hundreds of nanometres to centimetres.

As soon as the displacements become comparable with the characteristic pore length scale, reflections of the molecular trajectories at the pore walls break the uncorrelated character of the elementary diffusion steps and cause the MSD to grow more slowly than in the unrestricted spaces as schematically illustrated in Figure 6.11. It shows two typical patterns resulting from the two critically different morphologies of the pore spaces. One of them is related to diffusion in permeable porous matrices, while the second one exemplifies diffusion in closed regions with impermeable boundaries.

Irrespective of the pore space organization, if the molecular displacements are substantially shorter than the typical pore sizes, MSDs grow linearly with time as in the bulk liquid. During these diffusion times, only the molecules found at distances shorter than $l \approx \sqrt{2Dt_d}$ have a chance to reach the pore walls. All other molecules in the pore centre will diffuse freely. Depending on the pore surface-to-pore volume ratio, starting from some l deviations of the MSDs from the linear behavior can be observed and measured experimentally. Let us intentionally separate the entire ensemble of molecules in the pore into two parts: that in the shell of the thickness l adjacent to the pore walls and that in the remaining volume in the core part as illustrated in Figure 6.12. The MSD averaged over the entire ensemble of molecules in the pore volume can thus be represented as

$$\langle \Delta z^2(t_d) \rangle = \left(\frac{V - lS}{V} \right) \langle \Delta z_{\text{free}}{}^2(t_d) \rangle + \frac{lS}{V} \langle \Delta z_{\text{shell}}{}^2(t_d) \rangle \tag{6.34}$$

where S and V are, respectively, the surface area and pore volume. Let us further assume that the MSD of the molecules in the shell $\langle \Delta z_{\text{shell}}{}^2(t_d) \rangle$ is zero keeping in mind that reflections at the pore walls tend to reduce the displacement. It is

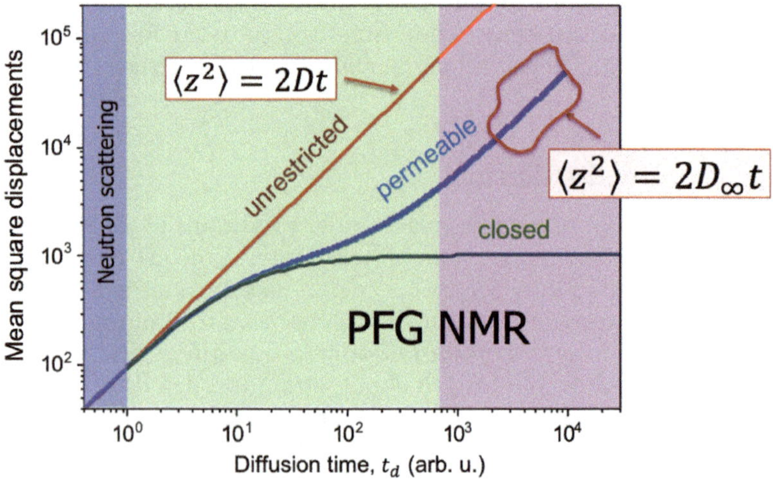

Figure 6.11 Mean-square-displacements as a function of diffusion time in bulk unrestricted fluids and in fluids confined in porous solids with interconnected and closed porosities.

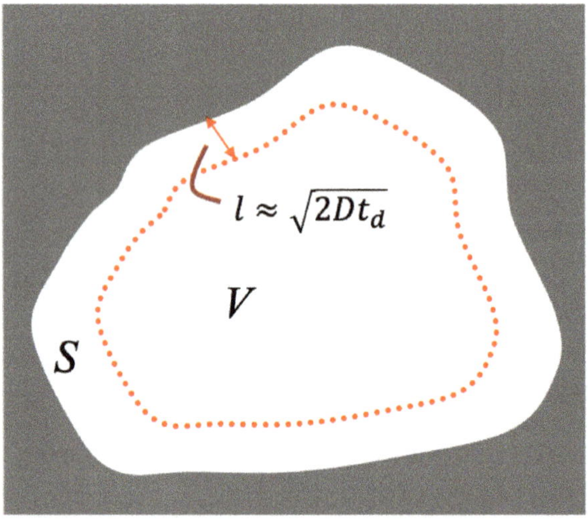

Figure 6.12 A closed pore with the surface area S and pore volume V. Only the molecules within the shell with thickness l can approach the pore walls by means of diffusion.

convenient to introduce, in the spirit of the Einstein equation, the effective or apparent diffusivity D_{eff} by dividing the MSDs in eqn (6.34) by $2t_{\mathrm{d}}$:

$$D_{\mathrm{eff}}(t_{\mathrm{d}}) = D\left(1 - \frac{S}{V}\sqrt{2Dt_{\mathrm{d}}}\right) \tag{6.35}$$

Eqn (6.35) highlights two important concepts. First, in confined spaces, in contrast to bulk fluids, the diffusivity is not constant, but is rather time-dependent.[62–64] Second, this time-dependency encodes important structural information. In this particular case, called in the literature the short-time diffusion behavior, it is the surface-to-volume ratio of the pore space that is probed in the experiments.[65] Remarkably, even with a very rough assumption of the immobile molecules in the shell, eqn (6.35) is found to be not very different from that obtained using exact calculations:[65]

$$D_{\mathrm{eff}}(t_{\mathrm{d}}) = D\left(1 - \frac{4}{9}\frac{S}{V}\sqrt{\frac{Dt_{\mathrm{d}}}{\pi}}\right) + O(t) \tag{6.36}$$

Figure 6.13 demonstrates the application of the short-time diffusion behavior for the estimation of the epidermal cell sizes in onion, resulting in values of the order of 110 μm by assuming a rectangular shape for the cells. Notably, the reflective boundary transitions may arise not only at hard walls, but also due to specific interactions. For example, strong adsorption potential in microporous crystals can give rise to reflective boundary conditions in the external boundaries of the crystals.

With further increasing diffusion time t_{d}, the behavior is strongly dictated by the architecture of the pore space. In permeable porous matrices with interconnected pores, statistics of the molecular trajectories at long

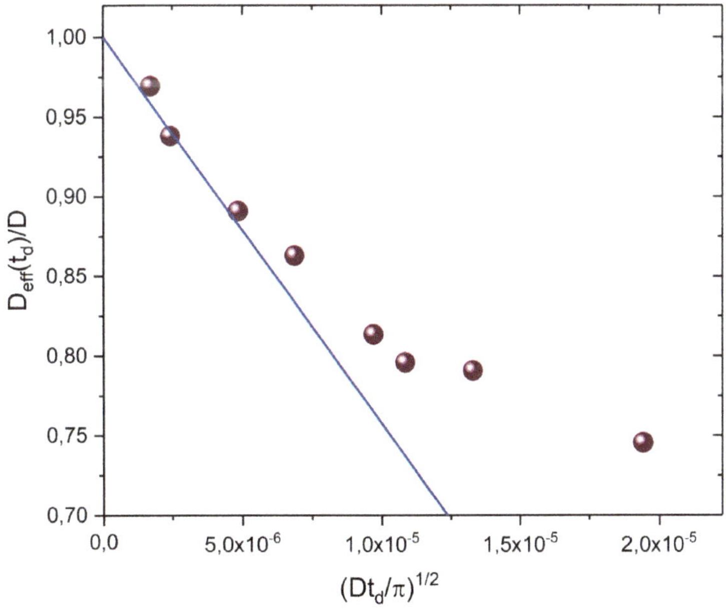

Figure 6.13 Short-time diffusion behavior of water in onion cells. The solid line shows the best fit of eqn (6.36) to the experimental data at short times with $S/V = 4.8 \times 10^{4}$ m^{-1}.

diffusion times once again become consistent with the Gaussian diffusion propagator given by eqn (6.14). In this regime, the MSDs recover their linear dependency on time. Respectively, the diffusivity at long times becomes independent of the diffusion time, but with the diffusivity lower than D in the unrestricted space. Typically, this diffusivity is called the long-range diffusivity and is denoted by D_∞. It plays an important role, revealing how much the matrix reduces the molecular transport. As an alternative parameter, the tortuosity factor τ defined as

$$\tau = \frac{D}{D_\infty} \tag{6.37}$$

is often used.

It can be shown that the time needed to approach the plateau value D_∞ is such that the molecules, starting their excursion close to the pore walls, forget information about their orientation by exploring *via* diffusion different pore wall orientations.[64] More precisely, this requires averaging out to zero the autocorrelation function $\langle \cos(\widehat{\boldsymbol{n}_0 \boldsymbol{n}_t}) \rangle$, where $\boldsymbol{n}_0$ and $\boldsymbol{n}_t$ are the normal unit vectors to the pore surface at the positions where molecules start their excursion and encounter the surface at time t, respectively. It turns out that if the entire surface area S needs to be explored before $\langle \cos(\widehat{\boldsymbol{n}_0 \boldsymbol{n}_t}) \rangle$ becomes zero, then D_∞ will as well tend to zero. This is the case of closed pores. In permeable or interconnected pores, $\langle \cos(\widehat{\boldsymbol{n}_0 \boldsymbol{n}_t}) \rangle$ averages out to zero more quickly by exploring only a part of the entire surface area S. The more interconnected is the pore space, the shorter is this time and the higher is D_∞. How $D_{\text{eff}}(t_\text{d})$ approaches its long-range value depends on the pore space geometry. An instructive illustration is presented in Figure 6.14.

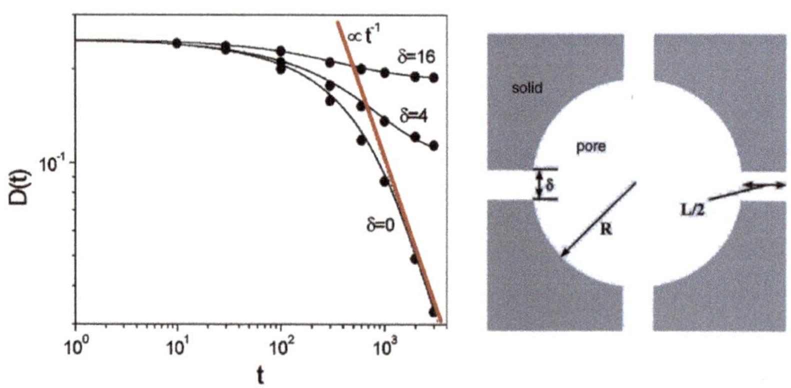

Figure 6.14 Time-dependent diffusivity as a function of time (in arb. u.) in a space formed by a periodic array of spheres interconnected by channels with diameter δ illustrated in the right panel. In closed spheres ($\delta = 0$) at long diffusion times the diffusivity follows $D \propto {t_\text{d}}^{-1}$ as shown by the solid line. In permeable structures, at long diffusion times D tends to a constant value, which is higher for more permeable structures (larger δ in this particular case). Reproduced from ref. 65 with permission from AIP Publishing, Copyright 2001.

In the closed cavities, the MSD cannot grow infinitely with increasing diffusion time and should saturate at a value of the order of a^2, where a is the cavity size. Respectively, as shown in Figure 6.14, D_{eff} follows t_{d}^{-1} dependency at long diffusion times. It is clear that, under these conditions, the diffusion propagator is very different from that given by eqn (6.14). Recall that the latter constituted the main assumption in deriving eqn (6.16) for the shape of the spin-echo diffusion attenuation function. Thus, in order to understand the echo attenuation behavior for closed cavities, one needs to start right from eqn (6.12). After a long excursion, a molecule diffusing in the cavity can be found anywhere within the volume V with the probability $\rho(z)$, where $\rho(z)$ is the density probability along the gradient axis. Thus, the ensemble average in eqn (6.12) can be found as the double integral over the initial and final positions of the molecules:

$$S(q, t_{\mathrm{d}}) = \int\int \rho(z_{\mathrm{i}})\rho(z_{\mathrm{f}})\mathrm{e}^{iq(z_{\mathrm{f}}-z_{\mathrm{i}})}\mathrm{d}z_{\mathrm{i}}\mathrm{d}z_{\mathrm{f}} \tag{6.38}$$

Because $\rho(z_{\mathrm{i}}) = \rho(z_{\mathrm{f}})$, $S(q, t_{\mathrm{d}})$ becomes independent of the diffusion time and represents the power spectrum of the density distribution:[66]

$$S(q) = \left|\int \rho(z)\mathrm{e}^{iqz}\mathrm{d}z\right|^2 = |F(q)|^2 \tag{6.39}$$

The function $F(q)$ is called the structure factor and defines the shape of the pore along the gradient direction. Obtaining the pore shape from eqn (6.39) is not directly possible because of cancelling phase factor. The problem can, however, be solved by designing an appropriate pulse sequence.[67–70]

In many cases, however, assessing the average size of the closed cavities is sufficient. It has been discussed in Section 6.2.1 that, irrespective of the diffusion propagator, the experimental data on $S(q,t_{\mathrm{d}})$ at low q-values are well captured by eqn (6.15). For example, in a closed cavity with radius a (sphere in 3D, circular region in 2D, or line in 1D) the MSDs at long diffusion times are equal to

$$\langle z^2 \rangle = \frac{2a^2}{2+d} \tag{6.40}$$

where d is the space dimension. Hence, the low-q data can be approached by

$$S(q, t_{\mathrm{d}}) = \exp\left\{-\frac{q^2a^2}{2+d}\right\} \tag{6.41}$$

to estimate the cavity size. Formally, the analysis can be done in terms of the effective diffusivities

$$D_{\mathrm{eff}}(t_{\mathrm{d}}) \equiv \frac{\langle z^2 \rangle}{2t_{\mathrm{d}}} = \frac{a^2}{(2+d)t_{\mathrm{d}}} \tag{6.42}$$

obtained from the spin-echo diffusion attenuations. This approach is widely used, *e.g.*, for the assessment of the emulsion droplet sizes.[71]

To demonstrate this, let us consider diffusion in a cylinder situated with its long axis perpendicular to the magnetic field gradient direction. This resembles then 2d-diffusion in a circular region. Hence, at long diffusion times the effective diffusivity should follow $D_{\mathrm{eff}}(t_{\mathrm{d}})=a^2/4t_{\mathrm{d}}$. Figure 6.15 shows experimental data for the diffusivities of hexadecane obtained in the channels of microporous silicon with a channel radius of 2.35 μm according to the SEM data. The sample was orientated such that the channel axes were perpendicular to the gradient direction. It is seen that, at long diffusion times, $D_{\mathrm{eff}}(t_{\mathrm{d}})$ is inversely proportional to the diffusion time as expected. The analysis yields a channel size of 2.4 μm, which is in excellent agreement with the SEM data.

In interconnected pore systems, the approach of the plateau value D_∞ depends strongly on the pore space morphology. Some particular solutions, for example, in a medium made of spherical cavities connected by narrow tubes or diluted impermeable spheres, can be found in the literature.[62,72] It has been claimed that the functional form found in ref. 62, namely

$$D_{\mathrm{eff}}(t_{\mathrm{d}})=D\left(C_1+C_2t^{-1}+C_3t^{-\frac{3}{2}}+\cdots\right) \tag{6.43}$$

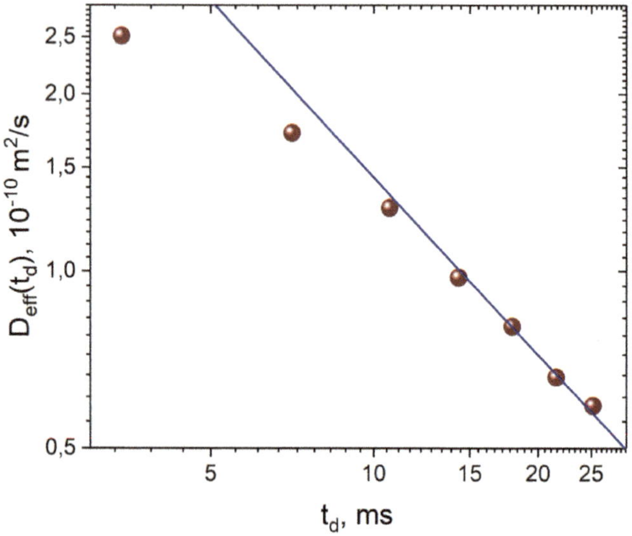

Figure 6.15 Long-time diffusion behavior of hexadecane in the macroporous silicon channels with a radius of 2.35 μm orientated perpendicular to the magnetic field gradient direction. The solid line represents the fit of eqn (6.42) with $d=2$ to the experimental data at long times yielding $a=2.4$ μm.

where D is the bulk diffusivity and C_i are the geometry-dependent constants, is also general for all permeable structures.[63] Because C_i are generally unknown, eqn (6.43) cannot be easily applied to obtain some structural details. As an alternative route, it has been shown that approaching of the long-time asymptote can be modelled as

$$\frac{D}{D-D_\infty}(D_{\text{eff}}(t_{\text{d}})-D_\infty)=\frac{{a_{\text{eff}}}^2}{2dt_{\text{d}}} \tag{6.44}$$

where D is the bulk diffusivity, d is the space dimension, and a_{eff} is a typical pore size.[64] Eqn (6.44) provides thus a simple means to estimate the pore sizes in permeable pore systems. If there is a broad distribution of a_{eff}, anomalous transient behaviors different from ${t_{\text{d}}}^{-1}$ may emerge.[73]

References

1. E. L. Cussler, *Diffusion: Mass Transfer in Fluid Systems*, Cambridge University Press, Cambridge, 3rd edn, 2009.
2. *Diffusion in Condensed Matter*, ed. P. Heitjans and J. Kärger, Springer, Berlin, 2005.
3. M. Cencini, A. Puglisi, D. Vergini and A. Vuöpiani, *A Random Walk in Physics*, Springer Nature, Switzerland AG, 2021.
4. P. T. Callaghan, *Translational Dynamics & Magnetic Resonance*, Oxford University Press, New York, 2011.
5. W. S. Price, *NMR Studies of Translational Motion*, University Press, Cambridge, 2009.
6. R. Kimmich, *NMR: Tomography, Diffusometry, Relaxometry*, Springer-Verlag, Berlin Heidelberg, 1997.
7. J. Kärger and W. Heink, *J. Magn. Reson.*, 1983, **51**, 1–7.
8. G. Zheng and W. S. Price, *eMagRes*, 1996, 561–574.
9. J. Kärger, D. M. Ruthven and R. Valiullin, *Adsorption*, 2021, **27**, 267–281.
10. A. Einstein, *Ann. Phys.*, 1905, **17**, 549–560.
11. M. J. Saxton, *Nat. Methods*, 2008, **5**, 671–672.
12. *Single Particle Tracking and Single Molecule Energy Transfer*, ed. C. Bräuchle, D. C. Lamb and J. Michaelis, Wiley-VCH, Weinheim, 2010.
13. M. Bauer, R. Valiullin, G. Radons and J. Kärger, *J. Chem. Phys.*, 2011, **135**, 144118.
14. M. H. Levitt, *Spin Dynamics: Basics of Nuclear Magnetic Resonance*, Wiley & Sons, Chichester, 2001.
15. E. O. Stejskal and J. E. Tanner, *J. Chem. Phys.*, 1965, **42**, 288–292.
16. R. W. Brown, Y.-Ch. N. Cheng, E. M. Haacke, M. R. N. Thompson, R. Venkatesan, *Magnetic Resonance Imaging: Physical Principles and sequence design*, John Wiley & Sons, New Jersey, 2rd edn, 2014.

17. G. H. Sorland and D. Aksnes, *Magn. Reson. Chem.*, 2002, **40**, S139–S146.
18. P. T. Callaghan and Y. Xia, *J. Magn. Reson.*, 1991, **91**, 326–352.
19. R. Kimmich and N. Fatkullin, *Prog. Nucl. Magn. Reson. Spectrosc.*, 2017, **101**, 18–50.
20. R. M. Cotts, M. J. R. Hoch, T. Sun and J. T. Markert, *J. Magn. Reson.*, 1989, **83**, 252–266.
21. G. Zheng and W. S. Price, *J. Magn. Reson.*, 2008, **195**, 40–44.
22. P. Z. Sun, J. G. Seland and D. Cory, *J. Magn. Reson.*, 2003, **161**, 168–173.
23. P. Galvosas, F. Stallmach and J. Kärger, *J. Magn. Reson.*, 2004, **166**, 164–173.
24. F. Stallmach and P. Galvosas, in *Annual Reports on NMR Spectroscopy*, ed. G. A. Webb, Academic Press, 1st edn, 2007, vol. 6, pp. 51–131.
25. G. H. Sorland, J. G. Seland, J. Krane and H. W. Anthonsen, *J. Magn. Reson.*, 2000, **142**, 323–325.
26. Y.-Q. Song, L. Venkataramanan, R. Kausik and N. Heaton, in *Diffusion NMR of Confined Systems: Fluid Transport in Porous Solids and Heterogeneous Materials*, The Royal Society of Chemistry, 2017, pp. 111–155.
27. Y. Q. Song, L. Venkataramanan and L. Burcaw, *J. Chem. Phys.*, 2005, **122**, 104104.
28. J. P. Gunther, L. L. Fillbrook, T. S. C. MacDonald, G. Majer, W. S. Price, P. Fischer and J. E. Beves, *Science*, 2021, **371**, eabe8322.
29. A. Shakhov, R. Valiullin and J. Kärger, *J. Phys. Chem. Lett.*, 2012, **3**, 1854–1857.
30. E. Yamamoto, T. Akimoto, A. Mitsutake and R. Metzler, *Phys. Rev. Lett.*, 2021, **126**, 128101.
31. J. Hou and E. Pearce, *Anal. Chem.*, 2021, **93**, 7958–7964.
32. H. Walderhaug, O. Soderman and D. Topgaard, *Prog. Nucl. Magn. Reson. Spectrosc.*, 2010, **56**, 406–425.
33. M. C. Miller, A. Klyosov, D. Platt and K. H. Mayo, *Carbohydr. Res.*, 2009, **344**, 1205–1212.
34. X. A. Guo, E. Laryea, M. Wilhelm, B. Luy, H. Nirschl and G. Guthausen, *Macromol. Chem. Phys.*, 2017, **218**, 1600440.
35. A. Chen, D. H. Wu and C. S. Johnson, *J. Am. Chem. Soc.*, 1995, **117**, 7965–7970.
36. R. Bernewitz, G. Guthausen and H. P. Schuchmann, *Magn. Reson. Chem.*, 2011, **49**, S93–S104.
37. S. J. Law and M. M. Britton, *Langmuir*, 2012, **28**, 11699–11706.
38. P. Stilbs, *J. Colloid Interface Sci.*, 1982, **87**, 385–394.
39. B. Antalek, *Concepts Magn. Reson.*, 2002, **14**, 225–258.
40. G. Pages, V. Gilard, R. Martino and M. Malet-Martino, *Analyst*, 2017, **142**, 3771–3796.
41. E. Fischer, R. Kimmich and N. Fatkullin, *J. Chem. Phys.*, 1997, **106**, 9883–9888.

42. R. Kimmich and N. Fatkullin, NMR 3D Analysis Photopolymerization, in *Advances in Polymer Science*, 2004, vol. 170, pp. 1–113.
43. D. Topgaard, in *Diffusion NMR of Confined Systems: Fluid Transport in Porous Solids and Heterogeneous Materials*, The Royal Society of Chemistry, 2017, pp. 226–259.
44. R. Valiullin, P. Kortunov, J. Kärger and V. Timoshenko, *J. Chem. Phys.*, 2004, **120**, 11804–11814.
45. R. Valiullin and A. Khokhlov, *Phys. Rev. E*, 2006, **73**, 051605.
46. A. C. Forse, S. A. Altobelli, S. Benders, M. S. Conradi and J. A. Reimer, *J. Phys. Chem. C*, 2018, **122**, 15344–15351.
47. C. M. Thiele, *Eur. J. Org. Chem.*, 2008, **2008**, 5673–5685.
48. P. Zeigermann, S. Naumov, S. Mascotto, J. Kärger, B. M. Smarsly and R. Valiullin, *Langmuir*, 2012, **28**, 3621–3632.
49. P. Levitz, *J. Phys. Chem.*, 1993, **97**, 3813–3818.
50. M. Knudsen, *Ann. Phys.*, 1909, **28**, 75.
51. J. Crank, *The Mathematics of Diffusion*, Clarendon Press, Oxford, 1975.
52. J. Jen, *J. Magn. Reson.*, 1978, **30**, 111–128.
53. J. R. Zimmerman and W. E. Brittin, *J. Phys. Chem.*, 1957, **61**, 1328–1333.
54. J. Kärger, *Ann. Phys.*, 1969, **24**, 1–4.
55. N. Moutal, M. Nilsson, D. Topgaard and D. Grebenkov, *J. Magn. Reson.*, 2018, **296**, 72–78.
56. M. Nilsson, E. Alerstam, R. Wirestam, F. Ståhlberg, S. Brockstedt and J. Lätt, *J. Magn. Reson.*, 2010, **206**, 59–67.
57. R. Valiullin, J. Kärger and R. Gläser, *Phys. Chem. Chem. Phys.*, 2009, **11**, 2833–2853.
58. H. S. Carslaw and J. C. Jaeger, *Conduction of Heat in Solids*, Clarendon Press, Oxford, 1946.
59. A. Yekta, J. Duhamel and M. A. Winnik, *J. Chem. Phys.*, 1992, **97**, 1554–1561.
60. J. Kärger, D. M. Ruthven and D. Theodorou, *Diffusion in zeolites and other nanoporous materials*, Wiley-VCH, Weinheim, 2012.
61. *Diffusion NMR of Confined Systems: Fluid Transport in Porous Solids and Heterogeneous Materials*, ed. R. Valiullin, The Royal Society of Chemistry, Cambridge, UK, 2017.
62. T. M. deSwiet and P. N. Sen, *J. Chem. Phys.*, 1996, **104**, 206–209.
63. P. N. Sen, *Concepts Magn. Reson., Part A*, 2004, **23A**, 1–21.
64. R. Valiullin and V. Skirda, *J. Chem. Phys.*, 2001, **114**, 452–458.
65. P. P. Mitra, P. N. Sen, L. M. Schwartz and P. Ledoussal, *Phys. Rev. Lett.*, 1992, **68**, 3555–3558.
66. D. G. Cory and A. N. Garroway, *Magn. Reson. Med.*, 1990, **14**, 435–444.
67. F. B. Laun, T. A. Kuder, W. Semmler and B. Stieltjes, *Phys. Rev. Lett.*, 2011, **107**, 048102.
68. F. B. Laun, L. Muller and T. A. Kuder, *Phys. Rev. E*, 2016, **93**.
69. S. Hertel, M. Hunter and P. Galvosas, *Phys. Rev. E*, 2013, **87**, 030802.

70. S. A. Hertel, X. D. Wang, P. Hosking, M. C. Simpson, M. Hunter and P. Galvosas, *Phys. Rev. E*, 2015, **92**, 012808.
71. M. L. Johns, *Curr. Opin. Colloid Interface Sci.*, 2009, **14**, 178–183.
72. A. M. Berezhkovskii, V. Y. Zitserman and S. Y. Shvartsman, *J. Chem. Phys.*, 2003, **119**, 6991–6993.
73. D. S. Novikov, E. Fieremans, J. H. Jensen and J. A. Helpern, *Nat. Phys.*, 2011, 7, 508–514.

CHAPTER 7

Monitoring of Lubricating Oil Degradation Via *Fast Field Cycling NMR Relaxometry*

E. ANOARDO*[a,b] AND E. M. ERRO[b]

[a] LaRTE – FaMAF, Universidad Nacional de Córdoba, Córdoba, Argentina;
[b] LaRTE – Instituto de Física Enrique Gaviola (IFEG) – CONICET, Córdoba, Argentina
*Email: eanoardo@unc.edu.ar

7.1 Introduction

Lubricating motor oils or internal combustion engine (ICE) lubricants are composed of a hydrocarbon mineral or synthetic blend (base oil), conveniently combined with specific chemicals aimed to prevent degradation and increase their lifespan.[1] Their main functions are to reduce friction and the cooling of moving mechanical parts that, in the absence of a lubricant, would be in direct contact. They also play an essential role in preserving the internal cleaning of the machinery. However, a critical role of lubricants is related to efficient energy handling.

Lubricants degrade due to different chemical and physical processes related to machine operation. Oxidation, nitration, sulphating and contamination are the most common chemical degrading factors, while thermal stress is the most important physical reason triggering chemical degradation. Such processes induce changes at the molecular level, usually reflected as hydrocarbon chain ruptures and branching. In addition, as a consequence of imperfect lubrication, metallic and soot micro/nanoparticles can be accumulated and retained by the lubricant fluid, even after circulating across filters.

New Developments in NMR No. 32
The Environment in a Magnet: Applications of NMR Techniques to Environmental Problems
Edited by Pellegrino Conte, Delia Francesca Chillura Martino and Paolo Lo Meo

Published by the Royal Society of Chemistry, www.rsc.org

In the case of internal combustion engines, mixing with gasoline/fuel and/or water is quite common.[2,3]

Real-time monitoring of the lubricant health is a topic of high interest with incipient results and moderate progress in the last few years.[4–8] In many cases, the machinery health can be inferred from the oil condition, although nowadays this can only be done in a reliable way by collecting a sample of the lubricant fluid and testing it in a laboratory (*i.e.* outside the machine) through different physicochemical and mechanical protocols. In any case, the tendency is to determine a precise moment for the oil replacement, based on reliable testing procedures, and not following a general prescription of the manufacturer (which may be overestimated to avoid incurring into a risky recommendation).[9] Similar machines may operate in different places, working conditions and regimes. Consequently, a given machine can be exposed to quite different stressing and degrading factors according to its operating circumstances. In this context, a precise monitoring of the lubricant health to prevent unnecessary or premature replacement turns into an essential feature for a positive environmental impact.[10]

During servicing and depending on the machine, more than 80% of the original fresh oil can be transformed into waste lubricating oil at the time of replacement. This waste oil contains an elevated concentration of contaminants like corrosive organic acids, remaining additives, metals, soot, carbon, and fuel, among others, which may be extremely hazardous to humans and the environment. Toxic, carcinogenic and ecotoxic properties are attributed to them in international regulations.[11] As a consequence of this fact, different strategies have been considered in the last few years: (i) the setup of a virtuous circular economy associated to waste lubricant recycling,[12,13] (ii) the incorporation of vegetal and biodegradable oils in lubricant formulations[14,15] and (iii) the implementation of more strict regulations concerning the handling, collection and storage of waste oil.[16]

Within the aforementioned context, different analytical tools have been used to determine lubricant health. Diverse methods and techniques were considered for both laboratory tests and real-time sensor technology. Some of the most commonly used are Fourier transform infrared (FTIR) spectroscopy,[17–21] electron spin resonance (EPR/ESR),[21–23] nuclear magnetic resonance (NMR) spectroscopy,[20,24–29] low-field (LF) and time-domain (TD) NMR[30–35] and differential scanning calorimetry (DSC).[36–39] Also tribological functional tests are usually considered to evaluate the performance of a lubricant formulation.[40–43]

Nuclear magnetic relaxation[44] has been shown to be an effective experimental technique to assess the molecular dynamics in fluids, polymers, liquid crystals, surfactants, and solids, among others. In particular, the fast field-cycling (FFC) NMR relaxometry technique has been successfully used in the last few years to characterize different material systems thorough the analysis of the spin–lattice relaxation within a wide Larmor frequency range.[45,46] In fact, this technique has been recently used to characterize the degradation properties of paraffinic hydrocarbons, lubricating oil bases and additivated lubricating oils, mainly due to thermal stress.[29,32,33,35,47]

In this chapter, we will summarize the progress made in the use of NMR relaxation and FFC-NMR relaxometry for the study and characterization of motor oil degradation.

7.2 A Brief History of FFC-NMR Relaxometry in Lubricants

Up to year 2005, NMR spectroscopy,[48,49] relaxation[50,51] and diffusion[52,53] experiments in mineral and synthetic bases and lubricants (and related systems) allowed understanding the essential features connecting the molecular motion with the molecular structure, macroscopic physical parameters (like viscosity), the temperature behaviour of the fluid and the main causes of degradation. Such features were also analysed by using computational models.[54,55]

The first experimental study of proton relaxation in an extended Larmor frequency domain using field-cycling NMR relaxometry was presented in 2005.[47] In this work, the lubricating oil was successfully modelled as a simple fluid. The spin–lattice relaxation rate dispersion (R_1) was explained by considering translational diffusion (Torrey's model) and molecular rotations (a Lorentzian term); see Figure 7.1.

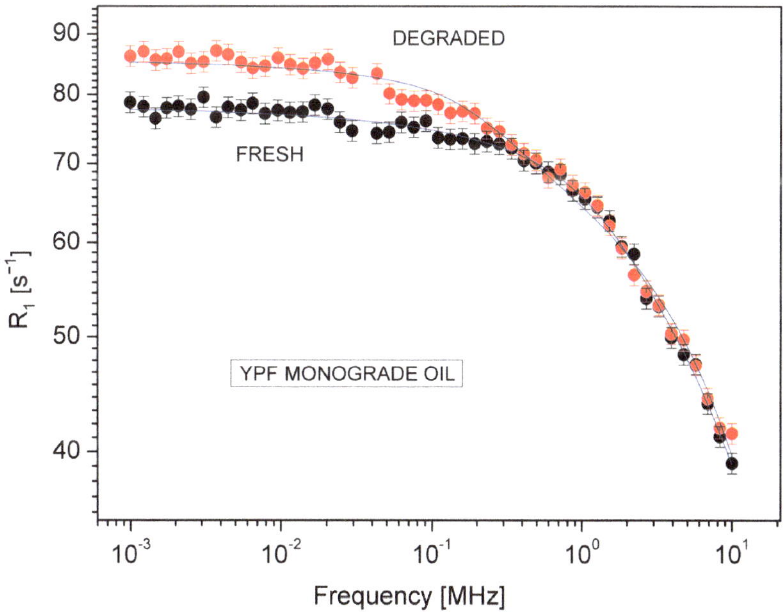

Figure 7.1 Relaxation rate dispersions of a fresh YPF monograde oil and the same oil after 5000 km of use in a diesel truck engine. Black lines show data fitting obtained after modeling the lubricant as a simple fluid by considering molecular translational diffusion and rotation as the main contributing nuclear spin–lattice relaxation mechanisms. Reproduced from ref. 47 with permission from IOP Publishing, Copyright 2005.

Eqn (7.1) describes the model used to explain the data shown in Figure 7.1:

$$R_1 = R_1^{R} + R_1^{D} \tag{7.1}$$

where the superscript "R" stands for molecular rotations and the superscript "D" for translational diffusion. Molecular rotations are represented here by a typical BPP (Bloembergen–Purcell–Pound) term, governed by a representative correlation time, τ. For the diffusional term, Torrey's model[56] was adopted in the limit $\omega\tau_D \ll 1$, where ω is the angular Larmor frequency and τ_D is the diffusional correlation time. From the experiments, it was observed that both correlation times tend to increase for the used oil, a fact that is consistent with the increase in the viscosity and the molecular changes that may occur during degradation of the fluid lubricant. Against the high sensitivity of NMR relaxation (particularly in the low-frequency regime), FFC-NMR relaxometry experiments in lubricating oils remained in suspense for more than 12 years. The reason of this inactivity in the field can be mainly attributed to the difficulty in unravelling the underlying processes at the molecular level in a complex system, like in the case of ICE lubricants after being exposed to different causes of degradation in an engine. Nevertheless, the proposed model was later used to analyze the relaxation rate in edible oils.[57]

In the following sections, we summarize the recent advances in the field. The main strategy to go deeper in the understanding of how FFC-NMR relaxometry can become a useful tool for lubricant analysis was to simplify the sample (paraffins and bases) and to expose them to a simple degrading factor at a time. As thermal stress is one of the main degrading processes, this was the first case considered.

7.3 Thermal Degradation of Paraffins

The name paraffin refers to linear alkanes, also known as normal alkanes or *n*-alkanes. These hydrocarbons are very simple in structure and lack functional groups. Even though the percentage of pure paraffin in lubricating oils may be small, they are excellent candidates to expose chemical transformation caused by the working conditions of internal combustion engines (heating, water and gasoline contamination, production of soot and metallic particles). The most common chemical changes include molecular oxidation and branching.

7.3.1 Paraffin Heat-induced Chemical Transformation

N-Octacosane ($C_{28}H_{58}$) is an average size paraffin of those contained in lubricating oils. In order to test the sensitivity of NMR relaxometry and individualize the molecular changes induced by heat, thermal degradation was carried out by heating an *n*-octacosane sample at 250 °C for 2 h. The relaxometric changes observed in the thermally treated paraffin are significant.

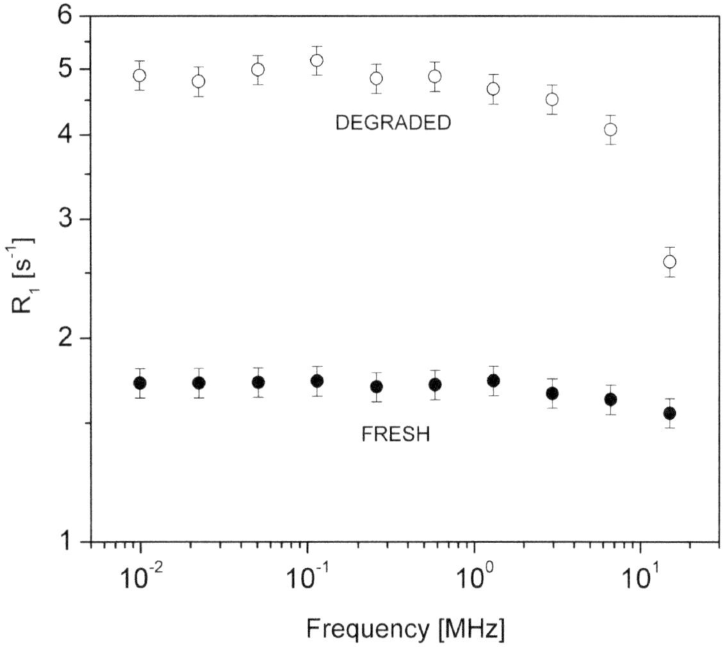

Figure 7.2 R_1 dispersion curves for *n*-octacosane in its pristine and thermally aged states. Measurements were performed at 70 °C.

Figure 7.2 shows the Larmor frequency dependence of the spin–lattice relaxation rate obtained for *n*-octacosane in its pristine state and after thermal aging, both measured at a temperature of 70 °C. It can be observed that the relaxation rate values measured for the thermally treated *n*-octacosane are larger than those obtained for the fresh (non-treated) sample. In addition, it can also be observed that the curve corresponding to the degraded sample shows a higher R_1 dispersion (clearly observable in the MHz regime). This result suggests that the chemical modifications promoted by thermal stress are responsible for a more efficient spin–lattice relaxation process.

Figure 7.3 shows the FTIR spectra of fresh (green) and degraded (red) *n*-octacosane. FTIR signals labeled as A, B, C and D show sensitivity to the thermal stressing process, that is, the heat-promoted chemical transformation in *n*-octacosane molecules was evidenced by changes in these signals. A, C and D are assigned to the stretching vibrations of methyl ($-CH_3$) and methylene ($-CH_2-$) groups. The changes observed in C and D indicate a decrease in the number of methylene groups, which is consistent with a molecular fragmentation process mediated by heat. Besides, the difference in the area ratio of A to C and that of A to D suggests a methyl group population increase. The B signal corresponds to the stretching vibrations of carbonyl groups ($-C{=}O-$) and appears only in the thermally treated sample. This change in the spectrum confirms that heat promotes oxidation.

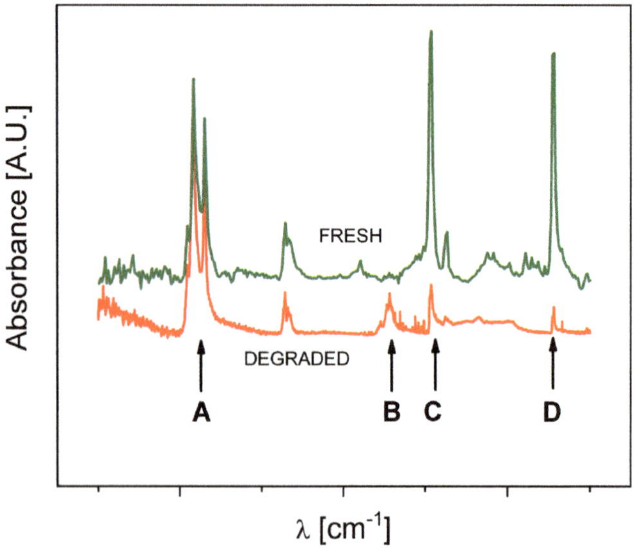

Figure 7.3 FTIR spectra of fresh *n*-octacosane (green) and thermally treated *n*-octacosane (red). New picks appearing in the red spectrum suggest the presence of an oxidative process and molecular fragmentation.

The data suggest that the linear molecule of *n*-octacosane is oxidized, fragmented and ramified[58] as a consequence of the thermal aging process.[35]

R_1 dispersion experiments in *n*-heptane (C_7H_{16}), *n*-octadecane ($C_{18}H_{38}$), oleic acid ($C_{18}H_{34}O_2$) and *n*-1-octadecene ($C_{18}H_{36}$) allowed comparing the relaxometric behavior for different molecular weights. The R_1 dispersion curves for the fresh *n*-alkanes (*n*-heptane, *n*-octadecane and *n*-octacosane) showed an increasing relaxation rate as a function of the molecular size (a situation that is also clearly observed in polymer melts[45]). This result suggests that the thermally treated *n*-octacosane shows an augmented relaxation efficiency as a consequence of an oxidation mediated chemical restructuring. Moreover, oleic acid shows that carboxylic acid groups cause a considerable increase in the relaxation rate values compared to those obtained for *n*-octadecane. The oxygens from carboxylic acid groups form hydrogen bond interactions with neighboring molecules, thus favoring molecular branching and providing a sort of structured liquid phase.[35]

7.3.2 Theoretical Model

The spin–lattice nuclear magnetic relaxation in liquids is generally interpreted in terms of intramolecular and intermolecular dipole–dipole interactions. The global relaxation process is modulated in time by molecular rotations and relative molecular translational diffusion.[44] So, as a first approach, a model to simulate proton spin–lattice relaxation dispersion of paraffins can be constructed as the sum of two terms: translational diffusion and molecular rotations (as it was previously assumed for a lubricant, see eqn (7.1)). Molecular rotations have been successfully described through a Lorentzian term. However, Torrey's diffusion was replaced by a force-free-hard-sphere (FFHS) model.[59,60] This model is based on the assumption that the molecules can be approximated by rigid spheres of uniform diameter that obeys Fick's diffusion law. The spins are centered in uniformly distributed molecules that can reach a unique minimum possible distance between them. That is, spin-pair interactions are not considered for inter-spin distances that are lower than this minimum distance. Therefore, an excluded volume effect is considered in the FFHS diffusion model, representing the potential of averaged forces between interacting spin pairs.[59] In fact, this is the main difference compared to Torrey's model.

At first sight, the use of a diffusion model for spherical molecules may turn inadequate for the description of translational diffusion of elongated molecules like fresh paraffins or lubricant oils. However, it seems to be that the break of the isotropic spherical symmetry is not affecting in a catastrophic way the use of this simplified model. In fact, it was also successfully used to describe diffusion in virgin rape oil.[61] Moreover, restructured branched paraffin molecules after degradation may approximate a closely spherical geometry. On this basis, a temptation appears to think this model is more precise for the description of degraded systems. However, this interpretation must be handled with care since other effects are present in the degraded sample which may affect the molecular diffusion (like soot and other residuals, for example).

In the Brownian diffusion limit (negligible small finite jumps), the diffusion term can be approximated by

$$R_1^{\mathrm{D}} = \frac{A_{\mathrm{D}}}{Dd}\{J[z(\omega)] + 4J[z(2\omega)]\} \tag{7.2}$$

where

$$A_{\mathrm{D}} = \frac{8}{45}\pi\gamma^4\hbar^2\left(\frac{\mu_0}{4\pi}\right)^2\rho$$

with γ being the nuclear gyromagnetic ratio, $\hbar$ the Planck constant divided by 2π, μ_0 the vacuum magnetic permeability constant and ρ the proton density. D is the self-diffusion coefficient, d the closest distance between two spins located in different molecules and

$$J(z) = \frac{1 + \frac{5}{8}z + \frac{z^2}{8}}{1 + z + \frac{z^2}{2} + \frac{z^3}{6} + \frac{z^5}{81} + \frac{z^6}{648}}.$$

In this last equation $z(\omega) \equiv \sqrt{\dfrac{2\omega d^2}{D}}$.

The contribution of molecular rotations to the spin–lattice relaxation rate R_1^R can be expressed by a Lorentzian function:

$$R_1^R = A_R\left[\frac{\tau_R}{1+\omega^2{\tau_R}^2} + \frac{4\tau_R}{1+4\omega^2{\tau_R}^2}\right] \tag{7.3}$$

where

$$A_R = \frac{3}{10}\gamma^4\hbar^2\left(\frac{\mu_0}{4\pi}\right)^2\frac{1}{r^6}.$$

Here r is the effective ^{1}H–^{1}H distance between two nuclei located in the same molecule and τ_R is the rotational correlation time.

Eqn (7.1)–(7.3) were used to explain the R_1 dispersions in paraffins, base and lubricating oils. The consistency of the model will be shown in the next section. For the application in pure paraffins, the reader can refer to ref. 35.

7.4 Thermal Degradation in ICE Lubricants

Since heat is a relevant feature in every internal combustion engine, it turns mandatory to assess how the molecular dynamics of lubricating oils becomes affected by its action.

7.4.1 Thermal Stress in Oil Bases

Experiments in base oils allow a better understanding of the efficiency of the additive package, once compared with the formulated lubricant (base plus additive pack). But, in turn, the analysis of the thermal degradation of the base itself facilitates the relaxometric characterization of the degradation process. In addition, since base oils do not contain any specific additives, it is usual to find interpretations considering them as an approximation to lubricating oils with depleted additives. However, although this assumption may make sense from the functional point of view, it seems to be hardly applicable in terms of a pure relaxometric analysis.

As an exemplary case, we consider a monograde SAE 40 (Normal 40) engine oil formulation from YPF (Yacimientos Petrolíferos Fiscales, Argentina) and its corresponding base (CB40). Figure 7.4 shows the spin–lattice relaxation dispersion rates obtained for the pristine and thermally degraded base oils. Black circles correspond to a pristine base oil sample, orange circles to a base oil sample after a thermal cycle at 90 °C (engine crankcase temperature) for 1000 h, and red circles correspond to a base oil sample after a cycle at 270 °C for 17 h. All measurements were performed at 30 °C. From this figure we may observe several interesting features. At first, we can see that, even without additives, the mineral base oil itself looks extremely resistant

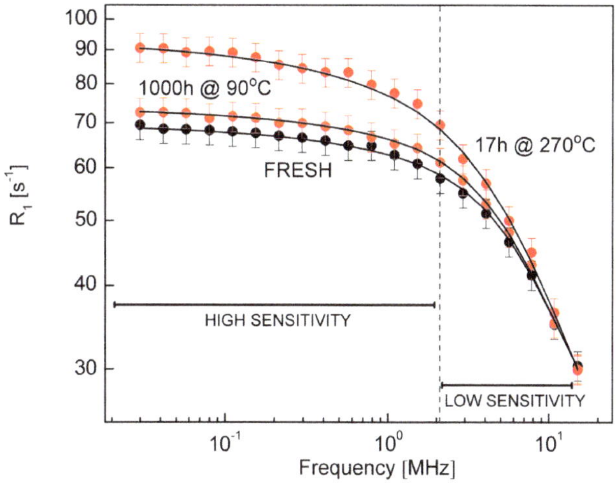

Figure 7.4 R_1 dispersions corresponding to mineral base oil under different thermal treatment conditions. Black circles correspond to a fresh sample, and orange and red to degraded samples (data from ref. 32). Straight black lines correspond to data fitting using the model described in Section 7.3.2. See text for further details.

after 1000 h at the typical crankcase temperature (90 °C). In fact, both fresh and degraded dispersions (black and orange symbols) are not strictly distinguishable within experimental errors, although a clear tendency is suggested by the data. This tendency tends to decrease if the experiment is repeated in the finalized lubricant after inclusion of the additive package. In contrast, just 17 h at a higher temperature (270 °C) is enough to create a noticeable degradation in the base oil, as shown in the figure. This fact suggests that the lubricating oil in an engine degrades mainly in the cylinders during combustion, where temperatures may be even higher. Another important feature is that the experiment becomes insensitive at Larmor frequencies higher than about 2 MHz. In fact, the highest sensitivity can be reached in the kHz frequency range and by running the test experiment at room temperature (or still better, by cooling the sample without freezing it).[29] Particularly, both R_1 and R_2 are extremely insensitive in the MHz frequency range unless measurements are carried out at temperatures around −10 °C. Figure 7.4 shows the performance of the model with fitting values that are consistent with the literature[30,62–64] (see Table 7.1 for details).

Figure 7.5 shows the relative contributions of the FFHS diffusional term (eqn (7.2)) and the Lorentzian term for the account of molecular rotations (eqn (7.3)). The diffusional term becomes dominant within the whole measured frequency range, while the rotational contribution becomes relevant only in the "sensitive" frequency interval (*i.e.*, for frequencies lower than 2 MHz).

Table 7.1 Parameters obtained after fitting the spin–lattice relaxation rate profiles using eqn (7.1)–(7.3) in the base oil corresponding to the YPF SAE40 formulation. For comparison, the sample was analyzed under three different thermal treatment conditions.

	CB40 base oil		
Parameters	Pristine	After 1000 h@90 °C	After 17 h@270 °C
d [m] $\times 10^{-10}$	2.4 ± 0.1	2.5 ± 0.1	2.7 ± 0.1
D [m^2 s^{-1}] $\times 10^{-12}$	9.2 ± 0.6	8.8 ± 0.6	5.1 ± 0.4
A_R [s^{-2}] $\times 10^8$	3.7 ± 0.4	3.6 ± 0.4	2.0 ± 0.2
τ_R [s] $\times 10^{-8}$	1.2 ± 0.2	1.3 ± 0.2	1.7 ± 0.4

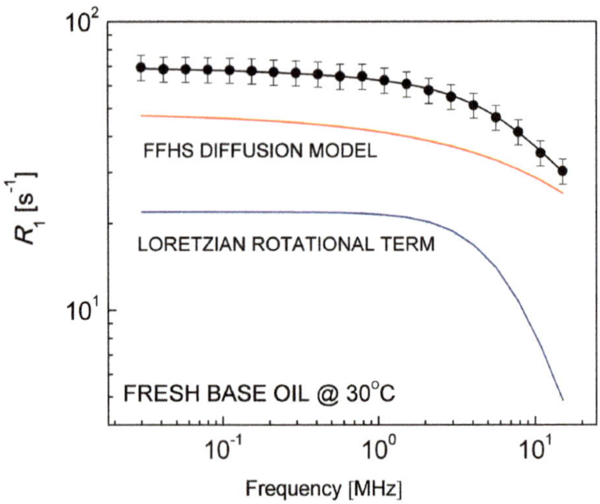

Figure 7.5 Relative contributions of diffusion and molecular rotations according to the model described in Section 7.3.2. Diffusion is dominating the relaxation rate in the whole measured frequency range, while rotations are only relevant for frequencies lower than about 2 MHz. Reproduced from ref. 32 with permission from Taylor & Francis, Copyright 2019.

Since the rotational contribution is not dispersive within the "sensitive" regime, within this Larmor frequency interval, the spin–lattice relaxation rate can be approximated by[32]

$$R_1(\nu_0) \cong R_0 - B\sqrt{\nu_0},$$

where $\nu_0 = \omega/2\pi$ is the Larmor frequency and the constant B is given by

$$B = \left(\frac{\mu_0}{4\pi}\right)^2 \rho\gamma^4\hbar^2 \left(\frac{8\sqrt{2}+2}{15}\right)\left(\frac{\pi}{D}\right)^{\frac{3}{2}}.$$

In this simplified model, the contribution from the molecular rotations to the relaxation is represented by the constant R_0. The diffusion

Table 7.2 Comparison of diffusion constants calculated from the model and its simplified version within the "sensitive" Larmor frequency interval.

	CB40 base oil		
Diffusion constant $\times 10^{-12}$ [$m^2 s^{-1}$]	Fresh	1000 h@90 °C	17 h@270 °C
Model	9.2 ± 0.6	8.8 ± 0.6	5.1 ± 0.4
Simplified model	9.1 ± 0.4	7.9 ± 0.3	5.3 ± 0.2

values obtained from this simplified model can be compared in Table 7.2.

Coming back to the consistency of the model, we observe that the values of the closest distance between two nuclei, d, did not show significant variations either as a function of the degradation level or as a function of the temperature. This result is consistent with the literature.[65] A slight decrease of the rotational amplitude (A_R) can be observed for degraded oils. This can only be explained as a slight increase in the effective distance r between the interacting spins. By considering that r appears in A_R to a sixth power, the net effect on r is hardly appreciable, *i.e.*, a minimum variation in r produces a noticeable change in A_R. This can be considered as an indicator that the molecular structure has changed in the degraded state. An increase of the rotational correlation time τ_R was observed for all aged samples (Table 7.1). This result is consistent with an increase in the viscosity and a lower molecular mobility.

The results here presented were analyzed by considering a mono-exponential relaxation scheme. However, due to the fact that the mineral base is a blend of paraffins of different molecular weights and other molecules that may contain aromatic or other functional groups, both spin–lattice and spin–spin relaxation processes may turn multiexponential. In such cases, we may think of a distribution of relaxation rates (or times) and an average representative value. This situation was considered for both bases and lubricants by using the gamma distribution approach.[29,30,33] Anyway, distributions tend to be narrow for relaxation experiments carried out at high Larmor frequencies ("insensitive" region) or/and high temperatures. In this limit, the average R_1 tends to the value obtained from a simple monoexponential fitting. It is worth mentioning that, even in cases where the distribution is not narrow, the general tendencies and the applicability of the model are preserved.

7.4.2 Aging of Automotive Lubricating Oils

In this section, we consider the case of fully formulated lubricants, that is, the base oil plus additive package. Figure 7.6 shows a comparative plot of the relative variation of the average relaxation rate (comparing fresh and degraded oils) for both the YPF mineral base (CB40) and the formulated

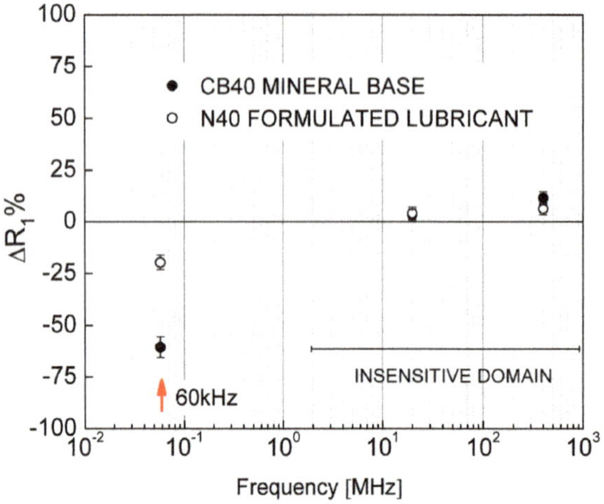

Figure 7.6 Mean R_1 relative variations (see eqn (7.4)) comparing the mineral base CB40 and the additivated oil N40 at different Larmor frequencies. Mean values of the relaxation rates were obtained by fitting the magnetization evolutions with a gamma distribution function. Reproduced from ref. 29 with permission from Elsevier, Copyright 2020.

lubricating oil (N40). From this figure, we learn the relevance of the Larmor frequency for the sensitivity of R_1 to oil degradation. In this plot the relative variation is defined as[29]

$$\Delta R_1\% = \frac{\left(\overline{R_1^{\mathrm{f}}} - \overline{R_1^{\mathrm{d}}}\right)}{\overline{R_1^{\mathrm{f}}}} \tag{7.4}$$

with $\overline{R_1^{\mathrm{f}}}$ being the mean value corresponding to the fresh and unused samples and $\overline{R_1^{\mathrm{d}}}$ being the corresponding mean value after a thermal degradation at 270 °C for 17 h. It can be seen from the figure that the experiments carried out at 20 MHz (using a fixed-field apparatus) are insensitive to degradation for both the mineral base and the additivated oil. No difference between the base and the formulated lubricant is observed. At 400 MHz (fixed-field apparatus) still fresh and degraded samples cannot be discriminated within experimental errors, although a bit higher sensitivity to degradation is observed. In contrast, a relative variation of the mean relaxation rate of the order of 60% can be observed for the base oil and around 20% for the additivated version. The data included in Figure 7.6 were obtained at a temperature of 30 °C. It is worth keeping in mind that the sensitivity of this experiment can be amplified by measuring at lower temperatures (but preserving the fluidity of the sample).

A similar analysis can be performed using R_2 relaxation. The presence of the secular term in R_2 provides certain sensitivity to the slow molecular dynamics, even within the insensitive Larmor frequency range for R_1. This sensitivity can also be amplified at low temperatures (−10 °C to 10 °C).

However, in any case, the relative differences between the base and the lubricating oils are insignificant within errors. That is, R_2 measured in the MHz frequency range has a much poorer performance than R_1 measured within the kHz Larmor frequency regime.[29]

From Figure 7.2, we observe that, for certain paraffins, R_1 shows a good sensitivity within the whole measured Larmor frequency range (10 kHz–15 MHz). Even in such cases, it is usually verified that the maximal sensitivity to degradation occurs for frequencies lower than 2 MHz. This suggests that, depending on the base's molecular blend and the used additives, the experiment can be completely insensitive in the MHz range (see Figure 7.1 for example) or be sensitive in the whole dynamic range (see Figure 7.2). In fact, an example of this last case was reported for a multigrade engine oil (Helix from Shell).[47] Figure 7.7 shows the relaxation rate dispersions for a fresh (0 km), a used (10 000 km) and an overused (20 000 km) multigrade Castrol Magnatec ICE lubricant formulation after use in a gasoline car engine subjected to mixed conditions (city, roads and dusty lanes).[35] The observed dispersions resemble the general characteristics of those reported for a Helix formulation from Shell years ago[47] after 5000 km (mainly dusty roads) but in another engine.

When analyzing Figure 7.7, it should be considered that different degrading factors were in play, not only thermal stress. Mixing with fuel and water, the presence of soot, dust, metals, paramagnetic impurities, *etc.* are playing an important role in the degradation. It is worth emphasizing that, after a centrifugation (9000 rpm for 20 minutes) of the degraded samples,

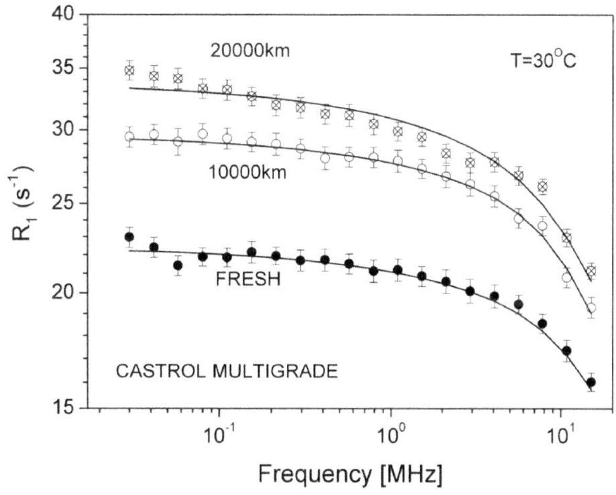

Figure 7.7 R_1 dispersion curves (symbols) and their corresponding fittings (solid lines) using the described model. Filled circles correspond to the fresh lubricating oil (0 km), and circles and crossed circles correspond to the used (10 000 km) and overused (20 000 km) lubricating oil samples, respectively. Experiments were performed at 30 °C. Original data are from ref. 35.

the supernatant did not show any difference in the relaxation dispersion profiles.[35] That is, the suspended solid particle concentrations are not influencing the relaxation properties in an observable way. This fact represents a big advantage if the interest is focused on the study of the molecular degradation of the lubricant itself. In many analytical techniques, used lubricants are cleaned by dilution with solvents and filtering mechanisms. Using solvents for NMR analysis may affect the results due to eventual residuals or interactions with the sample's molecules. Figure 7.7 and previous results obtained for a Helix formulation from Shell[45] suggest that the main changes in the relaxometric behavior occur after a runtime that can be much lower than the supposed useful life of the lubricant. However, this behavior may reverse depending on the engine conditions and the lubricant formulation.[33] A change of 45% in R_1 at low fields is observed in Helix after 5000 km, 32% after 10 000 km in Castrol Magnatec, and 20% between 10 000 km and 20 000 km. In this last case (Castrol), the mentioned changes in the relaxation dispersions imply a decrease in the diffusion constant of 22% after 10 000 km (used) and 41% after 20 000 km (overused). There is no significant change in the diffusion constant after centrifugation of the sample under the mentioned conditions. The rotational correlation time shows an increment of 30% under used conditions and 60% after overuse. There is no change in the correlation time after centrifugation. That is, in this case, we observe a linear variation of the parameters with the runtime of the lubricant in the engine.

Considering removal of suspended particles (an important part of them) through centrifugation, we may assume that variations in the diffusion constant and the rotational correlation time can be associated with changes in the chemical structure and oxidation of the hydrocarbonated chains. This assumption is in fact strongly supported by FTIR spectroscopy and other studies carried out in oil bases under controlled thermal stress.[32,35]

7.5 Concluding Remarks and Future Prospects

NMR relaxometry seems to be a highly sensitive method for the analysis of oils (lubricants, edible and others) that are subjected to different causes of stress. Our studies in ICE lubricants have confirmed so far that R_1 measurements at low field (Larmor frequencies lower than 2 MHz) and preferably at room temperature (or lower) meet optimal analytical conditions that are vastly superior to R_2 analytics in the MHz regime at low temperatures (−10 °C to 10 °C). The model we have been using, albeit simple, has been showing good consistency with pulsed field-gradient measurements of diffusion and other relaxation experiments performed on fixed-field instruments.[29]

A main advantage of the technique is the possibility for a fast acquisition of a relaxation dispersion profile. Due to the fact that oils are highly hydrogenated and the fact that most of such nuclei are subjected to an intense mobility, the FID signal arises with a very high signal-to-noise ratio. This essentially means that a fast scanning of the profile, even including a full receptor phase-cycle scheme, is feasible in a short time. Typical samples can be measured in less than 3 minutes without phase-cycling, which can be avoided if receptor offsets are properly settled. This fact constitutes a great motivation for further research in the field.

Applications in a short time include detection of adulteration and other forensic needs related to oils, fingerprint fast determination of an oil condition, testing efficiency and durability of additives, and quality determination of a base, among others. In any case, there is still some work pending to conclude about the limits of reliability of the technique and robustness of the model before we can define specific testing protocols for a full analytical package for oils based on FFC-NMR relaxometry.

References

1. *Synthetics, Mineral Oils and Bio-based Lubricants*, ed. L. R. Rudnick, CRC Press, Boca Raton, 3rd edn, 2020.
2. M. A. Al-Ghouti and L. Al-Atoum, *J. Environ. Manage.*, 2009, **90**, 98.
3. *Chemistry and Technology of Lubricants*, ed. R. M. Mortier, M. F. Fox and S. T. Orszulik, Springer, Dordrecht, 3rd edn, 2010.
4. J. M. Wakiru, L. Pintelon, P. N. Muchiri and P. K. Chemweno, *Mech. Syst. Signal Process.*, 2019, **118**, 108.
5. J. Zhao, D. Wang, F. Zhang, Y. Liu, B. Chen, Z. L. Wang, J. Pan, R. Larsson and Y. Shi, *ACS Nano*, 2021, **15**, 11869.
6. F. König, C. Sous, A. Ouald Chaib and G. Jacobs, *Tribol. Int.*, 2021, **155**, 106811.
7. P. Lu, H. E. Powrie, R. J. K. Wood, T. J. Harvey and N. R. Harris, *Tribol. Int.*, 2021, **159**, 106946.
8. A. Urban and J. Zhe, *Sens. Actuators, A*, 2022, **343**, 113671.
9. S. Yan, B. Ma, C. Zheng and J. Chen, *IEEE Access*, 2019, 7, 92110.
10. F. M. Pereira, J. A. Velásquez, J. L. Riechi, J. Teixeira, L. Ronconi, S. Riolfi Jr., E. L. Karas, R. A. Abreu and J. C. Travain, *Fuel*, 2020, **261**, 116418.
11. C. T. Pinheiro, M. J. Quina and L. M Gando-Ferreira, *Crit. Rev. Environ. Sci. Technol.*, 2021, **51**, 2015.
12. S. Widodo, D. Ariono, K. Khoiruddin, A. N. Hakim and I. G. Wenten, *Environ. Prog. Sustainable Energy*, 2018, **37**, 1867.
13. C. Sanchez-Alvarracín, J. Criollo-Bravo, D. Albuja-Arias, F. García-Ávila and M. R. Pelaez-Samariego, *Recycling*, 2021, **6**, 10.
14. F. J. Owuna, *Egypt. J. Pet.*, 2020, **29**, 251.
15. G. Öncü and E. Durak, *Proc. Inst. Mech. Eng., Part J*, 2021, **235**, 2144.
16. T. V. Sivanandhini, S. Unnikrishnan and V. Gedam, *Int. J. Sustainability Eng.*, 2021, **14**, 368.
17. H. E. Henderson, M. Fefer and A. Legzdins, *Am. Chem. Soc., Div. Pet. Chem.*, 1999, **44**, 284.
18. P. Mousavi, D. Wang, C. S. Grant, W. Oxenham and P. J. Hauser, *Ind. Eng. Chem. Res.*, 2005, **44**, 5455.

19. C. Besser, N. Dörr, F. Novotny-Farkas, K. Varmuzac and G. Allmaier, *Tribol. Int.*, 2013, **65**, 37.
20. V. Volli and M. K. Purkait, *Fuel*, 2014, **117**, 1010.
21. S. Zzeyani, M. Mikou, J. Naja, L. Bouyazza, G. Fekkar and M. Aiboudi, *Energy*, 2019, **180**, 206.
22. M. F. Fox, D. J. Picken, M. C. R. Symons and A. L. Thomson, *Tribol. Int.*, 1997, **30**, 417.
23. F. Cheng, T. Shibata, Y. Aoki and H. Hirata, *J. Magn. Reson.*, 2021, **332**, 107081.
24. K. Hosonuma, K. Yoshida and A. Matsunaga, *Wear*, 1985, **103**, 297.
25. B. Basu, M. P. Singh, J. S. Kapur, A. Nazakat, M. I. S. Sastry, S. K. Jain, S. P. Srivastava and A. K. Bhatnagar, *Tribol. Int.*, 1998, **31**, 159.
26. F. Owrang, H. Mattsson, J. Olsson and J. Pedersen, *Thermochim. Acta*, 2004, **413**, 241.
27. J. C. O. Santos, I. M. G. Santos, A. G. Souza, E. V. Sobrinho, U. J. Fernandes and A. J. N. Silva, *Fuel*, 2004, **83**, 2393.
28. M. A. D. Sousa Rios and S. E. Mazzetto, *J. Therm. Anal. Calorim.*, 2013, **111**, 553.
29. C. C. Fraenza, E. Förster, G. Guthausen, H. Nirlsch and E. Anoardo, *Tribol. Int.*, 2021, **153**, 106620.
30. E. Förster, H. Nirschl and G. Guthausen, *Appl. Magn. Reson.*, 2017, **48**, 51.
31. T. Rudszuck, E. Förster, H. Nirlsch and G. Guthausen, *Magn. Reson. Chem.*, 2019, **57**, 777.
32. E. M. Erro, C. C. Fraenza, L. J. Gerbino and E. Anoardo, *Mol. Phys.*, 2019, **117**, 983.
33. E. Förster, C. C. Fraenza, J. Küstner, E. Anoardo, H. Nirsch and G. Guthausen, *Mesurement*, 2019, **137**, 673.
34. L. F. Montes, E. C. S. Oliveira, A. C. Neto, S. M. C. Menezes, E. R. V. Castro and L. L. Barbosa, *Fuel*, 2019, **239**, 413.
35. E. M. Erro, L. J. Gerbino, C. C. Fraenza and E. Anoardo, *Magn. Reson. Chem.*, 2021, **59**, 447.
36. W. F. Bowman and G. W. Stachowiak, *Tribol. Int.*, 1996, **29**, 27.
37. C. D. Gamlin, N. K. Dutta, N. R. Choudhury, D. Kehoe and J. Matisons, *Thermochim. Acta*, 2002, **392**, 357.
38. J. C. O. Santos, I. M. García dos Santos, A. G. Souza, E. V. Sobrinho, V. J. Fernandes and A. J. N. Silva, *Fuel*, 2004, **83**, 2393.
39. I. Ahmad, J. Ullah, R. Khan and W. Ahmad, *Thermochim. Acta*, 2022, **713**, 179241.
40. S. Jahanmir, *Wear*, 1985, **102**, 331.
41. R. Bayat and A. Lethovaara, *Tribol. Int.*, 2021, **160**, 107002.
42. M. Ingram, I. Roots, T. Welham, M. Smeeth and C. Hamer, *Tribol. Trans.*, 2022, **65**, 1041.
43. A. V. Radulescu, I. Radulescu and F. Petrescu, *INCAS Bull.*, 2022, **14**, 125.
44. A. Abragam, *Principles of Nuclear Magnetism*, Oxford University Press, Oxford, 1st edn, 1961.

45. R. Kimmich and E. Anoardo, *Prog. Nucl. Magn. Reson. Spectrosc.*, 2004, **44**, 257.
46. *Field-cycling NMR Relaxometry: Instrumentation, Model Theories and Applications*, ed. R. Kimmich, New Developments in NMR Series, RSC, Cambridge, 2019.
47. M. Ballari, F. Bonetto and E. Anoardo, *J. Phys. D: Appl. Phys.*, 2005, **38**, 3746.
48. S. K. Naldu, E. E. Klaus and J. L. Duda, *Ind. Eng. Chem. Prod. Res. Dev.*, 1984, **23**, 613.
49. A. Adhvaryu, S. Z. Erhan, S. K. Sahoo and I. D. Singh, *Fuel*, 2022, **81**, 785.
50. S. Luff, NMR studies of the motion of molecules related in structure to commercial lubricants, PhD thesis, University of Surrey, Guilford – Surrey, 1995.
51. G. S. Kapur, A. S. Sarpal and A. K. Bhatnagar, *Lubr. Sci.*, 2002, **14**, 287.
52. G. S. Kapur and S. Berger, *Tribotest*, 2000, **6–4**, 323.
53. B. K. Sharma and A. J. Stipanovic, *Ind. Eng. Chem. Res.*, 2003, **42**, 1522.
54. M. Mondello and G. S. Grest, *J. Chem. Phys.*, 1995, **103**, 7156.
55. J. D. Moore, S. T. Cui, H. D. Cochran and P. T. Cummings, *J. Chem. Phys.*, 2000, **113**, 8833.
56. H. C. Torrey, *Phys. Rev.*, 1953, **92**, 962.
57. P. Conte, V. Mineo, S. Bubici, C. De Pascuale, F. Aboud, A. Macotta, D. Planeta and G. Alonzo, *Anal. Bioanal. Chem.*, 2011, **400**, 1443.
58. D. Rush, A. Jaeschke, J. A. J. Geenevasen, E. Tegelaar, J. Pureveen, M. D. Lewan, S. Schouten and J. S. S. Damsté, *Org. Geochem.*, 2014, **76**, 136.
59. L. P. Hwang and J. H. Freed, *J. Chem. Phys.*, 1975, **63**, 4017.
60. J. H. Freed, *J. Chem. Phys.*, 1978, **68**, 4034.
61. A. Rachocki and J. Tritt-Goc, *Food Chem.*, 2014, **152**, 94.
62. S. I. Mizushima and T. Simanouti, *J. Am. Chem. Soc.*, 1949, **71**, 1320.
63. D. C. Douglass and D. W. McCall, *J. Phys. Chem.*, 1958, **62**, 1102.
64. M. Hermansson, E. Johansson and M. Jansson, *J. Synth. Lubr.*, 1996, **13**, 279.
65. C. R. Jones, M. D. Greenhalgh, J. R. Bame, T. J. Simpson, R. J. Cox, J. W. Marshall and C. P. Butts, *Chem. Commun.*, 2016, **52**, 2920.

CHAPTER 8

MRI of Soil and Soil–Root Processes

A. POHLMEIER*[a] AND S. HABER-POHLMEIER[b]

[a] IBG-3, Research Centre Jülich, Germany; [b] IWS, University Stuttgart, Germany
*Email: a.pohlmeier@fz-juelich.de

8.1 Introduction

In the context of a growing world population, concomitant with a loss of acreage and an increasing climate change stress, there is an urgent need to increase crop productivity. At the same time, this increase must obey maximum sustainability in order to optimally use and conserve natural resources such as soil and water. One pathway followed is the adaptation of plants by either traditional breeding or genetic engineering.[1,2] At least as important is the improvement of the plant–soil interaction because plant productivity is strongly linked to the interaction with the soil as the main growth substrate. In this context, the main soil functions are anchoring of plants, supply with water and dissolved mineral nutrients, storage and buffering rainfall and artificial irrigation, and, last but not least, carbon sequestration. An important issue to be pointed out is that the interaction is not unidirectional because plants modify the surrounding soil in manifold ways. For example, growing roots break up solidified soil material, translocate water, and feed microorganisms such as symbiotic bacteria and mycorrhizal fungi.[3,4] All this makes the densely rooted soil zone a highly complex dynamic living system, whose thorough understanding is mandatory for improving sustainable plant production.

New Developments in NMR No. 32
The Environment in a Magnet: Applications of NMR Techniques to Environmental Problems
Edited by Pellegrino Conte, Delia Francesca Chillura Martino and Paolo Lo Meo

Published by the Royal Society of Chemistry, www.rsc.org

Many traditional experimental approaches are invasive or limited to visual inspection so there is a need for non-invasive methods with spatial resolution and monitoring of soil properties such as water content, pore size, and root distributions as well as their mutual interactions. Therefore, NMR methods are very suitable for these objectives because they are non-invasive and directly measure the amount and properties of the most important substance in the soil: water.

We organize this article as follows: since NMR is mostly known in the areas of chemical spectroscopy and medical imaging, we begin in Section 8.2 with some general remarks concerning the similarities and differences between spectroscopy, relaxometry, and imaging, and, in particular, the impact of relaxation in natural porous media on imaging. At the end of this chapter, we point out briefly that the important non-invasive methods such as MRI, X-ray CT, and neutron CT are complementary. In the second part of this article, Section 8.3, we present some exemplary work on the application of NMR in bare soil systems with a special focus on relaxometric characterization, also known as time domain NMR. Then we continue in Section 8.4 with imaging of root systems in soil as well as the mutual root–soil interaction before closing the article with a short summary.

8.2 Nuclear Magnetic Resonance

Nuclear magnetic resonance (NMR) is based on the interaction of nuclear spins having a magnetic moment with static and oscillating magnetic fields. A static magnetic field, B_0, which is applied in the z-direction, leads to Zeeman splitting of the energy levels proportional to the gyromagnetic ratio γ of the respective nucleus so that the energy levels are populated according to the Boltzmann distribution. A macroscopic magnetization develops which can be manipulated in many ways with magnetic radio-frequency fields. In soils and roots, the by far most relevant nucleus is that of the hydrogen atom in water, ^1H_2O, although some applications use so-called X-nuclei such as ^{19}F in surrogates of hydrocarbons and ^{23}Na for the investigation of salt migration.[5–7] If the system is irradiated with electromagnetic radiation of frequency $\upsilon_0 = \omega_0/2\pi = \gamma B_0/2\pi$, matching the difference of the Zeeman energy levels, the equilibrium Boltzmann distribution of the population of the Zeeman energy levels is disturbed. In most cases, the excitation consists of a radio-frequency pulse (rf-pulse) with a specific temporal shape in the time range of micro- to milliseconds so that the bandwidth, *i.e.* the spectral range $\Delta\omega$, is given by the inverse of the pulse length t_P:

$$\Delta\omega \cong \frac{2\pi}{t_P}. \tag{8.1}$$

The $\cong$ sign indicates that the exact bandwidth depends on the temporal pulse shape. For a "hard", rectangular pulse, it has the shape of a sinc function, which is the Fourier transform of the rectangular pulse. *Vice versa*, one frequently uses sinc- or Gauss-shaped rf-pulses yielding rectangular or normal frequency distributions, respectively, especially in imaging applications with slice selective excitation.[8] After pulsed excitation, the macroscopic magnetization in the excited volume evolves with time according to

$$M(t) = M_{0,\alpha} \exp(\mathrm{i}\omega t) \exp\left(-\frac{t}{T_2^*}\right)\left[1 - \exp\left(-\frac{t_\mathrm{R}}{T_1}\right)\right], \tag{8.2}$$

where $M_{0,\alpha}$ is the initial magnetization in the x–y plane perpendicular to the main magnetic field at the given flip angle α, T_2^* the effective transverse relaxation time describing the dephasing of transverse magnetization, t_R the repetition time between successive excitations, and T_1 the longitudinal relaxation time describing the re-establishment of the Boltzmann equilibrium along the z-axis. The term $\exp(\mathrm{i}\omega t)$ describes precessing magnetization in a plane normal to the static magnetic field, and the signal $M(t)$ is a complex quantity in this plane composed of a real and an imaginary part, equivalent to the magnitude and phase. Both parts can be evaluated during further image processing. This makes a significant difference to X-ray computed tomography (X-ray CT) and neutron computed tomography (neutron CT), which detect only the magnitude of the transmitted radiation (see below).

The relaxation times are controlled by magnetic dipole–dipole interactions in the excited spin system with fluctuating magnetic fields caused by molecular motions.[9,10] The apparent transverse relaxation rate is defined by the relaxation time T_2 and a contribution from the loss of phase coherence of transverse magnetization components, which precess at different frequencies in an inhomogeneous static magnetic field:

$$\frac{1}{T_2^*} = \frac{1}{T_2} + \frac{1}{T_{2,\mathrm{inhomo}}}. \tag{8.3}$$

An analogous term for longitudinal relaxation does not exist. In NMR spectroscopy with pulsed excitation, $M(t)$ is Fourier-transformed to yield the NMR spectrum, where the linewidth at half height is $\Delta\nu_{1/2} = 1/\pi T_2^*$. Especially for soil materials, the transverse relaxation time T_2^* can become very short due to the high contents of paramagnetic Fe, Mn and clay minerals so that linewidths of up to 20 kHz (24 000 ppm) were observed.[11,12] A consequence is that the excitation bandwidth should cover the entire spectrum unless one wants to excite only parts of the spectrum for chemical shift selective imaging or localized spectroscopy.[13–16]

8.2.1 Relaxation

The analysis of the NMR signal with respect to T_2 and T_1 or diffusion forms the wide field of NMR relaxometry, also known as time domain NMR. This is of particular interest since the relaxation behaviour informs about the local molecular dynamics of the fluid in porous systems, at solid–liquid interfaces and in gel phases. Common methods in soil science rely on saturation or inversion recovery for measuring T_1 and the CPMG echo train for determining T_2.[17–19] The latter profits from elimination of line broadening in time invariant magnetic field gradients by monitoring multiple Hahn echoes.[20] Here, Hahn echoes are created by an additional series of 180° radio-frequency (rf) pulses which refocus the dephasing magnetization components in stationary magnetic field gradients. The signal intensity is controlled by the echo time t_E which is twice the period between the centres of the initial excitation pulse and the first refocussing 180° pulse or the period between the centres of successive 180° pulses:

$$M(n_E t_E) = M_0 \exp\left(-\frac{n_E t_E}{T_2}\right)\left[1 - \exp\left(-\frac{t_R}{T_1}\right)\right], \tag{8.4}$$

where n_E denotes the number of equally spaced echoes. Please note that the signal depends no longer on T_2^* but only on T_2; see also eqn (8.3).

In natural porous media, some special processes control the relaxation mechanisms due to the interaction of water with the pore walls. They have been thoroughly studied experimentally and theoretically, driven by their huge economic importance in the context of oil, gas and water exploitation.[21–24] The most popular model is here the biphasic exchange model of Brownstein–Tarr, where the fluid phase in the pore is partitioned into a thin surface layer and a bulk phase with rapid mutual exchange by translational diffusion.[25,26] At the pore wall, relaxation is accelerated by partial immobilization prolonging the rotational correlation time and by the presence of local relaxation centres. These in turn are a consequence of the distortion of the magnetic field by susceptibility differences between the matrix and pores or the presence of paramagnetic ions at or near the surface. Both give rise to enhanced dipole–dipole interactions and transverse as well as longitudinal relaxation processes become faster.[27,28] $1/T_2$ in eqn (8.4) may additionally be accelerated by translational diffusion in external or local magnetic field gradients. In summary, the relaxation enhancement is described by

$$\frac{1}{T_1} = \frac{1}{T_{1,\text{bulk}}} + \rho_1 \frac{S}{V} \tag{8.5}$$

and

$$\frac{1}{T_2} = \frac{1}{T_{2,\text{bulk}}} + \rho_2 \frac{S}{V} + \frac{1}{T_{2,\text{diffusion}}}, \tag{8.6}$$

where S/V is the pore surface-to-volume ratio and the specific relaxivities $\rho_{1,2}$ in units of speed describe the local relaxation effects.[23] The diffusional acceleration term in eqn (8.6) depends on the local ratios of diffusion length, structural length and dephasing length in internal gradients[29] and it is identical to the classical Bloch–Torrey equation under conditions of short echo and long dephasing times:

$$\frac{1}{T_{2,\text{diffusion}}} = \frac{1}{12}\gamma^2 g^2 D_0 {t_E}^2. \tag{8.7}$$

Physical variables are the gyromagnetic ratio γ (for hydrogen atoms: $\gamma_H = 2.68 \times 10^8$ rad $T^{-1} s^{-1}$), the effective internal gradient g, the translational diffusion coefficient D_0 and the echo time t_E. Thus, the strength of internal gradients can be estimated from effective transverse relaxation times with the CPMG echo train method with different echo times.[30] The amplitude and phase of the echoes can also be modified by additional switching of pulsed magnetic field gradients to encode the signal for diffusion and flow. This methodology is termed pulsed field gradient NMR (PFG-NMR).[31] Please also note that NMR in the time domain has recently evolved to two-dimensional methods such as T_2–T_2, T_1–T_2, or T_2–D_0 correlation methods analogous to the well-established two-dimensional NMR spectroscopy.[32–36]

8.2.2 Imaging

The NMR signal (eqn (8.2) or (8.4)) can be modified by switching additional magnetic field gradients applied as pulses in defined spatial directions. For example, if a gradient of strength G_x is present in the x-direction of the Cartesian laboratory coordinate frame during the detection of a Hahn echo, the complex signal resolves contributions resulting from the spin density along that direction with frequencies $\omega(x) = \omega_0 + \gamma_H G_x x$. This type of encoding is termed frequency encoding. Fourier transformation yields the frequency distribution, which is proportional to the spatial origin of the spin density, *i.e.* one obtains a frequency encoded profile or, synonymously, a 1D image. If, on the other hand, a magnetic field gradient is applied for example in the y-direction *before* monitoring the signal, the phase of the signal is changed according to $\Delta\varphi(y) = \gamma_H G_y t y$, and this type of encoding is termed phase encoding. The third way of manipulation of the signal in space is the switching of a gradient during the excitation pulse. Analogous to the switching during the detection time, it modifies the frequency, but only for the excited region the resonance condition is met. For example, a sinc-shaped excitation pulse, applied simultaneously with a gradient along the z-direction, excites a slab normal to z with a thickness of $\Delta z = 2\pi/\gamma_H G_z t_p$. The three types of spatial encoding are combined orthogonally, often, but not necessarily, along each direction of the Cartesian coordinate system (see Table 8.1).

Please note that frequency as well as phase encoding determines the object in a reciprocal space, termed the k-space, with the coordinates $k_x = \gamma_H G_x t_{acq}$ and $k_y = \gamma_H G_y t_1$, where t_{acq} is the acquisition time and t_1 is the duration of the phase encoding gradient varied in steps of ΔG_y. For the frequency-encoded direction, n_{acq} complex data points are recorded in t_{dw} intervals during the acquisition time, *i.e.* $t_{acq} = n_{acq} t_{dw}$. Multi-dimensional discrete Fourier transformation (DFT) yields the real image with the resolutions $\Delta r_{x,y} = 2\pi/k_{max,x,y}$ and the field of view $FOV_{x,y} = 2\pi/\Delta k_{x,y}$. Before DFT, one can apply filters such as zero-filling, where the k-space is expanded by zeros which improves the Fourier transformation, or sine or sine^2 filters leading to smoothed images.

Table 8.1 Some frequently used MRI pulse sequences for soil and plant research.

Type	Example	Mode of encoding per direction[a] fe	pe	ss	Weighting	Total time[b]	Target Soil	Roots	^{23}Na
Spin echo	SEMS	1	1	1	$(t_E \ll T_2)$ and $(t_R > 5T_1)$: M_0	$n_{av}t_Rn_{pe}$	+	++	–
	SE3D	1	2	—		$n_{av}t_Rn_{pe1}n_{pe2}$			++
	MEMS	1	1	1			+	++	
	RARE	1	1	1		$n_{av}t_Rn_{pe}/n_E$	+–	++	–
	3DRARE	1	2	—		$n_{av}t_Rn_{pe2}n_{pe1}/n_E$			++
Gradient echo	FLASH	1	1	1	T_2^*, T_1		–	+	
Single point imaging	SPRITE	—	3	—	T_1	$n_{av}t_Rn_{pe1}n_{pe2}n_{pe3}$	++	+–	
Ultrashort	UTE	3	—	—	T_2^*, T_1	$n_{av}t_Rn_s$	+	+	
	ZTE	3	—	—			++	+	
	SWIFT	3	—	—			n.a.		

[a] fe: frequency encoding, pe: phase encoding, ss: slice selection.
[b] n_{av}: number of averages, n_{pe}: number of phase encoding steps, n_E: number of echoes, t_R: repetition time, n_s: number of spokes in the radial k-space.

Depending on the nature of the detected signal, one may discriminate between different types of imaging:[37] spin echo and multi-echo images encode the signal by a single Hahn echo or a CPMG echo train, respectively. Examples are SEMS (single-echo multi-slice), MSME (multi-slice multi-echo), and RARE (rapid acquisition with relaxation enhancement) or FSE (fast spin echo), where multiple lines in the *k*-space are addressed by an echo train with in-loop switching of a phase encoding gradient pulse. This accelerates the total time by a factor corresponding to the number of n_E echoes. Due to the compensation of the decay in stationary inhomogeneous fields, spin-echo based pulse sequences are most convenient for imaging roots and macropore water in soils as well as imaging dedicated properties such as relaxation time, diffusion, or velocity.

Omitting the time-consuming refocusing rf-pulse leads to the family of gradient echo sequences. An example is the popular FLASH sequence (fast low angle shot), which is fast but sensitive to T_2^*, and therefore not very convenient for soil systems due to the rapid T_2^* decay. The same is true for even faster sequences such as EPI (echo planar imaging), which encodes the *k*-space by multiple gradient echoes.

Much more convenient for soil systems are sequences with ultrashort detection times. The earliest are the single point imaging (SPI) and its successors, such as SPRITE (single point imaging with T_1 enhancement) and related derivatives.[38–41] Here the FID is directly probed by one single point a few μs after excitation, leading to one single data point in the *k*-space. They are time-consuming in the sense that all directions are scanned successively point-by-point and thus the total time scales with the product of all points per direction. Their great advantage is that all points are phase-encoded and no frequency artefacts occur.[40]

On the other hand, sequences with multiple detected points along the FID perform significantly faster by using a gradient during the decay of the FID and recording of multiple points per excitation. ZTE (zero t_E) switches the spatial encoding gradients before the excitation pulse, while UTE (ultrashort t_E) does it directly after excitation.[42,43] A similar pulse sequence is SWIFT (sweep imaging with Fourier transformation) with quasi-simultaneous excitation and detection,[44] which has not yet been applied to soil and plant systems. All these methods typically use small excitation angles α in combination with short repetition times t_R so that the object is at least in parts in partial saturation, and the image is frequently T_1-weighted. The extent of T_1-weighting can be assessed by calculation of the Ernst angle which yields the optimal intensity by a compromise between t_R, T_1 and α:

$$\cos(\alpha_{\mathrm{Ernst}}) = \exp\left(-\frac{t_R}{T_1}\right). \tag{8.8}$$

To reduce T_1-weighting, one should set the actual flip angle smaller than the Ernst angle; see also below.[45]

Further, sequences such as UTE and ZTE probe the *k*-space radially and therefore require re-gridding into the Cartesian *k*-space before digital

Fourier transformation. However, their great advantage is the monitoring speed due to the high number of acquisition points per excitation, at the expense of possible T_2^* weighting since the FID proceeds during acquisition and the risk of frequency encoding artefacts. Therefore, one should ensure that the acquisition time, $t_{acq} = n_{acq}/bw$, is significantly shorter than the FID decay time which has a second advantage of reducing the risk of frequency-shift artefacts since the external gradients are now stronger than the internal fields (see Section 8.3).

8.2.3 Image Contrast

Especially in porous media, the signal can be strongly influenced by relaxation properties. This is not necessarily a drawback since it allows creating contrast by various mechanisms. Contrast in an image is defined as the difference of pixel intensity (voxel intensity in 3D images) between adjacent regions with different physical or chemical properties. For example, a soil material has short T_2 and T_1 relaxation times and root tissue relaxes comparatively more slowly. This allows contrast enhancement by using longer echo times, which suppress the signal from the soil and therefore make the roots appear relatively bright. *Vice versa*, since T_1 of the roots is longer than that of soil, the root signal may be suppressed by short repetition times, t_R. Another option to create image contrast is by switching additional modules in a preparation period before the imaging part, also termed filters.[46] For example, one can prepare the magnetization by an inversion recovery module, *i.e.* an additional 180° pulse before the 90° excitation pulse. It will invert the equilibrium magnetization from the $+z$ to $-z$ direction, where it is subject to T_1 relaxation only since no transverse magnetization exists. Eqn (8.4) is modified as follows:

$$M(t_E, t_R, t_{inv}) = M_0 \exp\left(-\frac{t_E}{T_2}\right)\left[1-\exp\left(-\frac{t_R}{T_1}\right)\right]\left[1-2\exp\left(-\frac{t_{inv}}{T_1}\right)\right], \quad (8.9)$$

where t_{inv} is the inversion time. For simplification of the evaluation, one should set the middle term to unity by choosing $t_R \gg T_1$. Thus, rapid T_1 decaying compartments such as soil can be faded out if the inversion time is set so that the rightmost term is approximately zero, *i.e.* by $t_{inv} = -\ln(0.5)T_1$. Figure 8.1 illustrates this effect.

In Figure 8.1(b) roots are visible in the wet soil due to their higher water content but they appear blurry. Incorporating an inversion recovery pulse at an inversion time of 1.5 s, Figure 8.1(c) suppresses the signal from the soil almost completely and reveals fine details of the roots. Especially, the epidermis and the stele of the taproot in the centre are now clearly visible due to their shorter T_1. Also note the bright area around a lateral root near the centre, which is already closer to thermal equilibrium again due to a shorter T_1 time than the signal from the bulk soil. It is probably caused by exudation of mucilage by the root or presence of bacteria, which reduce local relaxation times.[47]

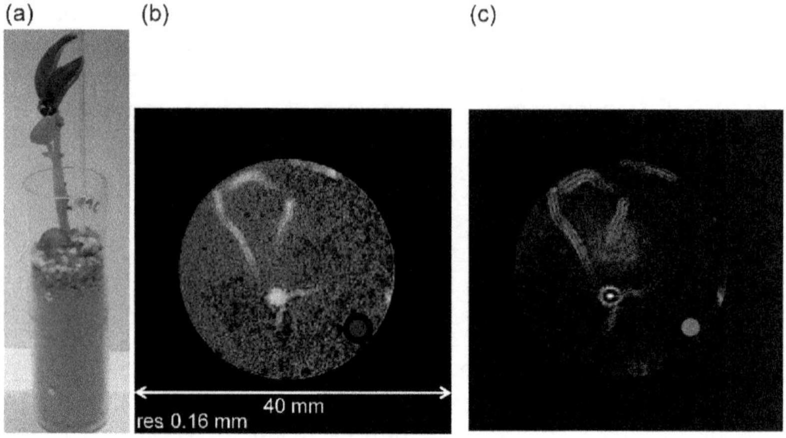

Figure 8.1 Axial cross section through a *Vicia faba* plant (B116) grown in sandy soil. Multi-slice single echo pulse sequence with $t_E = 3.4$ ms, $t_R = 9$ s, matrix size = 256 × 256, resolution = 0.16 mm × 0.16 mm, and slice thickness = 1.8 mm. The photograph in (a) shows the 10-day-old plant in the MRI cuvette in water-saturated sand and the marker tube. (b) Reference image without inversion recovery preparation. (c) With an additional 180° inversion recovery (IR) pulse with an inversion time of $t_{inv} = 1.5$ s.

8.2.4 MRI, X-ray CT and Neutron CT

The most important 3D non-invasive imaging techniques for high resolution soil and plant studies are MRI, μ-X-ray CT (μ-XRCT)[48–52] and neutron computed tomography (NCT).[53,54] Somewhat lower resolutions are obtained with electrical resistivity and impedance tomography methods, but these can be readily applied in the field.[55–57] The physical process for μ-XRCT is the absorption of X-rays by the electronic shells of atoms. Thus, the contrast relies on the electron density of the different materials. In soils, the mineral phase with dense quartz, clay or carbonate particles differs considerably from the less dense solution and gas phases. In addition, plant tissues like roots possess low density, which does not differ much from that of water so that roots are difficult to distinguish from the mineral phase. However, root imaging by μ-XRCT has become a standard method in recent times.[58,59] Its huge advantage is the very high resolution down to 1 micron in 3D so that even root hairs could be imaged.[48]

Neutron CT or neutron imaging has a lower resolution than X-ray CT. It is on the order of some tens of microns but offers very high sensitivity to all hydrogen atoms. The underlying physics is the scattering of neutrons by atomic nuclei, quantified by the so-called scattering cross section, which is a specific property of the individual nucleus.[53,60] In soil and plants, the contrast is primarily due to the different contents of ^{1}H atoms in different compartments, *e.g.* roots or soil, since the cross sections of other abundant atoms (^{12}C, ^{16}O, . . .) are relatively similar. However, a convenient method to enhance contrast in soil materials is the use of D_2O due to the high difference in the scattering cross sections of ^{1}H and ^{2}D nuclei.[61] Both μ-X-ray CT and neutron CT have a feature in common that one records 2D projections of the transmitted beam in real space while the object is rotated. The rotation angles are small in the range of 0.5° to 1° in most applications so that the total number of projections can be very large leading to high resolution. The 3D image is finally reconstructed by numerically filtered 3D back-projection techniques.[52] Some commonalities and differences are listed in Table 8.2.

The key physical information from X-ray CT, neutron CT and MRI is complementary, and their combination is highly valuable for investigating transient processes such as flow on the basis of structural information. Here the great advantage of MRI, namely its sensitivity to motion of liquids, is optimally combined with high resolution of solid structures by X-ray CT. For instance, a very recent study proved by means of flow MRI that micropores, which were not detectable by X-ray CT, carry about 1/3 of the total flow in a certain porous carbonate rock, since the X-ray CT resolution was limited to about 1 μm.[62] In earlier works, it was proven that infiltration into natural soils detected by MRI follows preferential flow pathways, visible in X-ray CT images as zones of reduced density[63] (see Section 8.3). In Section 8.4, we will point out the added value of the investigation of root–soil interactions by the combination of MRI with neutron CT.

8.3 Bare Soil

As pointed out above, the contrast and quality of MRI images rely to a greater extent on the relaxation properties of the materials. This fact motivated the groups of Hall and Cislerova in the 1990s to investigate the relaxation times of different soil materials, covering a wide texture range from clay loam to sand.[11] The linewidths at $B_0 = 2$ T, corresponding to the inverse relaxation times T_2^* (eqn (8.3)), showed a clear increase with the mass-based magnetic susceptibility, where the linewidths ranged from a few 100 Hz up to 20 000 Hz. The dependencies of the other relaxation times T_1 and T_2 were not so clear, although scattered trends of decreasing relaxation times with increasing contents of clay, iron oxide and exchangeable cations were

Table 8.2 Characteristics of 3D imaging techniques.

	MRI	μ-XRCT	Neutron CT
Principle	Resonance	Absorption	Scattering cross section
Active	Liquid phase (water) NMR active nuclei: ^{1}H, ^{23}Na, ^{19}F, ...	Heavy atoms	Hydrogens
Visibility	Pore space, cell liquid	Solid phase	Total hydrogens
Contrast	Water content, relaxation, diffusion, flow	Density	Hydrogen density
Typical resolution	30–300 μm	1–10 μm	20–50 μm

obvious. Later, they investigated the dependence of different types of NMR signals important for imaging application, namely FID, spin echo, inversion recovery (IR) and CPMG amplitudes, on their ability to reflect water content in fine sand and coarse-grained sandy loam soil.[64] Most reliable was the water content detected by the FID amplitude, whereas for the spin echo and CPMG amplitudes significant signal decays due to fast relaxation were observed. Especially for the "NMR-difficult" material, non-linear relations between the NMR signal and the water content became obvious. Based on these results, they then investigated infiltration processes in natural soil cores by combining MRI with XRCT[63] and discovered preferential water-flow pathways in less dense regions of soil cores as determined by XRCT images. Later, T_1 mapping and tracer tracking were added to this type of study, specifying the effect of entrapped air on the flow processes.[65]

Relaxation times are not only necessary for the interpretation of images, but also highly important for the characterization of the water relations in the pore system itself. The term water relation means in this context not the mere water content, but more specifically the interaction with the pore system. This implies the retention of water inside the porous medium in equilibrium with external pressure or suction and the hydraulic conductivity, which is also strongly influenced by the pore-size distribution and the internal connectivity.[66,67]

Water retention is usually described by the so-called water characteristic or water retention curve.[68] An example for a silt loam soil is depicted in Figure 8.2(b). These typically s-shaped curves plot the water content as a function of the logarithm of external pressure ψ or hydraulic head $h=\psi\rho^{-1}g^{-1}$ and may show hysteresis depending on whether a soil material is drying or wetting. The water retention curve reflects the pore-size distribution since narrow pores attract water more easily than large pores according to Young–Laplace's law of capillary suction.[69] Thus, different ranges in the water retention curve are classified as follows: the macropore region between a total water potential $\psi=0$ kPa and about -0.3 kPa where water flows freely in 1 to 10 mm wide pores. Mesopores between 1 mm and 30 μm are located between and within aggregates or individual sand grains and correspond to $\psi>-5$ kPa, and all smaller pores below 30 μm are termed micropores corresponding to $\psi<-5$ kPa.[70] This is the central range of the water retention curve where water is termed also as the capillary bound.[68] It is available for plants down to the permanent wilting point at $\psi=-1470$ kPa. The latter value corresponds to a capillary diameter of 0.2 μm, and pores below that size are termed clay pores.[66] A comparable classification into 5 categories has been proposed by Clothier,[71] which differs only in some threshold values for the pore diameter and defines the clay region as cryptopores with diameters <0.5 μm having $\psi<-0.6$ MPa. Transferring this system to the exemplary soil in Figure 8.2(b), we find a smaller fraction of 20% macropores, a wide range of meso-, micro- and ultramicropores (60%), and 20% cryptopores. These data correspond well to the textural composition of the soil which is 15% sand, 68% silt and 17% clay.[36]

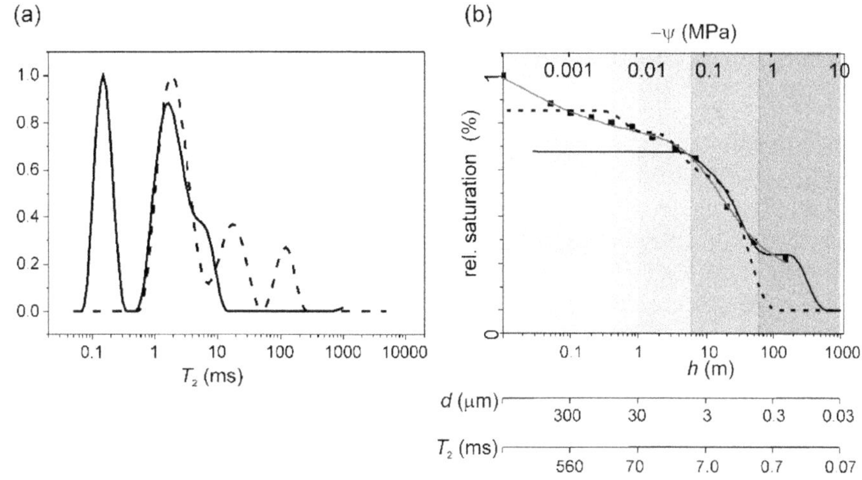

Figure 8.2 (a) T_2 relaxation spectra of a silt loam soil core at $(2\pi)^{-1}\gamma B_0 = 6.4$ MHz measured with $t_E = 0.15$ ms (----) and 0.05 ms (—). The usage of short t_E reveals the existence of a very fast mode at $T_2 = 0.2$ ms. (b) ■ Measured water retention curve. Black curve: WRC calculated from the microcoil with $t_E = 0.05$ ms. Grey curve: fit with a bimodal van Genuchten model. Broken curve: WRC calculated from the volume coil. Note that the large pores (25% of total porosity) are not represented due to the limited number of echoes, $n_E = 20\,000$. The shaded areas correspond to the pore diameter classification of Clothier:[71] macropores, >75 μm; mesopores, 30 to 75 μm; micropores, 5 to 30 μm; and ultramicropores, 0.5 to 5 μm. Smaller pores are termed cryptopores[71] or clay pores.[66]

Pore size distributions (PSD) in rocks and soils have been intensively studied by time-domain NMR, mostly by the analysis of T_2 relaxation spectra with eqn (8.6), in fewer cases also for T_1.[23,25–27] The crucial parameters are the surface relaxivities ρ_1 and ρ_2, which can be obtained from the ratio of the mean relaxation rates and inverse pore size S/V. The latter quantity is commonly obtained by gas adsorption or mercury intrusion. Alternatively, one can analyse water-retention curves with respect to the underlying PSD and match the NMR-derived PSD by variation of ρ_1 and ρ_2 and a model for the pore, for instance, an infinite cylinder. The procedure is as follows: first, the T_1 or T_2 distribution functions (Figure 8.2(a)) are integrated yielding the cumulative distribution function, which is proportional to the degree of saturation S_{NMR} of the pores with the corresponding relaxation time. Next, the T_1 or T_2 relaxation times are converted into equivalent radii using eqn (8.5) or (8.6), respectively, with a reasonable estimate of the surface relaxivities. Next, the equivalent radii are converted into hydraulic head parameters by the Young–Laplace equation for capillary rise:

$$h = \frac{2\sigma\cos(\varphi)}{rg\rho}, \tag{8.10}$$

where σ is the water–air surface tension (0.073 N m^{-1}), φ is the contact angle (usually 0° for a wetting fluid), and ρ is the density of water. Finally, one plots the relative saturation S_{NMR} *versus* h and compares this curve with the experimentally obtained water retention curve S_{WRC} (see Figure 8.2(b)) and fits ρ_1 or ρ_2 by matching of both curves. For the exemplary soil in Figure 8.2, we found $\rho_2 = 100$ µm s^{-1}, which is in the upper range of reported values.[30,72–76] Typical parameters of surface relaxivity ρ_2 of rocks and soils range from a few tens to a few hundreds of µm s^{-1}, the wide range of values being mainly due to the method used. Gas adsorption usually measures a high value of S/V since small gas molecules probe also small pores and even surface roughness. In this case, the corresponding surface relaxivities are small. In contrast, if the relevant S/V values are small as found by water retention curves, the corresponding surface relaxivities are high.[73,74,76]

NMR offers an alternative way to determine average pore sizes by diffusion measurement with pulsed gradient spin echo (PGSE). The background is that the diffusion is unrestricted on a diffusion length scale smaller than the pore size and becomes increasingly restricted for longer time scales.[77] The use of S/V values determined by this method leads to consistent values of the transverse surface relaxivities in the range between 2 and 7 µm s^{-1} at 6.4 MHz for porous media such as glass beads, purified and natural sands and a natural sandy loam soil.[75]

Analogous to 2D spectroscopy, 2D relaxation spectra of soil materials can also be obtained in terms of T_1–T_2 correlation or T_2–T_2 exchange diagrams. The latter method allows the assessment of the connectivity between pores of different size classes. In essence, CPMG echo trains are monitored in the 2nd dimension as a function of increasing numbers of echoes in a 1st, indirect dimension separated by an exchange period, τ.[32] The result is a correlogram, exemplarily shown in Figure 8.3(a). Subsequent 2D Laplace transformation yields a T_2–T_2 exchange map, where the signals on the diagonal are caused by water pools remaining in their original pores. More interesting features are off-diagonal peaks which indicate water pools that have exchanged between pores of different size classes. More precisely, for the example soil in Figure 8.3, the existence of off-diagonal peaks near $T_{2,\text{indirect}} = 2$ ms and $T_{2,\text{direct}} = 5$ ms proves the rapid exchange of water between clay-bound and capillary bound fractions.[36]

These few examples demonstrate the importance of determining also the rapidly relaxing components of soils to assess the entire water dynamics. For imaging, this means that one should measure the NMR signal as soon as possible after excitation to cover also the rapidly relaxing fractions of water in the soil matrix. As described in Section 8.2, a frequency artefact free method is the family of single point imaging pulse sequences, which avoids any frequency encoding by probing just a single point immediately after excitation. The cost is the long recording time or the limited resolution since each point in the *k*-space is addressed separately. However, 1D or 2D profiles can be scanned in a sufficiently short time. A compromise is the use of short repetition times t_R. Here one should be aware of the risk of T_1 saturation

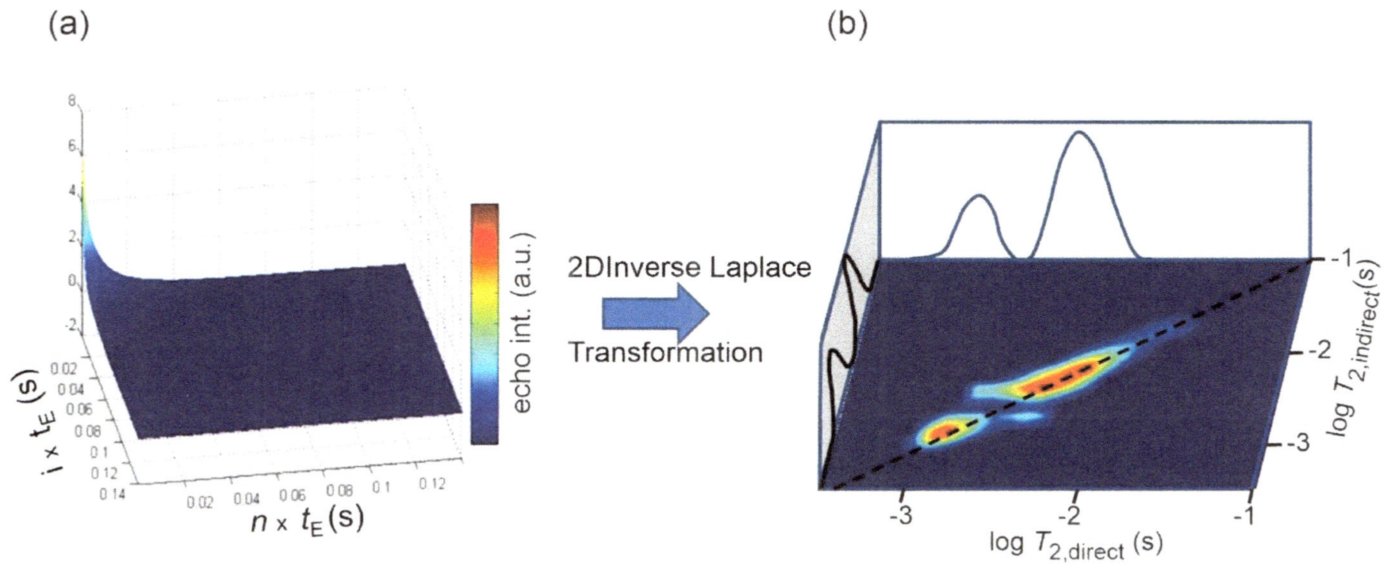

Figure 8.3 (a) T_2–T_2 relaxation correlation diagram of saturated silt loam soil from Merzenhausen, Germany. The mixing time was 4 ms and the echo time was 0.1 ms. (b) Exchange map obtained by 2D Laplace transformation of the correlation diagram. The curves on the left-hand and top sides are the projections along the indirect and direct detection axes, respectively. The signals on the main diagonal correspond to the transverse relaxation distribution functions and reflect water remaining inside the actual pore size class. The off-diagonal signals are caused by exchange between the pore size classes. Reproduced from ref. 36 with permission from John Wiley & Sons, Copyright 2010.

resulting in intensity loss for certain water pools. By proper choice of sequence design and scan parameters, this risk can be minimized as shown by the group of Balcom.[45] They used a centric scan SPRITE sequence where the centre of the *k*-space, which controls the absolute image intensity, is scanned after the establishment of thermal equilibrium at a flip angle smaller than the Ernst angle (see eqn (8.8)). It avoids the signal intensity decrease by changes in T_1 and the authors could obtain true water-content profiles for a horticultural planting soil even at low water contents.

One can also combine the single point imaging method with a preparatory module or filter that encodes the signal's sensitivity for T_2. This sequence is consequently termed the SE-SPI sequence.[78] Its usefulness was demonstrated in the example of T_2 relaxation time profiles in natural silt loam soil columns with a resolution of 2 mm.[41] For an initially saturated soil column, T_2 remained almost constant over the entire length with a slight decrease toward the surface. After 11 days of evaporation, a 15 mm deep secondary evaporation front was detected, below which a uniform T_2 profile had developed. Its mean value of about 1 ms indicates that water was mostly clay-bound (see Figure 8.4).

Such receding of the evaporation front below the soil–air interface is typical for porous materials in the so-called stage-II phase at lower water contents, where the further evaporation is no longer sustained by liquid transport in narrow capillaries or thin films, but only by vapour diffusion.

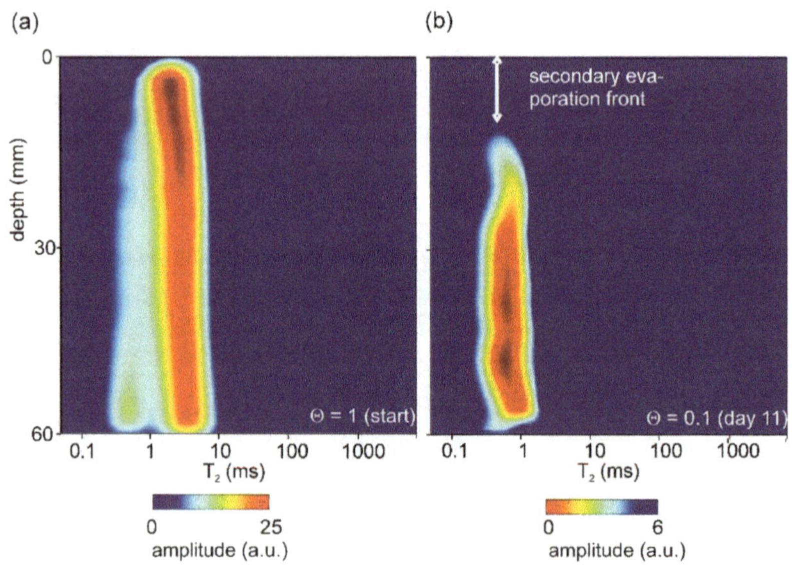

Figure 8.4 Vertical T_2 relaxation profiles of a silt loam soil column, obtained by an SE-SPI pulse sequence. (a) At the start of the evaporation at a saturation degree of 1.0 and (b) after 11 days at a saturation degree of 0.1. Reproduced from ref. 42 with permission from Springer Nature, Copyright 2016.

We could further prove this transition for the abovementioned soil by the combination of the experimental results with a soil physical model that includes also the vapour transport of water.[79]

The disadvantage of long acquisition times or reduced image dimensionality of the single point methods can be overcome by detection of multiple points per FID.[†] The image intensity is given by generalizing eqn (8.2):

$$M(t) = M_0 \exp\left(-\frac{t}{T_2^*}\right) \frac{\left[1 - \exp\left(-\frac{t_R}{T_1}\right)\right]}{\left[1 - \cos(\alpha)\exp\left(-\frac{t_R}{T_1}\right)\right]} \sin(\alpha). \quad (8.11)$$

The denominator is unity for $\alpha = 90°$ as typical for spin echo experiments and eqn (8.11) becomes identical to eqn (8.2) if the precession term is omitted. Since most pulse sequences with ultrashort detection times use short repetition times lower than T_1, they are inherently weighted by T_1 if the flip angle is not significantly smaller than the Ernst angle (eqn (8.8)).[45] On the other hand, by using $\alpha < \alpha_{\mathrm{Ernst}}$ and encoding times t which are significantly smaller than T_2^*, one obtains images of M_0 which reflect the true water content. Figure 8.5 demonstrates this effect by an example of a phantom composed of four cuvettes filled with saturated Kaldenkirchen soil material and two reference tubes filled with $CuSO_4$ solution in a mixture of H_2O/D_2O at a 0.33/0.67 volume fraction.

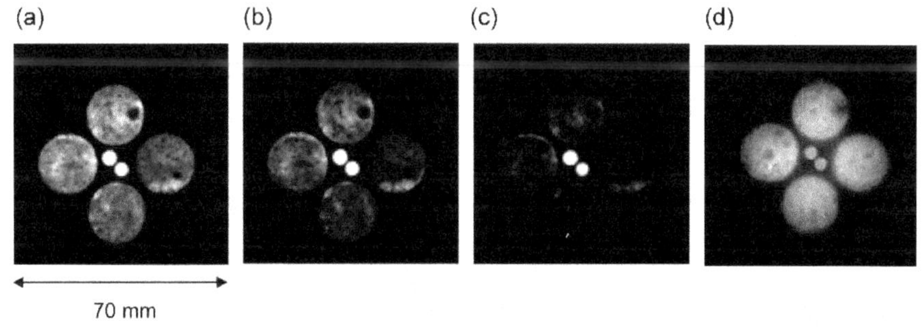

Figure 8.5 Images of a soil phantom with spin echo imaging (a–c) and ZTE (d). Echo times: 1.6, 3.2, and 4.8 ms, respectively. The acquisition bandwidth for ZTE was 300 kHz resulting in an acquisition time of $64 \times 3.3\ \mu s = 211\ \mu s$. The matrix size and in-plane resolution of all images were 128 and 0.55 mm, respectively. The image intensity of the soil domains in the spin-echo image is relatively low compared to the two marker tubes in the centre. In contrast, the image intensity of the ZTE image (d) is comparable to the marker tubes and thus reflects the true water contents. Reproduced from ref. 12, https://doi.org/10.3390/molecules26175130, under the terms of the CC BY 4.0 license, https://creativecommons.org/licenses/by/4.0/.

[†] It is sometimes misleadingly termed echo time, *e.g.* zero echo time, ZTE. No echo is created.

A spin echo sequence yielded echo intensities of the water from soil cuvettes of only 20 to 35% to that of the reference tubes, even with a short echo time of $t_E = 1.6$ ms. Regarding the fact that the saturated soil material has a porosity of $\varphi = 0.4$ cm^3 cm^{-3}, it means that a greater part of the soil water is invisible due to rapid T_2 relaxation (Figure 8.5(a)–(c)). In contrast, the ZTE image reflects the true water content (Figure 8.5(d)). The image appears slightly blurred which is due to the fact that the signal decayed with $T_2^* = 150$ μs, which is shorter than the acquisition time of 211 μs so that the intensity in the outer areas of the *k*-space already decayed. For objects with short T_2^*, the effective resolution is limited according to eqn (8.12):[80]

$$\Delta x, y = \frac{1}{\pi T_2^*} \frac{2\pi}{\gamma_H G_{x,y}}. \tag{8.12}$$

This T_2^* blurring can be reduced by increasing the receiver bandwidth, *i.e.* by decreasing the dwell time, so that the acquisition time becomes shorter than T_2^*.[12]

The image quality of objects embedded in the soil matrix may be also affected by frequency-shift artefacts. For instance, in the sample shown in Figure 8.6, a reference tube was placed directly in a layered soil phantom composed of sand, sandy loam and silt loam with increasing contents of clay and iron.[12] Using UTE, we could image only the marker tube correctly in the sand compartment at all bandwidths, whereas severe artefacts occurred in the soil compartments. The reason is that different magnetic susceptibilities at interfaces, here between the water tube and soil, shift the local Larmor frequencies and thus the apparent voxel positions after Fourier transformation. For instance, the 5 mm marker tube appears only as a thin line in the silt loam compartment in the partial image (ii) in Figure 8.6(b).

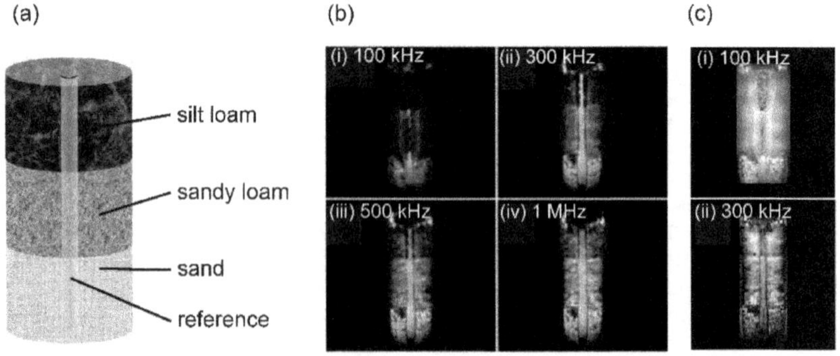

Figure 8.6 Imaging of a soil phantom (a) with UTE (b) and ZTE (c). The image quality is significantly influenced by the choice of the acquisition bandwidth. Too short bandwidths lead to artefacts and image blurring, which affect the image quality of an embedded object such as the reference tube with long T_2^*. Reproduced from ref. 12, https://doi.org/10.3390/molecules26175130, under the terms of the CC BY 4.0 license, https://creativecommons.org/licenses/by/4.0/.

An increase of the acquisition bandwidth also increases the read-out gradient strength so that it eventually overcompensates the internal magnetic field gradients ($G_{ro} \gg g_{int}$) and such frequency-shift artefacts become less pronounced.

Having determined these opportunities and limitations of UTE and ZTE, these methods of imaging were applied to a real-world problem, namely the detection of film flow in a natural soil core. Film flow can occur at high irrigation rates when the matrix is almost saturated. It is much faster than the mostly dominating matrix flow at lower saturation and may contribute to rapid displacement of solutes such as contaminants and fertilizers. The processes are rapid and the conducting structures such as wormholes or rotten root canals are 3D objects. Steady states developed immediately after start with ponding at the surface and the bottom of the wormhole, while thin liquid films at the wormhole walls occurred rapidly. The onset of film flow beyond an irrigation rate of 18 mm h^{-1} indicated that from this point on the hydraulic conductivity of the matrix did no longer sustain the downward flow and the additional pathway *via* the holes must be considered too (Figure 8.7).[12]

Such types of investigation are important for the assessment of the transport properties of soils under conditions of strong rain events, which

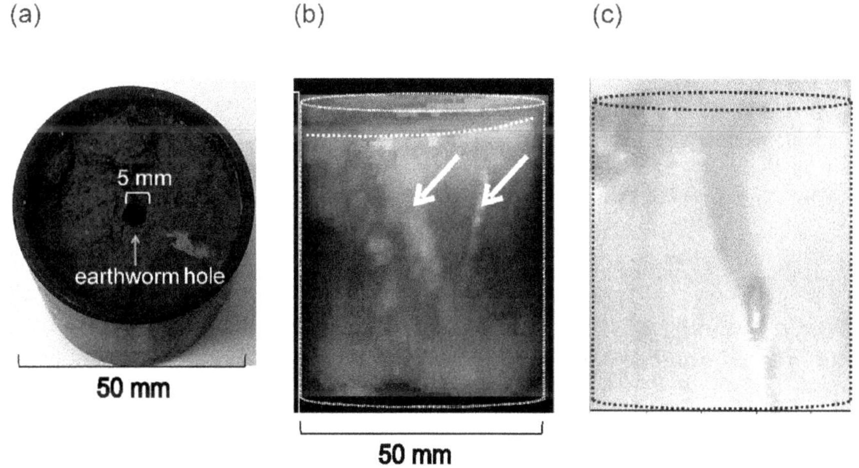

Figure 8.7 Natural soil core with a 5 mm wide earthworm hole, imaged with a ZTE sequence. (a) Photograph. (b) 3D maximum intensity projection showing the 5 mm wide wormhole and another narrow one, indicated by the arrows. The dotted curve shows the soil surface. (c) Central vertical slice 13 min after starting irrigation with a rate of 36 mm h^{-1}. Visible is ponding at the core surface and at the bottom of the wide wormhole plus film formation in the centre. The resolution is 0.55 mm × 0.55 mm × 0.7 mm. Reproduced from ref. 12, https://doi.org/10.3390/molecules26175130, under the terms of the CC BY 4.0 license, https://creativecommons.org/licenses/by/4.0/.

are supposed to occur with increasing strength and frequency due to global climate change.

In summary, natural soil cores and soil materials should be imaged using pulse sequences with ultrashort detection times. If 1D or 2D imaging (projections) at comparably low resolution is sufficient for the processes of interest, single point sequences are a good choice since they are frequency-artefact free. If soil cores should be investigated in 3D, at high resolution, and in a short time, 3D UTE and ZTE are preferably applied due to their short scan times. In these cases, one should determine T_1 and T_2^* first to determine the optimal Ernst angle and the lowest possible bandwidth for correct imaging.

8.4 Roots and Root–Soil Interactions

8.4.1 Root System Architecture Imaging

The investigation of dynamic and transient processes in soil–root systems, such as water uptake or the transport of tracers and solutes, requires the most accurate determination of the 3D structure of roots in soil. This can be performed by the combination of MRI with X-ray CT, which profits from high resolution and has been quantitatively evaluated by Metzner *et al.*[58] However, in larger plant growing containers, the resolution is practically limited by the quotient of cuvette diameter and the number of pixels, for instance, 50 mm/1024 = 50 microns. Since X-ray CT detects roots by the lack of a signal, at least two voxels are necessary for an unambiguous assignment and the effective resolution is about 100 microns. In contrast, MRI detects the roots by the presence of a signal because the high water content dominates the contrast with respect to the soil matrix. Furthermore, the significantly longer T_2 relaxation of roots favours the contrast to the soil when using T_2-weighted pulse sequences with a long echo time t_E. This makes the detection of roots by MRI with sub-voxel resolution possible, *i.e.* if a root fills only a part of an MRI voxel, it still contributes to a higher signal. Proving this point, we found that capillary tubes with a diameter of 0.1 mm are visible in MRI images obtained with a 0.6 T mobile low-field at nominal resolution after zero-filling of 0.16 mm.[47] The comparison with X-ray and neutron CT of identical plants validated the MRI approach to image the 3D root system architecture. In particular, a lupin plant was grown in a 30 mm wide glass cuvette filled with medium sand, regularly watered from bottom, and scanned after 21 days by the mobile MRI scanner, neutron CT and X-ray CT at the Helmholtz Centre Berlin. All three methods represented the root system in equal quality as demonstrated by the superposition of images in Figure 8.8.[47]

The good image quality of root systems measured by MRI opens the way for a detailed analysis and implementation in root water uptake models. To do this, it is necessary first to segment the roots from the soil in the data analysis procedure and in the second step to create a continuous root

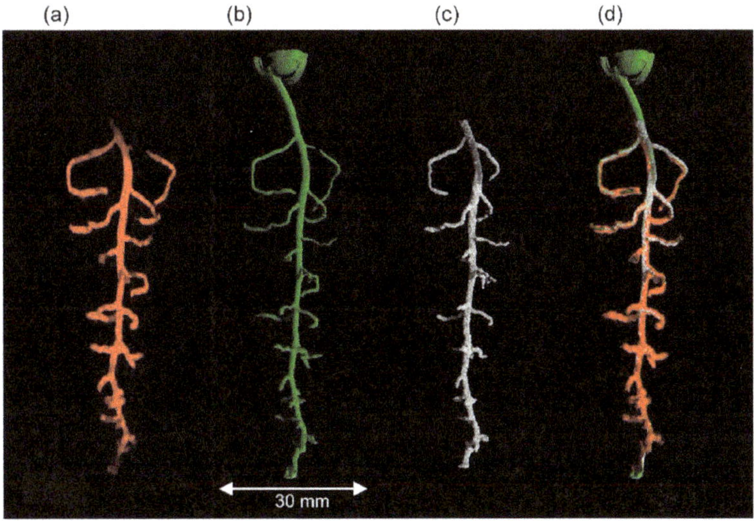

Figure 8.8 (a) MRI, (b) neutron CT and (c) X-ray CT images of a root system of a 21-day-old root system of broad bean grown in medium sand. (d) A superposition of all three images. Reproduced from ref. 47 with permission from John Wiley & Sons, Copyright 2019.

skeleton in the sense of a tree structure. Due to the high complexity, this must be done with the help of an automated algorithm. In the end, a 3D root model in terms of nodes and connections is obtained, which can then be integrated into a coupled soil–root–water uptake model.[81,82]

8.4.2 Root–Soil Interactions

The root system architecture is strongly coupled to one of their functions, namely the water uptake from soil. Therefore, the interaction of roots with soil has attracted the interest of MRI researchers as early as the 1980s. Thus, Bottomley *et al.* employed different MRI methods to detect the root system in an artificial soil, the uptake of $CuSO_4$ as a contrast agent, and the uptake of water during an illumination phase and subsequent recovery in a dark phase.[83] The group continued their work with an evaluation of the quality of root images in various natural and artificial soil materials. In almost all soils, the roots could be segmented from the nearly invisible soil matrix, but an iron content of more than 4% may have affected the image quality.[84]

The representation of the water content in the immediate vicinity of the roots is also important for the assessment of root function. Brown *et al.* investigated maize roots in an artificial growth substrate by a spin warp imaging pulse sequence with long echo and short repetition times resulting in T_2- and T_1-weighted single slice images with a resolution of 0.078 mm × 0.078 mm in 2 mm thick slices.[85] In addition to imaging the inner root tissues of the vascular cylinder and the cortex, they were able to distinguish the so-called rhizosphere, which extended to about 2 mm into the growth substrate, by its enhanced water content, possibly modified by relaxation. MacFall *et al.* continued such types of investigations in the same lab. They determined a distinct zone with reduced signal intensity around a loblolly pine seedling grown in moist sand and assigned it to a water depletion zone.[86,87] However, it could not be excluded that the reduced echo intensity in the rhizosphere was a result of the T_2 weighting due to the long echo time of 10 ms used in the single slice spin echo pulse sequence. The authors themselves discussed an exchange of water in the bulk soil and the rhizosphere, which is supposed to be bound more tightly.[87]

Since the relaxation times correlate with the water mobility, bound water relaxes faster and contributes to signal loss in T_2-weighted images. This effect is well known from many well logging investigations, where T_2 is chosen as a criterion to distinguish between free and bound water.[23] Similar depletion zones were later observed in barley roots in mini-lysimeters using sand as a substrate.[88,89] Also these authors used comparatively long echo times of 12 and 13 ms so that signal decay due to faster T_2 decay in the rhizosphere cannot be excluded. This view was later indirectly supported by neutron imaging experiments sensitive to the total hydrogen content in a sample (see also Section 8.2). In contrast to the MRI results mentioned above, the group of Carminati observed increasing water contents in the rhizosphere, which they assigned to the presence of gel substances exuded by the root.[54,90] These investigations could observe water depletion only if roots in soil were rewetted after drainage, during which the gel substances in the rhizosphere changed their properties from hydrophilic/hygroscopic to hydrophobic. Our group was able to confirm this type of behaviour. First results confirmed the persistence of zones of increased water content near rizinus roots as total moisture decreased when plants were initially well watered. This implies that roots in partially unsaturated soils attract water.[91,92]

To clarify these partial contradictory observations, we used a combination of methods of MRI and neutron CT in later experiments.[47] Here, neutron CT provided the water content in the rhizosphere with high accuracy and MRI relaxation mapping allowed conclusions about the corresponding local water dynamics. The underlying reasoning is the fact that the relaxation times reflect the presence of gel substances which contribute to the decrease of relaxation times since they restrict the local water mobility in a root-exudate polymer network. For the experimental study, we allowed the plants to grow for 16 days in 30 mm cuvettes made of boron-free fused silica, and we measured the T_1 distribution and spin-density maps around lupin roots in a saturated state (see Figure 8.9(c) and (d)). Subsequently the plants were transported to the Paul Scherrer Institute in Villigen, Switzerland (Figure 8.9(a)) and imaged again on-site with a transportable MRI instrument at the NEUTRA imaging facility. The T_1 map of an exemplary axial slice showed distinct zones with reduced T_1 times (Figure 8.9(c)) corresponding exactly to zones of increased water content under desiccation conditions (see Figure 8.9(d)–(f) and

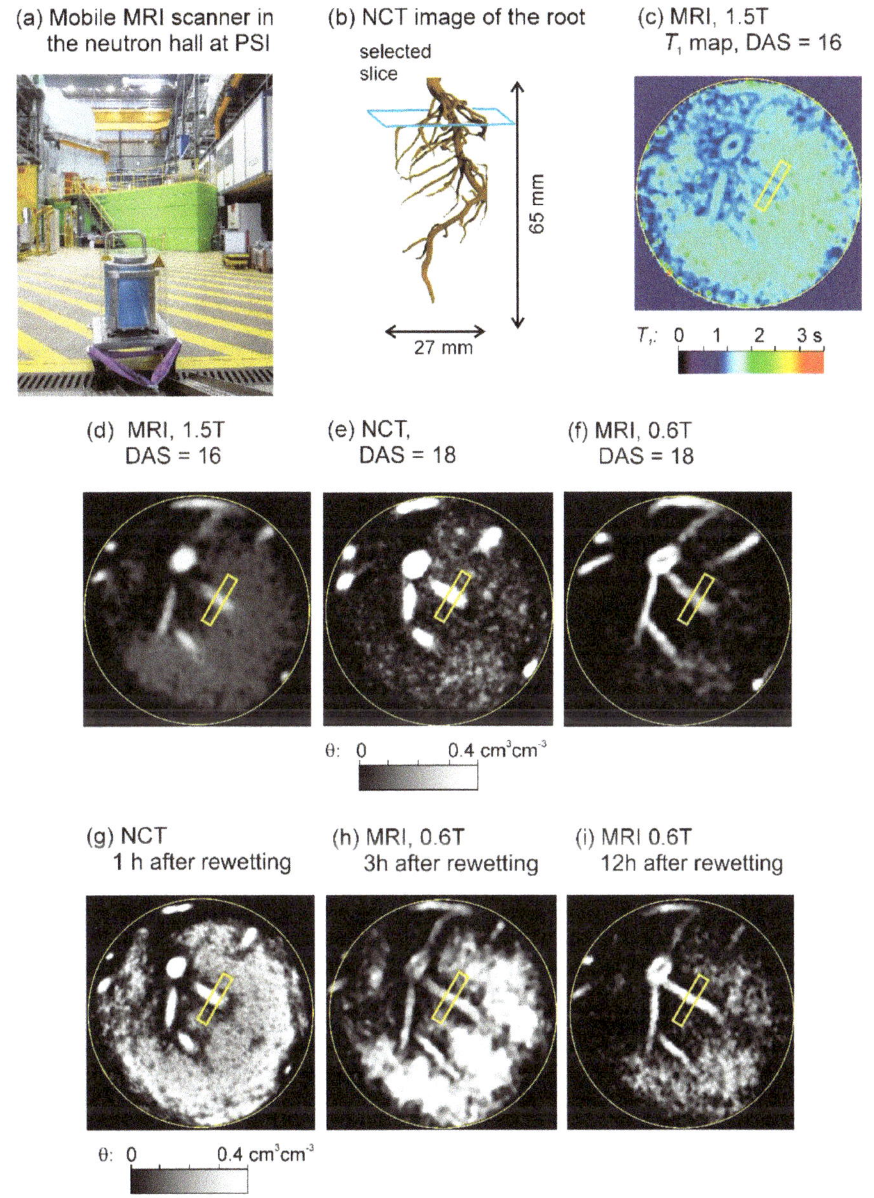

Figure 8.9 MRI and neutron CT investigation of the rhizosphere of broad bean (*Vicia faba*) grown in sand at the Paul Scherrer Institute in Villigen, Switzerland. (a) The transportable MRI scanner in the neutron hall at the PSI. (b) 3D neutron CT image of the root. (c–i) MRI and neutron CT axial cross sections through the top part of the root system. "DAS" is days after seeding. (f) The T_1 map showing distinct zones of reduced T_1 around roots corresponding to enhanced water content during desiccation (c–e) and reduced water content after re-wetting (g–i). Reproduced from ref. 47 with permission from John Wiley & Sons, Copyright 2019.

the discussion of the respective profiles in the original publication).[47] The situation reversed after renewed irrigation since these zones could not rewet immediately due to mucilage exudation or biofilm formation in the rhizosphere (Figure 8.9(g)–(i)).

Thus, these results confirm the change from hydrophilic to hydrophobic behaviour caused by gels in the rhizosphere depending on the hydraulic conditions. A possible explanation is that such a strategy helps roots to cope with draught stress, as temporary hydrophobicity protects against desiccation. However, such a generalization requires further investigation, in which MRI will be of great help, as sensitive changes in a complex environment require non-invasive 3D imaging.

8.4.3 MRI Solute: Na

MRI is not only capable of monitoring roots and the surrounding water content, but its versatility also makes it ideal for monitoring the accumulation of solutes. These can be naturally occurring solutes such as salt[7] or heavy metals,[93] but also artificial solutes that serve as a tracer.[94] To map the changes in NaCl concentration in the root system of tomato, the sensitivity of MRI to so-called X-nuclei can be used, for example, ^{23}Na. In general, the sensitivity of an NMR experiment is determined by the equilibrium magnetization, which is calculated according to Curie's law. Consequently, the relative signal intensity of a 1 M NaCl solution is only about 0.3% of that of pure water considering that the gyromagnetic ratio of ^{23}Na with $\gamma_{Na} = 70\,761 \times 10^7$ rad per Ts is only 26% of that of ^{1}H. In addition, ^{23}Na has a spin of 3/2 with a quadrupole moment, which is sensitive to electric field fluctuations. Therefore, the T_1 of aqueous $^{23}Na^+$ is much shorter than that of water. This is advantageous for MRI because short repetition times are possible without running into saturation and numerous experiments can be accumulated compensating for the lower sensitivity. If the echo time is then chosen to be much shorter than T_2, the signal intensity is proportional to the volumetric Na content in a voxel.

In preparation for the salt enrichment experiments, we first measured the relaxation properties of a 0.2 M NaCl solution in saturated sand at a water content of $\theta = 0.38$ cm^3 cm^{-3}. The relaxation times of $T_2 = 43$ ms and $T_1 = 50$ ms are almost identical and correspond to the values in aqueous solution. They indicate that there is no acceleration of the $^{23}Na^+$(aq) relaxation rate by the pore wall according to the Brownstein–Tarr model, unlike water. The next step was the determination of the lowest concentration that could be mapped in acceptable time. We have chosen a 3D spin echo sequence (see Table 8.1) with 1 mm isotropic resolution, acquisition bandwidth $b_w = 50$ kHz, $t_E = 2.6$ ms, repetition time $t_R = 0.3$ s, and up to 128 averages, resulting in a total acquisition time of up to 12 hours. With this setup, we mapped NaCl concentrations in sand down to 0.01 mol L^{-1}, independent of the water content, which is important for root–soil studies under unsaturated conditions.[7] It is noted that the use of $t_R > 5T_1$, *i.e.* $t_R > 250$ ms and $t_E \ll T_2$, leads to a spin-density image, $M_{0,xyz}$ (see eqn (8.4)), which is proportional to the *volumetric salt concentration, i.e.*

$$M_{0,xyz} \propto c_{xyz} \times \theta, \quad \text{for } t_R > 5 \times T_1 \text{ and } t_E \ll T_2, \tag{8.13}$$

but not to the salt concentration c_{xyz} in solution as in the case of paramagnetic contrast agents. Here, θ is the volumetric water content. Following this preparatory work, the accumulation of salt in the root system of tomatoes irrigated with 42 mM NaCl solution was studied. First, the root system was imaged in 3D in a ^{1}H rf-coil by a multi-slice multi-echo sequence and second the concentration of Na^+(aq) in a ^{23}Na coil by the 3D spin echo method described above. The results revealed salt accumulation in the neighbourhood of several roots with proceeding transpiration at a high rate of 0.54 cm d^{-1}, but also a partial uptake by the plants (Figure 8.10(c) and (d)). The enrichment patterns become clearer through the creation of correlation diagrams (see Figure 8.10(e) and (f)). Here, the concentration of NaCl in each voxel is plotted *versus* the voxel's distance to the next root by 3D correlation of the concentration maps with a distance transformation of the root images in Figure 8.10(a) and (b). The stronger enrichment in the high transpiration regime (Figure 8.10(e)) manifested in higher concentration immediately beyond the root surface. For a different scenario with a low transpiration of 0.3 cm d^{-1} (not shown here), lower accumulation and a weaker decay were observed compared to the high transpiration regime (Figure 8.10(f)). Studying such behaviour is an important step in evaluating plant strategies that deal with saline soils, as salt contributes to an increase in osmotic potential in the rhizosphere, which impedes water uptake.[95,96]

8.4.4 MRI Solute: Paramagnetic Tracers

Paramagnetic solutes in soil–root systems can be well imaged with T_1 sensitive pulse sequences, since in most cases they shorten the relaxation time *via* the interaction of the electron system with the surrounding water molecules. The concentration of paramagnetic substances in the solution decreases the inverse longitudinal relaxation time according to the equation:

$$\frac{1}{T_1} = \frac{1}{T_{1,\text{bulk}}} + \rho_1 \frac{S}{V} + a_1 c, \tag{8.14}$$

where a_1 is the specific relaxivity of the paramagnetic solute and c is its concentration in solution. A comparatively fast pulse sequence is a gradient-echo sequence using a short rf pulse with variation of the flip angle so that the image is strongly T_1-weighted if $t_E \ll T_2^*$. Moradi *et al.* used this method for the 2D mapping of Ni accumulation in the root systems of *Berkheya coddii* in glass-bead filled 2D rhizoboxes, known as a hypoaccumulator plant.[97] Ni^{2+} accumulated in an area on the order of 1 cm above the root surface, and fitting a transport model including Michaelis–Menten uptake indicated passive uptake rather than active uptake.

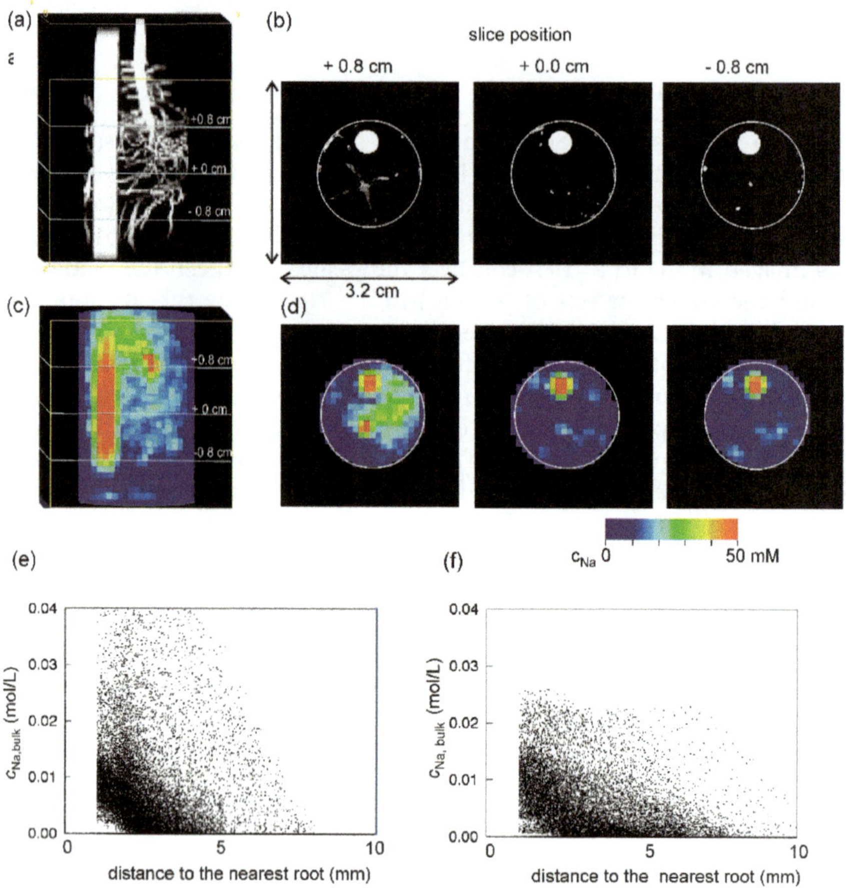

Figure 8.10 Imaging of NaCl enrichment in the root system of a tomato plant. (a) Root system of a 6-week-old tomato in wet medium sand. The white stripe is a marker tube for image correlation and Na calibration. (b) $^{1}H_2O$ images of axial cross sections at three selected depths. (c) Maximum intensity projection of the NaCl distribution in the soil matrix. (d) Na concentration maps at the three selected slices. (e) and (f) Distance correlation diagrams of Na^+(aq) concentration in the soil matrix with distance to the nearest root for (e) high transpiration (0.54 cm d^{-1}) and (f) low transpiration (0.30 cm d^{-1}) regimes. The higher transpiration leads to stronger enrichment of Na^+ in the neighbourhood of the roots. Reproduced from ref. 7, https://doi.org/10.1007/s11104-020-04628-8, under the terms of the CC BY 4.0 license, https://creativecommons.org/licenses/by/4.0/.

Unlike metal cations, which tend to adsorb to the mostly negatively charged surfaces of soil particles, anionic paramagnetic compounds are inert tracers, *i.e.*, they do not interact with the soil particles. This motivated us to use Gd-DTPA^{2-} (gadolinium-diethylenetriamine pentaacetate) as a tracer for monitoring transport in the soil–root compartment.[94] We have chosen a lupin plant in sand and a scenario of constant ratio of irrigation and percolation rates so that a certain fraction of solute remained in the system. The short T_2^* relaxation time of water in medium sand did not allow the use of a gradient echo sequence, and we opted for a spin echo sequence with inversion recovery (IR) preparation to map a wide range of T_1 and Gd-DTPA^{2-} concentrations. If, in addition to a series of scans with varying inversion times t_{inv}, a reference scan without inversion preparation is measured and the quotient of both measurements is formed, the influence of the T_2 relaxation and the varying water content ($\propto M_0$) can be eliminated. In this case, the signal intensity is given by

$$\frac{M(t_{\mathrm{inv}})}{M_{\mathrm{ref}}} = \frac{M_0 \exp\left(-\frac{t_{\mathrm{E}}}{T_2}\right)\left[1 - 2\exp\left(-\frac{t_{\mathrm{inv}}}{T_1}\right)\right]}{M_0 \exp\left(-\frac{t_{\mathrm{E}}}{T_2}\right)} = \left[1 - 2\exp\left(-\frac{t_{\mathrm{inv}}}{T_1}\right)\right]. \quad (8.15)$$

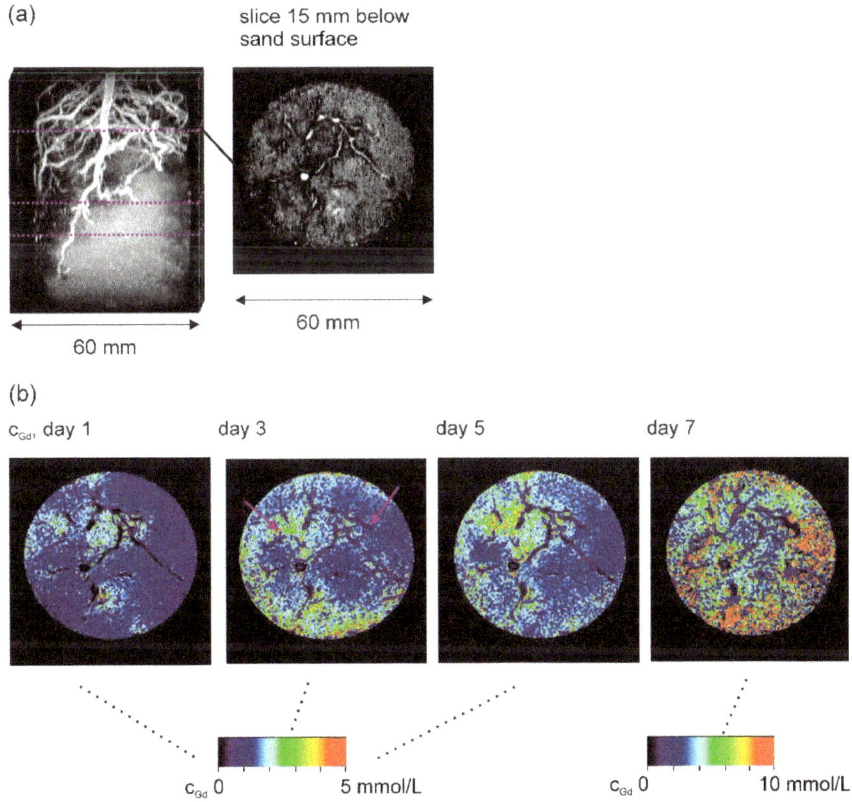

Figure 8.11 Tracer accumulation in a lupine root system. (a) 3D maximum intensity projection of the entire root system obtained with a fast spin echo 3D sequence. (b) Selected axial slice 15 mm below the soil surface. (c) Enrichment patterns of Gd concentration after 1, 3, and 5 days under continuous constant rate irrigation. (d) Enhanced concentration patterns due to stopping of irrigation on day 6 and desaturation over 1 day. Reproduced from ref. 94 with permission from John Wiley & Sons, Copyright 2017.

Furthermore, we showed that the division of the complex IR and reference images leads to the result that the signal is only included in the real part of the complex image, while the imaginary part contains just noise.[94] Using this method, we observed continuous accumulation of the tracer in the neighbourhood of the roots with increasing time and irrigation, and there were differences in the degree of accumulation, *i.e.*, some roots were more active than others that did not accumulate salt to the same extent (see Figure 8.11(b)).

Irrigation was terminated on day 6 so that the downward flux ceased immediately while transpiration continued. As a result, the local T_1 decreased accordingly since the solution concentration of Gd-DTPA^{2-} increased (see eqn (8.14)). Later, the results were combined with a fully coupled root–water uptake model proving that 76% of total transpiration proceeded *via* 3rd order roots.[98]

8.5 Concluding Remarks

In this chapter, we have outlined the development of NMR relaxometry and imaging (MRI) with the focus on clarifying the relations between mobile water and solid structures in soil and soil–plant systems with examples chosen from the literature as well as from our research. The underlying concept is the determination of relaxation times and the understanding of their origin because they control the image properties and image quality to a great extent. Many soil materials have relatively short relaxation times, which requires imaging pulse sequences with short detection times to recover all water fractions. We also showed that the interpretation profits from the combination of MRI with other 3D imaging methods such as X-ray and neutron CT, since the latter are sensitive to the static properties: solid structure and total hydrogen atom content, respectively, while MRI can image the dynamic properties of liquids in porous media. The versatility of MRI is also evident from its ability to map water content and salt concentration quasi-simultaneously. Thus, we could monitor salt enrichment around plant roots in soil either directly *via* X-nucleus imaging or indirectly by profiting from relaxometric imaging of paramagnetic contrast agents. In future, the combination of MRI with other such imaging methods will allow a more thorough understanding of soil–root interactions based on the interdependence of structures and transient processes.

Acknowledgements

The authors thank Bernhard Blümich for the critical reading and correction of the manuscript.

References

1. N. R. C. (US), in *Safety of Genetically Engineered Foods: Approaches to Assessing Unintended Health Effects*, Committee on Identifying and Assessing Unintended Effects of Genetically Engineered Foods on Human Health, National Research Council (US), Washington, DC, 2004.

2. M. F. Eckerstorfer, M. Engelhard, A. Heissenberger, S. Simon and H. Teichmann, *Front. Bioeng. Biotechnol.*, 2019, **7**, 26.
3. G. Neumann, T. S. George and C. Plassard, *Plant Soil*, 2009, **321**, 431–456.
4. V. Vives-Peris, C. de Ollas, A. Gomez-Cadenas and R. M. Perez-Clemente, *Plant Cell Rep.*, 2020, **39**, 3–17.
5. D. Courtier-Murias, E. Michel, S. Rodts and F. Lafolie, *Environ. Sci. Technol.*, 2017, **51**, 2602–2610.
6. M. J. Simpson, A. J. Simpson, D. Gross, M. Spraul and W. L. Kingery, *Environ. Toxicol. Chem.*, 2007, **26**, 1340–1348.
7. A. Perelman, N. Lazarovitch, J. Vanderborght and A. Pohlmeier, *Plant Soil*, 2020, **454**, 171–185.
8. P. T. Callaghan, *Principles of Nuclear Magnetic Resonance Microscopy*, Oxford University Press, Oxford UK, 1991.
9. W. I. Bakhmutov, *Practical NMR Relaxation for the Chemist*, Wiley, New York, 2004.
10. P. Conte, in *Fast Field Cycling Relaxometry*, ed. R. Kimmich, Royal Society of Chemistry, London, UK, 2018, pp. 229–254.
11. L. D. Hall, M. H. G. Amin, E. Dougherty, M. Sanda, J. Votrubova, K. S. Richards, R. J. Chorley and M. Cislerova, *Geoderma*, 1997, **80**, 431–448.
12. S. Haber-Pohlmeier, D. Caterina, B. Blümich and A. Pohlmeier, *Molecules*, 2021, **26**, 5130.
13. S. Lakshmanan, W. A. Maru, D. J. Holland, M. D. Mantle and A. J. Sederman, *Flow Meas. Instrum.*, 2017, **53**, 161–171.
14. T. Y. Tse, R. M. Spanswick and L. W. Jelinski, *Protoplasma*, 1996, **194**, 54–62.
15. A. Haase, J. Frahm, W. Hanicke and D. Matthaei, *Phys. Med. Biol.*, 1985, **30**, 341–344.
16. R. van Schadewijk, J. R. Krug, D. F. Shen, K. Gupta, F. J. Vergeldt, T. Bisseling, A. G. Webb, H. Van As, A. H. Velders, H. J. M. de Groot and A. Alia, *Sci. Rep.*, 2020, **10**, 971.
17. R. L. Vold, J. S. Waugh, M. P. Klein and D. E. Phelps, *J. Chem. Phys.*, 1968, **48**, 3831–3832.
18. H. Y. Carr and E. M. Purcell, *Phys. Rev.*, 1954, **94**, 630–638.
19. S. Meiboom and D. Gill, *Rev. Sci. Instrum.*, 1958, **29**, 688–691.
20. E. L. Hahn, *Phys. Rev.*, 1950, **80**, 580–594.
21. R. L. Kleinberg, in *Encyclopedia of NMR*, John Wiley & Sons, New York, 1996, pp. 4960–4969.
22. P. J. Barrie, *Annu. Rep. NMR Spectrosc.*, 2000, **41**, 265–316.
23. G. R. Coates, L. Xiao and M. G. Prammer, *NMR Logging Principles and Applications*, Halliburton Energy Services, Houston, 1999.
24. Y. Q. Song, *J. Magn. Reson.*, 2013, **229**, 12–24.
25. K. R. Brownstein and C. E. Tarr, *J. Magn. Reson.*, 1977, **26**, 17–24.
26. K. R. Brownstein and C. E. Tarr, *Phys. Rev. A: At., Mol., Opt. Phys.*, 1979, **19**, 2446–2453.
27. R. L. Kleinberg, W. E. Kenyon and P. P Mitra, *J. Magn. Reson., Ser. A*, 1994, 206–214.

28. R. L. Kleinberg, S. A. Farooqui and M. A. Horsefield, *J. Colloid Interface Sci.*, 1994, **158**, 195–198.
29. J. Mitchell, T. C. Chandrasekera and L. F. Gladden, *J. Chem. Phys.*, 2010, **132**, 244705.
30. L. R. Stingaciu, A. Pohlmeier, P. Blümler, L. Weihermüller, S. Stapf and H. Vereecken, *Water Resour. Res.*, 2009, **45**, W08412.
31. P. T. Callaghan, *Translational Dynamics And Magentic Resonance*, Oxford University Press, Oxford, UK, 2011.
32. K. E. Washburn and P. T. Callaghan, *Phys. Rev. Lett.*, 2006, **97**, 175502.
33. P. Callaghan, C. H. Arns, P. Galvosas, M. W. Hunter, Y. Qiao and K. E. Washburn, *Magn. Reson. Imaging*, 2007, **25**, 441–444.
34. Y. Q. Song, L. Zielinski and S. Ryu, *Phys. Rev. Lett.*, 2008, **100**, 248002.
35. J. Mitchell, J. D. Griffith, J. H. P. Collins, A. J. Sederman, L. F. Gladden and M. L. Johns, *J. Chem. Phys.*, 2007, **127**.
36. A. Haber, S. Haber-Pohlmeier, F. Casanova and B. Blümich, *Vadose Zone J.*, 2010, **9**, 893–897.
37. W. R. Nitz, *Eur. Radiol.*, 1999, **9**, 979–997.
38. B. J. Balcom, R. P. MacGregor, S. D. Beyea, D. P. Green, R. L. Armstrong and T. W. Bremner, *J. Magn. Reson., Ser. A*, 1996, **123**, 131–134.
39. F. Marica, Q. Chen, A. Hamilton, C. Hall, T. Al and B. J. Balcom, *J. Magn. Reson.*, 2006, **178**, 136–141.
40. C. E. Muir and B. J. Balcom, in *Annual Reports on NMR Spectroscopy*, ed. G. A. Webb, 2012, vol. 77, pp. 81–113.
41. S. Merz, A. Pohlmeier, B. J. Balcom, R. EnJilela and H. Vereecken, *Appl. Magn. Reson.*, 2016, **47**, 212–138.
42. M. Weiger and K. P. Pruessmann, *eMagRes*, 2012, 311–322.
43. M. D. Robson, P. D. Gatehouse, M. Bydder and G. M. Bydder, *J. Comput. Assisted Tomogr.*, 2003, **27**, 825–846.
44. D. Idiyatullin, C. Corum, J. Y. Park and M. Garwood, *J. Magn. Reson.*, 2006, **181**, 342–349.
45. I. V. Mastikhin, H. Mullally, B. MacMillan and B. J. Balcom, *J. Magn. Reson.*, 2002, **156**, 122–130.
46. B. Blümich, *NMR Imaging of Materials*, Clarendon Press, Oxford, 2000.
47. S. Haber-Pohlmeier, C. Tötzke, E. Lehmann, N. Kardjilov, A. Pohlmeier and S. Oswald, *Vadose Zone J.*, 2019, **18**, 180166.
48. S. D. Keyes, K. R. Daly, N. J. Gostling, D. L. Jones, P. Talboys, B. R. Pinzer, R. Boardman, I. Sinclair, A. Marchant and T. Roose, *New Phytol.*, 2013, **198**, 1023–1029.
49. S. D. Keyes, F. Gillard, N. Soper, M. N. Mavrogordato, I. Sinclair and T. Roose, *J. Biomech.*, 2016, **49**, 1802–1811.
50. S. Baraka-Lokmane, I. G. Main, B. T. Ngwenya and S. C. Elphick, *Mar. Pet. Geol.*, 2009, **26**, 39–56.
51. M. Halisch, H. Steeb, S. Henkel and C. M. Krawczyk, *Solid Earth*, 2016, 7, 1141–1143.
52. M. Ruf and H. Steeb, *Rev. Sci. Instrum.*, 2020, **91**, 113102.

53. N. Kardjilov, A. Hilger, I. Manke, R. Woracek and J. Banhart, *J. Appl. Crystallogr.*, 2016, **49**, 195–202.
54. A. B. Moradi, A. Carminati, D. Vetterlein, P. Vontobel, E. Lehmann, U. Weller, J. W. Hopmans, H. J. Vogel and S. E. Oswald, *New Phytol.*, 2011, **192**, 653–663.
55. S. Garre, M. Javaux, J. Vanderborght, L. Pages and H. Vereecken, *Vadose Zone J.*, 2011, **10**, 412–424.
56. A. Furman, A. Arnon-Zur and S. Assouline, in *Soil–Water–Root Processes: Advances in Tomography and Imaging*, ed. S. H. Anderson and J. W. Hopmans, 2013, vol. 61, pp. 223–245.
57. J. Vanderborght, J. A. Huisman, J. van der Kruk and H. Vereecken, in *Soil –Water–Root Processes: Advances in Tomography and Imaging*, ed. S. H. Anderson and J. W. Hopmans, 2013, vol. 61, pp. 247–282.
58. R. Metzner, A. Eggert, D. van Dusschoten, D. Pflugfelder, S. Gerth, U. Schurr, N. Uhlmann and S. Jahnke, *Plant Methods*, 2015, **11**, 17.
59. N. Koebernick, U. Weller, K. Huber, S. Schlueter, H.-J. Vogel, R. Jahn, H. Vereecken and D. Vetterlein, *Vadose Zone J.*, 2014, **13**, vzj2014.03.0024.
60. P. Vontobel, R. Hassanein, A. Carminati, A. Kaestner, P. Lehmann and A. Koliji, *Neutron Imaging for Soil Physics and Geology*, 2008.
61. M. Zarebanadkouki, Y. X. Kim and A. Carminati, *New Phytol.*, 2013, **199**, 1034–1044.
62. K. Karlsons, D. W. de Kort, F. O. Alpak, J. Dietderich, J. J. Freeman, M. Appel, M. D. Mantle, A. J. Sederman and L. F. Gladden, *Transp. Porous Media*, 2022, **143**, 373–396.
63. J. Votrubova, M. Cislerova, M. H. G. Amin and L. D. Hall, *Water Resour. Res.*, 2003, **39**, 1371.
64. J. Votrubova, M. Sanda, M. Cislerova, M. H. G. Amin and L. D. Hall, *Geoderma*, 2000, **95**, 267–282.
65. V. Jelinkova, M. Snehota, A. Pohlmeier, D. van Dusschoten and M. Cislerova, *Org. Geochem.*, 2011, **42**, 991–998.
66. T. J. Marshall, J. W. Holmes and C. W. Rose, *Soil Physics*, Cambridge University Press, Cambridge, UK, 1999.
67. M. B. Kirkham, *Principles of Soil and Plant Water Relations*, Elsevier, Burlington, MA, 2005.
68. W. A. Jury and R. Horton, *Soil Physics*, Wiley, New York, 2004.
69. E. H. D'Hollander, *Water Resour. Res.*, 1979, **15**, 107–112.
70. R. J. Luxmoore, *Soil Sci. Soc. Am. J.*, 1981, **45**, 671–672.
71. B. Clothier, in *The Encyclopedia of Soil Science*, ed. E. Chesworth, Kluwer, Dordrecht, 2004.
72. S. Falzone and K. Keating, *Vadose Zone J.*, 2016, **15**, vzj2015.11.0153.
73. M. Meyer, C. Buchmann and G. E. Schaumann, *Eur. J. Soil Sci.*, 2018, **69**, 393–406.
74. F. Jaeger, S. Bowe, H. Van As and G. E. Schaumann, *Eurasian J. Soil Sci.*, 2009, **60**, 1052–1064.
75. M. Duschl, P. Galvosas, T. I. Brox, A. Pohlmeier and H. Vereecken, *Water Resour. Res.*, 2015, **51**, 6549–6563.

76. L. R. Stingaciu, L. Weihermüller, S. Haber-Pohlmeier, S. Stapf, H. Vereecken and A. Pohlmeier, *Water Resour. Res.*, 2010, **46**, W11510.
77. P. P. Mitra, *Phys. Rev. E: Stat. Phys., Plasmas, Fluids, Relat. Interdiscip. Top.*, 1992, **68**, 3555–3559.
78. C. E. Muir and B. J. Balcom, *Magn. Reson. Chem.*, 2013, **51**, 321–327.
79. S. Merz, B. J. Balcom, R. Enjilela, J. Vanderborght, Y. Rothfuss, H. Vereecken and A. Pohlmeier, *Vadose Zone J.*, 2018, **17**, 160099.
80. P. T. Callaghan and C. D. Eccles, *J. Magn. Reson.*, 1988, **78**, 1–8.
81. L. Stingaciu, H. Schulz, A. Pohlmeier, S. Behnke, H. Zilken, M. Javaux and H. Vereecken, *Vadose Zone J.*, 2013, **12**, vzj2012.0019.
82. D. van Dusschoten, R. Metzner, J. Kochs, J. A. Postma, D. Pflugfelder, J. Buehler, U. Schurr and S. Jahnke, *Plant Physiol.*, 2016, **170**, 1176–1188.
83. P. A. Bottomley, H. H. Rogers and T. H. Foster, *Proc. Natl. Acad. Sci. U. S. A.*, 1986, **83**, 87–89.
84. H. H. Rogers and P. A. Bottomley, *Agron. J.*, 1987, **79**, 957–965.
85. J. M. Brown, P. J. Kramer, G. P. Cofer and G. A. Johnson, *Theor. Appl. Climatol.*, 1990, **42**, 229–236.
86. J. S. MacFall, G. A. Johnson and P. J. Kramer, *Proc. Natl. Acad. Sci. U. S. A.*, 1990, **87**, 1203–1207.
87. J. S. MacFall, G. A. Johnson and P. J. Kramer, *New Phytol.*, 1991, **119**, 551–560.
88. E. Segal, T. Kushnir, Y. Mualem and U. Shani, *Vadose Zone J.*, 2008, 7, 1024–1037.
89. E. Segal, T. Kushnir, Y. Mualem and U. Shani, *Vadose Zone J.*, 2008, 7, 1018–1026.
90. A. Carminati, C. L. Schneider, A. B. Moradi, M. Zarebanadkouki, D. Vetterlein, H. J. Vogel, A. Hildebrandt, U. Weller, L. Schuler and S. E. Oswald, *Vadose Zone J.*, 2011, **10**, 988–998.
91. A. Pohlmeier, A. M. Oros-Peusquens, M. Javaux, J. Lindenmair, M. I. Menzel, H. Vereecken and N. J. Shah, *Vadose Zone J.*, 2008, 7, 1010–1017.
92. A. Pohlmeier, A. M. Oros-Peusquens, M. Javaux, M. I. Menzel, H. Vereecken and N. J. Shah, *Magn. Reson. Imaging*, 2007, **25**, 579–580.
93. A. B. Moradi, S. E. Oswald, J. A. Massner, K. P. Pruessmann, B. H. Robinson and R. Schulin, *Eur. J. Soil Sci.*, 2008, **59**, 476–485.
94. S. Haber-Pohlmeier, J. Vanderborght and A. Pohlmeier, *Water Resour. Res.*, 2017, **53**, 7469–7480.
95. H. Jorda, A. Perelman, N. Lazarovitch and J. Vanderborght, *Vadose Zone J.*, 2018, **17**, 170029.
96. A. Perelman, H. Jorda, J. Vanderborght and N. Lazarovitch, *Plant Soil*, 2020, **447**, 55–71.
97. A. B. Moradi, S. Oswald, J. A. Massner, K. P. Pruessmann, B. H. Robinson and R. Schulin, *Geophys. Res. Abstr.*, 2008, **10**, EGU2008-A-08398.
98. A. Koch, F. Meunier, J. Vanderborght, S. Garre, A. Pohlmeier and M. Javaux, *J. Exp. Bot.*, 2019, **70**, 2797–2809.

CHAPTER 9

Using Magnetic Resonance Imaging to Study Contaminant Dynamics

R. H. MORRIS,*[a] M. I. NEWTON[a] AND N. NESTLE[b]

[a] Nottingham Trent University, Clifton Lane, Nottingham, NG11 8NS, UK;
[b] BASF SE, Carl-Bosch-Strasse 38, Ludwigshafen am Rhein 67056, Germany
*Email: rob.morris@ntu.ac.uk

9.1 Why Use MRI for Soil Contaminant Study

When we consider soil contaminants, there are three broad groups which are of critical importance: heavy or otherwise toxic metals and their compounds, oils and tars and finally solvents. Magnetic resonance imaging (MRI) offers information on spatial signal distribution and has the capability of identifying different substances through relaxation times and spin density. As the technique provides valuable insights into the internal processes within opaque samples with little to no external influence, we can use it to further our understanding of dynamic processes in ways that are not possible using destructive techniques. For example, the tracking of a contaminant along the length of a sample with traditional methods will often involve tapping into the sample to determine the time to breakthrough of a contaminant or estimating the residence time between the inlet and outlet. Nuances of in-sample interactions and flow paths are however lost in the near infinite possible paths which could have occurred in a tortuous porous system between two such measurement points. MRI is capable of resolving such systems and offers a glimpse inside the sample to reveal fluid

New Developments in NMR No. 32
The Environment in a Magnet: Applications of NMR Techniques to Environmental Problems
Edited by Pellegrino Conte, Delia Francesca Chillura Martino and Paolo Lo Meo

Published by the Royal Society of Chemistry, www.rsc.org

distributions from one or more phases, measurements of flow and streamlines and the resulting transport and mixing of solutes and contaminants.

Applying MRI to environmental samples is not however without challenges. The static magnetic field (B_0) which polarises the nuclei within the system, the time varying radio-frequency field (B_1) which manipulates them, and the gradient fields (B_g) which extend nuclear magnetic resonance (NMR) into imaging are all easily influenced by the presence of metals, their ions and compounds which can be abundant in environmental samples. At low concentrations, or for those compounds that have more favourable magnetic properties, their presence may cause artefacts which are correctable. At high concentrations, or for compounds with less favourable magnetic properties, they can render imaging all but impossible with a complete suppression of the signal.

In this chapter, we will consider five key areas which can help to build a complete picture of the internal process of contaminant dynamics in environmental samples. It should be noted that although, at first thought, direct measurements of kinetics and transport may seem to be the ultimate goal in studies of contaminants, vast gains in understanding can be made using flow measurements of proxy fluids. Thus, in this chapter, we will begin our journey briefly reviewing structural MRI which will allow us to characterise the properties of the soil media (either *in situ*, treated, cleaned or replicated) with a focus on direct application to contaminants. We will then consider similar systems experiencing flow for one or more fluidic or gaseous phases which provides an understanding of how contaminants may be transported. This can be extended to directly measuring the transport of simulated or real contaminants. Where the contaminants of interest are reactive, MRI can be used to explore how they react with their surroundings or break down. Most environmental systems contain significant populations of microorganisms which drive complex processes including the breakdown of contaminants. Many of these will secrete biofilms which can have a phenomenal impact on contaminant dynamics and are abundant in soils which will also be discussed here.

9.2 Structural MRI of Soils and Soil-like Systems

As briefly introduced in the previous section, MRI is a powerful tool for three-dimensional visualisation of static soil samples. The main areas of application of static imaging to support our understanding of soil contaminants are in gathering structural information for use in fluid dynamic simulations and determining structural properties such as hydraulic conductivity, porosity and water distribution non-invasively.

9.2.1 Initial Studies

Soil systems were initially investigated using MRI for studies exploring root systems *in situ* (for example, Omasa[1] and Rogers[2]). In 1993, Amin *et al.*[3]

began to explore the benefits of the technique for application directly to soil. They investigated the ability to map water distribution in both model samples prepared from glass beads of two diameters and two types of sand with different iron oxide contents (0.0009% and 0.09% w/w) and six different soil types including peat, clay, sandy and their mixtures. A demonstration of the artefacts caused by the presence of paramagnetic iron oxide can be seen in Figure 9.1 which shows a coronal view (top down) of tubes of aqueous solutions of copper sulphate placed in the dry versions of the different samples investigated. The top row are the model systems, whilst the second row are real soils with peat (e), the only one that does not cause significant signal attenuation. The extent of distortion and signal loss correlates directly with the iron content of the sample. Reducing distortion such as this is critical to a successful experiment. It should be noted that the images are not equally scaled and thus no comparison can be made between them regarding absolute signal levels. There are a myriad of modifications to pulse sequences and experimental hardware which can be applied to reduce the extent of the distortion, the key ones of which will be discussed here. The fine balancing act of magnetic resonance imaging is rarely made as clear as when imaging paramagnetic samples. The signal-to-noise ratio (SNR) of an MRI image is conventionally improved by increasing the static magnetic field, with an approximately linear increase in polarisation and gyromagnetic ratio leading to an up to square increase in the signal-to-noise ratio for small samples in narrow bore magnets (see ref. 4 for a comprehensive analysis of large, conductive samples which tend to increase closer to linearity once dielectric and inductive sample losses are accounted for. Enjilela *et al.* also provide an experimental analysis for rock cores in ref. 5).

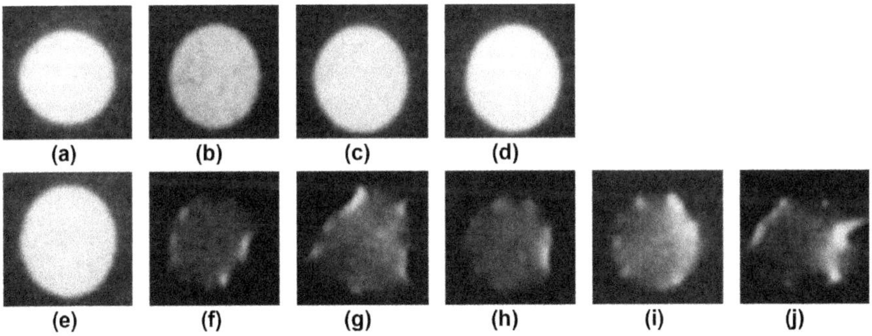

Figure 9.1 Coronal spin echo images of tubes of aqueous copper sulphate (visible) inserted in dry soil samples (not visible). The top row represents idealised samples of glass beads (average diameter of 300 μm in (a) and 3.5 mm in (b)) and brown and quartz sand ((c and d), respectively). The bottom row represents real soil samples: peat; woodland clay; farmland clay; sandy silt; clayey silt; and Cottenham sandy soil in (e–j), respectively. Reproduced from ref. 3 with permission from John Wiley & Sons, Copyright 1993.

However, the extent of susceptibility induced distortion also scales linearly which can negate this benefit for paramagnetic samples. Similarly in MRI it is common to reduce the acquisition bandwidth to increase the SNR by better excluding noise. However, the presence of field perturbations from susceptibility mismatch leads to a divergence in field strength and a corresponding broadening of the range of precession frequencies mandating an increase in the detection bandwidth to detect the entire proton population. This again worsens as the field increases and as the broadening of the bandwidth exceeds that of an imaging pixel. Increasing the gradient strength can help in some circumstances, but there are better gains to be had from pulse sequence modification[6] or imaging at lower field strengths.[7] The approach applied by Amin *et al.* was to utilise a pre-focused RF pulse.[8] This interesting class of shaped pulse minimises phase dispersion for a frequency selective pulse as is typically used for slice selection in typical MRI pulse sequences. They were initially developed for solvent suppression in NMR but have been applied in circumstances where gradient switching needs to be reduced or for rapid imaging, but are used here to minimise the artefact caused by phase dispersion from the susceptibility mismatch. They find that using a pre-focused 90° pulse followed by hard 180° pulses significantly reduces distortion permitting imaging of wet samples.

It is worth noting, at this point, the wide variability in reported parameters which impact the ultimate image quality. In the literature, they are broadly divided into two groups:

- water content, soil texture and the presence of paramagnetic species and
- water and iron contents.

Whilst there is little in the way of consensus in the general literature, studies that focus on real soil types tend to show that only those with significant water and minimal iron contents are clearly resolved, suggesting the latter though there are a vast number of other soil features such as clay content that contribute to the water holding capability and thus, as always, a vast number of parameters ultimately determine the overall quality and likelihood of successful imaging. Real soils are typically highly heterogeneous and contain a wide range of pore sizes. It is important to consider the size of samples needed in order to appropriately represent such distributions in experiments. The larger bore size of MRI compared to upright NMR spectroscopy tends to better facilitate this which is another area in which MRI offers significant benefit. It should be noted however that larger instruments also present additional challenges since gradient rise times and increased probe dead times result in longer echo times for most sequences and applications.

9.2.2 Evolution of the Technique and Physical Limitations

In 1997, Hall *et al.*[9] further explored static imaging of wet soils to better understand the range of pore sizes and levels of paramagnetic impurities which would afford imaging with MRI. A systematic evaluation of 23 soils from six countries was undertaken to understand what parameters are of most importance. MRI measurements were undertaken alongside traditional measures of particle size, organic matter content and chemical content of iron oxide and exchangeable cations of calcium, potassium and sodium. The authors discuss the process of matching spin echo pulse sequence parameters, namely the echo time and repetition time, to the properties of soil water and estimating the percentage of available signal based on echo decay. For example, on a modern MRI scanner, the minimum T_E for spin echo sequences without significant adjustment of many parameters is around 10 ms which would correspond to 60% of the maximum recoverable signal for soil water with $T_2 = 20$ ms which corresponds to a relatively clean (paramagnetically) soil. More challenging soil water might come with relaxation times as short as $T_2 = 1$ ms which for the same echo time will give a signal intensity of 0.005% of the maximum recoverable signal which will be completely obscured by the noise. It is possible to reduce the echo time somewhat by manipulating gradient switching times and pulse bandwidths, but using standard spin echo sequences, which are less affected by susceptibility artefacts, it is unlikely to be reduced beyond 4 ms on clinical imaging systems which would only achieve 8% of the maximum signal. Of critical importance is that the field strength within the pores between soil grains is also variable, leading to shortening of T_2^* towards the pore edge. This means that the theoretically achievable maximum is only observed in the largest of pores, for the least paramagnetic samples and in fully saturated pores. This identifies one of the significant limitations of MRI of soils for contaminant research which is the unintentional selection of water population based on pore size (though this gives rise to a useful technique in its own right for clean samples).

9.2.3 Application to Contaminant Dynamics

Once the challenges in acquiring useful signals are overcome, we can apply the technique to imaging for contaminant dynamics. Work by Chudek and Reeves in 1998[10] and 2001[11] investigated the capabilities of MRI to track the migration of diesel oil through estuarine soil samples. Figure 9.2 shows a two-dimensional cross-sectional image of an estuarine sample, initially mixed with diesel oil, and then after 18 and 36 hours showing the diffusion of the oil throughout the sample over time. The experimental parameters are chosen such that the signal from water in the samples is largely suppressed, thanks to long T_1 and short T_2 whilst the signal from the oil is collected efficiently. The images are scaled to provide bright pixels from the oil content. Whilst there is evidence of diffusion driving the oil in all directions throughout the sample, the authors note that the oil tends to accumulate in areas already saturated with oil. This is in agreement with previous findings suggesting that organics with limited water solubility will

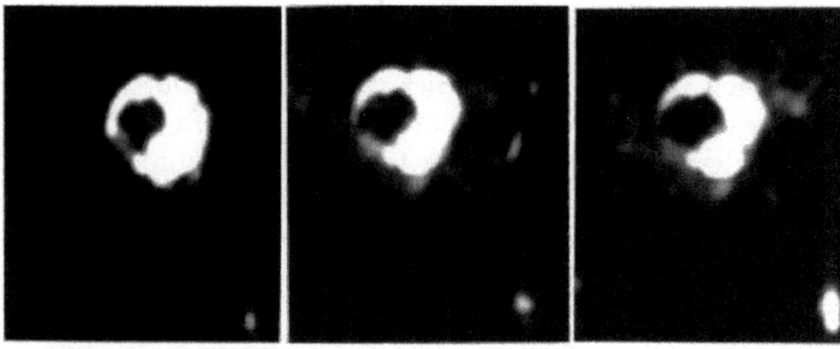

Figure 9.2 Diesel oil in estuarine sediment from left to right: immediately after mixing, after 18 hours and after 16 hours. Whilst there is some diffusion throughout the slice, most diesel oil accumulates in areas that are initially saturated. Reproduced from ref. 10 with permission from Springer Nature, Copyright 1969 Kluwer Academic Publishers.

tend to accumulate as wetting forces make it energetically favourable in comparison to creating interspersed regions.

A significant area of research which is supported by MRI of contaminants in soil is an understanding of how various contaminant removal processes work and a means to easily assess their efficacy. A frequently used group of removal techniques for volatile contaminants are soil vapour extraction (SVE), dual phase extraction (DPE) and air sparging (AS). These techniques all allow volatile compounds to be removed from soil by suction of vapour of the volatiles only from a borehole, suction of both vapour and liquid phases from a borehole, or increasing the evaporation of volatile compounds by pushing air over them between boreholes, respectively. Numerous contaminants can be removed using these techniques including petroleum and diesel hydrocarbons, organics such as benzene and toluene and some pesticides. They are often referred to in the literature as nonaqueous phase liquids (NAPL). Initial studies in this area were conducted on glass bead packs using octane or decane. The first study by Johns and Gladden[12] will be considered in the section on flow imaging. Such experiments were however also undertaken a little later using intermittent flow with imaging steps in between for example by Chu.[13] Here a model system combining elements of SVE and AS was explored using MRI. Four different columns containing one each of two different silica gel grain sizes and two mixed columns with larger beads forming a coarse angular column with a finer pore core are saturated with water and then drained before injecting decane from the base. The decane was then drained and volatilisation started using water saturated nitrogen gas. The flow was temporarily suspended for imaging which is correlated with traditional measurements of mass of decane in the column. The water signal is suppressed using an inversion recovery spin echo pulse sequence. Although linearity is seen for decane mass and signal intensity in all columns, the constant of proportionality differs, most likely owing to pore volume inhomogeneity issues already discussed. Example images for four experimental times are shown in Figure 9.3 for a heterogeneous column.

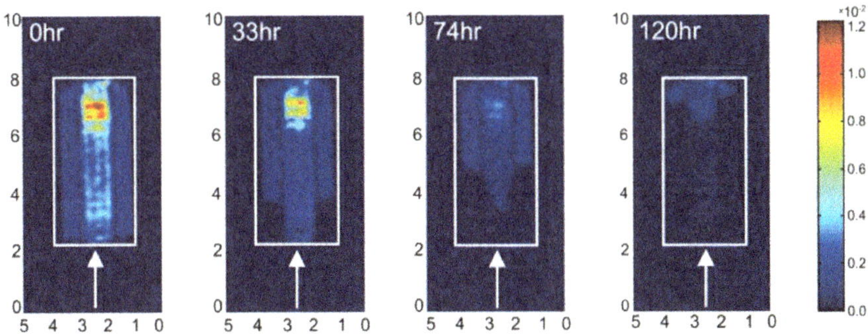

Figure 9.3 Centremost vertical cross-sectional images of annular columns with a fine core and a coarse shell. The core is initially above the residual saturation of decane. Horizontal and vertical scales are in centimetres. The colour scale is linearly proportional to the decane saturation. Reproduced from ref. 13 with permission from Elsevier, Copyright 2004.

Tracking the front in all columns during sparging suggests that below the residual decane saturation there is a moving volatilization front which moves steadily upward. In the heterogeneous cores, the shell progresses faster as expected given the pore size differences. The authors suggest that advection of volatilized decane from the shell, which is fed by diffusion from the internal column, is the mechanism of transport. Above residual saturation, there is instead a flow of liquid-phase decane to the shell. Such results imply that SVE rates will be enhanced at the start of the extraction process where there is a gradient in pore sizes from small to large and the contaminant is present above residual saturation. Following extraction to below the residual saturation, the diffusive process takes over reducing the extraction rate. Additional fluid phases may help the system to stay in the initial regime for longer enhancing extraction though in practical terms selection of an appropriate fluid which is miscible with the contaminant but with no additional environmental impact is likely to limit real-world applications.

9.2.4 Discrimination of Fluid Phases

The comprehensive study of these processes requires a means to discriminate between different fluid fractions. There are a number of approaches used to image a single phase at a time based on the inherent properties of relaxation, diffusion and alternate nuclei or doping.

9.2.4.1 Intrinsic Properties

There is some natural difference between relevant immiscible phases such as their relaxation times T_1 and T_2, their self-diffusion coefficient D_0 and their chemical shift. It is possible to discriminate between oil or other contaminants and water phases using each of these parameters which will be discussed in this section.

For example, for relaxation times, spin echo imaging can be used to exclude the aqueous phase of a water–octanol system in glass bead packs, thanks to their different T_1 values (3 s for water, 0.8 s for octanol) by varying T_R to exclude the water signal[14] or using inversion recovery to selectively invert and therefore remove the signal from each phase.[15] Using the inversion recovery method, the nulled phase can be tracked throughout the media. Various inversion times can be used to specifically interrogate the phase of interest.

An alternative method taking advantage of the intrinsic MR properties of the phases is to utilise chemical shift imaging. Here, as with the inversion recovery approach, it is possible to selectively remove one of the phases using a saturation pulse centred on the population's centre frequency or by spoiling the remaining signal with a gradient after the selected population has been inverted along z. This approach provides good differentiation between populations of oil and water which was first demonstrated using banded samples and a steady state free precession (SSFP) sequence by Blackband *et al.* in 1986[16] and extended to provided discrimination in mixed samples by Davies in 1994 as can be seen in Figure 9.4,[17] provided that there is no overlap of the peaks which is a reasonable assumption in glass bead packs but does not extend to samples with significant paramagnetic impurities. This approach has been employed by Lee and Lee[18] to separately resolve the signals from silicone oil and water in a glass bead pack initially saturated with the oil fraction. Using microimaging, they validated the method as a means to quantify fluid volumes of the individual fractions and also to monitor the motion of water through the beads confirming an exponential recovery function. The authors also use these data to produce

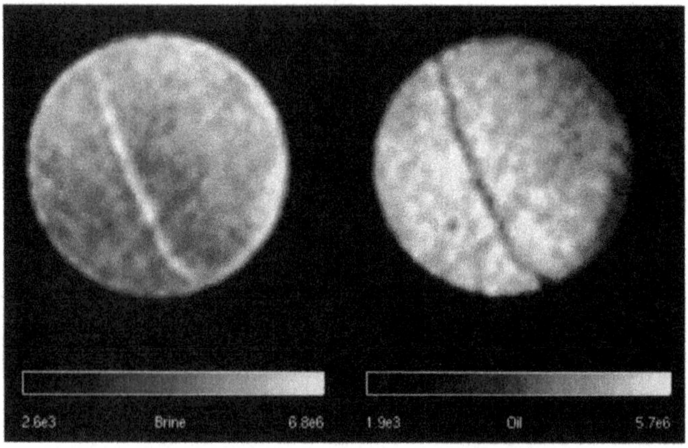

Figure 9.4 Demonstration of chemical shift imaging to show the location of brine (left image) in an oil saturated rock sample. The oil fraction is shown in the right-hand image. These images demonstrate the benefit of such an approach where the two fractions occupy similar volumes. Reproduced from ref. 17 with permission from Elsevier, Copyright 1994.

three-dimensional representations of the oil and water fractions (not shown), demonstrating increased water volume where there are larger pores.

Such methods continue to be employed, such as a T_1 encoding pulse and SPRITE imaging, as demonstrated by Zamiri *et al.* in 2022[19] for samples of shale as recovered. As a highly heterogeneous material with a wide range of pore sizes, shales present a range of imaging challenges; however, the distributions of T_1 values for oil and water do not overlap and thus populations can be easily segregated with an encoding pulse.

The final approach is to utilise the self-diffusion coefficient of the populations which experiences less overlap and is less sensitive to background gradients from susceptibility mismatch while retaining sensitivity to pore geometry. In 2007 Li *et al.*[20] explored the possibility of applying a diffusion editing pulse before a centric SPRITE sequence to select individual phases based on their self-diffusion coefficients. This sees the storage of desired population inverted along the *z*-axis using pulsed field gradient (PFG) preparation, followed by suppression of remaining signals before restoring the *z* magnetisation to the transverse plane. For application in realistic samples, it is necessary to reduce the influence of background gradients, which is most often achieved using one of the pulse sequences described in 1989 by Cotts *et al.*[21] The 13-interval sequence utilises stimulated echoes and suppresses gradient cross terms between the background gradient from susceptibility mismatch and the applied PFG (presuming that the diffusion encoding gradients are centred between refocusing RF pulses). The intensity of a given echo is then given by eqn (9.1) which allows calculation of the self-diffusion coefficient for an experiment where the gradient separation Δ or gradient amplitude g_a is varied and the echo amplitude recorded:

$$\ln\frac{I(g_a)}{I(0)} = \gamma^2 D_0\left[\delta^2\left(4\Delta + 6\tau - \frac{2}{3}\delta\right){g_a}^2 + \frac{4}{3}\tau^3{g_0}^2\right], \tag{9.1}$$

where I is the echo intensity, γ is the gyromagnetic ratio, D_0 is the self-diffusion coefficient, τ is the echo time, δ is the gradient duration and g_0 is the background gradient we are seeking to suppress.

In practice, a combination of these techniques may be needed to provide comprehensive information for real soil samples. In porous media where the mean free path is on the order of the pore size, the relaxation rates R_1 and R_2 ($1/T_1$ and $1/T_2$, respectively) are modulated by the surface-to-volume ratio of the pores which complicates the use of this approach. Fortunately, however, mean free paths in excess of 7 μm can be investigated using typical maximum delays for moderate gradient settings on commercial MRI systems, thus permitting investigation of relevant microporous systems at the expense of sub-10 μm pores. Sacrificing the background gradient suppression of the Cotts sequences or utilising higher gradient strengths allows for interrogation of smaller pores in cleaner samples. Using this strategy allows strong suppression of the water signal, only revealing the oil component and permitting tracking as the oil is displaced by water. Similar approaches with long echo times give hybrid T_2 and diffusion weighting and allow for good suppression of water signals.

Improvements to the methods described thus far continue to be made and the ability to undertake such experiments to be broadened. For example, in 2012, Yan *et al.*[22] demonstrated the ability to make quantitative measurements of oil quantity injected into water saturated glass bead packs using clinical MRI scanners with better than 2% agreement using a multi-spin-echo sequence to visualise the migration pathways, and in 2013,[23] as shown in Figure 9.5, contrast is demonstrated between water ($D = 2.3 \times 10^{-9}$ m^2 s^{-1}) and hexadecane ($D = 3.6 \times 10^{-10}$ m^2 s^{-1}) in clean quartz sand. If flow can be stopped, or one of the phases is naturally immobile, each of the phases can be selectively suppressed as with relaxation times for independent images of each.

More recently, an alternative, though more destructive method of suppressing the water fraction in water–NAPL silica columns has been presented,[24] in which the water is frozen in the presence of an NAPL with a lower freezing point. Solidifying the water phase is sufficient to suppress its signal, thanks to a decrease in T_2 of many orders of magnitude, leaving only the signal from the non-aqueous phase. As can be seen in Figure 9.6, this allows visualisation of contaminants that were previously obscured by the water signal. Simple subtraction provides the ability to estimate contaminant and water quantities. It should be noted that this technique will create multiple possible pathways for the local displacement of the NAPL phases owing to the expansion of water during freezing and the change in wettability which would not necessarily be encountered in the natural environment.

Most recently, the latest pulse sequence developments in magnetic resonance imaging have been applied to soil systems allowing for the study of water in pores with very short relaxation times. There are two sequences of particular interest: ultrashort echo time (UTE) and zero echo time (ZTE).

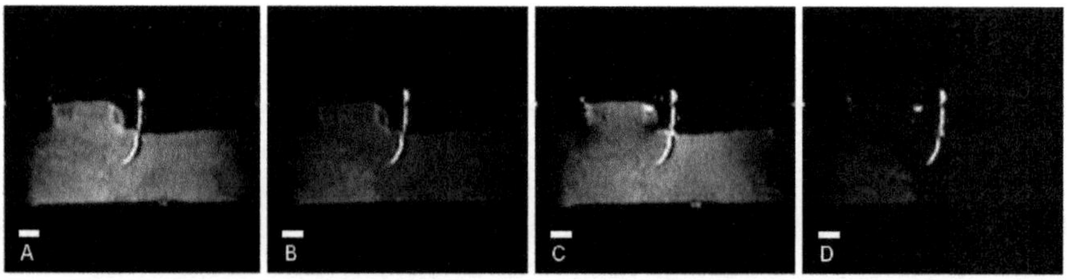

Figure 9.5 Diffusion contrast in a water-saturated quartz sand column (100–500 μm grain size) after injection of 35 ml hexadecane on the left-hand side. The white bar above the letters in the images corresponds to 2 cm. (A) A standard image showing minimal contrast from relaxation times. (B) An image with two bipolar pairs of gradients giving diffusion contrast. (C) The same column experiencing a flow at 0.3 mm s^{-1} without diffusion contrast and (D) that during flow with diffusion contrast. Reproduced from ref. 23 with permission from John Wiley & Sons, Copyright 2014.

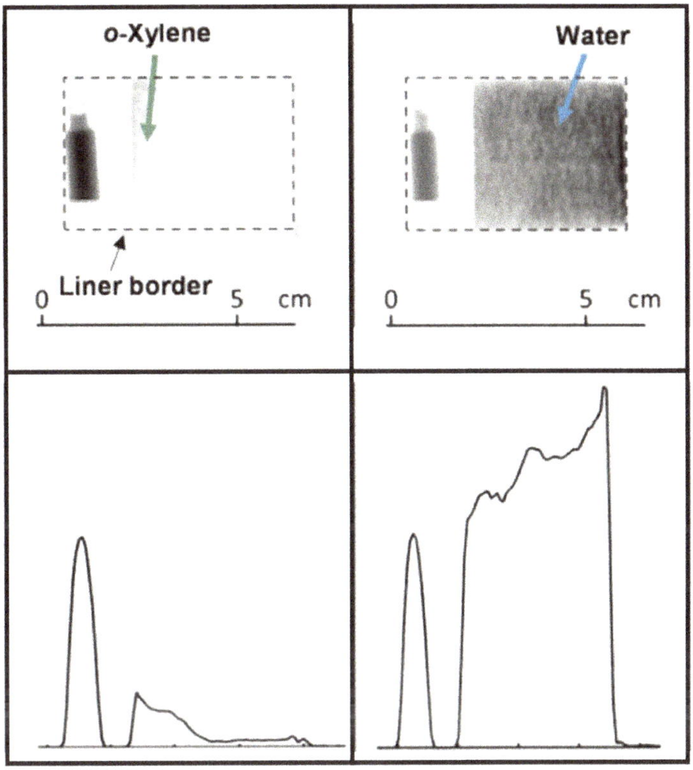

Figure 9.6 Images from the simulated soil core with 90% water and 10% *o*-xylene in Colorado silica sand. The top row are 2D images with a reference of *o*-xylene and the simulated core. The bottom row shows the 1D profiles of the same. The left-hand figures were collected at −20 °C suppressing the water signal, whilst the right-hand figures were collected at +20 °C representing both water and *o*-xylene. Reproduced from ref. 24 with permission from John Wiley & Sons, Copyright 2019.

Whilst pure phase encoded imaging techniques such as SPRITE and single point imaging (SPI) can use echo times as short as 500 μs and thus can resolve fluids with T_2^* on the order of a millisecond, UTE and ZTE offer echo times under 10 μs and rapid scan times at the expense of greater susceptibility artefacts. Although it had been suggested previously, it was not until 2012[25] when a number of practical issues regarding implementation were overcome that the technique really began to find application. The reader is directed to the original article for full detail on the sequence but in brief ultrashort echo times are delivered by utilising half RF pulses applied during slice selection, followed by a rapid refocusing and immediate application of the readout gradient, giving imaging echo times on the order of 150 μs. Further reduction is achieved by ramping down the slice selection gradient in tandem with the half RF pulse resulting in an echo time as short as a few μs (though practically digitisation cannot normally commence until some tens of microseconds later). Zero echo time is an alternative 3D imaging strategy where slice selection is abandoned and 3D spatial encoding gradients applied during the excitation pulse.[26] In practice, the trajectories are normally radial and thus benefit from relatively low noise, given that gradients can be increased by small steps in comparison to traditional imaging approaches. Recently these sequences were applied by Haber-Pohlmeier *et al.*[27] for use in soils to understand the passage of water in small pore volumes. They utilised ZTE and UTE imaging with high (300 kHz) and very high (1 MHz) receiver bandwidths, respectively, to minimise T_2^* blurring and overall improve the quality of the images as can be seen in Figure 9.7. This allowed them to resolve pore water at volumes as low as 100 μL. This represents an excellent imaging strategy for contaminant dynamics imaging in soils and is likely to see significant growth in the coming years.

9.2.5 Heteronuclear MRI

Whilst the vast majority of MRI studies utilise proton imaging, many other nuclei can be imaged using MRI. There is a wealth of literature detailing spectroscopic studies of pollutants in soil using other nuclei, but there are a limited number of publications utilising imaging. In part, this is due to the fact that for soil MRI we are already struggling with the SNR and reducing our sensitivity further has the potential to lose the signals in their entirety. Fluorine (^{19}F) however is almost as sensitive as proton imaging (around 80%) and possesses a similar gyromagnetic ratio. Fluorine is minimally present naturally in soils but is present in many contaminants of soils,

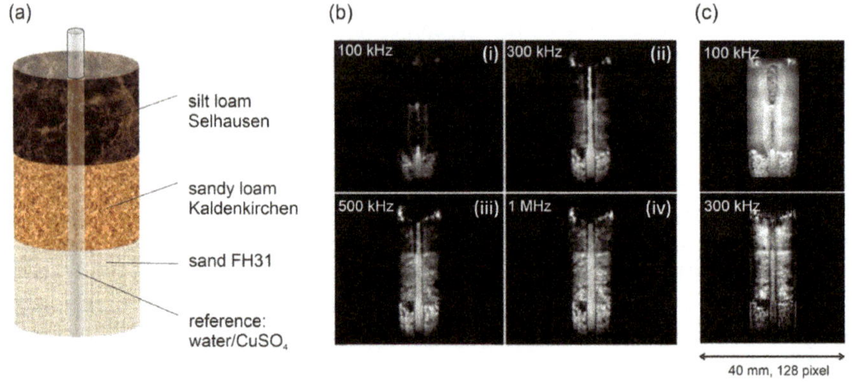

Figure 9.7 (a) Sketch of the composite soil sample. (b) Cross-sectional UTE images in the same orientation showing the three soils and the reference tube. Receiver bandwidth is shown on the images. Increasing bandwidth decreases T_2^* blurring artefacts which is of particular benefit to the silt loam portion of the images. (c) Cross-sectional ZTE images in the same orientation. Increasing the receiver bandwidth again improves image quality and reduces T_2^* blurring. Reproduced from ref. 27, https://doi.org/10.3390/molecules26175130, under the terms of the CC BY 4.0 license, https://creativecommons.org/licenses/by/4.0/.

particularly in the form of per- and polyfluoroalkyl substances (PFAS) which are growing in their presence due to the ubiquity of fluorine compounds in household and industrial effluents. Their specific and direct detection in natural samples is unfortunately not possible using MRI. However, MRI offers opportunities to monitor the migration patterns of PFAS through laboratory systems to better understand their interaction with such samples. At higher concentrations, the ability to specifically detect fluorinated tracers makes them useful for monitoring migration. It should be noted however that there is mounting pressure for PFAS to be banned as persistent organic pollutants (POP) in which case the use of some of the substances discussed in this section may not be available in the coming years. Whilst the approach has been used significantly more to track the uptake of contaminants by plants, it has been demonstrated in 2007 by Simpson *et al.* who utilised proton (^{1}H) and ^{19}F MRI to track the displacement of three fluorine based contaminants by water in real soil columns.[28] The three model contaminants explored were hexafluorobenzene, sodium fluoride and trifluralin. Whilst all three compounds were visible and could be monitored over time, hexafluorobenzene offered the highest SNR though this was primarily attributed to the greater concentration of fluorine atoms both from the perspective of fluorine nuclei per molecule, but also because sodium fluoride is water soluble and will be diluted into the sediment, whilst hexafluorobenzene is hydrophobic. Each contaminant exhibits a different behaviour within the soil system, with trifluralin for example undergoing strong binding with organic matter increasing the rate of signal loss. This proof of principle demonstrates the power of such an approach to observe contaminant dynamics in real systems as can be seen in the maximum intensity projection images in Figure 9.8. This also represents a significant area of opportunity for future research. There are studies applying multinuclear MRI to hydrogels[29] (using sodium and aluminium imaging) and catalysts,[30] the methods of which could be applied to explore both contaminant dynamics and remediation strategies in real and model soil systems.

9.2.5.1 *Manipulating Relaxation Times*

The alternative approach of doping offers a relatively straightforward approach where a dopant miscible in only one phase is used to increase the differences in relaxation times.

9.2.5.1.1 Spin–Spin Relaxation Time. The most popular dopant is manganese (typically Mn^{2+} ions, often in the form of water soluble manganese chloride) which was first demonstrated for differentiation of oil and water phases in Berea sandstones by Baldwin *et al.* in 1998.[31] Adding Mn^{2+} ions to water at concentrations ranging from 6 ppm to 4000 ppm before adding to samples containing oil, they were able increase the separation in the T_2 of the phases and thus image the different fractions at different echo times.

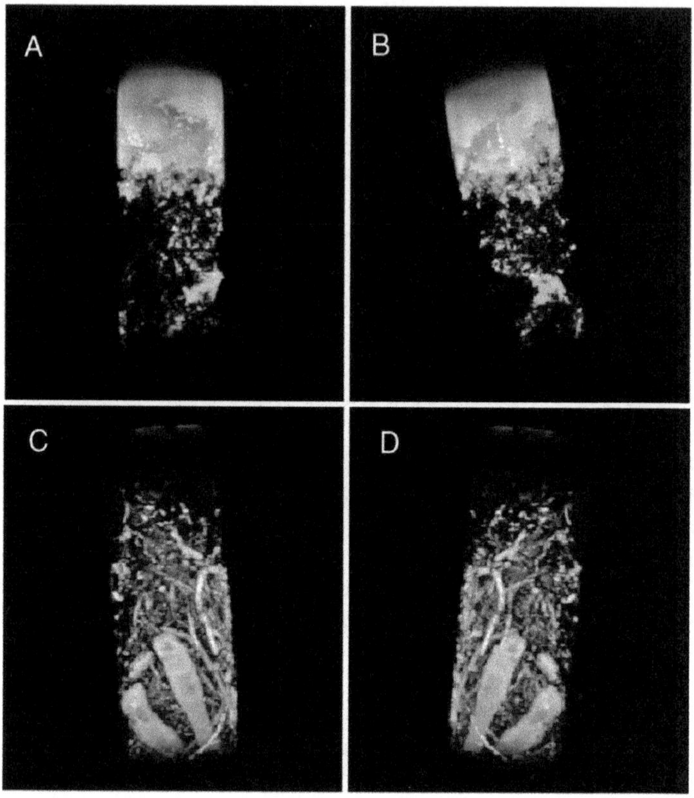

Figure 9.8 Hexafluorobenzene is added to a soil core and allowed to permeate for 24 hours. Multinuclear imaging is undertaken and a maximum intensity projection algorithm applied. (A and B) Fluorine images and (C and D) the water fractions. The left- and right-hand images are different rotations of the same images to provide a 3D view of the sample and the distribution of hexafluorobenzene and water. The images show that after 24 hours there has been minimal displacement of the water by the hexafluorobenzene despite its higher density. Reproduced from ref. 28 with permission from John Wiley & Sons, Copyright 2007 SETAC.

Such approaches are used widely in the literature either to enhance the separation of the relaxation times or for complete suppression of the signal from one of them. The suppression approach was used by Gaigalas *et al.* in 1989[32] and Guillot in 1991[33] to monitor the displacement of existing water volumes in a wet medium, with signal suppressed water, though care must be taken to avoid excessive signal loss with the dispersion of ions through the existing phase. In 2008, Nestle *et al.*[34] utilised Mn doping to study the distribution of *n*-octanol in quartz sand model sediments initially saturated with a Mn contrast agent solution. In this work, the Mn contrast agent was used to produce highly selective images of *n*-octanol, thanks to the very heavy T_2-weighting afforded by a turbo spin echo sequence. The effect of ethanol injections on the distribution of the *n*-octanol phase was studied and the results indicate many competing physical processes taking place including buoyancy and wetting which result in the formation of small droplets that then migrate throughout the sediment. More recently, this approach has been used to suppress the water signal during removal of a hydrocarbon fuel contaminant by the water flow experiment in polystyrene bead and sand packs[35] to better understand the influence of wetting. High resolution imaging was used to investigate trapped volumes and as expected it was found that water channelling led to retained oil volumes, the size of which was determined by capillary force and matrix wettability. For sand which is water wettable, the flow front travels as menisci with capillary force acting in the same direction as the flow, whilst in the oil wettable polystyrene pack a film-form configuration dominates trapping oil on the pore surface. This approach has also been extended to allow the mixing and chemical reactions to be monitored. In the work of Markale *et al.*,[36] glass bead columns are initially saturated with a solution of manganese(III) acetate providing a volume of Mn^{3+} ions which are less paramagnetic than Mn^{2+} we have seen thus far. A reducing agent, cyclohexanedione, and a volume of Mn^{2+} ions in the form of manganese(II) acetate are then introduced and the reduction in signal intensity owing to the conversion of Mn^{3+} to Mn^{2+} is monitored to better understand the dynamics of the process. A second series of experiments are also conducted where a fluorinated immiscible phase, visualised using fluorine imaging, is randomly distributed through the porous media before the reactants are introduced. Concentration gradients are well visualised using this technique and are used to monitor the progressing reaction front. This front is stretched and folded in three dimensions as it progresses through the porous media. This 3D imaging approach provides a wealth of information unavailable with 2D experiments and is used to validate lamella theory for reactive transport and observe the variable reaction rate as a function of saturation.

The spin–spin relaxation time can also be shortened using paramagnetic or superparamagnetic nanoparticles. These nanoparticles represent a potential contaminant in their own right but may also find application as a phase specific modifier in the same way we have seen with Mn^{2+}. There are three key studies published in 2015 which discuss the transport of nanoparticles in environmental systems. The first utilises gadolinium nanoparticles and investigates contaminant transport through gravels.[37] The interactions between nanoparticles and the matrix can be significant, as is the case in this study where the gravel comprises dolomite which creates an electrostatic attraction for the gadolinium nanoparticles without magnetic disruptions. This was compared to quartz gravels which do not have the same effect. As can be seen in Figure 9.9, the signal from water in proximity to the nanoparticles is bright allowing tracking of their location. A signal calibration can also be used to give concentration maps which are also shown. Unlike traditional breakthrough curve measurements, this imaging

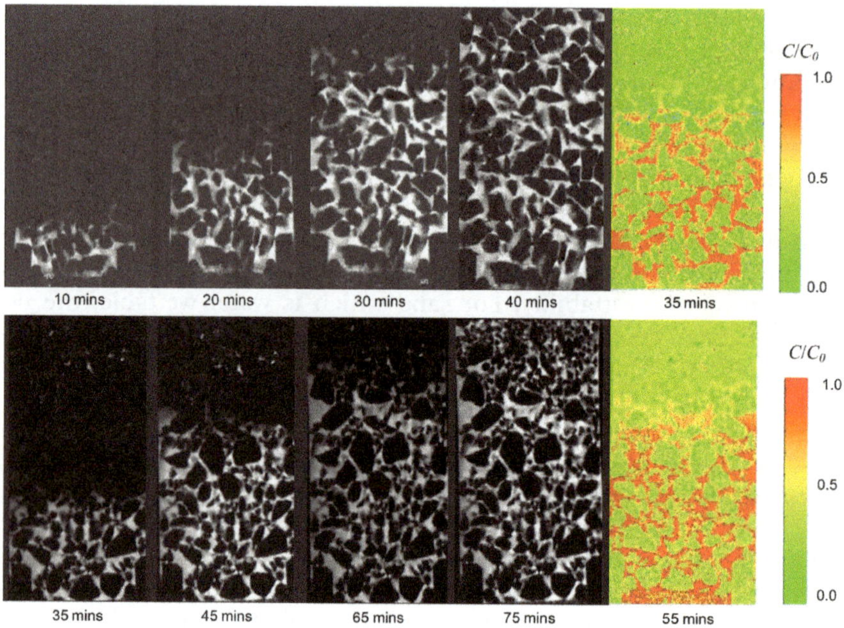

Figure 9.9 T_2-weighted images of gadolinium nanoparticles in quartz gravel (top) and dolomite gravel (bottom); coloured figures are concentration maps produced by calibration. The infiltration times are denoted below each panel. Reproduced from ref. 37, https://doi.org/10.1007/s13762-015-0767-4, under the terms of the CC BY 4.0 license, https://creativecommons.org/licenses/by/4.0/.

approach offers valuable insight into the surface interactions owing to electrostatic effects on nanoparticles.

Polystyrene coated superparamagnetic particles were used to investigate the effect of heterogeneous column packing by Baumann *et al.* in 2005.[38] A layer of coarse-grained silica is packed around a fine-grained core, in a similar arrangement to the Chu study discussed earlier[13] before introducing a pulse of tracer particles. The flow of this tracer front is tracked and the dispersion determined as a function of the packaging arrangement. As can be seen in Figure 9.10, when the shell and core have similar sizes (the left-hand side of the figure), the flow is mostly uniform, whilst when there is significant difference in sizes (the right-hand side of the figure), there is preferential flow along the outside of the column after initial introduction. Since the majority of real soil systems comprise a wide distribution of particle sizes, such studies are essential for understanding flows in practical applications.

The other studies both utilise iron oxide nanoparticles which are often used as a negative contrast agent in MRI since they promote rapid signal dephasing at very low concentrations. The first, also by Lakshmanan *et al.*,[39] investigates the transport of carboxyl functionalised nanoparticles through

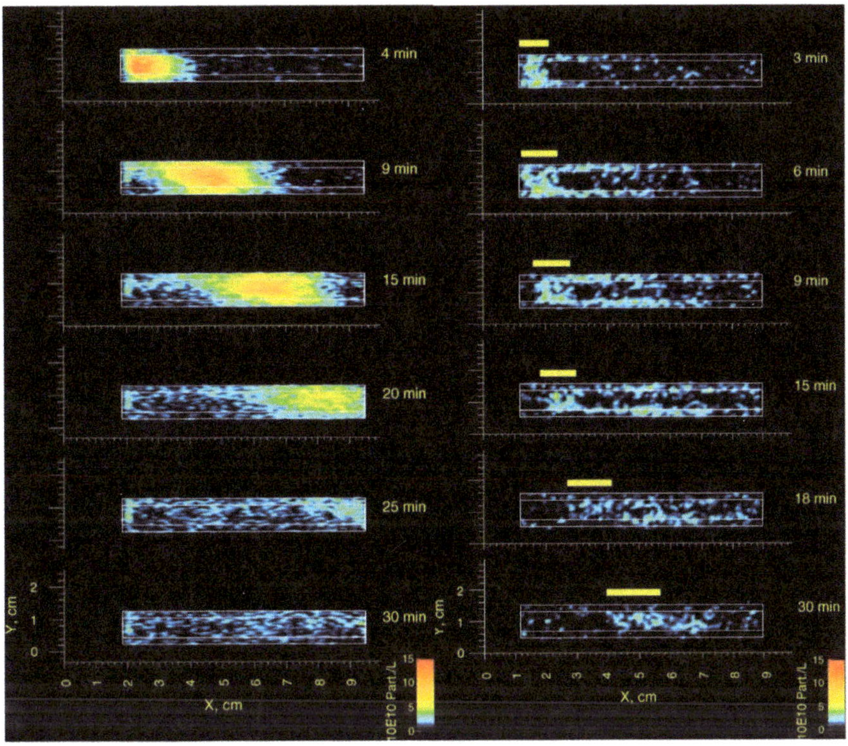

Figure 9.10 Figure showing the flow of polystyrene coated superparamagnetic particles through annular columns of silica gel with different grain sizes in the shell and core. The left-hand figure represents flow through a core 600–850 μm particle diameter, surrounded by 850–1000 μm particle diameter. The right-hand figure represents 250–600 μm particle diameter in the core, surrounded by a shell of particles with a diameter of 850–1000 μm. The boundaries of the core and shell for each set of results are shown with a white outline. Note that where the particles are similarly sized in the core and shell the flow progresses similarly through both, while for greater differences in diameters, the flow tends to pass through the higher diameter particle region. Reproduced from ref. 38 with permission from Elsevier, Copyright 2005.

sandstones which is then processed using a model to estimate the transport parameters through the length of the column providing information which can be scaled up to larger samples to develop improved remediation strategies. The second of the iron nanoparticle studies is by Cuny *et al.*[40] who explored the role of particle size and functionalisation in transport through quartz sand. As with the other studies, the MRI reveals information that is lost in the breakthrough curves, particularly with regard to different radial distributions of the contaminants despite similar breakthroughs, an example of which can be seen in Figure 9.11. The study confirms that surface functionalisation and the pH of fluid fraction have the greatest effect on the transport properties of the nanoparticles, and that sorption of natural

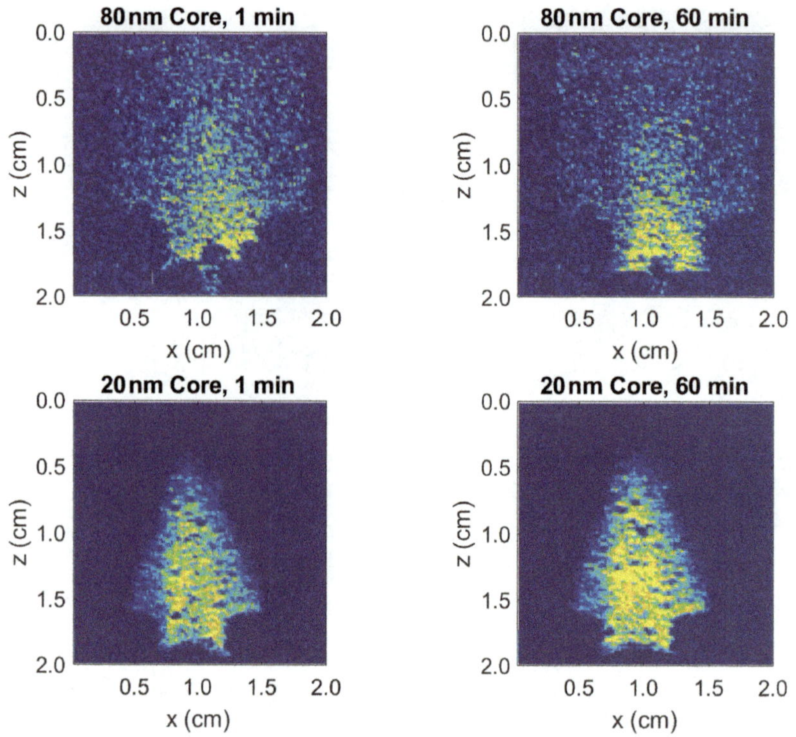

Figure 9.11 Difference images showing patterns of dispersion of superparamagnetic nanoparticles transported through water saturated porous media. Two particle sizes with different coatings are shown with 80 nm dextran coated particles on the top row and 20 nm uncoated particles on the bottom row. Similar breakthrough curves are found for both of these particles, despite significant differences in their distribution through the media. Reproduced from ref. 40 with permission from Elsevier, Copyright 2015.

organic matter (simulated with humic acid) on the particles has a strong influence on their zeta potential and increases their mobility which has significant implications for lab experiments when extended to environmental applications.

Similar iron oxide nanoparticles have been more recently applied[41] as tracer cores in microcrystalline cellulose as a toilet paper reference to understand breakdown in wastewater. The simulated wastewater is passed through a granular bed, left overnight, and an additional flow applied. A preferential flow path is created along which the particles are seen to deposit. The flow is sufficient to reorder the bed, but the second flow continues along a similar preferential path, leaving much of the column unused as can be seen in Figure 9.12. The tracers offer a useful means to assess the precise passage of particles which is not possible with similar resolution using any other technique and demonstrates the potential issue of preferential flow paths clearly.

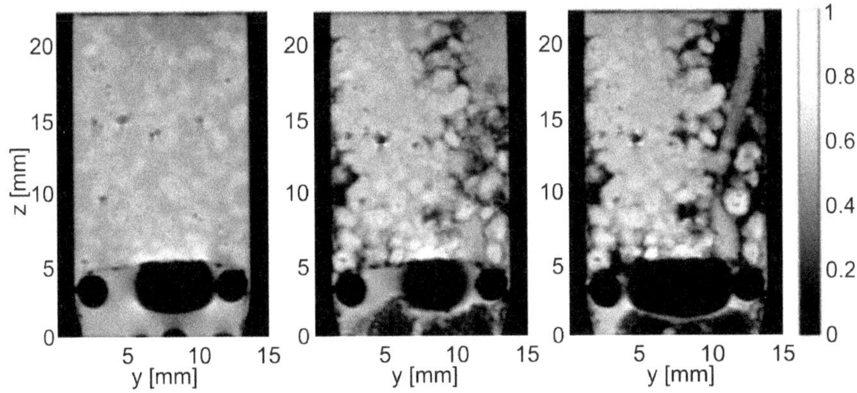

Figure 9.12 2D MRI images identifying the position of iron oxide nanoparticle cored microcrystalline cellulose particles in a granular sludge bed. Images from the left to right are before loading, moderate particle loading and additional particle loading. Images are T_2-weighted and show where particles have accumulated as areas of lower signal intensity. During flow, the granular bed is seen to reorder and a preferential flow channel is seen to appear on the right-hand side of the image. The particles are also seen to accumulate at the inlet and outlet. Reproduced from ref. 41 with permission from Elsevier, Copyright 2022.

The same process can be applied to shorten T_1, to provide variable contrast through the repetition time or preparatory inversion pulses. It should be noted that whilst T_2 can be readily shortened independently of T_1, the majority of compounds that shorten T_1 will also shorten T_2. The following discussion focuses on the use of such agents to influence T_1 but changes will inevitably also occur in T_2. Nickel ions (Ni^{2+}) can be used to reduce the value of T_1 creating signal enhancement for shorter repetition times as was demonstrated in 1991 by Pearl and Bendel:[42] two sand packs, separated vertically by a metallic barrier, are saturated with distilled water and a higher density aqueous sodium chloride solution doped with nickel chloride ($NiCl_2$). The barrier is then removed and periodic imaging commenced. Subtracting the image at time zero from later experimental times shows the migration of the Ni^{2+} ions through the packs. The behaviour of the phases is dependent on the orientation of the packs with respect to gravity. With the high density phase above, an unstable, convective front owing to the density difference moves downward under gravity; in the opposite configuration, the dominant process is diffusion with the upward migration of Ni^{2+} ions. These two processes can be seen in Figure 9.13 and provide a valuable visualisation tool for the migration of contaminants through aqueous systems. The growth of fingers in the unstable configuration was measured over time and correlated with permeability. As with the study of Guillot for Mn^{2+}, Ni^{2+} ions can be used to monitor the displacement of a water phase by water, which was demonstrated by Herrmann in 2002[43] to follow a flow front in three-dimensional space through a fine-grained quartz sand column.

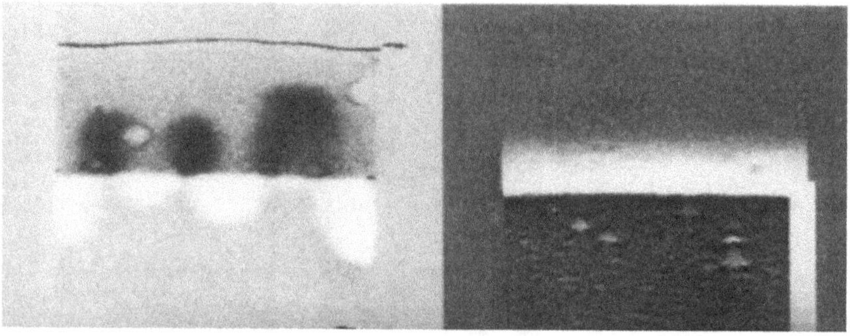

Figure 9.13 Two difference images demonstrating the different forms of transport in two horizontal sand packs saturated with distilled water or higher density aqueous sodium chloride solution doped with nickel chloride ($NiCl_2$). As difference images, the higher intensity region correlates to an increase in nickel ions. The left image is the high density phase above resulting in the formation of an unstable, convective front shown after 90 minutes. The right image is the high density phase below, and thus the dominant process is diffusion with the upward migration of Ni^{2+} ions. Reproduced from ref. 42 with permission from Springer Nature, Copyright 1993.

The shortening of T_1 opens up access to fast imaging sequences such as fast low angle shot (FLASH) which were used in this work with a 10° tip angle after a preparatory inversion pulse to suppress the undoped water signal. Tracking the front in this way can be used as a time-of-flight measurement to determine the velocity profile throughout the core, which is a simple means by which flow can be investigated which will be covered in detail in the following sections.

A similar approach in 2008 by Moradi *et al.*[44] found a linear correlation between T_1 and Ni^{2+} ion concentration down to 1.5 $mg\,l^{-1}$ using an inversion recovery approach. However this does not bring with it a high enough temporal resolution to track contaminant uptake. Instead, a variable flip angle approach is used to map T_1[45] by varying the flip angle instead of the inversion time, which is thus significantly faster yet retains quantitative mapping. This approach was validated by comparing absorption into an ion uptake gel to that into a glass bead pack. As expected, no concentration gradient was formed for the glass beads, whilst the gel created a depletion region after four hours, the concentration map of which can be seen in Figure 9.14.

The same tracer approach was used in 2011 to explore the influence of entrapped air in real soil samples.[46] Many real sample experiments of soil samples will begin with a saturation step which in some cases brings us one step further away from reality, particularly in soils which are not readily wet. Ni^{2+} tracer experiments were used to track flow paths during a recurrent ponded infiltration experiment where a volume of fluid is introduced on the top surface of the sample under investigation which is allowed to pass until the flow ceases. This process is then repeated. The technique is useful to understand the stability of results as determined from a single flow and to

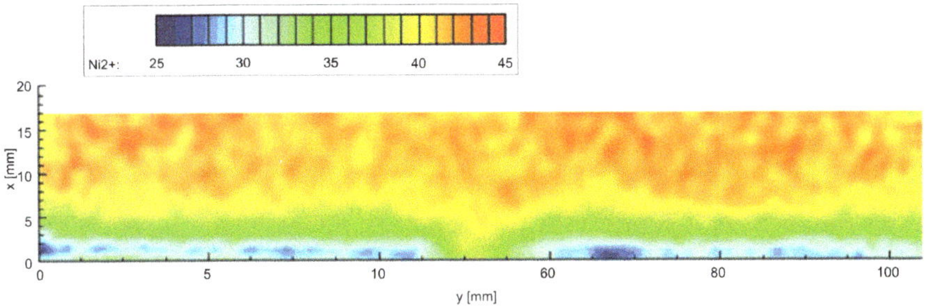

Figure 9.14 Averaged (over four 7 mm slices) Ni^{2+} distribution four hours after the start of the experiment using MRI. DGT resin strips are present, separated by a small gap at the base of a glass bead pack. The influence of the gap is visible in the centre of the figure at the base. Around this gap, the two-dimensional nature of the Ni^{2+} distribution is visible, whilst further away the uptake into the beads is mostly parallel to the gel surface showing a homogeneous depletion of Ni^{2+} along the gel surface. Reproduced from ref. 44 with permission from John Wiley & Sons, Copyright 2008 The Authors Journal compilation; Copyright 2008 British Society of Soil Science.

explore factors that lead to variability. It was found that whilst a preferential flow path was created during the first infiltration, during outflow a trapped air bubble was formed within the path which prevented its use during the subsequent infiltration. Although this bubble accounted for only a 2% decrease in used pore space, it resulted in a completely different flow path which is an essential result to keep in mind when processing data from single run experiments in real samples.

A 2003 study by Nestle *et al.*[47] also used FLASH imaging alongside spin echo (SE) imaging to explore the adsorption and remobilisation of heavy metals in sand columns in the head coil of a clinical MRI scanner. When such ions are adsorbed onto the surface of sediment, there can be a severe change in the relaxation contrast which is found in the sample. This offers opportunity to determine the extent of the adsorption but is also an important point to consider when undertaking general experiments in sediments with heavy metals. It needs to be taken into account that these effects may be strongly dependent both on the chemical nature of the sediment surface as well as on the details of the electron spin relaxation properties of the injected paramagnetic ion. For example, Gd^{3+} and Dy^{3+} were found to behave quite differently upon adsorption to quartz sand with Gd keeping a strong relaxivity but Dy turning almost silent. Furthermore, a long-time ageing effect due to rearrangement of the adsorbed ions could be observed for the case of Gd^{3+} as shown by Nestle *et al.* in 2007.[48,49] The combination of the two pulse sequences was beneficial beyond their individual contrasts, thanks to the improved SNR of the SE imaging technique offering accurate assessment of the shape of the propagating contaminant front at the expense of the ion concentration which was provided by the FLASH sequence despite the poor SNR. $Cr(NO_3)_3$ and $Gd(NO_3)_3$ are explored using this approach which was demonstrated with temporal and spatial resolutions of 1 min and 1 mm, respectively, offering plentiful opportunities for application.

It is also worth noting that, although not as frequently used, it is also possible to get T_1 and T_2 contrast images of the same system using different pulse sequences. For example, in the 2008 study of Nestle *et al.*,[34] which we mentioned in Section 9.2.5.1.1, where Mn doping was used to enable heavy T_2 weighting and thus suppression of a water phase, 3D FLASH imaging was also used to enable heavy T_1 weighting with high temporal resolution (an image every 11 seconds). Such an approach offers greater insight into the physical processes that occur.

9.3 Applications of MRI to Understand Physical Processes

In addition to resolving different phases, MRI is useful to understand the processes that underpin contaminant migration such as wetting. In 2000, Belliveau[50] *et al.* utilised the change in the transverse relaxation time T_2 to monitor the movement of pooled water (long T_2) into soil through a process of swelling and ingress through the opening of micropores represented as a shortening of T_2. This knowledge can be applied to the migration of contaminants such as pesticides and herbicides throughout soils. The challenges associated with obtaining sufficiently high signals in the presence of impurities have somewhat limited the application of relaxation measurements for measurement of swelling kinetics in soils such as in the work of Schaumann *et al.*[51] though it has been undertaken in bentonite clay, as done for example by Dvinskikh.[52] In this work, two-dimensional MRI is used to create one-dimensional profiles of compressed disks of hydrated, purified Na-ion exchanged bentonite clay placed underneath a pool of deionised water. Over the time course of the experiment, the water ingresses and swells the clay. To account for variability in the scanner and scan parameters, a similar sample is also prepared containing only water which is used as a reference signal. There is a significant change in the relaxation times as this process proceeds from tens of microseconds to seconds for T_2 and milliseconds to seconds for T_1. It is not possible in a single scan to accurately capture such a wide range of relaxation parameters so the authors undertook imaging experiments focused on free (long relaxation times) and confined water (short relaxation times) populations separately. Traditional spin echo images reveal useful, though not quantitative, spatial information on the propagation of a swelling front. However, T_1 mapping is useful to provide information on the clay concentration. A set of calibration samples were prepared for clay volume fractions from 0 to 0.2 for which T_1 values were obtained using standard inversion recovery imaging. The confined water however presents a more significant challenge. Owing to the significantly reduced T_2, the authors utilised constant time imaging (described in full by Bendel[53]) which has pure phase encoding in two dimensions, applied during the application of a 90° pulse and FID readout to allow for the acquisition of extremely short T_2. Unfortunately, the mean free path of the water diffusion through the clay matrix is on the order of the pellet size over the time course of the experiment and thus cannot be used quantitatively in isolation. The combination however of these two methods provides a comprehensive view of the swelling processes. Over the course of a 620 hour experiment, the authors were able to determine the profiles of clay volume fraction covering 5 orders of magnitude, thanks to the combination of these pulse sequences. Such an approach opens up a range of possible experiments with a wide range of relaxation times as is often the case in samples with paramagnetic impurities and other contaminants. Such an experiment was presented in 2009 by similar authors[54] where superparamagnetic iron oxide particles were included in the clay pellets to understand the influence of contaminants in these systems alongside results for different bentonite forms and different solvents. Of key interest for this chapter, the influence of solvent on the swelling of the clay is critical to understand the migration of pollutants through clay containing soil systems. A similar methodology was used as in the previous publication to produce several key findings. Firstly, distributed superparamagnetic particles throughout swollen clay are seen to sediment over the course of four hours, falling to the base of the system. However, if included in the clay pellet before swelling commences, they both reduce the speed at which swelling takes place and also remain in suspension for the duration of the experiment (22 days). The influence of the solvent on the swelling behaviour is demonstrated through the comparison of the swelling rate with different concentrations of calcium chloride in calcium and sodium exchanged bentonite. No difference was observed for the calcium exchanged bentonite, but in the sodium exchanged bentonite, the higher the concentration of calcium ions, the slower the rate of swelling, going from 100 mm over 80 days to 55 mm over 80 days when the concentration of calcium chloride is increased to 4 $mM\,L^{-1}$. This demonstrates the great influence of interactions between pollutants and soil components to understand their migration in soil systems. Correlating traditional measurements with those achievable with MRI is often needed to gain adoption of the technique within industry. The challenges associated with working with the short T_2 times often found in such systems were overcome with a novel approach in 2005 as demonstrated by Fagan *et al.*[55] who utilised continuous wave MRI. Although not as common as the typical pulsed MRI with which most facilities are familiar, the use of continuous wave MRI has been shown to be an effective alternative for many experiments. In this work, the diffusion of water throughout bentonite clay and kaolinite clay matrices was investigated to determine the transport behaviour in swelling and non-swelling clays. Packed columns of heavy water wetted clay are exposed to a water reservoir in the form of light water soaked laboratory tissue. The exchange between the two water pools is monitored using continuous wave MRI. Despite the difference in the molecular dynamics of the two types of clay, no strong effects on the long-range diffusive transport behaviour of the water molecules were observed, suggesting that tortuosity encountered by the molecules is similar in both systems and Fickian in nature. In 2013, Shamim *et al.*[56] demonstrated the correlation of T_2^{eff}, as measured using a multi-echo sequence

(a) N =1 (b) N =3 (c) N =7

Figure 9.15 MRI images of the cross-sectional view of the soil column for repeated drying–wetting cycles. N is the number of cycles which the soil has experienced. The signal intensity is proportional to the quantity of water in each voxel. As the number of cycles increases, the small fissures present initially expand and develop new areas of cracking which are filled with water. Reproduced from ref. 57, https://doi.org/10.1155/2022/9948658, under the terms of CC BY 4.0 license, https://creativecommons.org/licenses/by/4.0/.

along a glass bead column experiencing a flow of sterilised wastewater. The particulate clogging, free of the influence of biofilms, demonstrated a gradual decrease in T_2^{eff} starting from the inlet and moving to the outlet. The initial decrease in the relaxation time was homogeneous within the slices, thanks to the comparatively high mobility of the influent; however, as the overall hydraulic conductivity decreases, a gravitationally driven sedimentation results in pore space clogging from the base of the tube upwards over time. The traditional measures fail to identify the location of this occlusion which is essential to understanding contaminant transport in environmental systems. A more recent publication by Xu *et al.*[57] in 2022 applied repeated drying–wetting cycles and used MR imaging after each wetting cycle to observe the change in pore structure. Images after one, three and seven cycles are shown in Figure 9.15. As can be seen from these images, there is a significant increase in macropores as the number of drying–wetting cycles increases. This extends out from initial cracks which ultimately join as the porosity increases. Such information is not available using bulk measurement methods and highlights the benefit of MRI in such applications, revealing new information on the impact of repeating wetting–drying on soils. Such experiments provide an important piece of the puzzle in understanding how the structural and physical properties impact hydrodynamics.

9.4 MRI of Flow Through Soils

When we ask our students to consider what is needed for the best achievable MRI image, they will answer a strong, homogeneous, static magnetic field applied to image a stationary sample free of ferro- and paramagnetic species. We have already seen that we struggle with keeping our static field homogeneous because of the presence of magnetic species in most soil samples. In this chapter and those that follow, we will introduce a further challenge which is to undertake imaging for a sample that is not stationary.

In routine MRI, we consider the impact on our images from unavoidable flow as an artefact. However, relatively simple extensions to the pulse sequences of choice can intentionally encode information about flow in our signal bringing with it opportunities for qualitative and quantitative measurements.

In this section, we will see the power of flow MRI to determine how contaminants flow through soils and how this can be used to understand environmental impacts and inform strategies to reduce pollution.

9.4.1 Generalised Fluid Flow Used to Predict Contaminant Behaviour

In addition to the time lapse tracking of a contrast front which has been introduced in the previous section, it is possible to purposely encode displacement in the signal by applying an additional gradient. These gradients are often bipolar, in that they are first applied with a positive effect and then for an equal integral with a negative effect. In this way, there is no change to the overall phase for nuclei that have not moved, but those that have will gain (or lose) phase proportional to the distance that they have moved. The flow can then be encoded in one orthogonal direction at a time by changing the orientation of the applied gradient and in combination with the gradient duration can be used to calculate the flow velocity. Of particular importance for flow in real or model soils is that background gradients, such as those caused by paramagnetic impurities, are not necessarily cancelled out. To avoid this, the polarity of the gradient can be reversed and the difference between phases on those two measurements will give pure displacement encoding. Some understanding of maximum velocities is required in setting up such experiments as the gradients must be kept small enough that no more than 180° of phase shift is added whilst remaining strong enough that maximum sensitivity is retained.

9.4.2 Flow of Contaminant-like Fluids

The first studies utilising velocity encoding gradients in a porous medium system were conducted by Muller *et al.* in 1995[58] and Kutsovsky *et al.* in 1996.[59] The former utilising a velocity encoding gradient applied in three orthogonal directions with subsequent imaging to produce a Fourier encoded velocity and spatial distribution which was used to measure the flow of water through CNC produced 3D lacunar networks. The latter work imaged a 6 mm bead pack experiencing fluid flow, where thin slices were imaged with and without velocity encoding. There was sufficient resolution to measure the flow front within individual pores which was found to be parabolic as expected for relatively unrestricted flow. This was extended by Dijk *et al.*[60] three years later for rock systems saturated with seawater to produce three-dimensional flow maps. The first application of these techniques to complex porous media towards monitoring the flow of a contaminant-like fluid was reported in 2001 by Mantle *et al.*[61] who used the technique to validate lattice-Boltzmann simulations based on stationary structural imaging for both single-phase flow and two-phase, liquid–gas flow, which is applicable to the sparging remediation process as we have already seen in Section 9.2.3. The results also highlight the importance of understanding the wetting properties of the system under investigation as we also saw in an earlier section with the polystyrene *vs.* sand packs of Fannir.[35]

9.5 Reactive Contaminants and Their Kinetics

The extension of these flow processes to the study of dynamics, migration and kinetics was first published by Johns and Gladden in 1998[12] and again in 2000.[62] Pockets of octanol were deposited in a pack of glass ballotini whilst water was flowed throughout. The dissolution of the octanol pockets was then monitored using MRI, alongside velocity mapping of the water. Preferential flow paths are well identified and, as can be seen in Figure 9.16, pockets immediately adjacent to high flow areas are readily dissolved, whilst those proximal to slower flows still have some volume remaining. The volume labelled (2) remains unchanged by the end of the experiment as it is in an area of stagnant water. With over 80 citations, this approach has been well adopted, particularly to inform dissolution models. A 2007 study by Zhang *et al.* utilised approaches which we have met in other sections, namely paramagnetic tracing of the aqueous phase and fluorine imaging to monitor the dissolution dynamics of non-aqueous phase liquids into the water

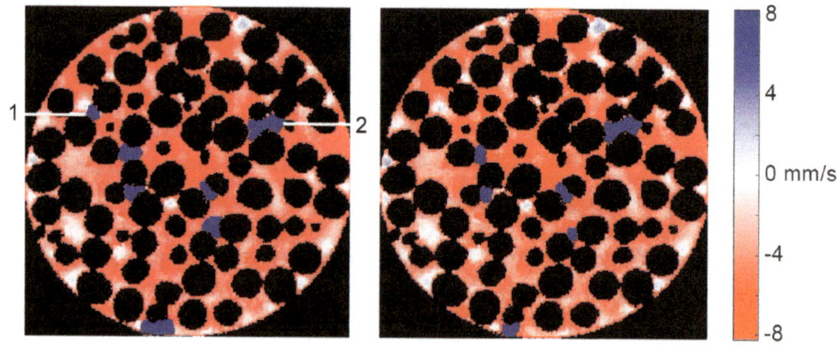

Figure 9.16 Velocity and hydrocarbon distribution maps of the multiphase system in glass beads. Hydrocarbon ganglia are shown in dark blue. The colour encodes the flow velocity. There is significant heterogeneity of the flow throughout the slice with stagnant areas and high velocity flow channels. The left image was collected 2.5 hours after starting dissolution, whilst the right-hand image was collected 34.5 hours later. Note the preferential dissolution of ganglia near areas of high aqueous phase flow (*e.g.* ganglia 1) in comparison to those near stagnant regions (*e.g.* ganglia 2). Reproduced from ref. 12 with permission from Elsevier, Copyright 1999 Academic Press.

fraction. T_2-weighted proton imaging with variable echo times is used to select water in different pore sizes of the heterogeneous porous media, whilst fluorine imaging reveals the pure NAPL phase. This approach highlights the fact that preferential flow paths through areas of higher permeability are responsible for more rapid dissolution. This grows increasingly significant as the concentration of NAPL is increased. Over time these preferential flows grow, increasing the relative flow through the contaminated areas culminating in high NAPL–water surface area.

9.6 Imaging of Biofilms in Soils

The final piece of the puzzle to gain a comprehensive understanding of real soil samples is the ability to determine the quantity, location and the form of biofilms present within. MRI offers us such an opportunity using a range of different methods. For example, since biofilms have an effect on the T_1 and T_2 of water, in and around the films, relaxation-time-weighted images or fitting can be used to identify areas with biofilms present.

An early publication in this area was *Selective imaging of biofilms in porous media by NMR relaxation* by Hoskins *et al.*[63] in 1999 where biofilms were selectively imaged using MRI. Transparent parallel plate cells with and without glass beads were used to compare biofilm growth to MR images with optical imaging. Lengthy imaging experiments are used to provide sufficient points for bi-exponential fitting to T_1 and T_2 to give components associated with the biofilm and in free water, yielding 4 images for each relaxation type (intensities Ioa and Iob and relaxation times T_1a or T_2a and T_1b or T_2b). These images are then sorted by Io:T_2 or Io:T_1 and a threshold used to assign the value of Io to pixels in either a biofilm or a water image. Results correlated with optical imaging for at least 75% of pixels in all samples and clearly delineate areas of the biofilm as can be seen in Figure 9.17.

Such an approach offers a useful demonstration of the ability of MRI to resolve the location of biofilms which can already give us a useful indication of areas that are likely to hold or break down pollutants. Of particular value in contaminant research is the work by Olsen *et al.*[64] who investigated bacterial chemotaxis (the motion of bacteria along chemical gradients) by labelling bacteria with a monoclonal antibody attached to 50 nm magnetite particles,[65] which was discussed in the paramagnetic labelling section previously. Imaging is undertaken in an aqueous saturated porous medium with increasing concentrations of trichloroethylene (TCE) which creates a concentration gradient along which *Pseudomonas putida* F1 is known to migrate. T_2-weighted spin echo images are collected and used to track the bacteria at 330 μm resolution. In the absence of TCE, the motion is purely diffusive, but as the concentration increases, the peak chemotactic velocity was seen to increase over two orders of magnitude. Whilst the average velocity could be determined with breakthrough curves, the use of MRI reveals the spatial variation of velocity, with maximal velocity found in the mid part of the column. In the later work of Cao *et al.*,[66] the water self-diffusion

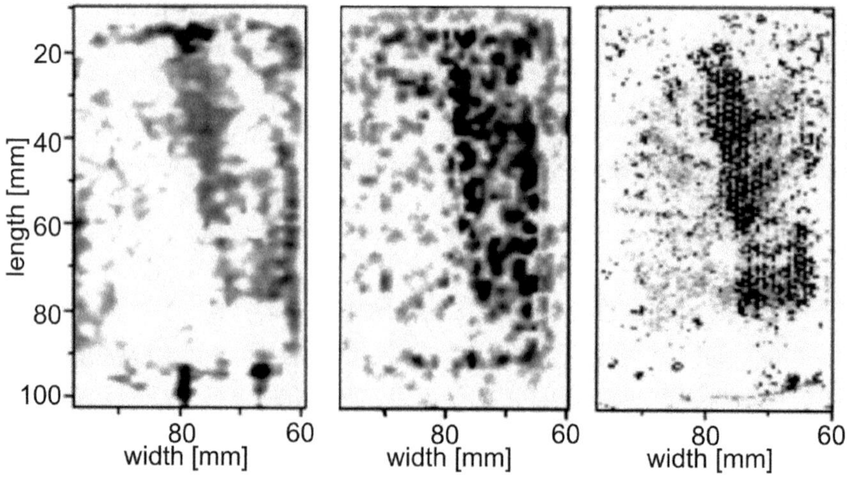

Figure 9.17 T_1 sorted biofilm image, T_2 sorted biofilm image, and optical image of the bead packed reactor experiencing biofilm growth. Reproduced from ref. 63 with permission from Elsevier, Copyright 1999.

coefficient of biofilms was measured in response to the presence of different ionic species, namely uranyl (UO_2^{2+}) and chromate (CrO_4^{2-}), modelling their behaviour in the presence of pollutants. It was found that while the overall mass of the biofilm remained unchanged in their presence, the chromate ions in particular induced significant spatial variation in the self-diffusion coefficient of water. Importantly, a stratified structure was seen to form with the coefficient reducing in the surface 100 μm and increasing in the lower 200 μm which has important implications for the effects of contaminants in soil systems, where diffusion through biofilms is a key mechanism of transport. The authors postulate that these ionic species interact with the polysaccharides in the biofilm matrix changing the structural and therefore hydrodynamic properties, resulting in the variation of the self-diffusion coefficient. In addition to the structural changes of the biofilms, the stratification is also likely driven by the cytotoxicity of the ionic species leading to a loss of the metabolic activity of the microorganisms leading to elimination of metabolic induced flow on a local level. In addition to the already mentioned impact on transport, this has a bearing on the nutrient availability to the biofilm which could negatively impact bioremediation efforts in soils.

A recent paper by Ranzinger *et al.*[67] investigates the impact of a changing water level on biofilms grown in soil proxies. The process of draining and flushing a bead pack during growth creates a different biofilm morphology as visualised by MRI. Particularly relevant was that the biofilm tended to first grow at the necks of the pores, which, whilst not unexpected based on previous studies, was able to completely entrap volumes of water in the pore spaces between regions of glass beads, between consecutive drainage and flushing cycles by day 9. Such an experiment cannot be achieved using traditional methods since external interrogation of the pore spaces would cause the held water to be drained.

9.7 Earth's Field Magnetic Resonance

Although, in comparison to high field and low field MRI, there is little in the way of earth's field MRI for soil science, there are significant future opportunities for its use and thus it is mentioned briefly here from the perspective specifically of imaging. Unlike traditional MRI, images of ground water, and potentially in the future contaminants, can be built using two approaches. The first is built up by moving and collecting signals from non-imaging systems, such as that disclosed in various Russian patents in the 1970 and 1980s, which detects the signals from subsurface protons after placing a large circular wire on the surface of the earth before transmitting an excitation pulse through it at the Larmor frequency to excite subsurface spins. This has been demonstrated in several applications by Yaramanci who, for example, in 2017[68] demonstrated the ability of multilocation earth's field NMR to identify the 2-dimensional location of subsurface groundwater which was verified with borehole logging. The alternative to produce an image is to use a 2D inversion approach such as that proposed and demonstrated in 2007 by Hetrich *et al.*[69] where overlapping surface coils are used and the signals from each combined to produce a subsurface water map in the vertical plane.

There are two major limitations of earth's field magnetic resonance: the first is the difficulty in applying the technique in an arbitrary area. While early experiments were undertaken in comparatively remote areas of minimally populated Siberia and Russia, attempts to undertake such experiments in built-up areas tend to have poor success rates. Proximity to power lines which themselves emit magnetic fields introduces a time varying component to the magnetic field which cannot be easily corrected for. Similar issues result from the presence of magnetic rocks in the subsurface which cause local variations in the geomagnetic field. Strategies to correct for these have been proposed, such as that of Trushkin in 1993[70] where the applied RF pulse has a frequency offset with respect to the Larmor frequency to introduce a modulation in the collected signal which varies with the presence of magnetic rocks, but is not lost. The second challenge is the inherently low polarisation owing to the small static magnetic field. This makes it highly challenging to produce time-efficient images. To date there have been several approaches to address this issue. The first, in 2016 by Grunewald,[71] attempts to increase the signal-to-noise ratio of the measurements by utilising broadband excitation pulses. Unilateral magnetic resonance is made challenging by the variable effectiveness of the B_1 field as a function of position, and thus a 90° excitation pulse only has the correct effect at a given distance from its surface. Moving closer to the coil, the efficiency increases which will result in a longer nutation, and moving further away, the efficiency is reduced, which will lead to a tip angle less than 90°. This problem is further exacerbated by inhomogeneities in B_0 which we have found in previous sections abound in soil samples. Thus, the

SNR can be improved by using an excitation pulse which can excite multiple frequencies or which is in other ways less sensitive to B_1 inhomogeneities. In an adiabatic excitation pulse, the pulse shape and frequency are modulated to produce complex magnetisation vectors, encompassing multiple conditions for the magnetic moments in terms of frequency and position in the B_1 field, to converge them to the x–y plane after its application. In particular, in this work they utilised a simple frequency swept pulse and the adiabatic half passage pulse. It was demonstrated that such an approach can produce 1D profiles of T_2^* in a subsurface aquifer system with similar results to on-resonance pulses. The gains in terms of the SNR are still somewhat lower than desired to achieve rapid subsurface mapping.

In laboratory earth's field NMR experiments, the lack of a signal due to the inherently low B_0 field is addressed by applying an additional pre-polarisation field to the system before the experiment takes place.[72] If a pre-polarisation pulse is applied parallel to the direction of the earth's magnetic field, the field should be removed steadily and traditional imaging can take place starting with a 90° excitation pulse at the Lamour frequency of the local earth's magnetic field. Alternatively, the field can be applied perpendicularly to the earth's magnetic field and removed suddenly, that is, non-adiabatically (referring to the polarisation pulse, not the excitation pulse discussed in the previous paragraph), leaving the magnetisation in the same plane as would follow the typical excitation pulse. This will generate an FID, and subsequent 180° pulses at the Lamour frequency of the local earth's magnetic field will generate echoes. Such an approach has been utilised by two groups for different environmental applications. The first is used to identify contaminants in sea ice which has direct applicability to contaminants in soil systems by Altobeli[73] building on their earlier work with

Figure 9.18 Earth's field MR system for positioning from the helicopter. The left image shows the suspended coil assembly carried by a Sikorsky S-92 helicopter. The right image shows the coil assembly tested without the helicopter using 30 jerry cans containing oil positioned above the coil on plastic buckets. Reproduced from ref. 73 with permission from Elsevier, Copyright 2019.

Conradi.[74] This system is designed to be flown under a helicopter on the surface of sea ice to produce maps that identify the location of subsurface oil spills. Utilising a large pre-polarisation coil and a smaller excitation and detection coil, along with adiabatic half passage pulses, and T_1 water suppression (by only waiting for the polarisation of the oil fraction), such a system has been demonstrated to have the capabilities to map oil hidden under a polystyrene ice surrogate in a large lake as can be seen in Figure 9.18.

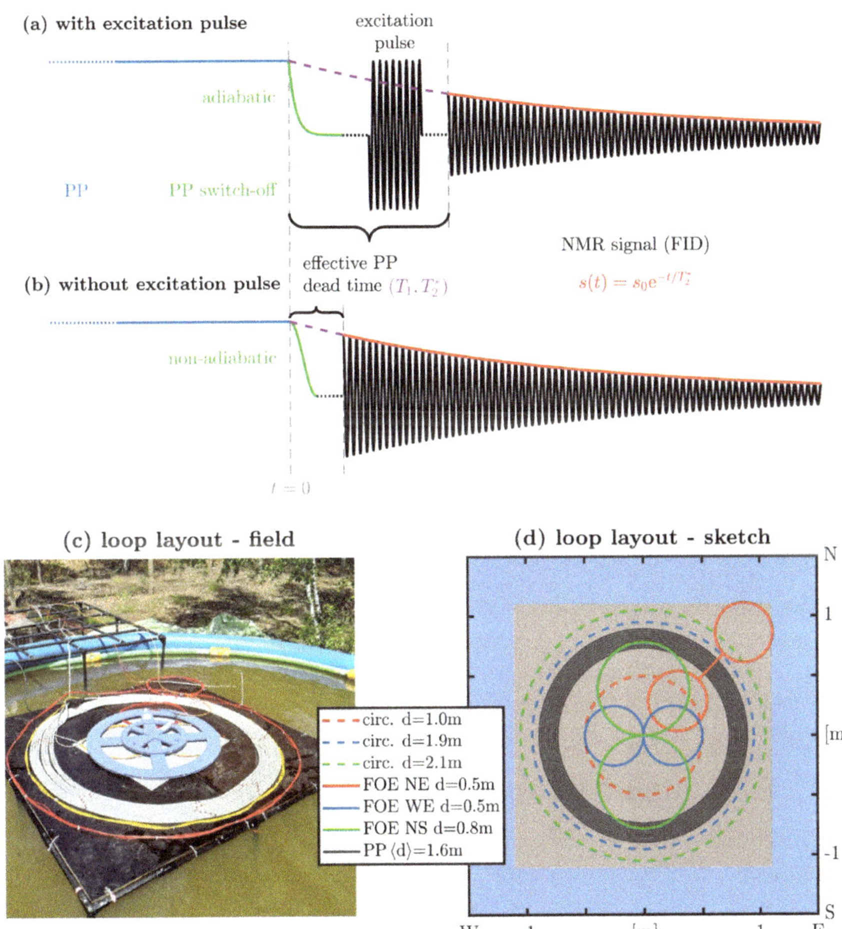

Figure 9.19 Large-scale earth's field MR. (a and b) Two possible excitation strategies, specifically (a) with a standard oscillating excitation pulse and (b) utilising the switch-off ramp of the pre-polarising field. The pre-polarisation pulse switch-off is shown in green and the FID envelope in red. The photograph (c) and sketch (d) of the experimental setup. In (d) the dashed lines refer to circular Rx-loops, while the solid lines refer to figure-of-eight loops, and the pre-polarisation coil is shown in grey. Reproduced from ref. 75, https://doi.org/10.1029/2021GL095371, under the terms of the CC BY 4.0 license, https://creativecommons.org/licenses/by/4.0/.

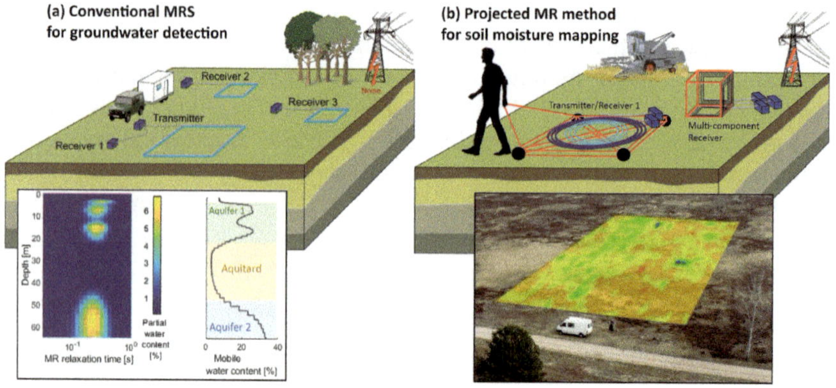

Figure 9.20 Schematic with data showing the conventional mapping approach on the left and a newly proposed projected imaging method on the right with moisture results overlaid on a field photograph in the inset. Reproduced from ref. 78 with permission from the authors.

However, there still exist challenges associated with limited SNR which have been most recently addressed in 2021 by Hiller *et al.*[75,76] where the inherent challenges of selective excitation pulses are addressed by eliminating them from the experiment with the previously discussed non-adiabatic prepolarisation pulse. An image and schematic of the coils used are shown in Figure 9.19. Furthermore, this approach eliminates the deadtime after the excitation pulse, facilitating access to the short T_2^* of water in realistic soils. They continue to make significant improvements to the technique, most recently by developing a concentric figure of eight coil for improved noise cancellation.[77]

This system has not been used to produce images in this publication, but the authors present mapping data in an online report,[78] showing the potential of the technique for groundwater soil measurements and investigation of contaminants which can be seen in Figure 9.20.

9.8 Conclusions and Future Directions

In this chapter, we have reviewed the historical and current state of the art of the application of magnetic resonance imaging to study contaminants in soils. Much of the work, particularly that which was undertaken before the 2020s, focuses on the application of imaging to soil proxies such as glass bead packs which have value in understanding the underlying hydrodynamics of soil samples but must be approached with caution for making broad conclusions about soil samples. We have shown how the different fluid fractions can be resolved using suppression and tracer techniques and seen the information which such experiments reveal, non-invasively, for a range of environmentally relevant samples.

Measurements in real soil samples using actual contaminants are understandably more rare owing to the challenges associated with achieving a sufficiently high SNR in the presence of strong paramagnetic impurities present in most real soil samples. Where this is achieved however the additional understanding we gain is significant, particularly in systems where the influence of the myriad of soil properties which impact soil hydro- and bio-dynamics is considered.

The next decade is likely to see significant application of imaging technology to real soil samples using ultrashort or zero echo time imaging which can help to overcome the inherently short T_2^* in the presence of paramagnetic impurities. Taking the measurement to the field, particularly with earth's field NMR, with recent developments using adiabatic half passage excitations or non-adiabatic prepolarisation pulses and no excitation, has extremely high potential for identifying contaminated soils and for monitoring remediation strategies and is another area which is likely to see good growth in application.

These advances are likely to be useful to resolve some of the ongoing areas of interest in soil contaminant research with a focus on soil remediation, biofuel conversion and wastewater treatment.

9.9 A Starting Point for Experiments

The main challenges involved in MRI for contaminant dynamics have been presented throughout this chapter. In this final section, we present a summary of strategies to employ as a starting point for undertaking imaging of samples of porous media with a view to monitoring contaminant dynamics. These will likely not be suitable as the final imaging parameters but are intended to serve as a springboard for optimising an individual experiment. The first decision encountered will almost always be the choice between acquired media and an artificial 'clean' medium which is free of impurities likely to cause disruption of the signal. As noted in previous sections, where artificial media are used, efforts to replicate the range of particle sizes and surface properties of media encountered in practice are rarely wasted and can greatly improve the applicability of the experiment to real-life scenarios. Where samples of media are used, paramagnetic impurities, air spaces and interfaces between air and liquid or liquid and liquid will often perturb the field gradients sufficiently to cause significant artefacts in the images. Where such impurities do not completely lose the signal, the artefacts can be somewhat minimised by using spin echo imaging instead of gradient echo imaging, setting large receiver bandwidths, strong gradients, short echo times, thin slices and choosing lower field instruments. As a starting point for a turbo spin echo imaging sequence, the bandwidth should be set to maximum, the echo time to the minimum value and any gradient noise reduction strategies disabled. Setting the slice thickness to be the smallest value which provides the signal serves to maximise the available signal.

Contrast can be achieved using the intrinsic properties of the sample such as relaxation times, diffusion coefficients or chemical shift imaging as appropriate. The choice is largely a personal preference in clean or artificial samples. In samples of real-life media, the choice is likely to be more limited. T_2^* will often be too short to provide much in the way of opportunities for quantitation, whilst T_1 or diffusion weighted images or maps can be applied in a wider range of circumstances. For multiphase systems, inversion recovery imaging or chemical shift imaging offers a simple means to interrogate individual fluid fractions. This is most readily achieved by measuring the T_1 of the individual fluid fractions and setting the inversion time to be equal to $T_1^* \ln(2)$ to selectively suppress each component. For diffusion, a similar approach can be used, where the evolution time is set to minimise the effects of restricted diffusion in pores. For example, for water at 2.2×10^{-9} m^2 s^{-1}, an appropriate evolution time is chosen according to $D = \frac{\xi^2}{2\tau}$ to exclude diffusion less than approximately 10 μm as $\frac{10 \times 10^{-62}/2.2 \times 10^{-9}}{2}$, where ξ is the mean free path and τ is the time which in this case is the evolution time giving a value of approximately 20 ms. The signal from the fraction with the higher diffusion coefficient will be suppressed at a lower gradient amplitude than the smaller diffusion coefficient and can thus be selectively suppressed where one phase has a diffusion coefficient more than around 5 times that of the other. Where pulse programming is available, the use of one of the Cotts[21] multiple interval sequences offers a simple way to reduce the influence of background gradients on measurements of diffusion.

Alternatively, an extrinsic contrast agent can be added to one of the fluid fractions allowing its passage to be tracked where the differences in intrinsic properties do not afford sufficient contrast. The choice of the contrast agent will initially depend on what is readily available, safe to handle, appropriate for the matrix of interest and most relevant to the process of interest. Each ionic species has a different sensitivity for use which will also in some way dictate an appropriate choice. The change in relaxation time as a function of contrast is shown in Figure 9.21 to help inform the choice. It is also important to consider the impact that this change in the relaxation rate will have on the final signal contrast since the expected behaviour is not linear. Again, to aid planning for such an experiment, the expected signal change, expressed as the ratio of the signal in the presence of the ions to the signal from distilled water, is shown in Figure 9.21(c).

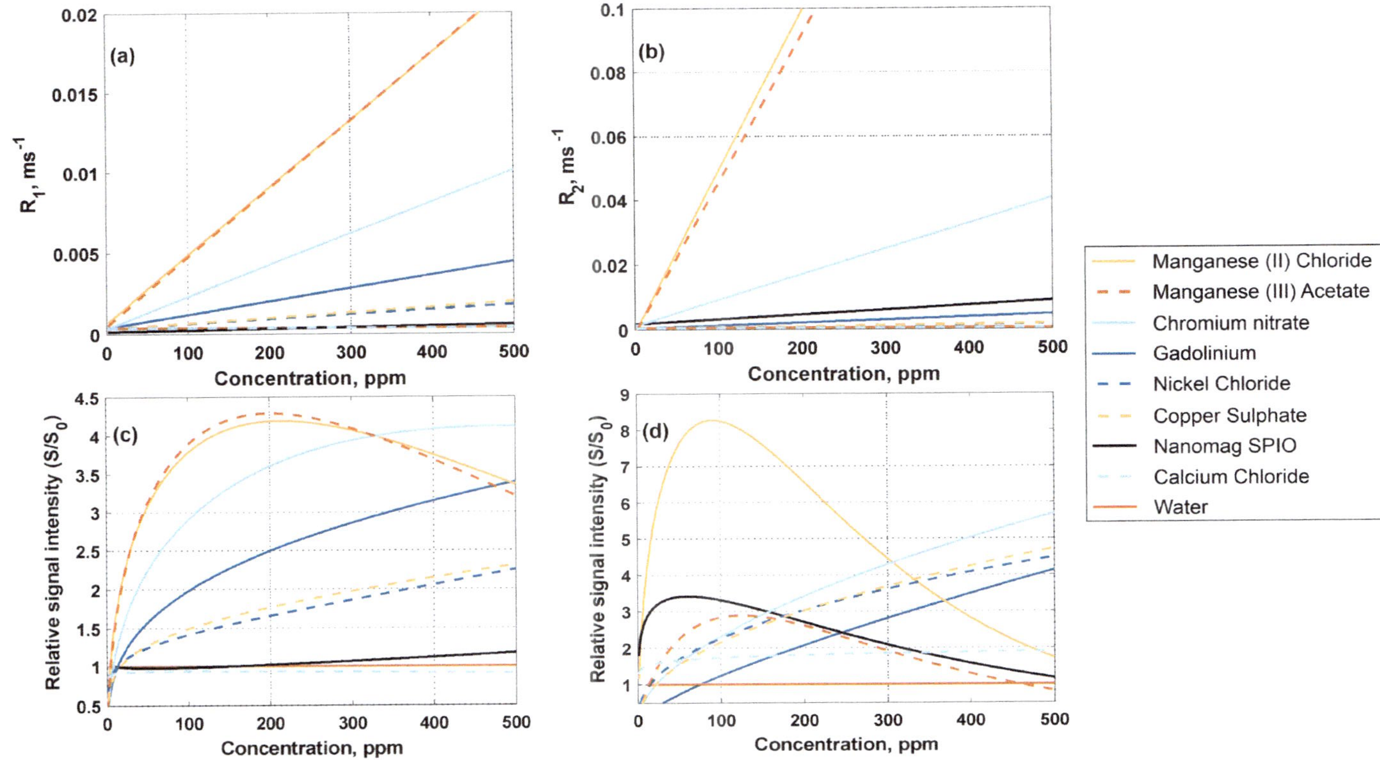

Figure 9.21 Comparison of relaxation rates for different ionic species over a range of concentrations at 1.5 T. (a) The change in R_1 as a function of concentration and (b) that for R_2. (c and d) The expected signal changes for a typical FLASH experiment ($T_E = 1.96$ ms, $T_R = 4.6$ ms, $\alpha = 15°$) and spin echo experiment ($T_E = 17$ ms, $T_R = 28$ ms) over the same range of concentrations. Reproduced from ref. 75, https://doi.org/10.1029/2021GL095371, under the terms of the CC BY 4.0 license, https://creativecommons.org/licenses/by/4.0/.

References

1. K. Omasa, M. Onoe and H. Yamada, Short Communication, *Environ. Control Biol.*, 1985, **23**(4), 99–102.
2. H. H. Rogers, H. H. Rogers and P. A. Bottomley, *Agron. J.*, 1987, **79**(6), 957–965.
3. M. H. G. Amin, R. J. Chorley, K. S. Richards, B. W. Bache, L. D. Hall and T. A. Carpenter, *Hydrol. Processes*, 1993, 7, 279–286.
4. D. I. Hoult and P. C. Lauterbur, *J. Magn. Reson.*, 1979, **34**, 425–433.
5. R. Enjilela, B. MacMillan, M. J. McAloon, O. V. Petrov, S. Vashaee and B. J. Balcom, *J. Magn. Reson.*, 2019, **307**, 106575.
6. P. T. Callaghan, *J. Magn. Reson.*, 1990, **87**, 304–307.
7. A. M. Coffey, M. L. Truong and E. Y. Chekmenev, *J. Magn. Reson.*, 2013, **237**, 169–174.
8. T. P. L. Roberts, T. A. Carpenter and L. D. Hall, *J. Magn. Reson., Ser. B*, 1993, **101**(1), 78–82.
9. L. D. Hall, M. H. G. Amin, E. Dougherty, M. Sanda, J. Votrubova, K. S. Richards, R. J. Chorley and M. Cislerova, *Geoderma*, 1997, **80**, 431–448.
10. J. A. Chudek and A. D. Reeves, *Biodegradation*, 1998, **9**, 443–449.
11. A. D. Reeves and J. A. Chudek, *J. Ind. Microbiol. Biotechnol.*, 2001, **26**, 77–82.
12. M. L. Johns and L. F. Gladden, *J. Colloid Interface Sci.*, 1999, **210**, 261–270.
13. Y. Chu, C. J. Werth, A. J. Valocchi, H. Yoon and A. G. Webb, *J. Contam. Hydrol.*, 2004, **73**, 15–37.
14. M. L. Johns and L. F. Gladden, *Magn. Reson. Imaging*, 1998, **16**(5–6), 655.
15. A. E. Fischer, J. Balcomt, E. J. Fordham, T. A. Carpenter and L. D. Hall, *J. Phys. D: Appl. Phys.*, 1995, **28**, 384–397.
16. S. Blackband, P. Mansfield, J. R. Barnes, A. D. H. Clague and S. A. Rice, Discrimination of Crude Oil and Water in Sand and in Bore Cores With NMR Imaging, *SPE Form. Eval.*, 1986, **1**(01), 31–34.
17. S. Davies, A. Hardwick, D. Roberts, K. Spowage and K. J. Packer, *Magn. Reson. Imaging*, 1994, **12**, 349–353.
18. B. H. Lee and S. K. Lee, *J. Hydrol.*, 2017, **553**, 637–650.
19. M. S. Zamiri, F. Marica, L. Romero-Zerón and B. J. Balcom, *Chem. Eng. J.*, 2022, **428**, 131042.
20. L. Li, F. Marica, Q. Chen, B. MacMillan and B. J. Balcom, *J. Magn. Reson.*, 2007, **186**, 282–292.
21. R. M. Cotts, M. J. R. Hdch, T. Sun and J. T. Markert, *J. Magn. Reson.*, 1989, **83**, 252–266.
22. J. Yan, X. Luo, W. Wang, F. Chen, R. Toussaint, J. Schmittbuhl, G. Vasseur and L. Zhang, *J. Pet. Sci. Eng.*, 2012, **86–87**, 237–245.
23. M. J. Simpson and A. J. Simpson, in *NMR Spectroscopy : A Versatile Tool for Environmental Research*, Wiley, 2014, p. 486.
24. S. Kiaalhosseini, B. Kohn, A. T. Watson, T. C. Sale and J. Blotevogel, *Groundwater Monit. Rem.*, 2019, **39**, 71–75.

25. P. M. Margosian, T. Takahashi and M. Takizawa, *Encyclopedia of Magnetic Resonance*, 2012, DOI: 10.1002/9780470034590.EMRSTM1247.
26. M. Weiger, K. P. Pruessmann and F. Hennel, *Magn. Reson. Med.*, 2011, **66**, 379–389.
27. S. Haber-Pohlmeier, D. Caterina, B. Blümich and A. Pohlmeier, *Molecules*, 2021, **26**(17), 5130.
28. M. J. Simpson, A. J. Simpson, D. Gross, M. Spraul and W. L. Kingery, *Environ. Toxicol. Chem.*, 2007, **26**, 1340–1348.
29. M. Raue, J. Martins, M. Küppers, T. Mang, B. Blümich and S. Stapf, *Prog. Polym. Sci.*, 2013, **140**, 35–43.
30. I. V. Koptyug, A. A. Lysova, A. V. Kulikov, V. A. Kirillov, V. N. Parmon and R. Z. Sagdeev, *Magn. Reson. Imaging*, 2005, **23**, 221–225.
31. B. A. Baldwin and W. S. Yamanashi, NMR Imaging of Fluid Dynamics in Reservoir Core, *Magn. Reson. Imaging*, 1988, **6**, 493–500.
32. A. K. Gaigalas, A. Chidester Van Orden, B. Robertson, T. H. Mareci and L. A. Lewis, *Nucl. Technol.*, 1989, **84**, 113–118.
33. G. Guillot, G. Kassab, J. P. Hulin and P. Rigord, *J. Phys. D: Appl. Phys.*, 1991, **24**, 763.
34. N. Nestle, A. Wunderlich and T. Baumann, *Eur. J. Soil Sci.*, 2008, **59**, 559–571.
35. J. Fannir, I. Panfilova, S. Leclerc and D. Stemmelen, *Mech. Ind.*, 2020, **1**(5), DOI: 10.1051/meca/2020058.
36. I. Markale, G. M. Cimmarusti, M. M. Britton and J. Jiménez-Martínez, *Environ. Sci. Technol.*, 2021, **55**, 8742–8752.
37. S. Lakshmanan, W. M. Holmes, W. T. Sloan and V. R. Phoenix, *Int. J. Environ. Sci. Technol.*, 2015, **12**, 3373–3384.
38. T. Baumann and C. J. Werth, *Colloids Surf., A*, 2005, **265**, 2–10.
39. S. Lakshmanan, W. M. Holmes, W. T. Sloan and V. R. Phoenix, *Chem. Eng. J.*, 2015, **266**, 156–162.
40. L. Cuny, M. P. Herrling, G. Guthausen, H. Horn and M. Delay, *J. Contam. Hydrol.*, 2015, **182**, 51–62.
41. F. Ranzinger, M. Matern, M. Layer, G. Guthausen, M. Wagner, N. Derlon and H. Horn, *Water Res.: X*, 2020, **7**, 100050.
42. Z. Pearl, M. Magaritz and P. Bendel, Nuclear Magnetic Resonance Imaging of Miscible Fingering in Porous Media, *Transp. Porous Media*, 1993, **12**, 107–123.
43. K.-H. Herrmann, A. Pohlmeier, S. Wiese, N. J. Shah, O. Nitzsche, H. Vereecken and S. Wiese, *J. Environ. Qual.*, 2002, **31**, 506–514.
44. A. B. Moradi, S. E. Oswald, J. A. Massner, K. P. Pruessmann, B. H. Robinson and R. Schulin, *Eur. J. Soil Sci.*, 2008, **59**, 476–485.
45. E. K. Fram, R. J. Herfkens, G. A. Johnson, G. H. Glover, J. P. Karis, A. Shimakawa, T. G. Perkins and N. J. Pelc, *Magn. Reson. Imaging*, 1987, **5**, 201–208.
46. V. Jelinkova, M. Snehota, A. Pohlmeier, D. van Dusschoten and M. Cislerova, *Org. Geochem.*, 2011, **42**, 991–998.

47. N. Nestle, A. Wunderlich, R. Niessner and T. Baumann, *Environ. Sci. Technol.*, 2003, **37**, 3972–3977.
48. T. Baumann, A. Wunderlich, M. Ebert and N. Nestle, *Diffus. Fundam.*, 2007, **5**, 1–19.
49. N. Nestle, A. Wunderlich and T. Baumann, *Magn. Reson. Imaging*, 2007, **25**, 576–577.
50. S. M. Belliveau, T. L. Henselwood and C. H. Langford, *Environ. Sci. Technol.*, 2000, **34**, 2439–2445.
51. G. E. Schaumann, E. Hobley, J. Hurraß and W. Rotard, *Plant Soil*, 2005, **275**, 1–20.
52. S. V. Dvinskikh, K. Szutkowski and I. Furó, *J. Magn. Reson.*, 2009, **198**, 146–150.
53. P. Bendel, M. Davis, E. Berman and G. W. Kabalka, *J. Magn. Reson.*, 1990, **88**, 369–375.
54. S. V. Dvinskikh and I. Furó, *Svensk Karnbranslehantering AB: Magnetic resonance imaging and nuclear magnetic resonance investigations of bentonite systems*, 2009.
55. A. J. Fagan, N. Nestle and D. J. Lurie, *Magn. Reson. Imaging*, 2005, **23**, 317–319.
56. M. F. Shamim, M. Bencsik, R. H. Morris and M. I. Newton, *Microporous Mesoporous Mater.*, 2013, **178**, 48–52.
57. X. T. Xu, D. Q. Liu, Z. X. Xian, F. Yang, W. Bin Jian, X. Xu and J. B. Huang, *Geofluids*, 2022, 9948658.
58. H. P. Müller, J. Weis and R. Kimmich, *Phys. Rev. E: Stat. Phys., Plasmas, Fluids, Relat. Interdiscip. Top.*, 1995, **52**, 5195.
59. Y. E. Kutsovsky, L. E. Scriven, H. T. Davis and B. E. Hammer, *Phys. Fluids*, 1996, **8**, 863–871.
60. P. Dijk, B. Berkowitz and P. Bendel, *Water Resour. Res.*, 1999, **35**, 347–360.
61. M. D. Mantle, A. J. Sederman and L. F. Gladden, *Chem. Eng. Sci.*, 2001, **56**, 523–529.
62. M. L. Johns and L. F. Gladden, *J. Colloid Interface Sci.*, 2000, **225**, 119–127.
63. B. C. Hoskins, L. Fevang, P. D. Majors, M. M. Sharma and G. Georgiou, Selective Imaging of Biofilms in Porous Media by NMR Relaxation, *J. Magn. Reson.*, 1999, **139**(1), 67–73.
64. M. S. Olson, R. M. Ford, J. A. Smith and E. J. Fernandez, *Environ. Sci. Technol.*, 2004, **38**, 3864–3870.
65. J. L. Sherwood, J. C. Sung, R. M. Ford, E. J. Fenandez, J. E. Maneval and J. A. Smith, *Environ. Sci. Technol.*, 2003, **37**, 781–785.
66. B. Cao, P. D. Majors, B. Ahmed, R. S. Renslow, C. P. Silvia, L. Shi, S. Kjelleberg, J. K. Fredrickson and H. Beyenal, *Environ. Microbiol.*, 2012, **14**, 2901–2910.
67. F. Ranzinger, K. Schröter, H. Horn and M. Wagner, *Water*, 2021, **13**, 2456.
68. U. Yaramanci, in *SEG Global Meeting Abstracts*, 2017.
69. M. Hertrich, M. Braun, T. Günther, A. G. Green and U. Yaramanci, in IEEE Transactions on Geoscience and Remote Sensing, 2007, vol. 45.

70. D. V. Trushkin, O. A. Shushakov and A. V. Legchenko, *Appl. Magn. Reson.*, 1993, **5**(3), 399–406.
71. E. Grunewald, D. Grombacher and D. Walsh, *Geophysics*, 2016, **81**, WB85–WB96.
72. G. S. Waters, *Nature*, 1955, **176**(4484), 691–691.
73. S. A. Altobelli, M. S. Conradi, E. Fukushima, J. Hodgson, T. J. Nedwed, D. A. Palandro, A. Peach, N. J. Sowko and H. Thomann, *Mar. Pollut. Bull.*, 2019, **144**, 160–166.
74. M. S. Conradi, S. A. Altobelli, N. J. Sowko, S. H. Conradi and E. Fukushima, *J. Magn. Reson.*, 2018, **288**, 95–99.
75. T. Hiller, S. Costabel, R. Dlugosch and M. Müller-Petke, *Geophys. Res. Lett.*, 2021, **48**, e2021GL095371.
76. T. Hiller, S. Costabel, T. Radić, R. Dlugosch and M. Müller-Petke, *Vadose Zone J.*, 2021, **20**, e20138.
77. T. Hiller, S. Costabel, R. Dlugosch, T. Splith and M. Müller-Petke, *Magn. Reson. Lett.*, 2023, **3**, 140–149.
78. Magnetic resonance for detecting ground and soil water, https://www.openaccessgovernment.org/magnetic-resonance-for-detecting-ground-and-soil-water/84182/ (accessed 22 February 2023).

CHAPTER 10

NMR Relaxation in Porous Media for Environmental Applications

A. MAVRIDIS[a] AND C. D'AGOSTINO*[a,b]

[a] Department of Chemical Engineering, The University of Manchester, Oxford Road, Manchester, M13 9PL, UK; [b] Dipartimento di Ingegneria Civile, Chimica, Ambientale e dei Materiali (DICAM), Alma Mater Studiorum – Università di Bologna, Via Terracini, 28, 40131 Bologna, Italy
*Email: carmine.dagostino@manchester.ac.uk

10.1 Introduction

The study of fluids in porous media is an active, interdisciplinary field, relevant to several areas of science and engineering, with applications to environmental science, food science, catalysis, and petroleum engineering as well as in many other fields. In the context of environmental sciences, topics of interest include contaminant transport in soil and underground aquifers, biofilms, carbon geo-sequestration, hydrogen storage, geothermal energy, *etc.* Modelling processes involving the behaviour of fluids in porous media through computer simulations is commonplace nowadays. However, accurate mathematical descriptions of such phenomena can be immensely complicated, or even unfeasible. For example, for systems involving the flow of multiple fluid phases, there is no analytical method that could predict the relative permeabilities of the different phases, and thus experimental data are necessary.[1] Furthermore, knowing all the details about the geometry and the composition of a solid porous network is unrealistic, and thus inaccuracies in computer models are inevitable.

New Developments in NMR No. 32
The Environment in a Magnet: Applications of NMR Techniques to Environmental Problems
Edited by Pellegrino Conte, Delia Francesca Chillura Martino and Paolo Lo Meo

Published by the Royal Society of Chemistry, www.rsc.org

Nuclear magnetic resonance (NMR) is a non-destructive and non-intrusive technique, mostly famous for spectroscopy and medical imaging applications, but it has also been extensively used for studying the behaviour of fluids within porous media. Experimental NMR techniques for such applications were largely developed by the oil industry for well logging purposes, to identify and characterize fluids in subsurface geological formations, as well as the pore geometries of the rocks which contain them.[2] Generally, NMR relaxation methods can be used to study the dynamics of fluids within porous media, such as in soil, wood, underground aquifers, biological systems, *etc.* By measuring the relaxation times of the hydrogen nuclei within the sample, researchers can gain insight into the pore size distribution, fluid mobility, and the interactions between the fluid and the solid matrix. These methods are often applied both in laboratory studies and *in situ*, and may also be combined with other NMR techniques, such as spectroscopy, pulsed field gradient diffusivity, or imaging measurements, to provide a more complete picture of the examined systems.

Climate change and other environmental issues are nowadays almost universally recognised as the most important challenges humanity has to confront. Many of the processes found in natural environments directly involve porous media, and thus studying such systems in detail can be of paramount importance.

10.2 Theory of NMR Relaxation in Porous Media

The nuclear magnetic resonance effect relies on the intrinsic magnetic properties of subatomic particles. Nuclei of certain atoms, such as hydrogen, have a magnetic spin and therefore possess an angular momentum, and this angular momentum results in a net magnetic moment. When these nuclei are exposed to an external magnetic field, they will precess around an axis parallel to that magnetic field. The angular frequency of this precession is called the "Larmor frequency" (ν_0) and is proportional to the magnitude of the magnetic field experienced by the nucleus, calculated as

$$\nu_0 = \frac{\gamma B_0}{2\pi}, \tag{10.1}$$

where γ is the gyromagnetic ratio, a characteristic constant for each type of spin-bearing nucleus, and B_0 is the magnitude of the external magnetic field.

If the nuclei have a spin of $\frac{1}{2}$ (such as ^{1}H nuclei), due to the Zeeman effect, the spins may be observed in one of two allowed energy states, one "parallel" to the external magnetic field (low energy) and one "anti-parallel" (high energy state). Nuclei with larger spin values have access to more energy states in between; however, we are not concerned with such cases in this chapter, as we focus exclusively on hydrogen. A slight majority of the spins will always favour the low-energy state, and this creates a net magnetization of the sample, often represented as $\overrightarrow{M_0}$, parallel to $\overrightarrow{B_0}$ (with M_0 and B_0 denoting the magnitudes of the vectors, respectively).

When an electromagnetic square pulse composed of frequencies close to the Larmor frequency of a group of nuclei is applied perpendicularly to the axis of their precession, some of these nuclei will absorb energy from the pulse due to resonance, and their original alignment will be temporarily disturbed. Effectively, this causes $\vec{M}_0$ to be rotated by an angle which depends on the magnitude and the duration of the pulse. After the pulse is terminated, the nuclei relax back to their original alignment, by releasing their energy excess to their immediate environment. This relaxation process can generate an electrical signal, by inducing a current through a coil placed around the sample. This signal is called free induction decay (FID), and its frequency depends on all the underlying Larmor frequencies present within the sample, which vary according to their chemical environment and the strength of the magnetic field. This process is schematically shown in Figure 10.1.

Two distinct relaxation processes can be distinguished here. In a cartesian coordinate system where the z axis is parallel to the external magnetic field, longitudinal relaxation describes the rate at which the z component of the net magnetization vector returns to its original value, whereas transverse relaxation describes the rate at which the xy component of the net magnetization vector decays to zero.

In order to experimentally evaluate these relaxation processes, sequences of consecutive radiofrequency pulses are applied repeatedly on the sample. The most well-known pulse sequences for relaxation experiments are the inversion recovery (IR) sequence for longitudinal relaxation measurements and the Carr–Purcell–Meiboom–Gill (CPMG) sequence for transverse relaxation measurements, as described in Figures 10.2 and 10.3, respectively.

Due to the nature of those pulse sequences, measuring T_2 is significantly faster than measuring T_1. Consequently, T_2 measurements are most commonly used in cases where the total experiment time is important, *e.g.* in dynamic samples which change over time, samples with low signal-to-noise ratios where many repetitions are required, and for NMR-logging applications where the whole apparatus must be on the move during measurements.[3]

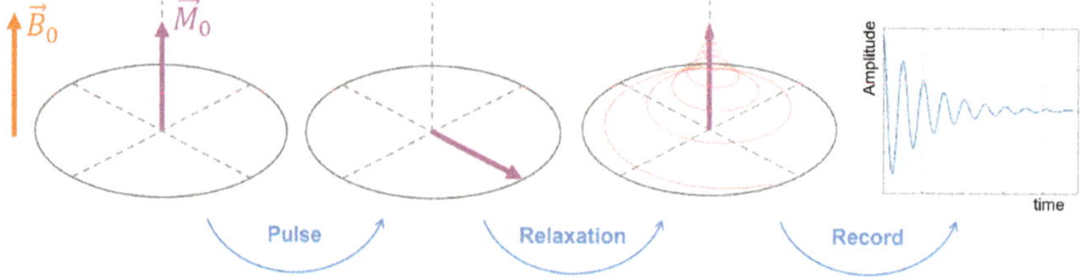

Figure 10.1 Graphical illustration of a simplified version of an NMR experiment involving a single radiofrequency pulse tipping the net magnetization by 90°.

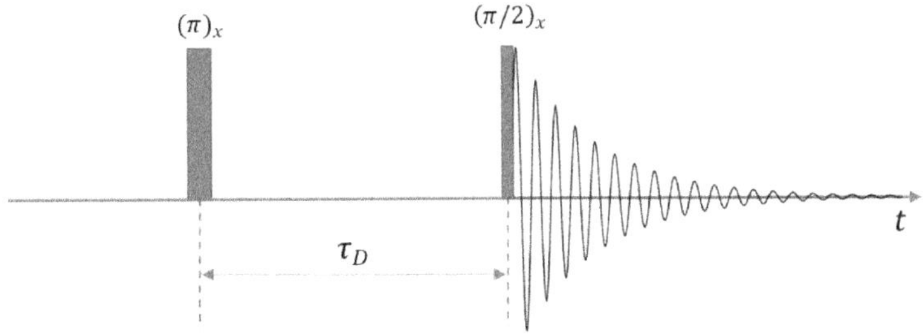

Figure 10.2 The inversion recovery pulse sequence. The initial pulse rotates the net magnetization by 180° about the x axis, and after the delay time (τ_D) the magnetization is rotated again by 90° to be placed on the xy plane, making its magnitude detectable by the instrument. The sequence is repeated for several increasing delay times in order to look at the decay of longitudinal magnetization over time.

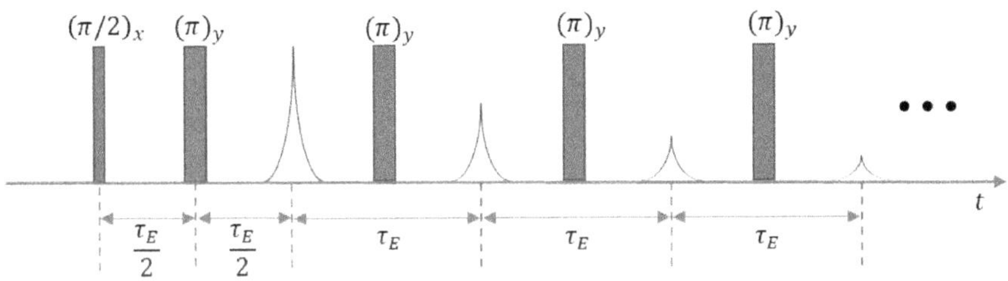

Figure 10.3 The CPMG echo sequence. The initial pulse rotates the net magnetization by 90° about the x axis, after which the spins begin to dephase and lose coherence. A 180° pulse rotating the magnetization about the y axis is then used to make the spins realign, resulting in the formation of a detectable signal (an echo) occurring at a predetermined time interval after the first pulse, which is named "echo time" (τ_E). What follows is a periodic series of 180° pulses, separated by τ_E, generating a large number of consecutive echoes with decaying intensities, known as an "echo train".

For nuclei in fluid molecules, both longitudinal and transverse processes occur at exponential rates according to the Bloembergen, Purcell and Pound (BPP) theory,[4] and thus exponential models are used to fit the experimental results and quantify the relaxation rates in fluids. The data points obtained from the IR and CPMG pulse sequences mentioned above can be approximated by eqn (10.2) and (10.3), respectively:

$$M(t) = M_0\left[1 - 2\exp\left(-\frac{t}{T_1}\right)\right], \tag{10.2}$$

$$M(t) = M_0 \exp\left(-\frac{t}{T_2}\right), \tag{10.3}$$

where $M(t)$ is the magnitude of the detected magnetization at each time step, t, and T_1, T_2 are the relaxation time constants which characterize the rates of longitudinal and transverse relaxation processes, respectively.

In the case of fluids in porous media, NMR relaxation can provide a plethora of information regarding both the behaviour of the fluid itself, as well as the pore spaces, which restrict it. Due to differences in the magnetic susceptibility between the solid network of a porous medium and the liquid contained within, the external magnetic field in different areas within the sample will be distorted in generally unpredictable ways. Therefore, due to the magnetic field gradients present within the sample, there will be a wide range of Larmor frequencies for all the nuclei, overwhelming chemical shift information, which could otherwise be obtained by NMR spectra in bulk liquids. Therefore, NMR spectra are usually not useful when it comes to fluids in porous media, and we have to solely rely on information provided by relaxation instead. Due to the reduced mobility of the fluid molecules and the potential presence of paramagnetic materials in the solid matrix, the nuclei residing very close to the solid surface experience significantly faster relaxation rates, compared to the nuclei in the bulk liquid. As a result, nuclei in smaller pores will generally have shorter relaxation times than nuclei residing in larger pores, since they encounter the solid surfaces more frequently. Thus, the relaxation time distribution for a sample can be directly correlated to the pore size distribution. For this correlation to be accurate, the sample needs to be in the "fast diffusion regime", according to the Brownstein–Tarr model,[5] defined by the condition

$$\rho \ll D\frac{S}{V}, \tag{10.4}$$

where D is the self-diffusion coefficient of the fluid, S/V is the surface-to-volume ratio of the porous network, and ρ is the surface relaxivity of the pore walls, defined for both longitudinal and transverse relaxation processes as

$$\rho_{1,2} = \frac{\lambda}{T_{1,2}}, \tag{10.5}$$

where λ is the thickness of the surface layer within which the nuclei experience surface-enhanced relaxation. Several methods can be used to estimate this surface relaxivity parameter, as reviewed by Ge *et al.*[6] For a fluid in the fast diffusion regime, the relaxation rate can be approximated as

$$\frac{1}{T_{1,2}} \approx \rho_{1,2}\frac{S}{V}. \tag{10.6}$$

In this case, for a porous medium with n different environments characterized by different sizes (*e.g.*, micropores, mesopores, *etc.*), there would be n different modes, or relaxation components, and the final signal would be expressed as a sum of relaxation rates. For example, for a T_2 relaxation experiment, we would write[7]

$$M(t) = M_0 \sum_{i=1}^{n} I_i \exp\left(-\frac{t}{T_2}\right) + \varepsilon(t), \tag{10.7}$$

where I_i is the relative intensity of the ith mode and $\varepsilon(t)$ is the noise which superimposes the pure signal. Effectively, in the fast diffusion regime each pore would be characterized by a single relaxation rate since relaxation would occur almost exclusively on the pore surfaces. This is because relaxation happens relatively slowly on the bulk phase, and most of the nuclei relax once they reach a surface, by diffusing through the pore space. In contrast, in the slow diffusion regime, each pore would be characterized by a multi-exponential relaxation, since both bulk and surface contributions would be significant.[7]

In order to uncover all the n different T_2 components hidden behind the final detectable signal, as well as their corresponding intensities, numerical inversion methods are generally used. This inversion process is an ill-posed problem with many potential solutions, many of them being not physically reasonable. Considerable effort has gone into developing appropriate algorithms to reach the desired results.[8–11] These methods involve a minimization problem solved by least-squares methods, followed by a regularization with smoothness constraints. For a more detailed explanation of the inversion process, readers are also referred to Dunn *et al.*[12]

It should be noted that if the aforementioned n different environments within the porous system are well connected, the diffusing molecules would be able to shuttle between them, and the final relaxation would only reflect the average S/V of the whole medium, instead of the S/V of those individual environments.[13] NMR has been used extensively as a method for estimating the pore sizes of porous structures for the last three decades.[14,15] This is achieved by looking at the relaxation time distributions of a fluid contained within the pores and using appropriate models to convert the surface-to-volume ratio into characteristic pore sizes (*e.g.*, diameters of cylindrical or spherical pores). Also, in order to assert that the relaxation time distribution of a fluid in a porous medium is congruent to the pore size distribution, uniform relaxivity across the solid matrix must often be assumed. Even though this supposition has no reason to be physically accurate, several experiments have shown that this is very often a sensible assumption to make.[12] One of the most significant advantages of such NMR methods is non-intrusiveness, as the measurement leaves the sample unchanged, as opposed to mercury porosimetry or nitrogen adsorption methods, which require the material to be dried and flooded with the appropriate fluid in each case.

Furthermore, it can also be shown analytically that the ratio between T_1 and T_2 relaxation times is directly proportional to the strength of interactions between the fluid and the solid matrix due to the differences in the dependences of T_1 and T_2 on surface diffusion correlation times.[16] Therefore, relaxation measurements can also be generally used to compare the wettability of different fluids in a confined environment by comparing their respective T_1/T_2 ratios.[17]

10.3 Applications

10.3.1 Soil

Soils are complex mixtures of minerals, organic matter, liquids and gases, as well as living organisms, that support plant growth, filter and purify water, and provide habitat for a wide range of species. It is needless to say that soil plays an essential role in environmental sciences, as it acts as the foundation for terrestrial ecosystems, while it can also be considered an ecosystem on its own.[18] Furthermore, it serves as an important component of the Earth's carbon cycle, as it can store large amounts of CO_2 within its porous matrix, regulating the Earth's climate. Therefore, understanding the properties and processes that govern soil formation, structure, and function is necessary for successfully managing natural resources, mitigating the effects of climate change and preserving biodiversity.

The size of soil pore spaces plays an important role in determining the physical and chemical properties of the whole system, as well as the various processes that occur within it. For example, pore size can affect water retention and drainage, gas exchange, and the movement of nutrients and other chemicals through the soil. Therefore, pore size can influence the activity of microorganisms, as well as the survival and growth of plants. In agriculture, understanding the pore size distribution of a soil can help with developing strategies for soil management, such as irrigation and tillage practices, which can improve soil quality and crop productivity.

A distinct advantage of using NMR methods to study the pore size distribution of soil samples is the ability to monitor the swelling of both clay and organic matter within the soil.[19] Jaeger *et al.*[20] found that, for such experiments, it is better to avoid using a single relaxivity parameter for the whole sample, as using one value for micropores and another one for mesopores significantly increases the accuracy of such estimations. Stray field NMR has also been used for measuring pore size distributions in soils.[21] The advantages of stray field instruments include the ability to study samples that are not easily accessible to traditional NMR techniques, such as samples with complex geometries or sizes too large to fit into conventional laboratory instruments. Such devices may also be used for *in situ* applications.

The water content of soil is a key determinant of plant growth and productivity, and changes in the water content can affect the distribution and abundance of different plant species. NMR relaxation measurements can be used to monitor the moisture content of soil *in situ*[22] by looking at the volume under the T_2 distribution peaks, producing reliable results, as illustrated in Figure 10.4.

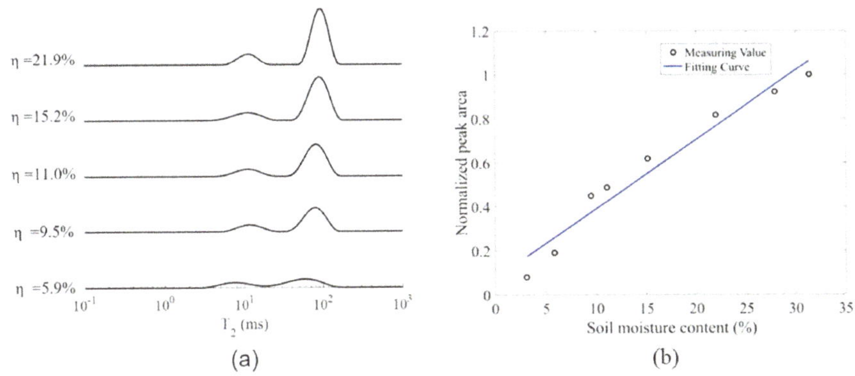

Figure 10.4 (a) T_2 relaxation distributions of soil as a function of moisture content (η). (b) Area under the long-time, high T_2 component of the T_2 relaxation distributions as a function of moisture content. Reproduced from ref. 22 with permission from IOP Publishing, Copyright 2022.

Such measurements may also be used to quantify proportions of free *versus* bound water within the porous structures of the soil. In Figure 10.4(a), two peaks can be distinguished for all of the different moisture content values, one with higher T_2 than the other. Low T_2 values usually correspond to the nuclei of molecules residing in more restricted spaces, which can be interpreted as water bound in small pores, whereas mobile water in larger pore spaces is characterized by longer relaxation times. In this example, during the drying process, the area of the low T_2 peak decreased only by 16.7% during the drying process, while the area of the high T_2 peak decreased by 80.8%, indicating that the decrease in free water content was much more significant than the decrease in bound water content.[22]

Water retention curves (WRC) are frequently used to describe the water holding capacity of a soil. They are determined experimentally and illustrate the relationship between water content and water potential within a soil. As an alternative to conventional methods of determining those curves, NMR relaxation methods have been used by Costabel and Yaramanci.[23] The caveat is that those methods were only successful for predicting WRC in sandy soils, whereas for soils rich in clay and silt with a broad distribution of constrictions in their pore spaces, the NMR methods had to be complemented by additional measurements in order to be reliable.

Soil wetting and drying is a critical process in environmental science, as it affects a wide range of ecological and hydrological processes, impacting the transfer of both nutrients and contaminants. Furthermore, it can affect the physical properties of the soil, such as its structure and porosity, which can in turn affect its ability to retain water and support the growth of plants and other organisms. Understanding the dynamics of soil wetting and drying might be crucial for predicting and managing the response of ecosystems to changing conditions such as climate change, land-use changes and soil degradation.

Wetting processes of dried soils were studied using NMR relaxation by Todoruk *et al.*[24] They used natural as well as artificial soil samples of various organic and mineral component compositions and observed the way water is penetrating different pore sizes when added drop by drop on top of the dried samples, simulating rainfall conditions. For the T_2 distributions acquired at different time steps, they generally observed peaks of relatively high relaxation times at early times of the wetting, shifting towards lower T_2 values later on, implying that the bigger pores get filled before the micropores. They attribute this effect to opening of micropores due to gel swelling, and it was only observed on soils composed of both organic and mineral components. As for soil drying processes, Tian and Wei[25] used T_1 distributions to monitor drying in clay soil samples using stepwise temperature increases, where they also observed a dependency of surface relaxivity on water saturation levels. Cycles of drying and wetting in soils have also been studied using NMR relaxometry,[26,27] revealing a decrease in soil water retention capacity after each of the first few iterations. For both of these studies, this reduction in water capacity reached a steady state after 4–5 cycles, and no further reduction was observed.

10.3.2 Contaminants

Contaminants existing in porous environments, both the vadose zone and aquifers, play an important role in environmental sciences, as they can often have detrimental effects on the environment and human health, and thus they need to be carefully monitored and controlled. Monitoring contaminants in porous media can be challenging due to the complex and heterogeneous nature of the subsurface structures. NMR relaxation can be used to identify the presence of contaminants in porous media, as well as their dynamics and absorption to the pore surfaces.

When two different phases are present within a porous medium, the wetting phase will have its NMR relaxation behaviour strongly affected by the solid surface interactions, as opposed to the non-wetting phase. This principle can be used for the detection of hydrocarbon contaminants in water-wet rocks, as two separate relaxation behaviours can be observed due to the wetting differences between the fluid phases.[28] Bryar *et al.*[29] reported that sorbed oil on sand grain surfaces can have an effect on the NMR relaxation distributions of an aqueous phase, by affecting both the surface relaxivity and the surface area of the pore network. Fay *et al.*[30] studied different water-saturated sediment samples contaminated with gasoline, diesel, crude oil, and trichloroethylene in using T_2 distributions. They found that, depending on each contaminant–sediment combination, the relaxation times for the contaminant and the aqueous phases might be very close, and thus distinguishing them exclusively by relaxation methods would be unfeasible. Clay-containing sands are an example of porous media where this separation was favourable, since water in clay tends to relax rapidly compared to the non-aqueous contaminants. This effect has been used by Bryar and Knight[31] and Hertzog *et al.*[32] to accurately quantify the amounts of various non-aqueous contaminants in sand–clay mixtures. An example illustrating the consistency of such measurements is shown in Figure 10.5.

In situ detection and quantification of hydrocarbon contaminants has also been implemented by Fay *et al.*[33] using NMR logging tools to detect gasoline and diesel leakage directly underground, close to the source of the contamination. In this case, the contaminants could be identified and quantified with an accuracy of around one percent volume fraction for slit sediments, where water and hydrocarbon relaxation times had significant differences. In contrast, contaminants in sand could not be detected due to the overlap of their relaxation times and that of water.

NMR methods may also be used to study the presence of non-liquid contaminants indirectly. Grunin *et al.*[34] sampled soil in several different proximities from a busy highway and studied those samples using NMR relaxometry to look for signs of pollution. They found that relaxation times in samples collected closer to the highway are notably decreased due to increased concentrations of metal ions, and as a result the heavy traffic nearby. Furthermore, this fast relaxation effect was also attributed to fungal growth due to radiotropism, as it is known that radiation sources have frequently been transported across that same highway. Mireiter *et al.*[35] also showed that the concentration of Fe^{3+} can even be quantified in soil samples using NMR relaxation rates, as long as the soil grain sizes are larger than 100 μm.

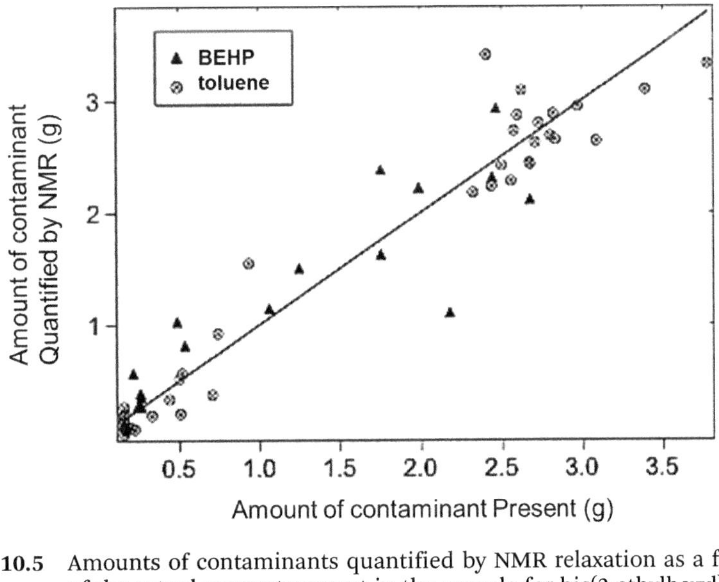

Figure 10.5 Amounts of contaminants quantified by NMR relaxation as a function of the actual amount present in the sample for bis(2-ethylhexyl)phthalate (BEHP) and toluene in sand–clay mixtures. Reproduced from ref. 22 with permission from John Wiley & Sons, Copyright 2008.

10.3.3 Biofilms

Biofilms are defined as "aggregates of microorganisms in which cells that are frequently embedded within a self-produced matrix of extracellular polymeric substance (EPS) adhere to each other and/or to a surface".[36] Effectively, the formation of biofilms is a survival mechanism for the microorganisms that create them since they protect them from environmental threats and enable efficient sharing of nutrients between them. In porous media, such as soil and rocks, biofilms can form within the pores and crevices of the material due to microbial activity in aquatic environments. They can be harnessed for bioremediation and bioconversion processes, making them a valuable tool in sustainable technologies, but they can also facilitate biofouling, where excessive accumulation of biofilms may cause several problems, such as clogging in tubes or filters. Thus, biofilms can have a significant impact on the environment, and studying them is important for understanding and managing the effects of pollution and contamination.

The gel-like environment created biofilms which substantially affect the mobility of the hydrogen-bearing molecules, thus causing a considerable shift in NMR relaxation times. MRI has been used extensively for the study of biofilms,[37] but simple relaxation measurements can also provide a lot of information on biofilm behaviour in porous media.

Jaeger *et al.*[38] studied biofilm growth on natural soil samples and suggested the following general mechanisms to explain the changes in relaxation times during these growth processes:

- Changes in spin relaxation mechanisms due to
  i. Dissolution of organic or paramagnetic substances into the solution.
  ii. Intrusion of water into the soil organic matter leading to exchange between bound and free protons.
  iii. Production and release of EPS and bacterial association within biofilms increasing the amount of water being restricted by the gel structures.
- Changes in pore size distribution due to
  i. Changes in the packing and the total volume of soil samples.
  ii. Swelling of soil organic matter leading to shrinking of pore spaces.
  iii. Production of EPS creating a secondary network of smaller pores between soil particles or bacteria.

Bayer *et al.*[39] performed experiments on glass beads and natural soil samples where they found relaxation time distribution peaks to be consistently shifted towards lower times after the biofilm growth process. Sanderlin *et al.*[40] also conducted a similar study of biofilm growth on unconsolidated sand porous media and reported similar results.

Biofouling has been investigated by Codd *et al.*[41] by performing relaxation measurements on biofouled and non-biofouled glass bead packs, as well as natural rock core samples. They utilized T_2–T_2 correlation methods, which can provide information about exchange of molecules between different pore sizes, to assess the effects of biofouling in those exchange processes. These experiments utilize two consecutive CPMG pulse sequences, separated by a time interval, denoted as exchange or storage time. During the storage time, only T_1 relaxation takes place since the net magnetization vector is on the *z* axis and there is no transverse component. In the meantime, Brownian diffusion takes place and the hydrogen-bearing molecules might end up relaxing while residing in a different environment after that time interval, thus yielding a different T_2 distribution for the second CPMG sequence, which can be denoted as T_2'. Looking at two-dimensional T_2–T_2' distribution maps for different storage times (Figure 10.6), it is possible to identify the time scales at which these exchange processes happen and distinguish the different pathways between these distinct environments.

If all the peaks fall close to the diagonal, it means that T_2 and T_2' are equal, and thus there is no movement between different environments, while the presence of any non-diagonal peaks indicates exchange of nuclei between these environments. This exchange process can also be quantified and compared for different exchange times by looking at the respective volumes of the non-diagonal peaks in each case.

Biofilms in some cases may also induce precipitation of calcite minerals, which can fill up pore spaces. Such a process of microbially induced calcite precipitation has been studied by Kirkland *et al.*[42] using an NMR logging tool in a controlled, laboratory setting. They found that the relaxation times were significantly altered through the process. The majority of the nuclei relaxed at much slower rates due to a layer of calcite shielding the fluid molecules from the surface paramagnetic sites, while some nuclei displayed very fast relaxation due to confinement in the micropores between $CaCO_3$ crystals.

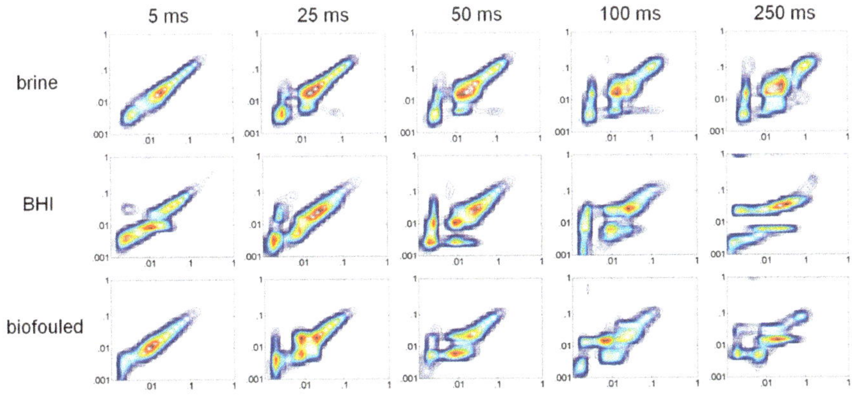

Figure 10.6 T_2–T_2 maps for five different exchange times for a Berea rock core sample with brine, brain heart infusion (BHI) and biofilm. Reproduced from ref. 41 with permission from Elsevier, Copyright 2011.

As for *in situ* detection of biofilms, Kirkland *et al.*[43] used well logging probes to monitor the growth of biofilm directly underground, for a period of 15 days, by observing the decrease in relaxation times over the time period of the experiment. The soil samples were inoculated on the first day. The bacteria were fed by substrate injections during days 2–10, starved during days 11–14 and then removed using groundwater and a bleach solution on the 15th day. As shown in Figure 10.7, the initial relaxation time distributions are bimodal in both cases, and they reflect the pore size distribution of the soil sample.

After the biofilm growth begins, the two peaks merge into a single one, a phenomenon that is most likely attributed to chemical exchange of protons between the aqueous phase and the EPS substances.[40] After the biofilms were removed at the end of the experiment, the original bimodal relaxation behaviour was restored for both cases.

Comparable to biofilms, mucilage is a thick, gel-like substance, often produced by some species of plants, which may be relevant to environmental science in several ways. For example, it can play a role in seed dispersal, water retention, and erosion control, and can also act as a natural adhesive, binding soil particles together and forming aggregates that improve soil

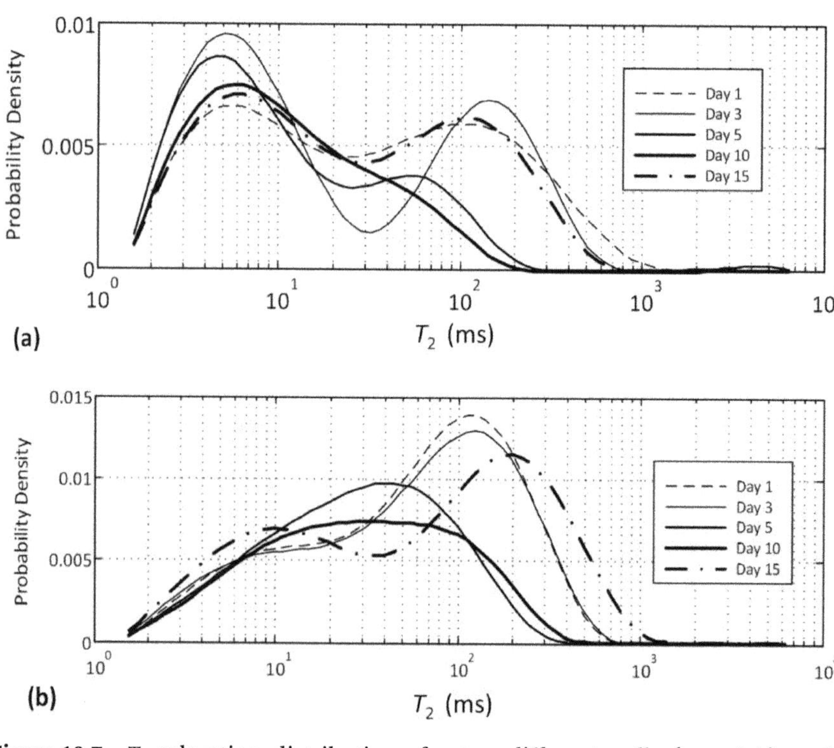

Figure 10.7 T_2 relaxation distributions for two different wells (a and b) at five different time steps. Reproduced from ref. 43 with permission from American Chemical Society, Copyright 2015.

structure and fertility. Brax *et al.*[44] used a combination of NMR relaxation and electron microscopy to investigate the ways in which mucilage polymer networks can be arranged within pore spaces made of glass beads, according to the sizes of the pores. To do that, they used the reduction of bulk relaxation times due to the presence of the mucilage gel, as found in a previous study,[45] which can carry information about the concentrations of mucilage polymers.

10.3.4 Carbon Storage

Through the Paris Agreement, the United Nations pledged their commitment to "hold the increase in the global average temperature to well below 2 °C above pre-industrial levels and pursuing efforts to limit the temperature increase to 1.5 °C above pre-industrial levels".[46] In order to achieve the desired 1.5 °C target, the IPCC estimates that the total anthropogenic CO_2 emissions will have to be reduced down to net zero by the year 2050.[47] Since many industrial processes will inevitably continue to rely on fossil fuels in the foreseeable future, some unavoidable emissions will have to be compensated by permanently removing harmful substances from the atmosphere. Carbon geo-storage (or carbon geo-sequestration, both abbreviated as CGS) is one of the most promising methods to deal with carbon emissions which are currently very hard or impossible to avoid. It involves injecting CO_2 in depleted oil reservoirs or deep saline aquifers, where it can be stored in a supercritical state for geological time scales.

As promising as the concept may sound, CGS processes may involve some serious hazards which need to be meticulously considered before the implementation of such projects at industrial scales. Examples of risks involve leaking of the injected CO_2 back to the surface, and induced seismicity caused by the intense pressures at the injection site.[48] Apart from technical difficulties, public acceptance for such projects is often low, and this is something that needs to be addressed accordingly for each implementation, according to the area's local population and the scale of the project.[49] CGS technology has already been successfully implemented in recent years, largely used for enhanced oil recovery methods. However, the amount of CO_2 sequestered so far is nowhere near the amounts required to reach our climate change mitigation targets. It is estimated that a cumulative total of 197 megatons of CO_2 have been stored through CGS, between 1996 and 2020.[50] This might be a considerable amount, but it only represents a small fraction of global CO_2 emissions, which are currently estimated to be around 36 000 megatons per year.[51] However, there are several CGS projects under development, and more are expected to be built in the future as governments and industry work to reduce their greenhouse gas emissions, making CGS a robust pathway to meet current climate goals.

NMR studies of carbonates before and after CO_2 injection have been performed by Adebayo *et al.*,[52] where the changes in pore size distribution were monitored by NMR relaxometry and validated by X-ray CT scans. Zhao *et al.*[53] and Gholami and Raza[54] also studied the pore size distributions of sandstone cores using T_2 relaxation methods before and after CO_2 injections. Both of the aforementioned studies found that the pore size distributions change considerably after exposure to CO_2, but the changes seem to depend on the characteristics of the individual rock samples in each case. An example is shown in Figure 10.8, where Zhao *et al.*[53] support that small pores are formed after CO_2 injection, through reactions of CO_2 with feldspar and carbonate minerals present within the rocks. Similar NMR studies have also been performed for coal formations,[55] which may also have potential for CGS on a large scale.

Connolly *et al.*[56] characterized the capillary trapping of CO_2 using T_2 distributions after repeated drainage–imbibition cycles and compared the behaviour of CO_2 with the behaviour of nitrogen using the same processes.[57] The same also used NMR relaxation to analyze the dissolution of paramagnetic minerals like iron and manganese from the rock into the bulk brine phase during a supercritical CO_2 injection process, as paramagnetic components within a solution are known to make both T_1 and T_2 relaxation processes significantly faster. Baban *et al.*[58] used relaxation data to assess the wettability of a sandstone sample before and after CO_2 injection and found that the rock became less hydrophilic after the process. Cai *et al.*[59] and Baban *et al.*[60] both used T_1–T_2 NMR methods combined with mass balance to successfully monitor the residual CO_2 in an EOR process involving three fluid phases (brine–hydrocarbon–CO_2) under reservoir conditions of high pressure and temperature.

Regarding the induced seismicity hazard of CGS processes, Zhang *et al.*[61] used NMR T_2 relaxation distributions to help assess the mechanical properties of limestone after CO_2 injection, suggesting that the rock becomes weaker due to rock dissolution, which might lead to earthquakes or rock layer collapse.

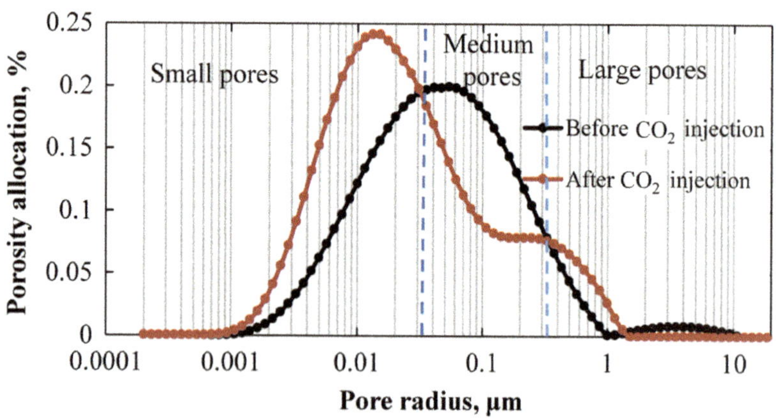

Figure 10.8 Pore size distributions of a tight sandstone sample obtained through NMR relaxation before and after CO_2 injection. Reproduced from ref. 53, https://doi.org/10.1002/ese3.663, under the terms of the CC BY 4.0 license, https://creativecommons.org/licenses/by/4.0/.

10.3.5 Hydrates

Clathrate hydrates consist of ice-like structures, where guest molecules (such as light hydrocarbons, CO_2, H_2, *etc.*) are trapped, or enclathrated, within cage structures made up of host, water molecules. These hydrates can form in subsea sedimentary rocks and permafrost regions, in conditions that can be found in various regions of the planet, but can also form in man-made systems such as pipelines and oil drilling equipment. Many studies have suggested hydrates might be a vast source of hydrocarbons for the years to come,[62] but in the context of environmental science, hydrates are of particular interest due to their potential impact in several environmental issues.

Methane hydrates are essentially a form of natural gas, trapped within ice-like structures in the ocean floor and permafrost regions. The sum of all these hydrate-bearing formations contains immense amounts of methane, and as the climate gets warmer, the stability of these hydrates might be at risk. Melting of methane hydrate formations can lead to all that methane being released into the oceans and the atmosphere, leading to significant impacts on ocean life and the Earth's climate, as methane is a much more potent greenhouse gas than CO_2. Additionally, the melting of methane hydrates can also disrupt ocean currents, leading to sea level rise and coastal erosion. Furthermore, the destabilization of methane hydrates could potentially trigger underwater landslides,[63] which could in extreme cases lead to tsunamis. Thus, understanding and mitigating these risks would require a better understanding of hydrate structures.

Since CO_2 is one of the common guest molecules in hydrate formation, it has been speculated that hydrates could be used for large-scale CO_2 sequestration. The use of hydrates for CGS is still in the early stages of research, but they have been shown to be a promising option due to their high storage capacity and the potential for formation *in situ*. However, there are significant challenges associated with the formation and stability of hydrates, as well as potential environmental impacts that need to be considered when evaluating the risk of such developments. A promising method to combine CO_2 storage and methane production from hydrate formations through exchange processes has been proposed by several researchers,[64–66] where injected CO_2 replaces the methane within the hydrate cells. This concept has already been applied to a very limited extent in a few case studies,[67,68] but more research is required before moving to large-scale production, to ensure enough efficiency and environmental safety, since the underlying mechanisms are not clear enough yet.

Several laboratory NMR relaxation studies have been performed in recent years, where the formation rate of hydrates encaging methane[69–71] and CO_2[72] has been quantified by looking at the evolution of T_2 distributions during the formation process. This is possible due to the changes in relaxation time for protons in water molecules when water is converted to hydrate since the reduced mobility of the water molecules makes relaxation times extremely short in the solid phase. As shown in Figure 10.9, when methane is

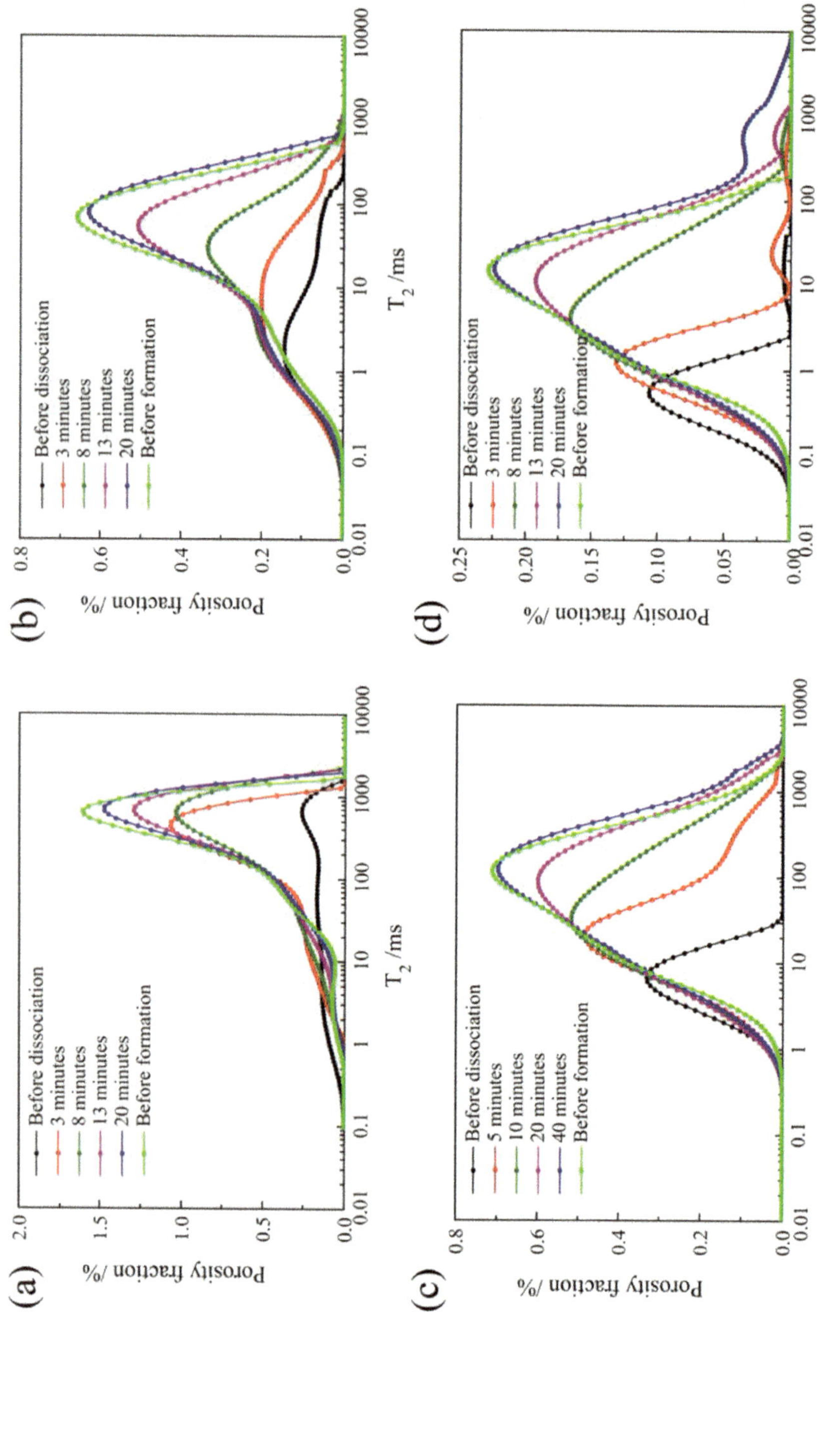

Figure 10.9 T_2 distributions of four different water-saturated sandstone samples at different stages of hydrate formation. Reproduced from ref. 71 with permission from John Wiley & Sons, Copyright 2018.

injected in rock samples and hydrates start to form, the intensity of the relaxation time components attributed to the larger pores starts to diminish. This indicates that hydrates form in larger pores, while it seems that the smaller pores are left unchanged during the process.

Liu *et al.*[73] used X-ray CT scans in combination with relaxivity measurements to estimate relaxivity values for hydrate-bearing sandstones for the accurate determination of pore size distributions. NMR relaxation has also been used to investigate the observed "memory effect",[74] where the formation rate of a hydrates depends strongly on the thermal history of such a system.

For NMR logging applications, the relaxation response of hydrogen nuclei within hydrates is often too fast to measure, but NMR logs in combination with other methods, such as neutron and density porosity, can still be used successfully to assess the porosity of hydrate bearing formations.[75] Another challenge is that hydrates might be difficult to distinguish from permafrost, since ice has similar physical properties to hydrates.[76] A remotely operated submarine vehicle equipped with an NMR logging probe capable of performing CPMG measurements has also been used by Kleinberg *et al.*,[77] revealing the potential of NMR relaxation methods to study hydrate formations *in situ*. Measurements made on the seafloor were congruent with laboratory measurements, validating the reliability of this novel setup; however, the limitations are the same as those in NMR logging, and applications might be limited. MRI has also been used extensively to study hydrates in laboratory scales,[78–81] providing useful images of the hydrate structures and the processes of hydrate formation and decomposition.[82] However, MRI instruments tend to be delicate in comparison to relaxometers, since they require precise magnetic field gradients, and thus they are not suitable for field use with currently available technologies. Furthermore, resolution limitations prevent the study of materials with pore diameters below the scale of a few hundred micrometers.[82]

Overall, NMR relaxation can be a powerful method for the study of hydrate behaviour in porous media, and while there are some limitations due to the nature of the hydrate protons, they can be overcome by combining NMR relaxation with other techniques such as X-ray CT and MRI.[82]

10.3.6 Marine Life

It is widely known that increased atmospheric temperatures would lead to an increase in the ocean temperature as well, potentially having severe consequences on ocean ecosystems. As an example, corals are important for several reasons in the context of environmental science. They play a critical role in supporting the biodiversity of coral reefs, which are among the world's most diverse and productive ecosystems. Furthermore, coral reefs act as a natural barrier, protecting coastlines from wave damage and erosion.

Fantazzini *et al.*[83] used NMR relaxation measurements on coral skeleton samples collected from three sites across the Italian coast characterized by different temperatures, to monitor the effects of those variations in temperature on the corals, validating their results by comparing them to mercury intrusion porosimetry and scanning electron microscopy imaging of the same samples. They found an increased fraction of larger pores appearing at increased temperatures, which could reduce their resistance to mechanical stresses. Reduced mechanical strength in corals can cause coral reefs to collapse under environmental stress, destroying entire ecosystems which are important to marine life as a whole.

Apart from the well-known effects of global warming, another important, but often overlooked consequence of increased atmospheric CO_2 concentrations is ocean acidification. Increasing levels of CO_2 dissolve into ocean waters, reducing the pH and making marine environments overall more acidic. In a follow-up study, Fantazzini *et al.*[84] studied 74 corals of similar age collected from areas of different acidity caused by the CO_2 concentration gradient produced by a volcanic vent on the seabed. NMR relaxometry combined with several other methods confirmed that there is a clear correlation between increased acidity and increased skeletal macro-porosity (Figure 10.10).

Additional coral species have also been studied by similar methods in a more recent study,[85] which yielded equivalent results. The scope of NMR relaxation in the context of marine sciences is still limited, but new challenges that arise due to climate change might increase the need for novel methods to study those delicate ecosystems of the utmost importance to the planet.

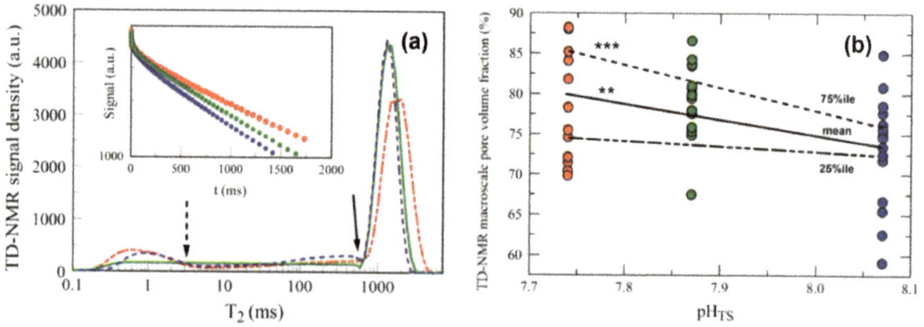

Figure 10.10 (a) Examples of T_2 distributions for three different water-saturated coral skeletons obtained from regions of different acidity. The inset shows the corresponding relaxation curves. Dashed and solid arrows represent the cutoff values from nanopores to micropores, and from micropores to nanopores, respectively. (b) Macroscale pore volume fractions of several samples plotted against total scale pH. Reproduced from ref. 84, https://doi.org/10.1038/ncomms8785, under the terms of the CC BY 4.0 license, https://creativecommons.org/licenses/by/4.0/.

10.4 Conclusions

Overall, NMR relaxation has been proven to be a very powerful tool for research involving fluids in porous media. Its non-invasive nature and its sensitivity to the ever-abundant hydrogen atom make it applicable to numerous different situations. However, it should be noted that, to fully utilize its potential, NMR relaxometry often has to be combined with other methods, in order to turn the measurements from qualitative to quantitative descriptions of the studied systems. Furthermore, researchers must often rely on empirical cut-off values and make idealistic assumptions, such as negligible pore connectivity and uniform surface relaxivity, in order to infer the desired results. Small pore sizes and paramagnetic impurities can make relaxation too fast to detect, and in some cases different liquids within the same medium have similar relaxation times, making them indistinguishable. Therefore, the reliability of relaxation measurements is highly dependent on the nature of the porous medium and the fluids contained within, as well as on the judgement of the individual researcher.

Nevertheless, despite these limitations, there are vast advantages to NMR relaxation methods, and there is still progress to be made when it comes to refining the algorithms used for data processing and the models used to interpret the final results. Due to the abundance of porous media in natural systems, and the ever-growing demand for research in environmentally relevant areas, NMR relaxation can be expected to be the protagonist in many studies to come for the foreseeable future.

References

1. R. O. Baker, H. W. Yarranton and J. L. Jensen, in *Practical Reservoir Engineering and Characterization*, Elsevier, 2015, pp. 239–295.
2. Y.-Q. Song, *J. Magn. Reson.*, 2013, **229**, 12–24.
3. R. L. Kleinberg, C. Straley, W. E. Kenyon, R. Akkurt and S. A. Farooqui, in *68th Annual Technical Conference and Exhibition of the Society of Petroleum Engineers*, SPE, Houston, TX, USA, 1993, pp. 3–6.
4. N. Bloembergen, E. M. Purcell and R. V Pound, *Phys. Rev.*, 1948, **73**, 679–712.
5. K. R. Brownstein and C. E. Tarr, *Phys. Rev. A: At., Mol., Opt. Phys.*, 1979, **19**, 2446–2453.
6. X. Ge, M. T. Myers, J. Liu, Y. Fan, M. A. Zahid, J. Zhao and L. Hathon, *Mar. Pet. Geol.*, 2021, **126**, 104934.
7. J. Mitchell, T. C. Chandrasekera, M. L. Johns, L. F. Gladden and E. J. Fordham, *Phys. Rev. E: Stat., Nonlinear, Soft Matter Phys.*, 2010, **81**, 026101.
8. L. Venkataramanan, Y.-Q. Song and M. D. Hurlimann, *IEEE Trans. Signal Process.*, 2002, **50**, 1017–1026.
9. Y.-Q. Song, L. Venkataramanan, M. D. Hürlimann, M. Flaum, P. Frulla and C. Straley, *J. Magn. Reson.*, 2002, **154**, 261–268.

10. M. D. Hürlimann, L. Venkataramanan and C. Flaum, *J. Chem. Phys.*, 2002, **117**, 10223–10232.
11. G. C. Borgia, R. J. S. Brown and P. Fantazzini, *J. Magn. Reson.*, 1998, **132**, 65–77.
12. K. Dunn, D. Bergman and G. LaTorraca, *Nuclear Magnetic Resonance: Petrophysical and Logging Applications*, Pergamon, 2002.
13. K. R. McCall, D. L. Johnson and R. A. Guyer, *Phys. Rev. B: Condens. Matter Mater. Phys.*, 1991, **44**, 7344–7355.
14. S. Davies and K. J. Packer, *J. Appl. Phys.*, 1990, **67**, 3163–3170.
15. R. Kleinberg, *Magn. Reson. Imaging*, 1994, **12**, 271–274.
16. C. D'Agostino, J. Mitchell, M. D. Mantle and L. F. Gladden, *Chem. – Eur. J.*, 2014, **20**, 13009–13015.
17. A. Valori and B. Nicot, *Petrophysics – SPWLA J. Form. Eval. Reserv. Descr.*, 2019, **60**, 255–263.
18. J.-F. Ponge, *Biol. Fertil. Soils*, 2015, **51**, 645–648.
19. M. Meyer, C. Buchmann and G. E. Schaumann, *Eur. J. Soil Sci.*, 2018, **69**, 393–406.
20. F. Jaeger, S. Bowe, H. Van As and G. E. Schaumann, *Eur. J. Soil Sci.*, 2009, **60**, 1052–1064.
21. N. R. A. Bird, A. R. Preston, E. W. Randall, W. R. Whalley and A. P. Whitmore, *Eur. J. Soil Sci.*, 2005, **56**, 135–143.
22. C. Wan, W. He, P. Guo, J. Wu, Y. He and Z. Xu, *Meas. Sci. Technol.*, 2022, **33**, 075902.
23. S. Costabel and U. Yaramanci, *Water Resour. Res.*, 2013, **49**, 2068–2079.
24. T. R. Todoruk, C. H. Langford and A. Kantzas, *Environ. Sci. Technol.*, 2003, **37**, 2707–2713.
25. H. H. Tian and C. F. Wei, *Water Resour. Res.*, 2020, **56**(10), DOI: 10.1029/2020WR027537.
26. L. Kong, H. M. Sayem and H. Tian, *Can. Geotech. J.*, 2018, **55**, 208–216.
27. X. T. Xu, D. Q. Liu, Z. X. Xian, F. Yang, W. Bin Jian, X. Xu and J. B. Huang, *Geofluids*, 2022, **2022**, 1–15.
28. S. A. Hedberg, R. J. Knight, A. L. MacKay and K. P. Whittall, *Water Resour. Res.*, 1993, **29**, 1163–1170.
29. T. R. Bryar and R. J. Knight, *Geophysics*, 2003, **68**, 942–948.
30. E. L. Fay and R. J. Knight, *Geophysics*, 2016, **81**, EN87–EN97.
31. T. R. Bryar and R. J. Knight, *Water Resour. Res.*, 2008, **44**(2), DOI: 10.1029/2006WR005635.
32. R. C. Hertzog, T. A. White and C. Straley, *J. Environ. Eng. Geophys.*, 2007, **12**, 293–306.
33. E. L. Fay, R. J. Knight and E. D. Grunewald, *Geophysics*, 2017, **82**, EN81–EN92.
34. L. Grunin, E. Nikolskaya and J. Edwards, *Air, Soil Water Res.*, 2013, **6**, ASWR.S12406.
35. I. Mitreiter, S. E. Oswald and F. Stallmach, *Open Magn. Reson. J.*, 2010, **3**, 46–51.
36. M. Vert, Y. Doi, K.-H. Hellwich, M. Hess, P. Hodge, P. Kubisa, M. Rinaudo and F. Schué, *Pure Appl. Chem.*, 2012, **84**, 377–410.

37. M. P. Herrling, S. Lackner, H. Nirschl, H. Horn and G. Guthausen, in *Annual Reports on NMR Spectroscopy*, Academic Press Inc., 2019, vol. 97, pp. 163–213.
38. F. Jaeger, E. Grohmann and G. E. Schaumann, *Plant Soil*, 2006, **280**, 209–222.
39. J. V Bayer, F. Jaeger and G. E. Schaumann, *Open Magn. Reson. J.*, 2010, **3**, 15–26.
40. A. B. Sanderlin, S. J. Vogt, E. Grunewald, B. A. Bergin and S. L. Codd, *Environ. Sci. Technol.*, 2013, **47**, 987–992.
41. S. L. Codd, S. J. Vogt, J. A. Hornemann, A. J. Phillips, J. E. Maneval, K. R. Romanenko, L. Hansen, A. B. Cunningham and J. D. Seymour, *Org. Geochem.*, 2011, **42**, 965–971.
42. C. M. Kirkland, S. Zanetti, E. Grunewald, D. O. Walsh, S. L. Codd and A. J. Phillips, *Environ. Sci. Technol.*, 2017, **51**, 1537–1543.
43. C. M. Kirkland, M. P. Herrling, R. Hiebert, A. T. Bender, E. Grunewald, D. O. Walsh and S. L. Codd, *Environ. Sci. Technol.*, 2015, **49**, 11045–11052.
44. M. Brax, M. Köhne, E. Kroener and G. E. Schaumann, *Geoderma*, 2019, **340**, 269–278.
45. M. Brax, C. Buchmann and G. E. Schaumann, *Microporous Mesoporous Mater.*, 2018, **269**, 47–50.
46. C. A. Horowitz, *Int. Leg. Mater.*, 2016, **55**, 740–755.
47. IPCC, *Global Warming of 1.5 °C*, 2018.
48. V. Vilarrasa, J. Carrera, S. Olivella, J. Rutqvist and L. Laloui, *Solid Earth*, 2019, **10**, 871–892.
49. S. Krevor, H. de Coninck, S. E. Gasda, N. S. Ghaleigh, V. de Gooyert, H. Hajibeygi, R. Juanes, J. Neufeld, J. J. Roberts and F. Swennenhuis, *Nat. Rev. Earth Environ.*, 2023, 1–17.
50. Y. Zhang, C. Jackson and S. Krevor, *Environ. Sci. Technol. Lett.*, 2022, **9**, 693–698.
51. P. Friedlingstein, M. O'Sullivan, M. W. Jones, R. M. Andrew, L. Gregor, J. Hauck, C. Le Quéré, I. T. Luijkx, A. Olsen, G. P. Peters, W. Peters, J. Pongratz, C. Schwingshackl, S. Sitch, J. G. Canadell, P. Ciais, R. B. Jackson, S. R. Alin, R. Alkama, A. Arneth, V. K. Arora, N. R. Bates, M. Becker, N. Bellouin, H. C. Bittig, L. Bopp, F. Chevallier, L. P. Chini, M. Cronin, W. Evans, S. Falk, R. A. Feely, T. Gasser, M. Gehlen, T. Gkritzalis, L. Gloege, G. Grassi, N. Gruber, Ö. Gürses, I. Harris, M. Hefner, R. A. Houghton, G. C. Hurtt, Y. Iida, T. Ilyina, A. K. Jain, A. Jersild, K. Kadono, E. Kato, D. Kennedy, K. Klein Goldewijk, J. Knauer, J. I. Korsbakken, P. Landschützer, N. Lefèvre, K. Lindsay, J. Liu, Z. Liu, G. Marland, N. Mayot, M. J. McGrath, N. Metzl, N. M. Monacci, D. R. Munro, S.-I. Nakaoka, Y. Niwa, K. O'Brien, T. Ono, P. I. Palmer, N. Pan, D. Pierrot, K. Pocock, B. Poulter, L. Resplandy, E. Robertson, C. Rödenbeck, C. Rodriguez, T. M. Rosan, J. Schwinger, R. Séférian, J. D. Shutler, I. Skjelvan, T. Steinhoff, Q. Sun, A. J. Sutton, C. Sweeney, S. Takao, T. Tanhua, P. P. Tans, X. Tian, H. Tian, B. Tilbrook, H. Tsujino,

F. Tubiello, G. R. van der Werf, A. P. Walker, R. Wanninkhof, C. Whitehead, A. Willstrand Wranne, R. Wright, W. Yuan, C. Yue, X. Yue, S. Zaehle, J. Zeng and B. Zheng, *Earth Syst. Sci. Data*, 2022, **14**, 4811–4900.

52. A. R. Adebayo, M. E. Kandil, T. M. Okasha and M. L. Sanni, *Int. J. Greenhouse Gas Control*, 2017, **63**, 1–11.
53. J. Zhao, P. Wang, Y. Zhang, L. Ye and Y. Shi, *Energy Sci. Eng.*, 2020, **8**, 2286–2296.
54. R. Gholami and A. Raza, *Fuel*, 2022, **324**, 124781.
55. X. Sun, Y. Yao, D. Liu and Y. Zhou, *Int. J. Coal Geol.*, 2018, **188**, 38–50.
56. P. R. J. Connolly, S. J. Vogt, M. Mahmoud, C. N. Y. Ng, E. F. May and M. L. Johns, *Water Resour. Res.*, 2019, **55**, 10466–10478.
57. P. R. J. Connolly, S. J. Vogt, S. Iglauer, E. F. May and M. L. Johns, *Water Resour. Res.*, 2017, **53**, 7917–7932.
58. A. Baban, A. Al-Yaseri, A. Keshavarz, R. Amin and S. Iglauer, *Int. J. Greenhouse Gas Control*, 2021, **111**, 103435.
59. M. Cai, Y. Su, Y. Hao, Y. Guo, D. Elsworth, L. Li, D. Li and X. Li, *Fuel*, 2021, **305**, 121606.
60. A. Baban, A. Keshavarz, R. Amin and S. Iglauer, *Fuel*, 2023, **332**, 126000.
61. Y. Zhang, M. Sarmadivaleh, M. Lebedev, A. Barifcani, R. Rezaee, N. Testamantia and S. Iglauer, in *Day 3 Thu, March 24, 2016*, OTC, 2016, pp. 22–25.
62. G. J. Moridis, T. S. Collett, R. Boswell, M. Kurihara, M. T. Reagan, C. Koh and E. D. Sloan, in *Society of Petroleum Engineers - Unconventional Reservoirs Conference 2008*, OnePetro, 2008, pp. 30–71.
63. M. Parlaktuna, in *Submarine Landslides and Tsunamis*, Springer Netherlands, Dordrecht, 2003, pp. 163–169.
64. D.-Y. Koh, H. Kang, J.-W. Lee, Y. Park, S.-J. Kim, J. Lee, J. Y. Lee and H. Lee, *Appl. Energy*, 2016, **162**, 114–130.
65. W.-N. Wei, B. Li, Q. Gan and Y.-L. Li, *Fuel*, 2022, **312**, 122873.
66. G. G. Tsypkin, *Energy*, 2021, **233**, 121115.
67. H. Zhu, T. Xu, Y. Yuan, Y. Xia and X. Xin, *Appl. Energy*, 2020, **275**, 115384.
68. R. Boswell, D. Schoderbek, T. S. Collett, S. Ohtsuki, M. White and B. J. Anderson, *Energy Fuels*, 2017, **31**, 140–153.
69. Z. Cheng, S. Wang, N. Xu, W. Liu, Y. Zhao, J. Zhao, L. Jiang and J. N. Zheng, *Fuel*, 2021, **301**, 121021.
70. Y. Ji, J. Hou, G. Cui, N. Lu, E. Zhao, Y. Liu and Q. Du, *Fuel*, 2019, **251**, 82–90.
71. X. Ge, J. Liu, Y. Fan, D. Xing, S. Deng and J. Cai, *J. Geophys. Res.: Solid Earth*, 2018, **123**, 3339–3346.
72. J. Zheng, L. Yang, S. Ma, Y. Zhao and M. Yang, *AIChE J.*, 2020, **66**, 16820.
73. L. Liu, Z. Zhang, C. Liu, N. Wu, F. Ning, Y. Zhang, Q. Meng and C. Li, *Energy Fuels*, 2021, **35**, 6144–6152.
74. Z. Wen, Y. Yao, W. Luo and X. Lei, *Fuel*, 2021, **299**, 120922.
75. D. Murray, R. Kleinberg, B. Sinha, M. Fukuhara, O. Osawa, T. Endo and T. Namikawa, in *SPWLA 46th Annual Logging Symposium 2005*, OnePetro, 2005.

76. S. Prensky, in *SPWLA 36th Annual Logging Symposium 1995*, OnePetro, 1995.
77. R. L. Kleinberg, C. Flaum, C. Straley, P. G. Brewer, G. E. Malby, E. T. Peltzer, G. Friederich and J. P. Yesinowski, *J. Geophys. Res.: Solid Earth*, 2003, **108**, 2137.
78. B. A. Baldwin, A. Moradi-Araghi and J. C. Stevens, *Magn. Reson. Imaging*, 2003, **21**, 1061–1069.
79. J. Zhao, Q. Lv, Y. Li, M. Yang, W. Liu, L. Yao, S. Wang, Y. Zhang and Y. Song, *Magn. Reson. Imaging*, 2015, **33**, 485–490.
80. J. Zhao, L. Yang, K. Xue, W. Lam, Y. Li and Y. Song, *Chem. Phys. Lett.*, 2014, **612**, 124–128.
81. M. Yang, Z. R. Chong, J. Zheng, Y. Song and P. Linga, *Renewable Sustainable Energy Rev.*, 2017, **74**, 1346–1360.
82. B. Liu, L. Zhan, H. Lu and J. Zhang, *Water*, 2022, **14**, 330.
83. P. Fantazzini, S. Mengoli, S. Evangelisti, L. Pasquini, M. Mariani, L. Brizi, S. Goffredo, E. Caroselli, F. Prada, G. Falini, O. Levy and Z. Dubinsky, *Environ. Sci. Technol.*, 2013, **47**, 12679–12686.
84. P. Fantazzini, S. Mengoli, L. Pasquini, V. Bortolotti, L. Brizi, M. Mariani, M. Di Giosia, S. Fermani, B. Capaccioni, E. Caroselli, F. Prada, F. Zaccanti, O. Levy, Z. Dubinsky, J. A. Kaandorp, P. Konglerd, J. U. Hammel, Y. Dauphin, J.-P. Cuif, J. C. Weaver, K. E. Fabricius, W. Wagermaier, P. Fratzl, G. Falini and S. Goffredo, *Nat. Commun.*, 2015, **6**, 7785.
85. F. Prada, L. Brizi, S. Franzellitti, S. Mengoli, S. Fermani, I. Polishchuk, N. Baraldi, F. Ricci, Q. Palazzo, E. Caroselli, B. Pokroy, L. Giorgini, Z. Dubinsky, P. Fantazzini, G. Falini, S. Goffredo and K. E. Fabricius, *Sci. Rep.*, 2021, **11**, 19927.

CHAPTER 11

Characterization of Cyclodextrin Cross-linked Polymers Used in Environmental Applications by Solid-state NMR Spectroscopy: a Historical Review

D. LACALAMITA,[a] S. BERTINI,[b] C. MONGIOVÍ,[a] C. COSENTINO,[b] N. MORIN-CRINI,[a] G. TORRI,[b] M. FOURMENTIN,[c] A. NAGGI,[b] S. FOURMENTIN,[d] M. GUERRINI[b] AND G. CRINI*[a]

[a] Université de Franche-Comté, UFR Sciences et Techniques, Chrono-environnement, UMR 6249, 16 route de Gray, 25000 Besançon, France; [b] Istituto di Chimica e Biochimica G. Ronzoni, 81 *via* G. Colombo, 20133 Milano, Italy; [c] Université du Littoral Côte d'Opale, Laboratoire de Physico-Chimie de l'Atmosphère (LPCA, UR 4493), ULCO, 59140 Dunkerque, France; [d] Université du Littoral Côte d'Opale, Unité de Chimie Environnementale et Interactions sur le Vivant (UCEIV, UR 4492), ULCO, 59140 Dunkerque, France

*Email: gregorio.crini@univ-fcomte.fr

11.1 Introduction

Cyclodextrins (CDs) are commercial substances produced at industrial level by cyclodextrin glycosyltransferase (CGTase) using starches or their derivatives as substrates through enzymatic degradation.[1] These molecules are

New Developments in NMR No. 32
The Environment in a Magnet: Applications of NMR Techniques to Environmental Problems
Edited by Pellegrino Conte, Delia Francesca Chillura Martino and Paolo Lo Meo

Published by the Royal Society of Chemistry, www.rsc.org

cone-shaped oligosaccharides composed of six or more D-glucopyranoside units linked in α(1–4), like in amylose (Figure 11.1). Typical CDs have six, seven, or eight glucose units, known as α-cyclodextrin (α-CD), β-cyclodextrin (β-CD), and γ-cyclodextrin (γ-CD), respectively.[2–9] Among them, β-CD and its derivatives are the most studied and used, based on their cheapness, availability, and particular structure and properties.[10–12] Indeed, CDs are cyclic oligosaccharides with a toroidal structure and remarkable amphiphilic properties. Their central cavity is hydrophobic and their external part is hydrophilic due to the presence of numerous hydroxyl groups. The molecular structure and characteristics of β-CD molecules are schematized in Figure 11.1.

Figure 11.2 compares the cross polarization magic angle spinning (CPMAS) and carbon-13 nuclear magnetic resonance NMR (^{13}C NMR) spectra of a powder sample of β-CD. The ^{13}C NMR spectrum obtained in D_2O (Figure 11.2(b)) shows the presence of 6 signals attributed to the 6 carbons of the glucose units (see Figure 11.1(b)). This spectrum is consistent with the fact that the β-CD molecule is a cyclic oligomer in which all the 7 glucose units are identical. The CPMAS spectrum of β-CD obtained in the solid state (Figure 11.2(a)) highlights the different degrees of crystallinity of the sample as revealed by the number of resonances, particularly for the anomeric C-1 signals and the carbons in position 6 on the glucose unit.[13]

Due to their cage-like supramolecular structure, β-CD molecules can interact with the structures of various molecules and ions, forming inclusion complexes through host–guest interactions, either in the solid state or in aqueous solution (Figure 11.3). They are also able to form non-inclusion complexes *via* other specific interactions and processes (self-assembly, aggregation). So, CDs are recognized as being among the most important supramolecular host molecules with a large range of applications. Native CDs, CD derivatives and CD-based polymers, in soluble or insoluble forms, find important uses in chemistry, biology and polymer science, *e.g.* in synthesis, materials science, catalysis, supramolecular chemistry, molecular recognition, and surface modification, separation science (chromatography, biotechnology), foods and beverage industry, cosmetics and personal care products, pharmacy and medicine (drug delivery systems, biomedicine), textile industry, wastewater treatment, and air filtration.[14–29] CDs and their derivatives have been also successfully employed to create innovative materials such as nanoparticles, nanosponges, membranes, fibers, and metal organic frameworks.[30–38]

Among the many materials available, cyclodextrin cross-linked polymers, marketed by several companies, are attracting increasing interest for the complexation of environmental pollutants.[39–50] Indeed, these polymeric architectures have been proposed as biosorbents to treat wastewater contaminated by classical pollutants, such as metals,[51,52] polycyclic aromatic hydrocarbons (PAHs),[47] dyes,[53–55] chlorophenols,[56,57] cresols[58] and pesticides,[59] or by emerging pollutants such as phthalates,[47] pharmaceuticals (hormones,

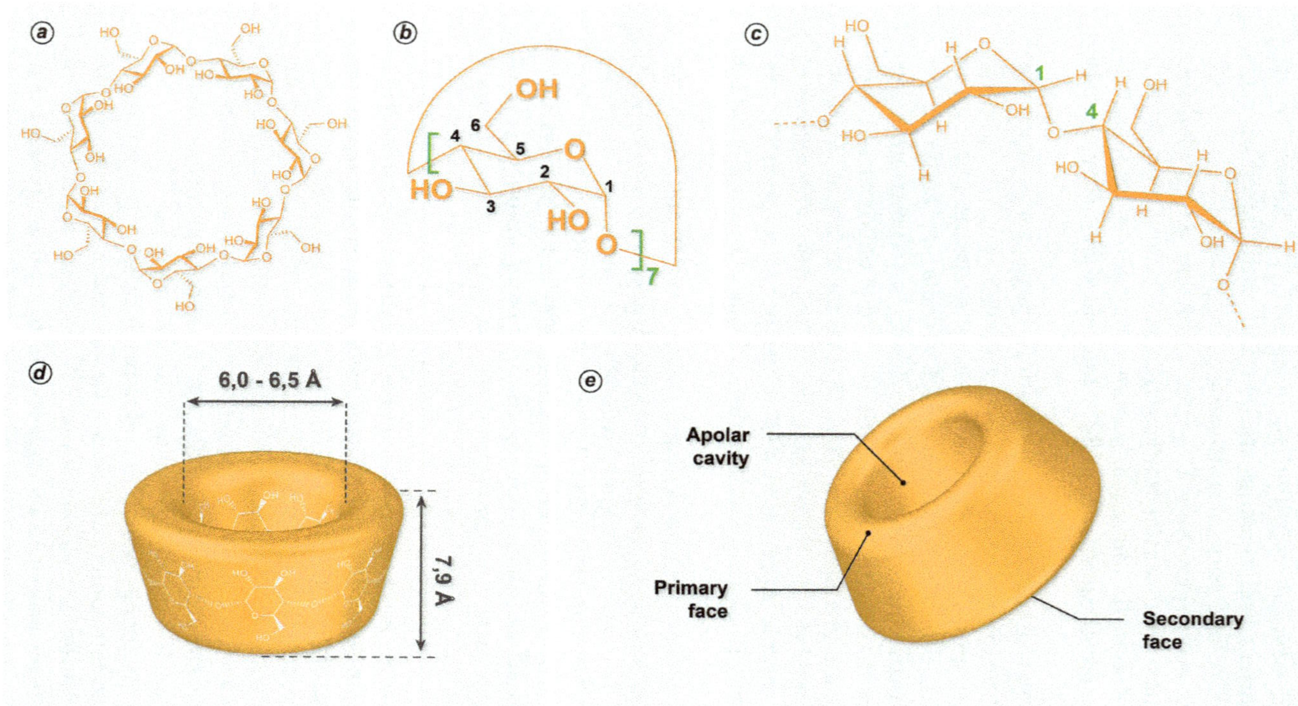

Figure 11.1 Schematic representation of (a) the general chemical structure of β-cyclodextrin (β-CD), (b) the ring glucose, (c) the α-(1–4)-glycosidic linkage and the C-1(D) chair conformation between two glucopyranose units (the number indicates the numbering system employed to describe the ring glucose), (d) the dimensions, and (e) the 3D structure of β-CD. The hydroxyl groups are located at the edges of the CD torus, which itself consists of the sugar backbone and glycosidic oxygen bridges, with primary and secondary OH groups located at the top and bottom of the structure, respectively. β-CD has 21 hydroxyl groups present at positions 2, 3 and 6 of the glycoside unit, and the 7 primary OH groups are free to rotate and therefore the most reactive. The free rotation primary OH groups reduce the diameter of the upper side and give a truncated cone shape, while the secondary OH groups form strong hydrogen bonds that stiffen the lower part of the CD.

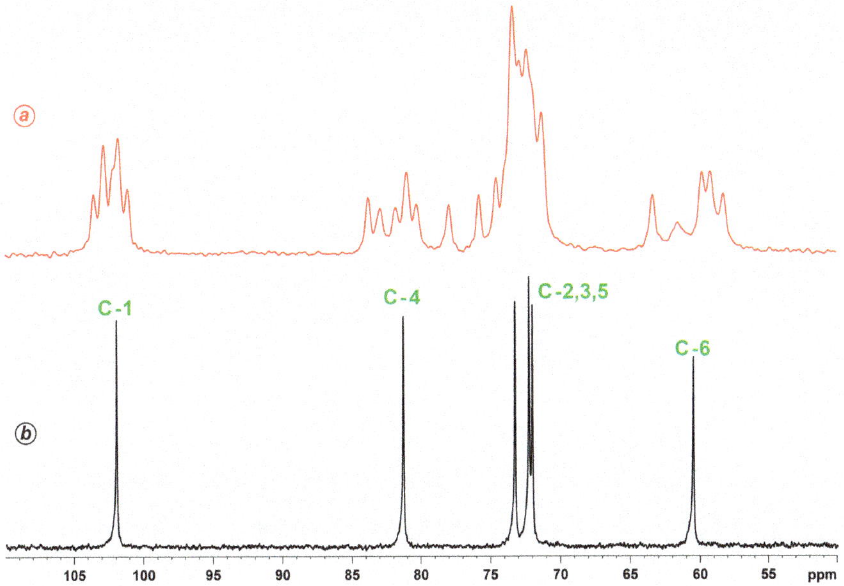

Figure 11.2 Comparison of (a) solid-state cross polarization magic angle spinning (CPMAS) and (b) carbon-13 nuclear magnetic resonance (^{13}C NMR) spectra obtained in D_2O of a β-cyclodextrin sample. The solid-state NMR spectrum was recorded with a Bruker CXP-300 spectrometer operating at 75.47 MHz and 303 K (the powder was placed in a zirconium rotor with a 7 mm diameter and 21 mm height). The ^{13}C NMR spectrum in D_2O was measured at 100.61 MHz on a Bruker AMX 400 spectrometer fitted with a 10 mm broadband probe. The ^{13}C NMR spectrum shows the 6 signals attributed to the 6 carbons of the 7 glucose units, while the CPMAS spectrum highlights the different degrees of crystallinity of the sample in the solid state (spectra recorded in 1994; source: Ronzoni's archives).

antibiotics),[60,61] bisphenols,[62] alkylphenols,[63,64] per- and polyfluoroalkyl substances,[65,66] thiophenic substances,[67] rare earth elements,[68] and mycotoxins.[69] An important advantage of CD cross-linked polymers is that, compared to other conventional commercial adsorbents/sorbents, they are able to interact specifically with pollutants present as traces in complex solutions.[70,71] All results are mainly explained by the presence of CD molecules. Indeed, in cross-linked polymers, CD cavities retain their ability to form inclusion complexes although their mobility is greatly reduced. However, although numerous results have been published, many questions remain not only on the chemical structure of the macromolecular networks of these functional polymers but also on the biosorption mechanisms and the CD role when these cross-linked polymers are used as biosorbent materials. Solid-state NMR can help answer these questions.

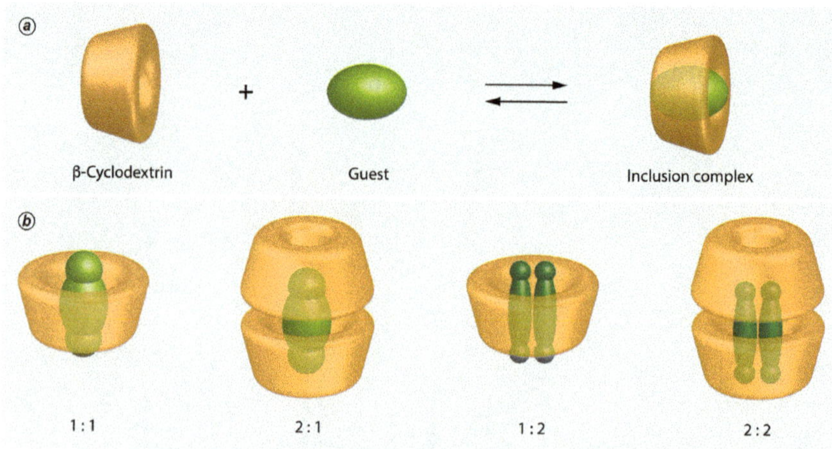

Figure 11.3 Schematic representation of (a) the formation of an inclusion complex between β-cyclodextrin (host) and a solute (guest) and (b) the main stoichiometries obtained between cyclodextrins and solute molecules. The main recognized characteristic of cyclodextrins is the formation of inclusion complexes (molecular encapsulation) with various guests and various stoichiometries reported in the literature, with most inclusion complexes having a cyclodextrin : guest stoichiometry of 1 : 1.

This chapter aims to summarize and discuss the role of solid-state NMR spectroscopy in the structural characterization of cross-linked cyclodextrin polymers from a historical point of view. First, we give a brief overview of their synthesis and their use as biosorbents in oriented-biosorption processes. Then, we present the interest of using solid-state NMR to characterize these complex macromolecular structures using the state of the art of the literature and the results of our research. Finally, the interactions involved in biosorption processes and characterized by NMR are also discussed.

11.2 β-Cyclodextrin Cross-linked Polymers for Wastewater Treatment

11.2.1 Synthesis of Cross-linked Polymers Prepared by Direct Cross-linking of β-Cyclodextrin

Due to their particular structure and characteristics, native CDs are versatile molecules for polymer science and in the design and the construction of macromolecular networks. Indeed, CDs are compatible with a range of polymerization techniques and grafting reactions to obtain multifunctional materials for target applications. The presence of a distinct reactivity of their numerous OH groups on the glycoside unit (with their anisotropy and polar spatial distribution) is often cited as the key element. There are three hydroxyl environments in these oligomers (Figure 11.1): (1) the primary hydroxyls at the C6 position are the most nucleophilic and basic, and hence the most reactive with free rotation; (2) the secondary hydroxyls at the C3 position are the least reactive because they are most inaccessible due to steric hindrance and hydrogen bonding, and (3) those at the C2 position are the most acidic.[72–77] Secondary OH groups are also relatively rigid. The presence of these OH groups and their different reactivity dictate the behaviour of various chemical reactions.

Among chemical modification techniques, cross-linking reactions using a cross-linking agent play an important role in creating CD-based materials. Indeed, CDs can easily be used to obtain cross-linked polymers under mild synthesis conditions (chemistry in water at moderate temperatures and atmospheric pressure). The main cross-linking agents used to produce gels and resins are isocyanates (*e.g.* toluene-2,4-diisocyanate and 1,6-hexamethylene diisocyanate), anhydrides (pyromellitic dianhydride), carbonyls (diphenyl carbonate, 1,1′-carbonyl diimidazole), acrylates, acrylamides, carboxyls, and especially epoxide (epichlorohydrin) derivatives.[78–81] Indeed, chemical cross-linking using epichlorohydrin or 1-chloro-2,3-epoxypropane is the method used on a large scale to produce water-insoluble commercial CD polymers. It is also the most commonly used cross-linker in polysaccharide chemistry and in the pharmaceutical industry.

The cross-linking reaction of CD molecules using epichlorohydrin has been developed in the mid-1960s by Solms' group[82] at the Research Laboratory of the Nestlé group in Vevey (Switzerland) and later by the Dutch group of Niels Wiedenhof[83] (Laboratory of General Chemistry, Eindhoven). This chemical reaction with CDs as with other oligosaccharides and polysaccharides is well-known and relatively easy.[41] The reaction mechanism was proposed in the 1960s by Jerald L. Hofmann[84] (University of Louisville, Kentucky, USA), adopted by József Szejtli[85–87] (Chinoin Chemical and Pharmaceutical Works, Budapest, Hungary) in the 1980s and described in detail for the first time by Bernard Sébille[88] (*Université de Paris XII*, France) in 1997 for water-soluble CD polymers. The same year, Torri and Crini's group (G. Ronzoni Institute for Chemical and Biochemical Research, Milano, Italy) demonstrated for the first time the cross-linking reaction mechanism and the structure of water-insoluble epichlorohydrin polymers by solid-state NMR spectroscopy. These results were presented at the IXth European Carbohydrate Symposium at Utrecht (The Netherlands, July 6–11, 1997).[42]

Figure 11.4 shows the chemical reaction between β-CD and a cross-linking agent such as epichlorohydrin in a basic medium to give a CD cross-linked polymer. During the cross-linking reaction, a number of CD rings are interconnected and a three-dimensional macromolecular network is formed. Indeed, the numerous primary and secondary –OH groups available on CD molecules are active sites capable of forming an important number of linkages, allowing the rapid formation of a CD polymer. However, it is important to note that the cross-linking reaction is random, difficult to control, due to the fact that many primary and secondary hydroxyl functions may be involved in the reaction, resulting in complicated macromolecular

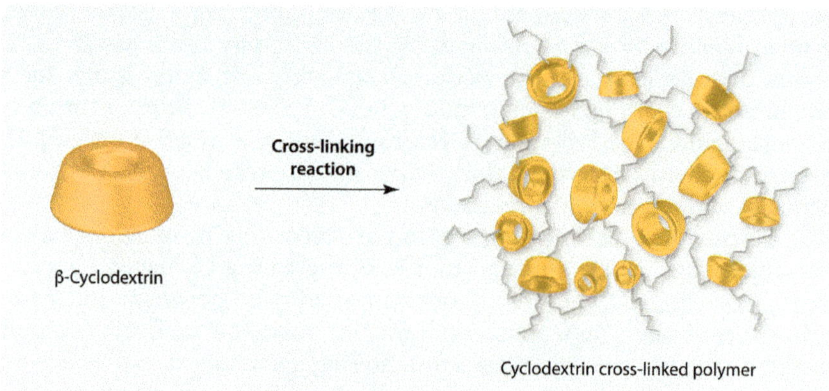

Figure 11.4 Chemical reaction between β-cyclodextrin and a cross-linking agent in a basic medium to give a cyclodextrin cross-linked polymer. During the cross-linking reaction, a number of CD rings are interconnected and a three-dimensional polymer network is formed. Another possible reaction is the self-polymerization of the cross-linking agent. The structure of a CD–epichlorohydrin polymer was proposed for the first time in 1972 by Jerald L. Hofmann (University of Louisville, Kentucky, USA).[42]

structures. Nevertheless, some parameters such as the degree of cross-linking and the morphology of the particles can be controlled.

The primary product of the reaction of CD molecules with a cross-linking agent in an alkaline medium is a heterogeneous mixture of various CD glyceryl ethers (Figure 11.4).[41,42] However, this is not a true polymer but an A–B copolymer, with A being the CD monomer and B the epichlorohydrin monomer. This was first suggested in the 1970s by Hoffman and taken up by Szejtli in the 1980s.[42] Changes in the relative mole ratio of the CD monomer to epichlorohydrin monomer alter the repeating structure of the monomer units from an A–B type copolymer to an A–B_n type copolymer, the latter containing epichlorohydrin-rich domains which are hydrophilic in nature. Indeed, in the presence of epichlorohydrin, another possible reaction is the self-polymerization of the cross-linking agent. This bifunctional cross-linking agent contains two reactive groups, which can form bonds not only with CD molecules (cross-linking step) but also with themselves *via* a polymerization step. The structure of a CD polymer is then similar to the Sephadex® resin, prepared by cross-linking dextran with epichlorohydrin (Figure 11.5). Both types of polymers are used in gel filtration, for example. However, unlike Sephadex® resin, CD polymers allow interactions not only with their macromolecular network but also with their cavity *via* inclusion complexes.

By adapting the experimental conditions of the synthesis (*e.g.* the amounts of the different reagents, the CD/epichlorohydrin molar ratio, NaOH concentration, reaction temperature, reaction time) and the reagents (type of CD, cross-linking spacer), several insoluble three-dimensional macromolecular networks with different characteristics in terms of cross-link

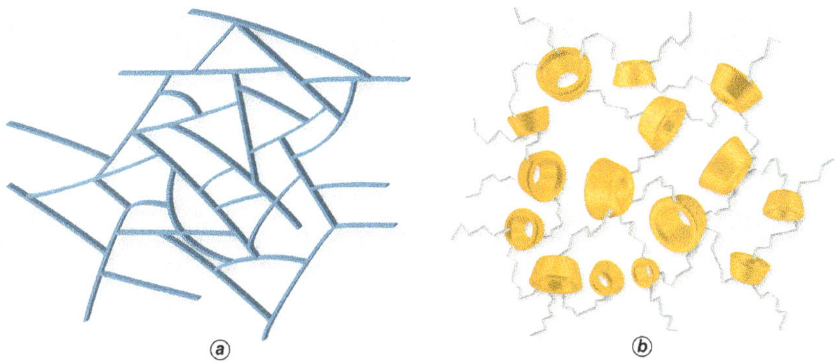

Figure 11.5 Comparison of the structure of (a) a cross-linked dextran resin (Sephadex®) and (b) a cyclodextrin cross-linked polymer. The two commercial products are obtained using epichlorohydrin as the cross-linking agent, with Sephadex® being prepared from the polysaccharide dextran.

density, structure, biosorption behaviour and chemical properties can be obtained. In general, the resulting cross-linked materials are either gels (irregularly shaped particles) or resins (spherical beads), depending mainly on the degree of cross-linking.

11.2.2 Chemical Functionalization of β-Cyclodextrin Cross-linked Polymers

In polymer science, chemical functionalization allows the introduction of reactive ligands to the surface of a material in order to improve its performance, to confer new functionalities, and to broaden its potential applications. The development of new functionalized materials based on cross-linked CDs and the study of their physical and chemical properties are a particularly active field of research, in line with societal and industrial demands in health and environmental technologies.

In principle, as already pointed out, the distinctive characteristics of CDs and the presence of numerous OH groups, at positions 2, 3 and 6 on their glycoside ring, with their different reactivity allow a certain selectivity in the chemical modifications of native CDs, CD derivatives, and materials containing CDs. There are two main relatively simple methods of preparing cross-linked and functionalized CD polymers. The first one is to carry out the cross-linking in a single step in the presence of a mixture of cross-linking agents carrying ionic groups, such as 2,3-epoxypropyltrimethylammonium chloride, or in the presence of ammonia (Figure 11.6). The second method, carried out in two separate steps, is the post-modification of a cross-linked CD polymer by grafting specific groups onto the hydrophilic CD molecule, using for example classical organic chemistry reactions such as carboxymethylation or amination in the presence or absence of an ionic reagent. All these chemical functionalizations of the CD polymer can then lead to

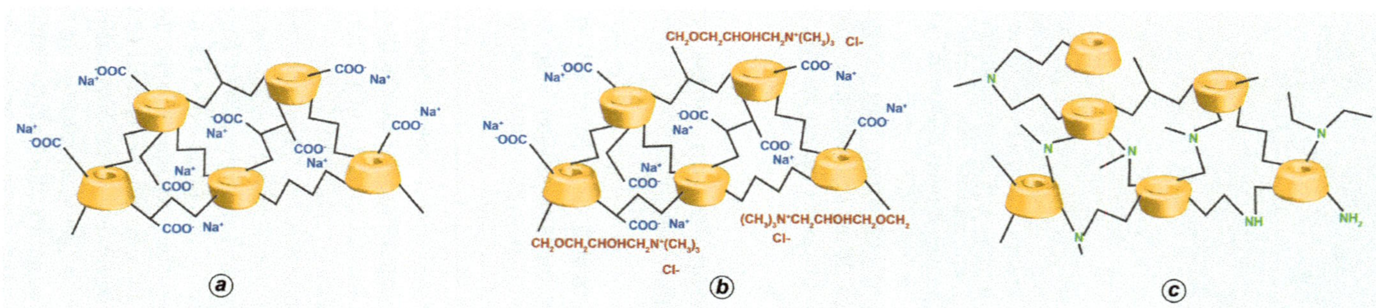

Figure 11.6 Possible structure of functionalized cyclodextrin polymers obtained by (a) carboxymethylation, (b) carboxymethylation in the presence of 2,3-epoxypropyltrimethylammonium chloride, and (c) amination.

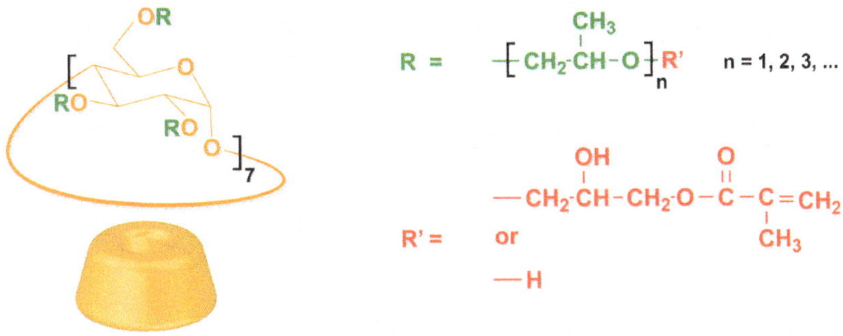

Figure 11.7 Structure of BETAW7MAHP®, a cyclodextrin monomer (2-hydroxy-3-methacryloyloxy-propyl-β-cyclodextrin (average number of double bonds per cyclodextrin unit: 2.5).

significant changes in its chemical properties, which can then make it amphoteric and/or amphiphilic in nature. Thus, several materials with different physical and chemical properties can be obtained. These chemical modifications can be demonstrated using various analytical and spectroscopic techniques, such as NMR. It is important to note that the selective derivatization of a CD polymer is not an easy task, which explains the relatively high prices, with some companies offering materials on demand (there is no industrial route yet). In addition, the grafting of ligands onto a CD polymer is generally random and the degree of substitution is low.

Other preparation methods of gels or resins include the use of CD monomers or so-called molecularly imprinted polymer techniques. For example, Janus *et al.* were the first to propose the synthesis of CD cross-linked polymers using a CD monomer such as 2-hydroxy-3-methacryloyloxy-propyl-β-cyclodextrin ready to polymerize (Figure 11.7). The authors characterized these polymers using several spectroscopic techniques (IR, Raman, mass, solid-state NMR).[89,90] The molecular mobilities of these materials were also analysed in terms of nuclear relaxation parameters such as ^{13}C spin–lattice relaxation (T_1) and H spin–lattice relaxation in the rotating frame ($T_{1\rho}$).

11.2.3 A Brief and Recent Review of the Literature on the Use of β-Cyclodextrin Cross-linked Polymers as Biosorbents

Water insoluble cross-linked-epichlorohydrin CD polymers are now produced on an industrial scale, with the β-CD polymer being the most widely produced and popular. These materials have been proposed as biosorbents to remove conventional (metals, dyes, *etc.*) and emerging (hormones, drugs, *etc.*) pollutants present in wastewater using biosorption-oriented processes, usually using a batch method. This analytical technique consists of bringing the CD polymer into contact with the solution to be treated for a certain time and under predefined conditions. The analysis of the solution before and after biosorption can then demonstrate the interest in using CD polymers to treat environmental pollutants. The results in terms of chemical efficiency (biosorption capacity, abatement or performances) are then explained by the presence of particular interactions due to the presence of CD molecules. Indeed, in these materials, the CD cavities retain their ability to form inclusion complexes although their mobility is greatly reduced. However, other interactions may also be involved depending on the type of cross-linked polymer used as the biosorbent. To explain the mechanisms of biosorption, other various interactions have been proposed in the literature, the main ones being described in Figure 11.8.

It is currently accepted that the purification performance of CD polymers is due to the presence of both the cavities of the CDs and the cooperative effects between these oligomers and the cavities induced by the particular structure of the three-dimensional macromolecular networks. A detailed discussion of this theory has been published in a comprehensive review by Crini *et al.*[41] The schematic representation given in Figure 11.9 shows the

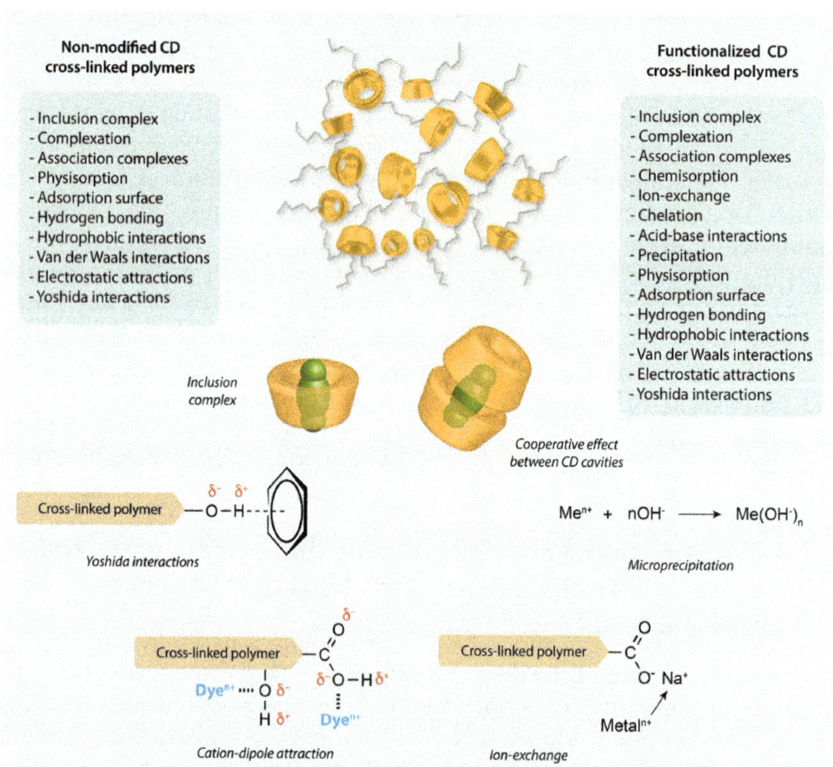

Figure 11.8 Main interactions reported in the literature to explain the mechanisms of biosorption depending on the type of cyclodextrin cross-linked polymers used as biosorbents to remove pollutants present in wastewater.

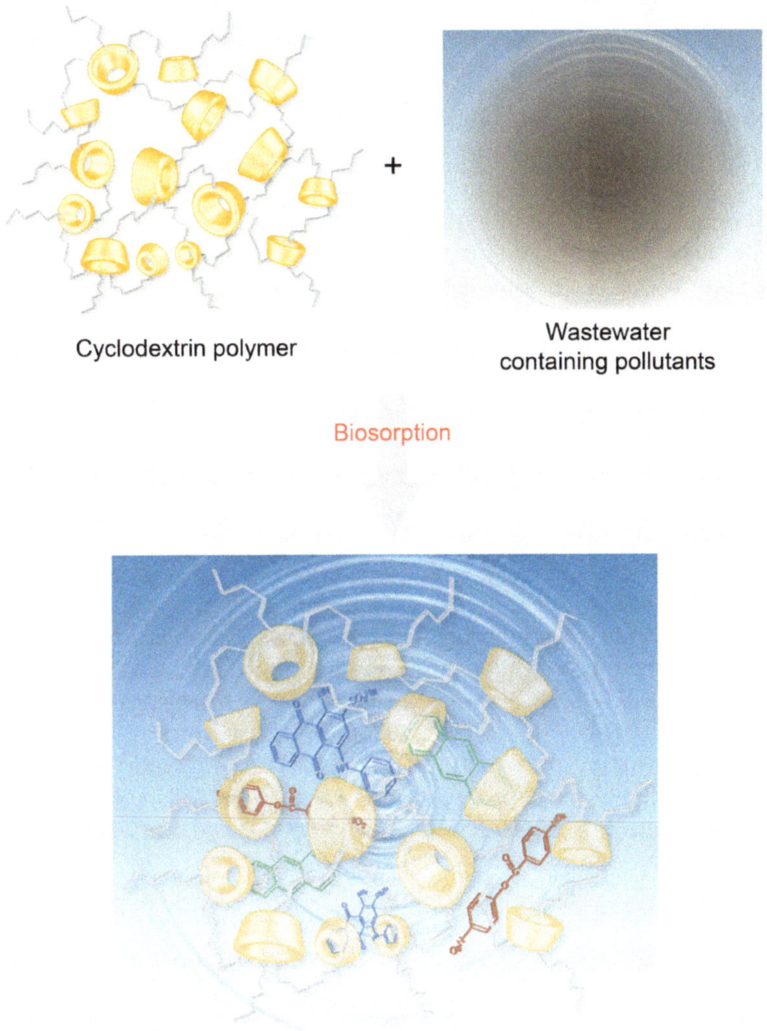

Figure 11.9 Use of a cyclodextrin polymer as a biosorbent to remove pollutants present in wastewater. This schematic representation shows the cooperative effect between cyclodextrin cavities and the role of the 3D polymer network during the removal of pollutants such as dyes and polycyclic aromatic hydrocarbons by a cyclodextrin polymer.

cooperative effect of CD cavities and the role of the 3D polymer network during the removal of pollutants such as dyes and PAHs.

The literature on the use of CD polymers in wastewater treatment is extensive. As examples, we have selected some recent reference works on the subject, focusing on examples of pollutants that pose an environmental challenge.[91–95] Hundreds of thousands of tons of pesticides, including herbicides, insecticides and fungicides, are used every year in Europe, about 90% of them in agriculture. This use of phytosanitary products, although strictly controlled, leads to the contamination of our environment and water resources. Nevertheless, some substances such as the herbicide atrazine, which is now banned (since 2004 in Europe), are regularly found in the drinking water supplied to our taps. Atrazine degrades very slowly in the environment and leaches into the groundwater very frequently. Its elimination in trace amounts is also a challenge in the field of water treatment, although methods such as adsorption on activated carbon are used. Recently, Cosma's group[91] proposed α-, β- and γ-CD epichlorohydrin copolymers to remove atrazine. Their results proved the chemical efficiency of these polymers, their ease of use and regeneration, while being in line with the principles of a circular and sustainable economy. Coupled with advanced oxidation techniques to both partially degrade the organic substances and remove their mutual interactions, the use of CD polymers could be an alternative to treat the residual presence of pesticides in wastewater. Similar findings were published by Xie's group[92] who proposed CD polymers with magnetic properties to facilitate their recovery when used in biosorption processes.

Other substances such as per- and polyfluoroalkyl substances (PFASs) are of particular interest to the scientific community and health organisations. Due to their widespread occurrence (in human and animal blood and urine and in the environment), bioaccumulation, persistence, toxicity, and associated health effects, PFASs are considered emerging contaminants. This class of compounds is a large and complex group of manufactured chemicals, consisting of a chain of linked carbon and fluorine atoms with high stability due to the carbon–fluorine bond. PFASs are used in a variety of industries, including construction, automotive, electronic, aerospace and military domains. In our daily lives, due to their resistance to heat, water, oil and grease, products based on these molecules are used to prevent food from sticking to kitchen utensils, to make clothes and carpets stain-resistant or to create more effective fire-fighting foams. Some molecules are also authorized in food packaging and food processing equipment. PFASs as contaminants are found in wastewater and groundwater. Recently, Ching *et al.*[93] proposed the use of CD polymers for PFASs. The removal capacities of these substances in groundwater were high but depended on the type and density of the cross-linkers as well as the loading of the ligands grafted onto the materials. The authors also pointed out that CD polymers had potential for regeneration and easy reuse in proof-of-concept experiments.

Fenyvesi *et al.*[94] published the first study on the industrial-scale use of CD–epichlorohydrin polymers to sequestrate endocrine disruptors from wastewater treatment plant effluents. These polymers are effective in removing substances such as estrogen-like endocrine disrupting chemicals (estrogens, hormones) and drugs (antibiotics, anti-inflammatories) that are found in trace amounts in complex mixtures. Endocrine disruptors are exogenous synthetic products, which alter the functions of the endocrine system and cause various health problems by interfering with the synthesis, metabolism, or cellular responses of natural estrogens. These substances are the subject of numerous studies on the development of techniques capable of eliminating them effectively. The results of Fenyvesi *et al.*[94] have also indicated the environmental benefits obtained after CD polymer biosorption: the decrease in pollutant concentrations in effluents resulted in a significant reduction of water toxicity. Tang *et al.*[95] also reported that more than 99% estrogens present in water can be effectively removed by β- and γ-CD polymers within 1 min, with the latter having a better biosorption capacity due to its higher inclusion ability demonstrated by NMR techniques. Solid-state ^{13}C NMR was also used to characterize the structure of polymers and their effects on performances.

Alkylphenols (APs) are also endocrine disruptors that are particularly studied.[62] These synthetic substances are used in numerous industrial, agricultural, and household applications as detergents, emulsifiers, wetting agents, dispersants, antistatic agents, demulsifiers, stabilizers, or solubilizers. In Europe, APs are strictly controlled and have been classified as priority substances because they may mimic or interfere with the body's hormones and they are linked with developmental, reproductive, brain, immune, and other problems. However, the existing water and wastewater treatment plants have not been designated for these emerging contaminants. Conventional treatments such as biodegradation, sand filtration, carbon adsorption and/or chemical oxidation in place are not effective in their removal. In contrast, CD polymers are versatile biosorbents able to retain alkylphenols present at trace concentrations.[62]

11.3 History of NMR Structure Determination of β-Cyclodextrin Cross-linked Polymers

There are several analytical and spectroscopic methods[85,87,96,97] for characterizing CD polymers (Figure 11.10), among which NMR spectroscopy plays an important role. Indeed, NMR spectroscopy can provide information on their structure and their interaction with other chemicals, such as environmental pollutants, and drugs. Torri and Crini's group was the first in the late 1990s to study cross-linked polymers of CDs in the water-insoluble form using solid-state NMR techniques.[98]

The first important result was the demonstration of the existence of two types of structures in cross-linked CD copolymers, depending mainly on the degree of cross-linking, with different molecular mobility. This was demonstrated for the first time by solid-state NMR measurements. As already pointed out, changes in the relative mole ratio of CD (monomer A) to epichlorohydrin (monomer B) altered the repeating structure of the monomer units from an A–B to an A–B_n type copolymer; the latter type contains epichlorohydrin rich domains. Using cross polarization magic angle spinning with dipolar decoupling (CPMAS) and high-resolution magic angle spinning (HRMAS) spectra and relaxation techniques (^{13}C spin–lattice relaxation, H spin–lattice relaxation in the rotating frame), Torri and Crini's group

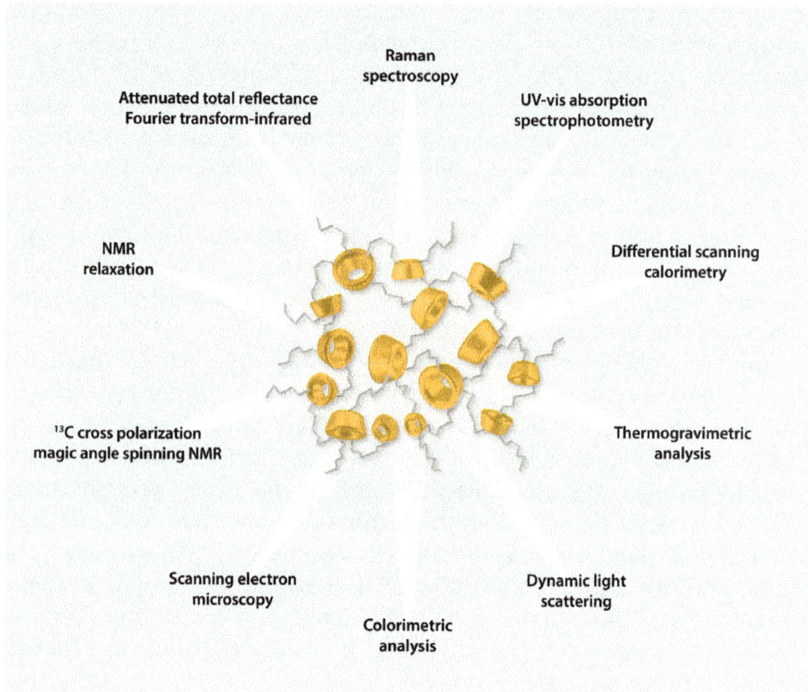

Figure 11.10 Main techniques for characterization of cyclodextrin cross-linked polymers.

showed that, in the materials, two kinds of structures existed: CD cross-linked by epoxide (epichlorohydrin) used as a cross-linking agent due to the reaction between the various groups of CD molecules and the epoxide, and polymerized epichlorohydrin due to the homopolymerization of epoxide with itself, this reagent being very reactive in a basic medium. The authors also pointed out that the polymer structure was complicated with numerous units interconnected. This was mainly explained by the important reactivity of the hydroxyalkyl groups in the C-6 position in the glycoside units.[13,42]

The second important result was related to the mobility and the degree of crystallinity of the cross-linked polymers. When the degree of cross-linking was high, the sample was mostly amorphous with an important rigidity and an important decrease of swelling properties, and cross-linking was not homogeneous. This was demonstrated using relaxation parameters such ^{13}C spin–lattice relaxation (T_1) and H spin–lattice relaxation in the rotating frame ($T_{1\rho}$). Another result concerned the hydrophilic/hydrophobic and chemical properties of the materials introduced by the functionalization, which could also be demonstrated by NMR.

All these data were later confirmed by the Spanish group of José Ramon Isasi[99–103] at Navarra University, the Canadian group of Lee D. Wilson[104–107] at the University of Saskatchewan, the Turkish group of Mustafa Yilmaz[108] at Selçuk University, and the Italian group of Andrea Mele at Politecnico di Milano.[109–111]

11.4 NMR Characterization of β-Cyclodextrin Cross-linked Polymers

Figure 11.11 compares the CPMAS spectra of a β-CD sample and a β-CD cross-linked polymer. After the cross-linking reaction, the CPMAS spectrum obtained is typical of a polysaccharide solid with an amorphous structure. Four broad signals at 101, 80, 72 and 61 ppm are observed due to the cross-linking of the glucose units, while the signals of polymerized epichlorohydrin are completely hidden. This spectrum allows only one signal to be attributed with certainty because there was a high degree of signal overlap in the range 50–90 ppm. Indeed, the resonance at 101 ppm due to the anomeric C-1 is a well-defined peak. The first assumption was that the signals from the polymerized epichlorohydrin were completely hidden by the peaks at carbons C-2,3,4, and 5 of the CD units.

The effect of the degree of cross-linking on the resolution of the CPMAS and ^{13}C NMR spectra of β-CD cross-linked polymers is shown in Figure 11.12.

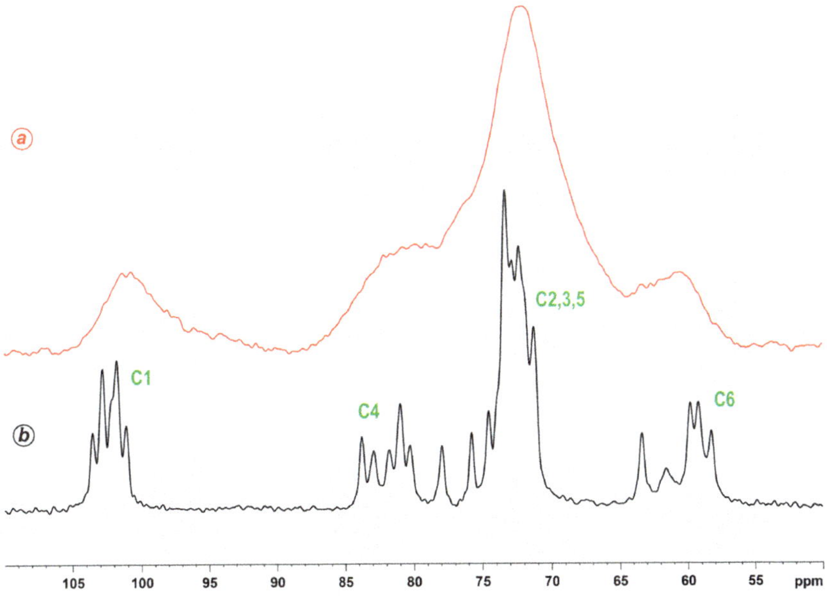

Figure 11.11 Comparison of the CPMAS spectra of a β-cyclodextrin sample (a) and a β-cyclodextrin cross-linked polymer (b). Solid-state NMR spectra were recorded with a Bruker CXP-300 spectrometer operating at 75.47 MHz and 303 K; the powder was placed in a zirconium rotor with a 7 mm diameter and a 21 mm height (spectra recorded in 1994; source: Ronzoni's archives). After the cross-linking reaction, the CPMAS spectrum is typical of a polysaccharide solid with an amorphous structure.

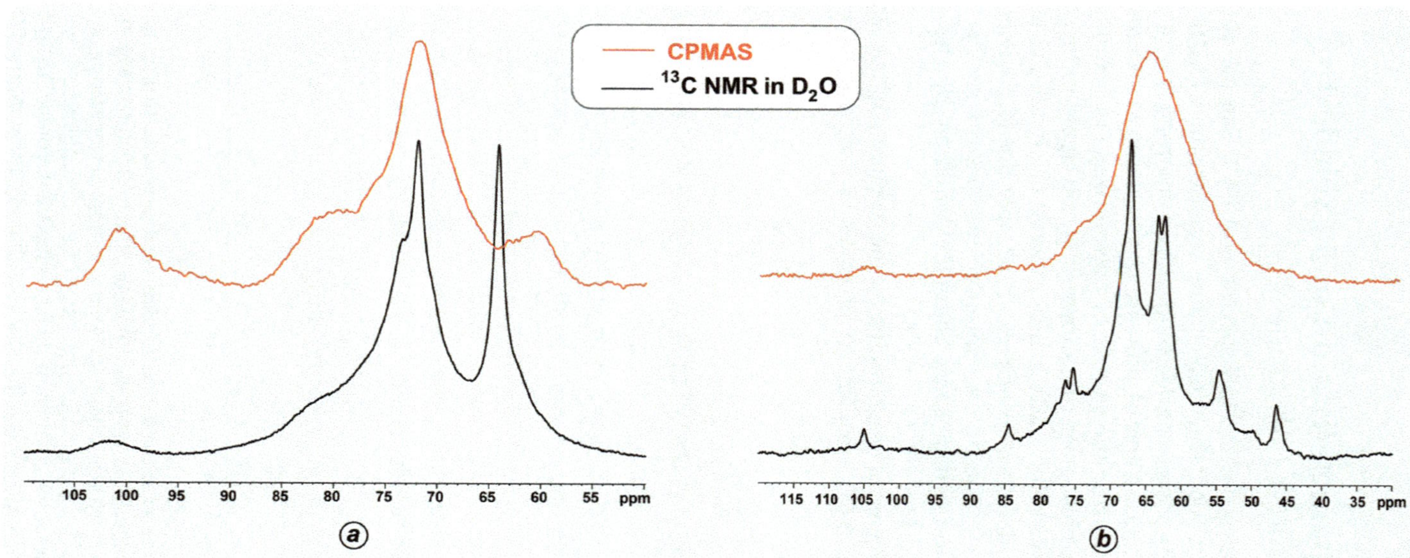

Figure 11.12 Effect of the degree of cross-linking on the resolution of the CPMAS and ^{13}C NMR spectra in D_2O of β-cyclodextrin cross-linked polymers: (a) low and (b) high cross-linking degrees. Solid-state NMR spectra were recorded with a Bruker CXP-300 spectrometer operating at 75.47 MHz and 303 K. The ^{13}C NMR spectrum in D_2O was measured at 100.61 MHz on a Bruker AMX 400 spectrometer fitted with a 10 mm broadband probe (spectra recorded in 1994; source: Ronzoni's archives).

When the degree of cross-linking increases, the resolution decreases in the CPMAS spectra due to the rigidity of the system and the amorphous structure. However, the resolution increases in the spectra recorded in solution, as revealed by the number of resonances corresponding to several CHOH and CH_2OH groups (Figure 11.12). This demonstrates the mobility of the polymerized epichlorohydrin grafted onto the surface of the cross-linked polymer. As the degree of cross-linking increases, the number of grafted epichlorohydrin chains also increases.

The CPMAS technique allows observation of both the rigid and mobile components in solid polymers, whereas the MAS technique principally shows signals arising from the mobile component. Figure 11.13 compares the MAS spectra without dipolar decoupling and ^{13}C NMR in the D_2O spectra of a β-CD cross-linked polymer. For the dry polymer sample, a very broad band typical of solids is obtained, while for the hydrated sample, sharp resonances are detected in the MAS spectrum. In particular, two hydroxyalkyl group peaks are visible at 62 and 64 ppm attributed to the CH_2OH of the glucose unit and the CH_2OH of the polymerized epichlorohydrin, respectively. However, the peak due to anomeric carbon does not appear on the MAS spectrum due to its rigidity (Figure 11.13). It is also very interesting

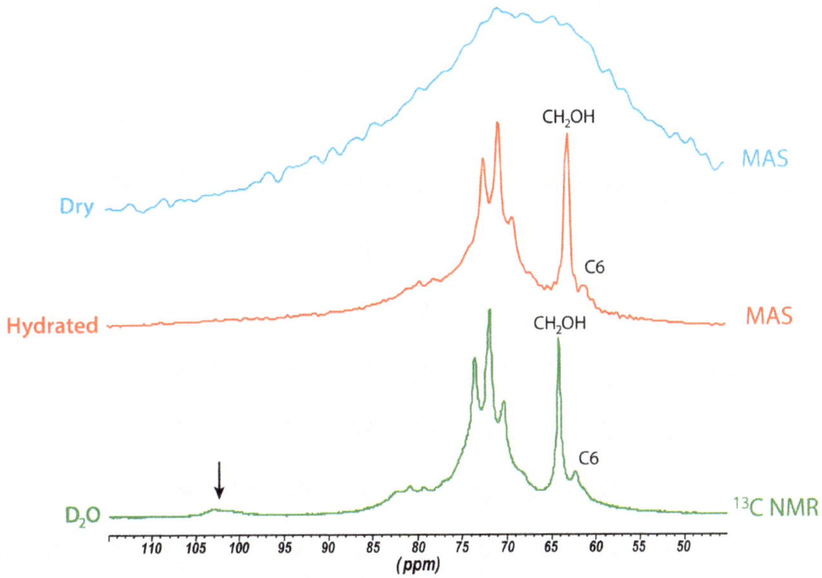

Figure 11.13 Comparison of MAS spectra without dipolar decoupling and ^{13}C NMR spectra in D_2O of a β-cyclodextrin cross-linked polymer. The MAS spectra were recorded with a Bruker Avance-300 spectrometer operating at 75.47 MHz and 303 K equipped with a 4 mm broadband CPMAS probe head. The ^{13}C NMR spectrum in D_2O was measured at 100.61 MHz on a Bruker AMX 400 spectrometer fitted with a 10 mm broadband probe (spectra recorded in 1994; source: Ronzoni's archives).

to note that the MAS spectrum is very similar to the spectrum recorded in solution, and in the last one, the anomeric signal at 103 ppm is now evident (Figure 11.13). The interaction of epichlorohydrin with the solvent (D_2O) is comparable with that observed in solution.

To confirm these results, we also recorded dipolar dephasing CPMAS spectra using different delay times for a hydrated polymer sample (Figure 11.14). As the dipolar dephasing time increases, the spectra showed a decrease in the intensity of the signals arising from the most rigid parts (anomeric carbons and carbons in position 4). For the hydrated sample, the dipolar dephasing

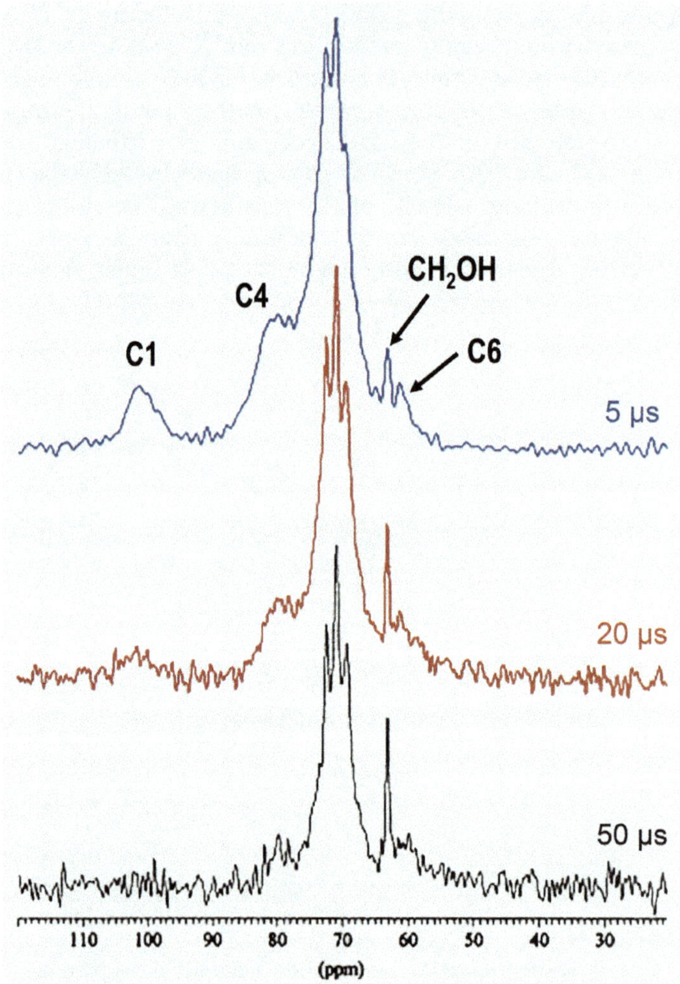

Figure 11.14 Comparison of dipolar dephasing CPMAS spectra using different delay times for a hydrated sample of a β-cyclodextrin cross-linked polymer. CPMAS spectra were recorded with a Bruker CXP-300 spectrometer operating at 75.47 MHz and 303 K (spectra recorded in 1994; source: Ronzoni's archives).

CPMAS spectra showed results comparable to those obtained with dipolar decoupling MAS techniques.

In NMR spectroscopy, the use of the distortionless enhancement by the polarization transfer (DEPT) sequence is interesting because it can distinguish between a methyl, methylene, methane or quaternary carbon present in a molecule or a macromolecule. The DEPT spectrum then permits the assignments of the peaks in the ^{13}C NMR spectrum and also of the broad peaks in the CPMAS spectrum. Figure 11.15 compares the ^{13}C NMR and DEPT-135 spectra of a β-CD cross-linked polymer in D_2O. Several methylene (CH_2) peaks are visible on the DEPT spectrum, in particular those at 62 and 64 ppm attributed to the CH_2OH of the glucose unit and the CH_2OH of the polymerized epichlorohydrin, respectively. The data of this spectrum confirm the mobility of the polymerized epichlorohydrin grafted onto the surface of the cross-linked polymer.

Figure 11.16 compares the CPMAS and MAS spectra of dry and hydrated samples for β-CD and the β-CD cross-linked polymer. For the hydrated β-CD sample, addition of water to the sample decreases the resolution as expected. After cross-linking, the CPMAS spectrum of the dry sample shows the peaks of a disordered system, while for hydrated samples two interesting

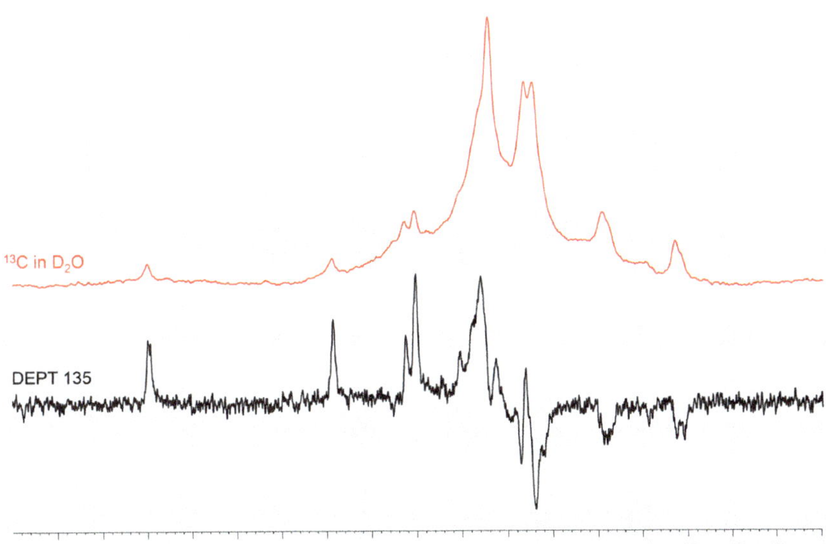

Figure 11.15 Comparison of the ^{13}C NMR and DEPT-135 spectra of a β-cyclodextrin cross-linked polymer in D_2O. The DEPT (distortionless enhancement by polarization transfer) spectrum confirms the presence of two signals attributed to the CH_2OH of the polymerized epichlorohydrin and to the hydroxymethyl of the glucose unit. The NMR spectra in D_2O were measured at 100.61 MHz on a Bruker AMX 400 spectrometer fitted with a 10 mm broadband probe (spectra recorded in 1995; source: Ronzoni's archives).

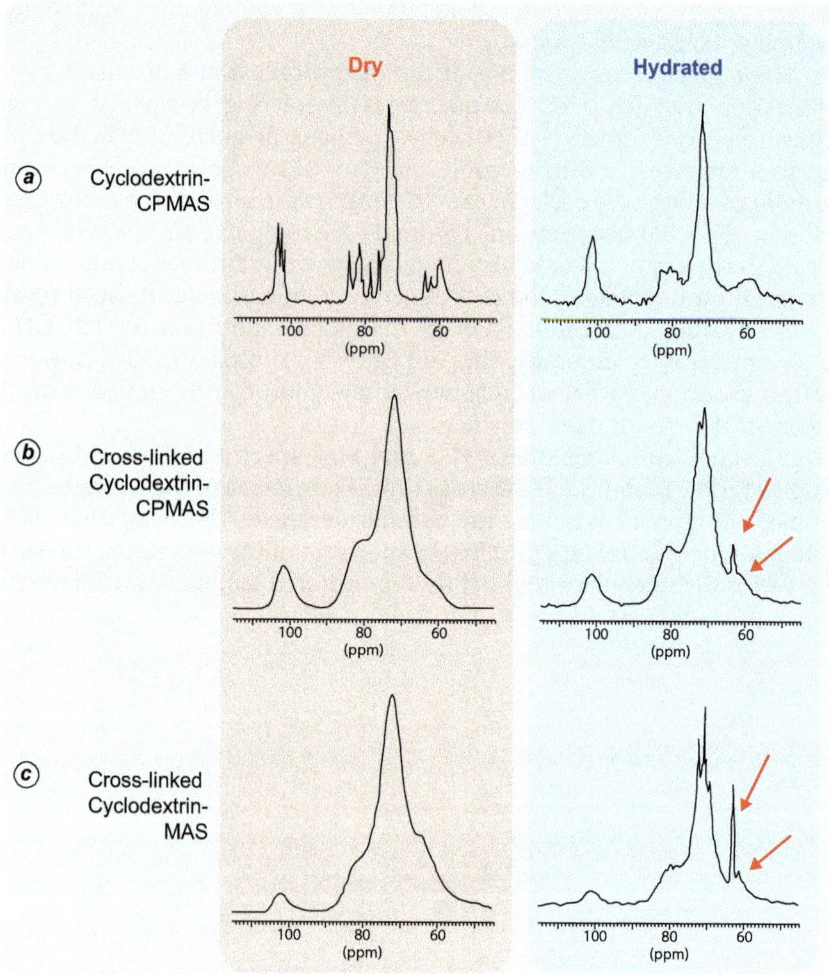

Figure 11.16 Comparison of the CPMAS and MAS spectra of dry and hydrated samples for β-cyclodextrin (a) and the β-cyclodextrin cross-linked polymer (b and c). CPMAS and MAS spectra were recorded on a Bruker Avance-300 spectrometer equipped with a 4 mm broadband CPMAS probe head; the spinning of the sample was 4000 Hz (spectra recorded in 1994; source: Ronzoni's archives).

signals appear, both in CPMAS and MAS spectra, and in particular a peak at 64 ppm due to the mobility of the hydroxymethyl group of the epichlorohydrin terminal residue.

This assignment was confirmed by studying the effect of the temperature on the CPMAS and MAS spectra of a β-CD cross-linked polymer (Figure 11.17). This variable temperature study was also conducted to overcome the overlap problem of the peaks and to confirm that the signals of polymerized epichlorohydrin were completely hidden by CD peaks. The sample temperature was raised from 30 °C to 110 °C at 20 °C increments. For each temperature, the CPMAS and MAS spectra were recorded. The results in Figure 11.17 showed that, when the samples are heated, significant changes in the resolution of both CPMAS and MAS spectra were observed due to the increase in the mobility of the CH_2OH groups at the surface of the samples with temperature. The two signals at 62 and 64 ppm clearly appeared in the CPMAS (110 °C) and MAS (30 °C) spectra. At 110 °C, the two resonances in the CPMAS spectrum, which appeared upfield to the main signal located at 70 ppm, were due to a significant increase in the mobility of the terminal CH_2OH groups. These signals are attributable to the CH_2OH at position C-6 in the glucose unit of β-CD and to the CH_2OH group of the epichlorohydrin terminal residue. The change in resolution at higher temperature reflected not only the increase of the molecular mobility of the polymer but also the temperature dependence of the chemical shifts. It is interesting to note that the MAS spectrum of the sample at 30 °C presented different ratios of the signals like the CPMAS spectrum at the same temperature. This was due to the MAS technique allowing the observation of the mobile components. So, the two signals at 62 and 64 ppm were present at a temperature of 30 °C and can be attributed to the CH_2OH groups. Finally, in the MAS spectra, the signals of the C-1 carbon of CD became very weak, whereas resolution improvement occurred in the high-field part of the spectra. All these data confirmed the presence of two components with different molecular motions in the sample. The two signals can be attributed either to the CH_2OH at position C-6 in the CD unit or to the CH_2OH group of the epichlorohydrin terminal residue. However, it was difficult to choose between these two assumptions according only to these spectra. After reaching the 110 °C temperature, the sample was cooled and the spectra were recorded. The resulting CPMAS and MAS spectra were identical to those recorded during the heating cycle, demonstrating the stability of the sample over the temperature range studied and the homogeneous nature of the polymer. However, this temperature (110 °C) was not sufficient for a better resolution.

With the use of proton NMR spectroscopy, it is possible to obtain information about the structure and dynamic behaviour of oligomers such as β-CD. However, at that time it was difficult to record a proton spectrum because, in solids with limited mobility, strong dipole–dipole interactions are predominant, causing a line broadening of up to some tens of kHz. Chemical shift anisotropy also leads to broad proton NMR spectra that are unexploitable. A possible method was the use of combined rotation and multiple spectroscopy (CRAMPS). However, this technique when used with a CPMAS probe head did not produce results in contrast to high-resolution magic angle spinning with gradients (HRMAS technique). By spinning the sample at the magic angle and using a HRMAS probe head, the classical 2D liquid experiments such as heteronuclear single quantum correlation (HSQC) $^{1}H/^{13}C$ can be applied with or without gradients. Figure 11.18 shows the first published HSQC HRMAS spectrum of a cross-linked polymer, clearly demonstrating the presence of two signals at 62 and 64 ppm attributed to

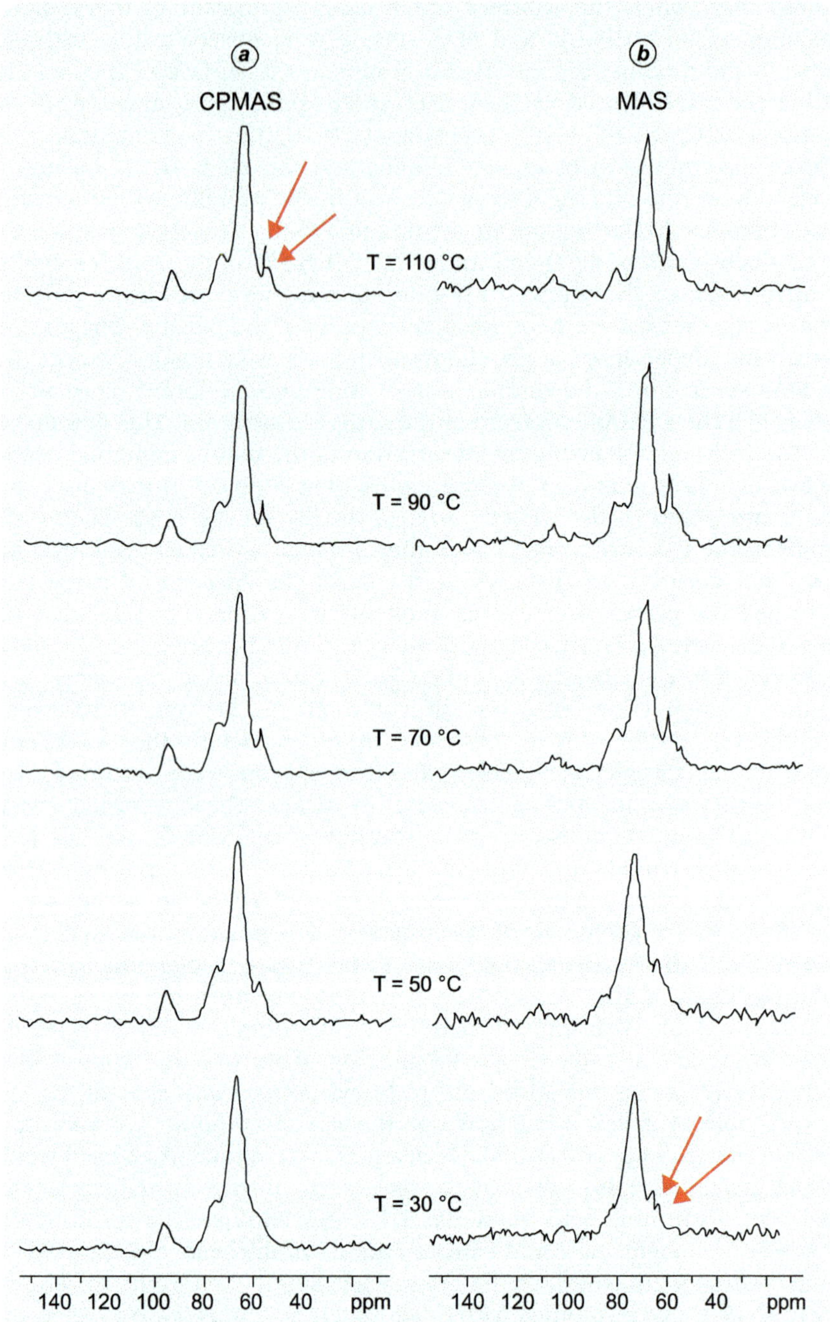
a
CPMAS
b
MAS
T = 110 °C
T = 90 °C
T = 70 °C
T = 50 °C
T = 30 °C
140 120 100 80 60 40 ppm
140 120 100 80 60 40 ppm

the polymerized epichlorohydrin and the signal to the CH_2OH of the glucose unit (with the intensities 1/3). Four other signals between 69 and 72 ppm corresponding to several CHOH and CH_2OH groups are also present.[98]

Double quantum filtered homonuclear correlation spectroscopy (DFQ-COSY), total correlation spectroscopy (TOCSY), heteronuclear multiple bond correlation (HMBC) $^{1}H/^{13}C$, and nuclear Overhauser enhancement spectroscopy (NOESY) of cross-linked CD polymers were conducted. In the NOESY spectrum (Figure 11.19), a correlation between the anomeric protons at 5.2 ppm corresponding to H-1 and the signals at 3.6 ppm corresponding to the H-2, H-3, and H-5 protons of the glucose unit of CD was observed. For the first time, MAS NMR combined with proton HRMAS and multidimensional techniques has provided structural information on cross-linked CD polymers.

The ^{13}C NMR spectra of raw and carboxymethylated β-CD cross-linked polymers with low reticulation are shown in Figure 11.20. These spectra clearly show the grafting of carboxylic groups onto the surface of the polymer with the appearance of additional bands at 175 ppm (–COOH group) and 47 ppm (–CH_2). Again, the DEPT sequence allows us to distinguish between three types of CH_2OH: those attributed to the glucose unit in position 6, those introduced by the carboxymethylation reaction, and those of the terminal chain of the polymerized epichlorohydrin (Figure 11.21). Figure 11.22 compares the ^{13}C NMR spectra of the carboxymethylated β-CD cross-linked polymer having different percentages of grafting. These data show a correlation between the intensity of the peaks and the percentage of grafting.

11.5 Usefulness of NMR Techniques to Characterize the Biosorption Mechanism

Liquid-state NMR spectroscopy is a powerful tool and a routine method for the characterization of CD derivatives and their inclusion complexes. For example, this technique is used to elucidate the geometrical accommodation of the guest within the CD cavity, to determine the formation constant of each inclusion complex, and of course to provide direct evidence of the

Figure 11.17 Effect of the temperature on the CPMAS and MAS spectra of a β-cyclodextrin cross-linked polymer. Significant changes in resolution are observed when the samples were heated due to a temperature dependence of the chemical shifts and an increase of the molecular mobility. At 110 °C, two sharp resonances at 62 and 64 ppm are evident and appear upfield to the main signal located at 70 ppm. These signals are attributable to the CH_2OH at position C-6 in the glucose unit of β-cyclodextrin and to the CH_2OH group of the epichlorohydrin terminal residue. CPMAS and MAS spectra were recorded on a Bruker Avance-300 spectrometer equipped with a 4 mm broadband CPMAS probe head; the spinning speed of the sample was 4000 Hz (spectra recorded in 1994; source: Ronzoni's archives).

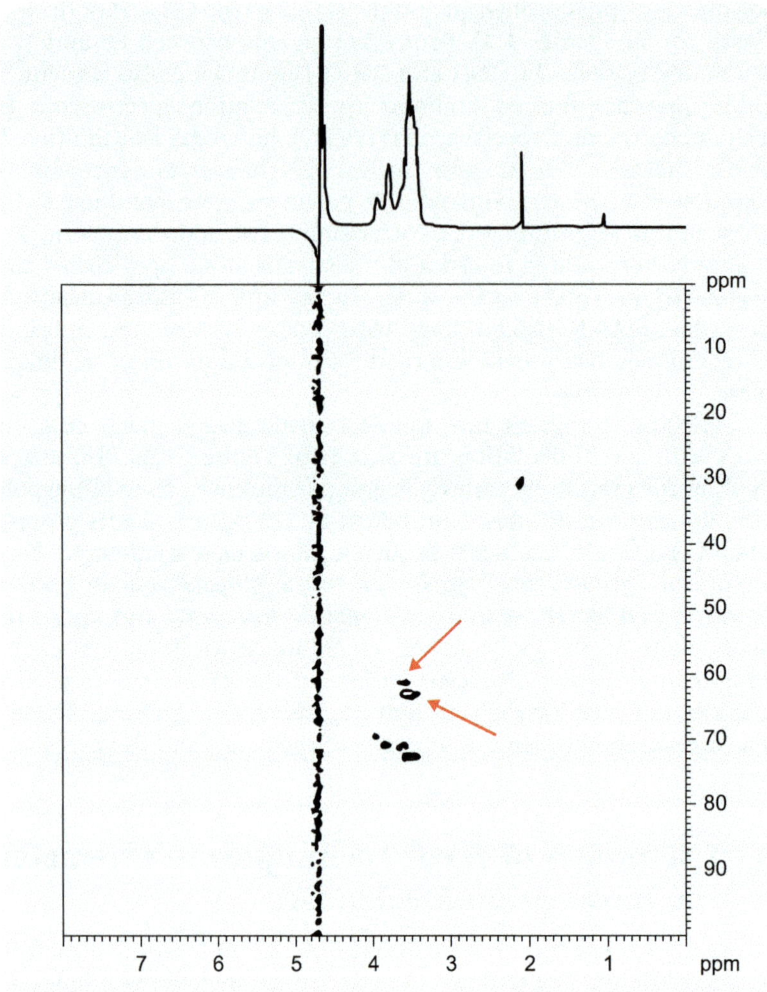

Figure 11.18 Heteronuclear single quantum correlation (HSQC) spectrum of a β-cyclodextrin cross-linked polymer using a HRMAS probe head in D_2O. The two most important signals at 62 and 64 ppm are attributed to the CH_2OH of the glucose unit and to the CH_2OH of the polymerized epichlorohydrin, respectively (with the intensities 1/3). Four other signals between 69 and 72 ppm corresponding to several CHOH and CH_2OH groups are also present. The NMR experiment was recorded with a Bruker Avance-400 spectrometer equipped with a 4 mm $^{13}C/^{2}H/^{1}H$ probe head provided with a gradient coil. The HSQC spectrum was acquired on 1024 points in the F2 dimension with 128 scans and 5.2 kHz spectral width, and the recycle delay was 3 s. In the F1 dimension, 256 increments were used with a spectral width of 10 kHz (spectra recorded in 1996; source: Ronzoni's archives).

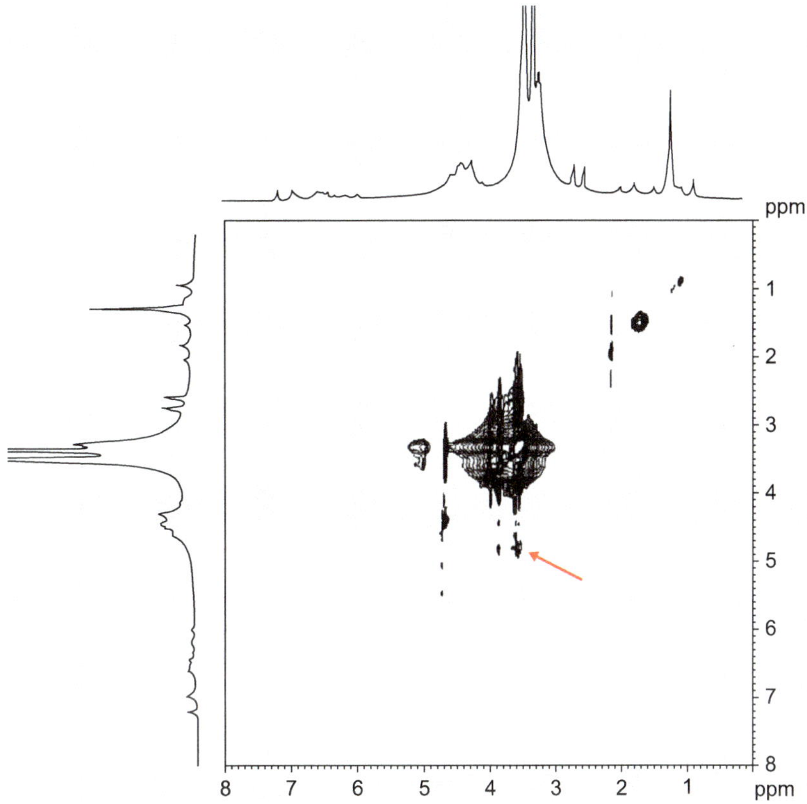

Figure 11.19 Nuclear Overhauser enhancement spectroscopy (NOESY) spectrum of a β-cyclodextrin cross-linked polymer. The NMR experiment was recorded with a Bruker Avance-400 spectrometer equipped with a 4 mm $^{13}C/^{2}H/^{1}H$ probe head provided with a gradient coil: the spectrum was registered on 2048 points in the F2 dimension with 96 scans and 4.4 kHz spectral width; the recycle delay was 1 s and a mixing time of 200 ms was used. Two hundred and fifty-six experiments were done in the F1 dimension using the same spectral width as in F2 (spectra recorded in 1996; source: Ronzoni's archives).

inclusion of the guest in the CD cavity. A large number of results using various NMR techniques (*e.g.* COSY, HMQC, ROESY and DOSY techniques, NMR titrations) have been published in the literature. ^{13}C, ^{15}N, ^{19}F and ^{31}P solid-state NMR spectroscopy has also already proved to be a valuable technique to the pharmaceutical industry, for example, in studies of inclusion complexes involving drugs and native CDs and CD derivatives. However, for cross-linked CD polymers, the data published in the literature are more scarce.[110–112]

Torri and Crini's research group was the first to demonstrate that solid-state NMR spectroscopy was also useful for studying the interactions

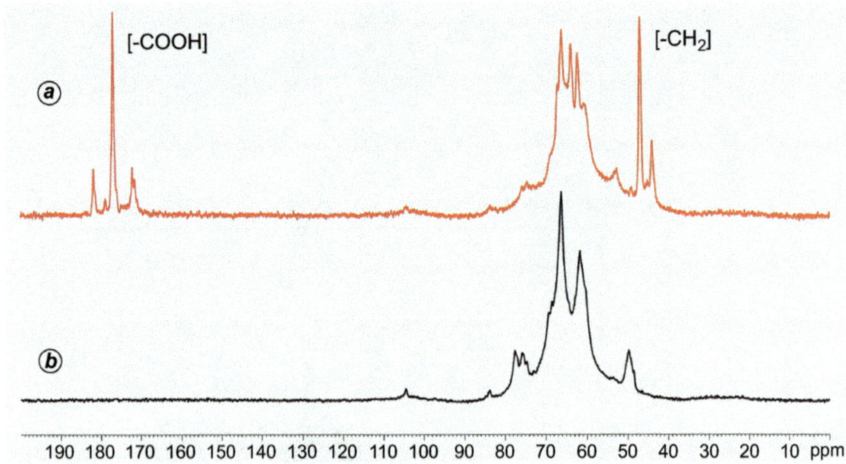

Figure 11.20 Comparison of the ^{13}C NMR spectra of (b) raw and (a) carboxymethylated β-cyclodextrin cross-linked polymers with low reticulation. These spectra clearly show the grafting of carboxylic groups onto the surface of the polymer with the appearance of additional bands at 175 ppm (–COOH group) and 47 ppm (–CH_2). The NMR spectra in D_2O were measured at 100.61 MHz on a Bruker AMX 400 spectrometer fitted with a 10 mm broadband probe (spectra recorded in 1997; source: Ronzoni's archives).

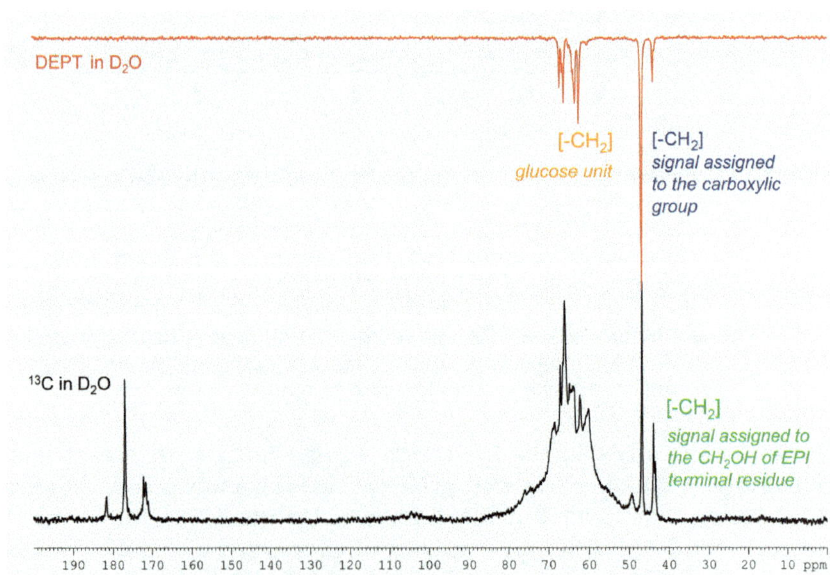

Figure 11.21 Comparison of the ^{13}C NMR and DEPT-135 spectra of a carboxymethylated β-cyclodextrin cross-linked polymer in D_2O. The DEPT spectrum permits the assignments of the three methylene groups, in particular that of the carboxylic groups at 47 ppm. The NMR spectra in D_2O were measured at 100.61 MHz on a Bruker AMX 400 spectrometer fitted with a 10 mm broadband probe (spectra recorded in 1997; source: Ronzoni's archives).

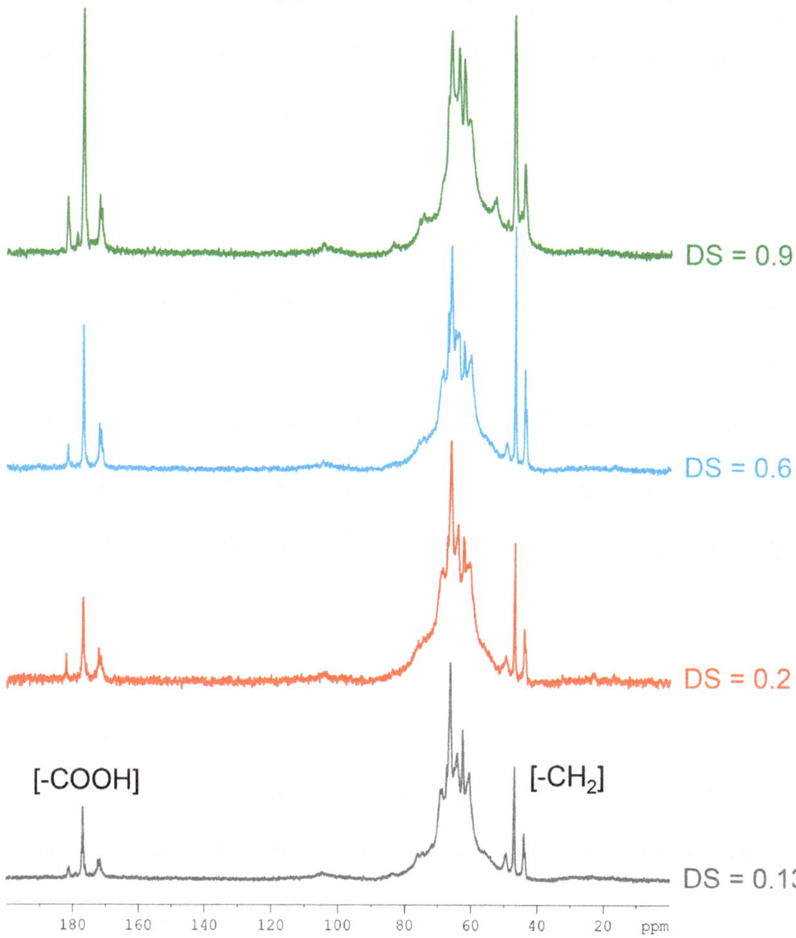

Figure 11.22 Comparison of the ^{13}C NMR spectra of the carboxymethylated β-cyclodextrin cross-linked polymer having different percentages of grafting in D_2O (DS corresponds to the average number of carboxymethyl groups per anhydroglucose unit). These data show a correlation between the intensity of the peaks and the percentage of grafting. The NMR spectra in D_2O were measured at 100.61 MHz on a Bruker AMX 400 spectrometer fitted with a 10 mm broadband probe (spectra recorded in 1999; source: Ronzoni's archives).

between the macromolecular network of a CD polymer and the encapsulated pollutants.[13,39,41,98] Figure 11.23 shows the proton HRMAS spectra in D_2O of a β-CD cross-linked polymer before and after biosorption of β-naphthol. The 1D proton spectrum of the insoluble polymer swollen in D_2O was obtained using a HRMAS probe head and a CPMG sequence with presaturation. The D_2O solvent was added to the sample directly inside the

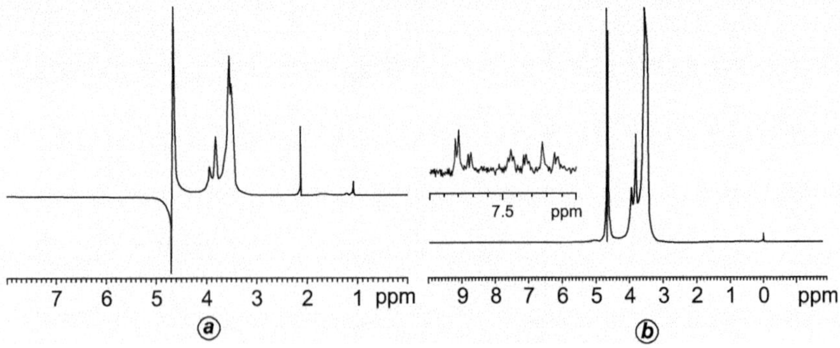

Figure 11.23 Proton high-resolution magic angle spinning with gradients (HRMAS) spectra in D_2O of a β-cyclodextrin cross-linked polymer (a) before and (b) after biosorption of β-naphthol (conditions: 100 μg adsorbed per g of polymer). The NMR experiments were conducted with a Bruker Avance-400 spectrometer equipped with a 4 mm $^{13}C/^{2}H/^{1}H$ probe head provided with a gradient coil; the spinning speed of the sample was 4000 Hz (spectra recorded in 1996; source: Ronzoni's archives).

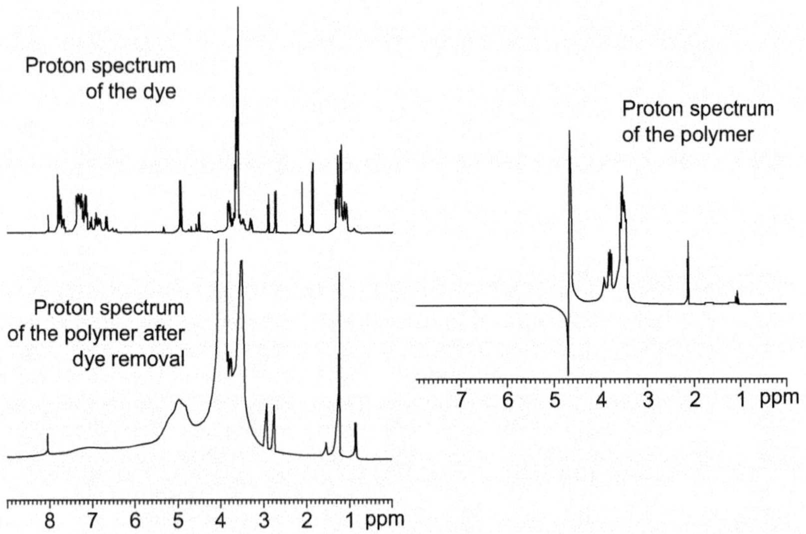

Figure 11.24 Proton high-resolution magic angle spinning with gradients (HRMAS) spectra in DMF of a β-cyclodextrin cross-linked polymer before and after biosorption of a dye (conditions: 200 mg adsorbed per g of polymer). The NMR experiments were recorded with a Bruker Avance-400 spectrometer equipped with a 4 mm $^{13}C/^{2}H/^{1}H$ probe head provided with a gradient coil; the spinning speed of the sample was 4000 Hz (spectra recorded in 1996; source: Ronzoni's archives).

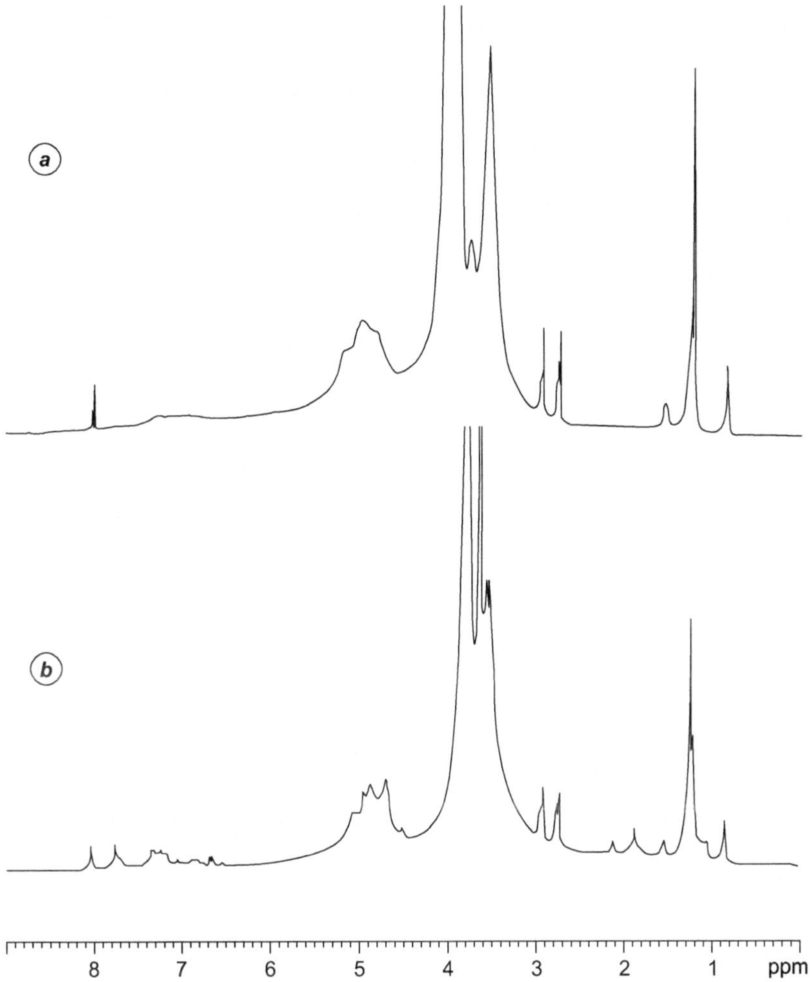

Figure 11.25 Proton high-resolution magic angle spinning with gradients (HRMAS) spectra in DMF of a β-cyclodextrin cross-linked polymer after (a) biosorption of a dye and (b) after adding a small quantity of the same dye (source: Ronzoni's archives). The NMR experiments were recorded with a Bruker Avance-400 spectrometer equipped with a 4 mm $^{13}C/^{2}H/^{1}H$ probe head provided with a gradient coil; the spinning speed of the sample was 4000 Hz (spectra recorded in 1996; source: Ronzoni's archives).

rotor in order to introduce some degree of mobility into the cross-linked polymer and to lock the spectrometer during the experiment. Before biosorption, at least four resonances are visible in the spectrum. Running the same experiment with a liquid probe head, only a broad peak, including the water signal, was observed. After biosorption of beta-naphthol by the polymer, we can observe the presence of several peaks in the range 7–8 ppm due to the aromatic protons of beta-naphthol complexed. Unfortunately, the low amount of contaminant adsorbed limited the study to the 1D proton experiment.

Figure 11.24 shows another example obtained with an industrial textile dye. This figure compares the 1D spectra of the textile dye, and the polymer before and after dye removal. In these experiments, dimethylformamide (DMF) was added to the sample directly inside the rotor. The large peaks in the proton HRMAS spectrum in DMF used to swell the sample obtained after dye removal are attributed to the aromatic and aliphatic protons of the dye molecules adsorbed by the polymer. To confirm the presence of an inclusion complex between the polymer and the dye, a small amount of dye was added to the complex (Figure 11.25). The resulting spectrum is the sum of the proton spectrum of the non-complexed dye and the proton spectrum of the complexed one. Unfortunately, the difficulties of observing the signal of the dye prevented us from analysing the interactions between the polymer and the dye. Other examples can be found in the study by Crini *et al.*[98]

11.6 Conclusion

At the end of the 1990s, for the first time, NMR spectra of insoluble β-cyclodextrin polymers obtained using CPMAS, MAS and HRMAS spectroscopy were reported by our group. The results demonstrated that these techniques were useful for characterizing cross-linked materials having a limited mobility. Two kinds of structures, cross-linked cyclodextrin and polymerized epichlorohydrin, were highlighted. In a biosorption process using cyclodextrin polymers, HRMAS spectroscopy also proved to be a useful technique to determine the relationship between their retention behaviour and molecular structure, and to analyse the interactions involved in the mechanism. Currently, solid-state NMR techniques are routinely used to characterize all types of cyclodextrin-based materials.[33,41,113]

Abbreviations

APs	Alkylphenols
CDs	Cyclodextrins
CPMAS	Cross polarization magic angle spinning
DEPT	Distortionless enhancement by polarization transfer
DMF	Dimethylformamide
HRMAS	High-resolution magic angle spinning
HSQC	Heteronuclear single quantum correlation
MAS	Magic angle spinning
NOESY	Nuclear Overhauser enhancement spectroscopy
PFASs	Per- and polyfluoroalkyl substances
PHAs	Polycyclic aromatic hydrocarbons

Acknowledgements

The authors are grateful to the late Professor Benito Casu (1927–2016), one of the pioneers in the dissemination of knowledge on NMR of cyclodextrins in the 1960s and 1970s, for initiating the results we present in this chapter. Professor Casu, as Director of the G. Ronzoni Institute, had suggested to Grégorio Crini and Giangiacomo Torri in the mid-1990s to open a new research theme on cyclodextrin polymers. Grégorio Crini would like to thank the *Région Bourgogne Franche-Comté* and the FEDER (Fineau project) for financial support and the research grants awarded to postdoctoral Dario Lacalamita, doctoral student Chiara Mongioví, and guest researcher Cesare Cosentino.

References

1. G. Crini, *Chem. Rev.*, 2014, **114**, 10940.
2. D. French, The Schardinger dextrins, in *Advances in Carbohydrate Chemistry*, ed. D. L. Wolfrom, Academic Press Inc., New York, 1957, vol. 12, pp. 189–260.
3. G. V. Caesar, The Schardinger dextrins, in *Starch and its Derivatives*, ed. J. A. Radley, Chapman and Hall Ltd., EC4, London, 4th edn, 1968, ch. X, pp. 290–305.
4. K. Kainuma, Starch oligosaccharides: Linear, branched and cyclic, in *Starch – Chemistry and Technology*, ed. R. L. Whistler, J. N. BeMiller and E. F. Paschall, Academic Press Ltd., London, 1984, ch. V, pp. 125–152.
5. R. J. Clarke, J. H. Coates and S. F. Lincoln, *Advance in Carbohydrate Chemistry*, 1988, vol. 46, p. 205.
6. S. Li and W. C. Purdy, *Chem. Rev.*, 1992, **92**, 1457.
7. J. Szejtli, *Chem. Rev.*, 1998, **98**, 1743.
8. D. Armspach, G. Gattuso, R. Koeniger and J. F. Stoddard, Cyclodextrins, in *Bioorganic Chemistry: Carbohydrates*, ed. S. M. Hecht, Oxford University Press Inc., New York, 1999, pp. 458–488.
9. J. Szejtli, *Pure Appl. Chem.*, 2004, **76**, 1825.
10. H. Viernstein and P. Wolschann, *ScienceAsia*, 2020, **46**, 254.
11. B. G. Poulson, Q. A. Alsulami, A. Sharfalddin, E. F. El Agammy, F. Mouffouk, A. H. Emwas, L. Jaremko and M. Jaremko, *Polysaccharides*, 2022, **3**, 1.
12. A. Cid-Samamed, J. Rakmai, J. C. Mejuto, J. Simal-Gandara and G. Astray, *Food Chem.*, 2022, **384**, 132467.
13. G. Crini, C. Cosentino, S. Bertini, A. M. Naggi, G. Torri, C. Vecchi, L. Janus and M. Morcellet, *Carbohydr. Res.*, 1998, **308**, 37.
14. N. Morin-Crini, S. Fourmentin, E. Fenyvesi, E. Lichtfouse, G. Torri, M. Fourmentin and G. Crini, *Environ. Chem. Lett.*, 2021, **19**, 2581.
15. M. Petitjean, I. X. García-Zubiri and J. R. Isasi, *Environ. Chem. Lett.*, 2021, **19**, 3465.

16. Y. Liu, T. Lin, C. Cheng, Q. Wang, S. Lin, C. Liu and X. Han, *Molecules*, 2021, **26**, 1090.
17. A. Dodero, G. Schlatter, A. Hébraud, S. Vicini and M. Castellano, *Carbohydr. Polym.*, 2021, **264**, 118042.
18. Z. Liu, L. Ye, J. Xi, J. Wang and Z. Feng, *Prog. Polym. Sci.*, 2021, **118**, 101408.
19. C. Dang, V. Jiranek, D. K. Taylor and K. L. Wilkinson, *Molecules*, 2020, **25**, 910.
20. F. M. Bezerra, M. J. Lis, H. B. Firmino, J. G. Dias da Silva, R. D. C. S. Curto Valle, J. A. Borges Valle, F. A. P. Scacchetti and A. L. Tessaro, *Molecules*, 2020, **25**, 3624.
21. V. Kadam, I. L. Kyratzis, Y. B. Truong, L. Wang and R. Padhye, *J. Appl. Polym. Sci.*, 2020, **137**, e49228.
22. G. Crini, S. Fourmentin, N. Morin-Crini and M. Fourmentin, *Techniques de l'ingénieur Collection Recherche et Innovation*, 2019, **232**, 1.
23. G. Crini, S. Fourmentin, É. Fenyvesi, G. Torri, M. Fourmentin and N. Morin-Crini, *Environ. Chem. Lett.*, 2018, **16**, 1361.
24. C. Folch-Cano, M. Yazdani-Pedram and C. Olea-Azar, *Molecules*, 2014, **19**, 14066.
25. J. D. Heidel and T. Schluep, *J. Drug Delivery*, 2012, **2012**, 262731.
26. C. O. Mellet, J. M. G. Fernández and J. M. Benito, *Chem. Soc. Rev.*, 2011, **40**, 1586.
27. S. Tan, K. Ladewig, Q. Fu, A. Blencowe and G. G. Qiao, *Macromol. Rapid Commun.*, 2014, **35**, 1166.
28. A. Matencio, A. R. Pedrazzo, A. Difalco, S. Navarro-Orcajada, Y. K. Monfared, I. Conesa, A. Rezayat, J. M. López-Nicolás and F. Trotta, *Polymers*, 2021, **13**, 4226.
29. S. Y. Qie, Y. Hao, Z. J. Liu, J. Wang and J. Wi, *Acta Chim. Sin.*, 2020, **78**, 232.
30. F. Caldera, M. Tannous, R. Cavalli, M. Zanetti and F. Trotta, *Int. J. Pharm.*, 2017, **531**, 470.
31. A. P. Sherje, B. R. Dravyakar, D. Kadam and M. Jadhav, *Carbohydr. Polym.*, 2017, **173**, 37.
32. Y. Han, W. Liu, J. Huang, S. Qiu, H. Zhong, D. Liu and J. Liu, *Pharmaceutics*, 2018, **10**, 271.
33. M. T. Sikder, M. M. Rahman, M. Jakariya, T. Hosokawa, M. Kurasaki and T. Saito, *Chem. Eng. J.*, 2019, **355**, 920.
34. I. Krabicová, S. L. Appleton, M. Tannous, G. Hoti, F. Caldera, A. Rubin Pedrazzo, C. Cecone, R. Cavalli and F. Trotta, *Polymers*, 2020, **12**, 1122.
35. G. Utzeri, L. Verissimo, D. Murtinho, A. A. C. C. Pais, F. X. Perrin, F. Ziarelli, T. V. Iordache, A. Sarbu and A. J. M. Valente, *Molecules*, 2021, **26**, 1426.
36. G. Utzeri, P. M. C. Matias, D. Murtinho and A. J. M. Valente, *Front. Chem.*, 2022, **10**, 18.
37. I. Roy and J. F. Stoddart, *Acc. Chem. Res.*, 2021, **54**, 1440.

38. S. Iravani and R. S. Varma, *Appl. Sci.*, 2022, **12**, 4182.
39. N. Morin-Crini and G. Crini, *Prog. Polym. Sci.*, 2013, **38**, 344.
40. B. Gidwani and A. Vyas, *Colloids Surf., B*, 2014, **114**, 130.
41. N. Morin-Crini, P. Winterton, S. Fourmentin, L. D. Wilson, E. Fenyvesi and G. Crini, *Prog. Polym. Sci.*, 2018, **78**, 1.
42. G. Crini, *Environ. Chem. Lett.*, 2021, **19**, 2383.
43. S. Safapour, M. Mazhar, M. Nikanfard and F. Liaghat, *Int. J. Environ. Sci. Technol.*, 2021, **19**, 5753.
44. B. Tian, S. Hua, Y. Tian and J. Liu, *Environ. Sci. Pollut. Res.*, 2021, **28**, 1317.
45. K. Köse, M. Tüysüz, D. Aksüt and L. Uzun, *Environ. Sci. Pollut. Res.*, 2022, **29**, 182.
46. C. Belenguer-Sapiña, E. Pellicer-Castell, A. R. Mauri-Aucejo, E. F. Simó-Alfonso and P. Amorós, *Nanomaterials*, 2021, **11**, 7.
47. T. F. Cova, D. Murthino, R. Aguado, A. A. C. C. Pais and A. J. M. Valente, *Polysaccharides*, 2021, **2**, 16.
48. E. Fenyvesi and T. Sohajda, *Environ. Sci. Pollut. Res.*, 2022, **29**, 20085.
49. X. Qin, L. Bai, Y. Tan, L. Li, F. Song and Y. Wang, *Chem. Eng. J.*, 2019, **372**, 1007.
50. J. Qu, Y. Yuan, Q. Meng, G. Zhang, F. Deng, L. Wang, Y. Tao, Z. Jiang and Y. Zhang, *J. Hazard. Mater.*, 2020, **400**, 123142.
51. Z. Duan, Y. Li, M. Zhang, H. Bian, Y. Wang, L. Zhu and D. Xia, *J. Cleaner Prod.*, 2020, **256**, 120308.
52. I. Mallard, D. Landy and S. Fourmentin, *Appl. Sci.*, 2020, **10**, 4679.
53. J. A. Pellicer, M. I. Rodríguez-López, M. I. Fortea, V. M. Gómez-López, D. Auñón, E. Núñez-Delicado and J. A. Gabaldón, *Polymers*, 2020, **12**, 188.
54. Y. Tu, G. Xu, L. Jiang, X. Hu, J. Xu, X. Xie and A. Li, *Chem. Eng. J.*, 2020, **382**, 123015.
55. H. Yamasaki, A. Odamura, Y. Makihata and K. Fukunaga, *Polym. Bull.*, 2023, **80**, 3265.
56. F. J. Peñas, A. Romo and J. R. Isasi, *J. Polym. Environ.*, 2022, **30**, 1189.
57. M. Wang, G. Li, C. Xia, X. Jing, R. Wang, Q. Liu and X. Cai, *Chem. Eng. J.*, 2021, **411**, 128489.
58. R. Romita, V. Rizzi, J. Gubitosa, J. A. Gabaldón, M. I. Fortea, T. Gómez-Morte, V. M. Gómez-López, P. Fini and P. Cosma, *Chemosphere*, 2021, **283**, 131238.
59. T. Gómez-Morte, V. M. Gómez-López, C. Lucas-Abellán, I. Martínez-Alcalá, M. Ayuso, S. Martínez-López, N. Montemurro, S. Pérez, D. Barceló, P. Fini, P. Cosma, J. P. Cerón-Carrasco, M. I. Fortea, E. Núñez-Delicado and J. A. Gabaldón, *J. Hazard. Mater.*, 2021, **402**, 123504.
60. K. Hemine, N. Łukasik, M. Gazda and I. Nowak, *J. Hazard. Mater.*, 2021, **418**, 126286.

61. M. Erdos, R. Hartkamp, T. J. H. Vlugt and O. A. Moultos, *J. Phys. Chem.*, 2020, **124**, 1218.
62. G. Crini, C. Bradu, M. Fourmentin, C. Cosentino, A. R. L. Ribeiro and N. Morin-Crini, *Environ. Sci. Pollut. Res.*, 2022, **29**, 171.
63. L. Xiao, C. Ching, Y. Ling, M. Nasiri, M. J. Klemes, T. M. Reineke, D. E. Helbling and W. R. Dichtel, *Macromolecules*, 2019, **52**, 3747.
64. M. J. Klemes, L. P. Skala, M. Ateia, B. Trang, D. E. Helbling and W. R. Dichtel, *Acc. Chem. Res.*, 2020, **53**, 2314.
65. Z. Duan, S. Wei, H. Bian, C. Guan, L. Zhu and D. Xia, *Sep. Purif. Technol.*, 2021, **254**, 117643.
66. F. Nkinahamira, A. Alsbaiee, Q. Zeng, Y. Li, Y. Zhang, M. Feng, C. P. Yu and Q. Sun, *Water Res.*, 2020, **181**, 115857.
67. V. Mohos, Z. Faisal, E. Fliszár-Nyúl, L. Szente and M. Poór, *Environ. Sci. Pollut. Res.*, 2022, **29**, 210.
68. T. F. G. G. Cova, D. Murthino, A. A. C. C. Pais and A. J. M. Valente, Cyclodextrin-based materials for removing micropollutants from wastewater, *Curr. Org. Chem.*, 2018, **22**, 2146.
69. P. Kumari, P. Singh, A. Singhal and A. Gautam, Cyclodextrin-based nanostructured materials for sustainable water remediation applications, *Environ. Sci. Pollut. Res.*, 2020, **27**, 32432.
70. W. C. E. Schofield, C. D. Bain and J. P. S. Badyal, *Chem. Mater.*, 2012, **24**, 1645.
71. G. Crini, *Prog. Polym. Sci.*, 2005, **30**, 38.
72. X. K. Yao, P. Huang and Z. H. Nie, *Prog. Polym. Sci.*, 2019, **93**, 1.
73. J. Liu, B. Tian, Y. Liu and J. B. Wan, *Int. J. Mol. Sci.*, 2021, **22**, 13516.
74. R. W. M. Krause, B. B. Mamba, F. M. Bambo and T. J. Malefetse, Cyclodextrins polymers: Synthesis and application in water treatment, in *Cyclodextrins: Chemistry and Physics*, ed. J. Hue, Transworld Research Network, Kerala, India, 2011, ch. 9, pp. 1–25.
75. B. Tian and J. Liu, *New J. Chem.*, 2020, **44**, 9137.
76. M. A. Przybyla, G. Yilmaz and R. Becer, *Polym. Chem.*, 2020, **11**, 7582.
77. F. Seidi, Y. Jin and H. Xiao, *Carbohydr. Polym.*, 2020, **242**, 116277.
78. G. Crini and M. Morcellet, *J. Sep. Sci.*, 2002, **25**, 789.
79. G. Crini, *Bioresour. Technol.*, 2003, **90**, 193.
80. G. Crini and H. N. Peindy, *Dyes Pigm.*, 2006, **70**, 204.
81. G. Crini, *Bioresour. Technol.*, 2006, **97**, 1061.
82. J. Solms and R. H. Egli, *Helv. Chim. Acta*, 1965, **48**, 1225.
83. C. L. Wiedenhof, J. N. J. J. Lammers and C. L. van Panthaelon van Eck, *Die Stärke*, 1969, **21**, 119.
84. J. L. Hoffmann, *Anal. Biochem.*, 1970, **37**, 209.
85. J. Szejtli, *Cyclodextrins and their inclusion complexes*, Akadémiai Kiadó, Budapest, 1982.
86. J. Szejtli, *Starch/Stärke*, 1984, **36**, 429.
87. J. Szejtli, *Cyclodextrin Technology*, Kluwer Academic Publishers, Dordrecht, 1988.

88. E. Renard, A. Deratani, G. Volet and B. Sébille, *Eur. Polym. J.*, 1997, **33**, 49.
89. L. Janus, G. Crini, V. El-Rezzi, M. Morcellet, A. Cambiaghi, G. Torri, A. Naggi and C. Vecchi, *React. Funct. Polym.*, 1999, **42**, 173.
90. L. Janus, B. Carbonnier, M. Morcellet, G. Ricart, G. Crini and A. Deratani, *Macromol. Biosci.*, 2003, **3**, 198.
91. R. Romita, V. Rizzi, P. Semeraro, F. Gubitosa, J. A. Gabaldon, M. I. F. Gorbe, V. M. G. López, P. Cosma and P. Fini, *Environ. Technol. Innovation*, 2019, **16**, 100454.
92. X. J. Hu, Y. Hu, G. Z. Xu, M. Li, Y. T. Zhu, L. Jiang, Y. Z. Tu, X. Q. Zhu, X. C. Xie and A. Li, *Environ. Res.*, 2020, **180**, 108796.
93. C. Ching, M. J. Klemes, B. Trang, W. R. Dichtel and D. E. Helbling, *Environ. Sci. Technol.*, 2020, **54**, 12693.
94. É. Fenyvesi, K. Barkács, K. Gruiz, E. Varga, I. Kenyeres, G. Záray and L. Szente, *J. Hazard. Mater.*, 2020, **383**, 121181.
95. P. X. Tang, Q. M. Sun, Z. Suo, L. D. Zhao, H. Q. Yang, X. Xiong, H. G. Pu, N. Gan and H. Li, *Chem. Eng. J.*, 2018, **344**, 514.
96. P. Mura, *J. Pharm. Biomed. Anal.*, 2014, **101**, 238.
97. P. Mura, *J. Pharm. Biomed. Anal.*, 2015, **113**, 226.
98. G. Crini, M. Bourdonneau, B. Martel, M. Piotto, M. Morcellet, T. Richert, J. Vebrel, G. Torri and N. Morin, *J. Appl. Polym. Sci.*, 2000, **75**, 1288.
99. A. Romo, F. J. Peñas and J. R. Isasi, *J. Colloid Interface Sci.*, 2004, **279**, 55.
100. A. Romo, F. J. Peñas, X. Sevillano and J. R. Isasi, *J. Appl. Polym. Sci.*, 2006, **100**, 3393.
101. I. X. García-Zubiri, G. González-Gaitano and J. R. Isasi, *Thermochim. Acta*, 2006, **444**, 57.
102. I. Vélaz, J. R. Isasi, M. Sanchez, M. Uzqueda and G. Ponchel, *J. Macromol. Macrocyclic Chem.*, 2007, **57**, 65.
103. A. Romo, F. J. Peñas, J. R. Isasi, I. X. García-Zubiri and G. González-Gaitano, *React. Funct. Polym.*, 2008, **68**, 406.
104. L. D. Wilson, M. H. Mohammed, R. Guo, D. Y. Pratt, J. H. Kwon and S. T. Mahmud, *J. Agromedicine*, 2010, **15**, 105.
105. M. H. Mohammed, L. D. Wilson and J. V. Headley, *Carbohydr. Polym.*, 2010, **80**, 186.
106. M. H. Mohammed, L. D. Wilson, D. Y. Pratt, R. Guo, C. Wu and J. V. Headley, *Carbohydr. Polym.*, 2010, **87**, 1241.
107. D. Y. Pratt, L. D. Wilson, J. A. Kozinski and A. M. Mohart, *J. Appl. Polym. Sci.*, 2010, **116**, 2982.
108. E. Y. Ozmen and M. Yilmaz, *J. Hazard. Mater.*, 2007, **148**, 303.
109. I. Mallard, D. Baudelet, F. Castiglione, M. Ferro, W. Panzeri, E. Ragg and A. Mele, *Beilstein J. Org. Chem.*, 2015, **11**, 2785.
110. M. Ferro, F. Castiglione, N. Pastori, C. Punta, L. Melone, W. Panzeri, B. Rossi, F. Trotta and A. Mele, *Beilstein J. Org. Chem.*, 2017, **13**, 182.

111. A. Mele, F. Castiglione, L. Malpezzi, F. Ganazzoli, G. Raffaini, F. Trotta, B. Rossi, A. Fontana and G. Giunchi, *J. Inclusion Phenom. Macrocyclic Chem.*, 2011, **69**, 403.
112. P. Lo Meo, G. Lazzara, L. Liotta, S. Rielaa and R. Noto, *Poym. Chem.*, 2014, **5**, 4499.
113. S. E. Z. Syeda, D. Nowacka, M. S. Khan and A. M. Skwierawska, *Polymers*, 2022, **14**, 2341.

CHAPTER 12

NMR Techniques for the Evaluation of Biochar Characteristics

P. CONTE,*[a] D. F. CHILLURA MARTINO,[b] A. SPINELLA,[c] C. LIBRICI,[a] P. BAMBINA,[a] E. PIACENZA[b] AND P. LO MEO[b]

[a] Department of Agricultural, Food, and Forrest Sciences – University of Palermo, v.le delle Scienze ed. 4, 90128 Palermo, Italy; [b] Department of Biological, Chemical and Pharmaceutical Sciences and Techniques (STEBICEF) – University of Palermo, v.le delle Scienze ed. 17 "S. Cannizzaro", 90128 Palermo, Italy; [c] Advanced Technologies Network Center (ATeN Center), v.le delle Scienze ed. 18, 90128 Palermo, Italy
*Email: pellegrino.conte@unipa.it

12.1 What Biochar Is: A Definition

Since the early 2000s, terms such as char, charcoal, and biochar have been used interchangeably to define a carbonaceous material obtained by thermal degradation under reductive conditions of biomasses derived from either plants or animals.[1] However, definitions are fundamental in science, as they set the limits within which a researcher must operate to fulfil the objectives of his/her study.[2] Therefore, starting already from the first decade of the 2000s, these terms were tentatively distinguished. In particular, while Bao *et al.*[3] and He *et al.*[4] reported that char is the carbonaceous material produced to obtain new materials in industrial chemistry, Knicker[5] assessed charcoal as the carbonaceous material incorporated into soils after forest fires. In this sense, biochar must refer to the carbonaceous material that is

New Developments in NMR No. 32
The Environment in a Magnet: Applications of NMR Techniques to Environmental Problems
Edited by Pellegrino Conte, Delia Francesca Chillura Martino and Paolo Lo Meo

Published by the Royal Society of Chemistry, www.rsc.org

deliberately applied to soils as an amendment or a fertilizer.[6] Due to the increasingly important meaning of the term "sustainability", which cannot be uniquely defined, the European Biochar Certificate (EBC) reported that "*biochar is a porous, carbonaceous material that is produced by pyrolysis of plant biomasses and is applied in such a way that the contained carbon remains stored as a long-term C sink or replaces fossil carbon in industrial manufacturing. It is not made to be burnt for energy generation*".[7]

The latter definition is broad and accounts for all the concepts of sustainability given in many different areas. Indeed, it fulfils the sustainability concept related to the ability to manage "*a resource for maximum continuing production, consistent with the maintenance of a constantly renewable stock*" when it deals with biological resources;[8] it answers to "*the ability of a system to maintain productivity in spite of a major disturbance*" when one considers agriculture;[8] it achieves the "*transition from a global energy system based on consuming depletable fossil fuels to a sustainable system based on non-depletable fuels*"[8] when accounting for the energy field; finally, it satisfies the concept of sustainability given in ecology that refers to the conservation of the micro- and macro-environments where flora and fauna can survive.[8] The EBC also suggests the type of biomasses mandatory to produce biochar (Table 12.1) and what must be the specific chemical–physical characteristics

Table 12.1 Different types of biomasses that must be used for the production of biochar according to the EBC.[7] A complete list is available on the EBC website.

Origin	Feedstock
Local waste collection services with waste separation	Biodegradable waste, biodegradable waste with kitchen waste, biodegradable waste with kitchen waste and leftovers
Garden wastes	Leaves, flowers, vegetables, roots, pruning from trees, vines and bushes, clippings from nature conservation measures, hay, grass
Agriculture and forestry	Harvest leftovers, straw, used straw, husks and grain dust, grain, feedstuffs, pruning from biomass plantations grown for energy or biomass uses (renewable resources), pruning from trees, vines and bushes, seeds and plants, bark, chipping, wood, sawdust, wood shaving, wood wool
Kitchens and canteens	Kitchen, canteen, and restaurant leftovers
Vegetable productions	Materials from washing, cleaning, peeling, centrifuging and separation
Waterway maintenance (vegetable material)	Raked material, flotsam, fishing residues, harvested material, water plants
Materials from food and packaging	Seasoning residues; residues from potatoes, corn, rice, and starch production; residues from dairy processing; fruit and grain residues; marc; residues from beer production
Textiles	Cellulose, cotton, vegetable fibers, hemp, wool leftovers, and wool dust
Paper	Paper fiber sludge
Biogas plants	Fermentation residues

Table 12.2 Technical parameters for biochars established by the European Biochar Certificate.[7] The list is incomplete. However, it is available on the EBC website.

Parameter	Value
C	35–95%
H/C	<0.7
O/C	<0.4
Heavy metal content	Pb < 150 g Mg^{-1}; Cd < 1.5 g Mg^{-1}; Cu < 100 g Mg^{-1}; Ni < 50 g Mg^{-1}; Hg < 1 g Mg^{-1}; Zn < 400 g Mg^{-1}; Cr < 90 g Mg^{-1}
pH, bulk density, water, and ash content	There are not fixed values. They must be measured and indicated
Polycyclic aromatic hydrocarbons (PAH)	PAH content (sum of the EPA 16 priority pollutants) must be <6 mg kg^{-1} (<4 mg kg^{-1} for organic agriculture)
Polychlorinated biphenyls (PCB)	<0.2 mg kg^{-1}

of any carbonaceous material obtained by thermal degradation under oxygen-starved conditions to consider as biochar (Table 12.2).

As an example, Table 12.1 reports that biochar can be produced only by using organic wastes coming from forest care, garden management, food production, agriculture, the textile industry, and biogas plants, to name a few. In other words, biochar cannot be obtained by pyrolyzing long-living plants or any other system with a long life cycle (*i.e.*, longer than human life). Under these circumstances, any sustainability concept cannot be satisfied, thereby making biochar application environmentally useless.

Table 12.2 shows that the amount of organic carbon in biochars must vary between 35% and 95% of dry matter. The width of this interval depends on biomass feedstock and pyrolysis temperature.[1] For instance, the carbon content of pyrolyzed straw or wood and nutshells is included in the range of 40–50% or 70 and 90%,[7] respectively. Due to the environmental relevance of biochar (see below), this material must not supply contaminants. Therefore, the EBC[7] also suggests the maximum amount of pollutants that any biochar must contain (*e.g.*, Table 12.2) to prevent any environmental contamination derived from biochar application in the different environmental compartments.

12.2 Biochar Environmental Relevance

One of the most important environmental concerns that humankind is facing nowadays is the increase of the Earth's surface temperature due to the greenhouse effect. This issue originated from introduction of greenhouse gases such as carbon dioxide (CO_2), water, and all the volatile chemical species having a non-null electrical dipolar moment[9] into the atmosphere. Among these, CO_2 is the most worrying gas because it lasts in the atmosphere for at least a couple of centuries, while the others remain there for up to 20 years.[10] Therefore, it is the atmospheric persistence of CO_2 that makes this gas very effective for the greenhouse effect.

The increase of carbon dioxide in the atmosphere is mainly due to anthropic activities related, just as examples, to agriculture and fuel production for energy achievement. Therefore, any action devoted to reducing (or even null) CO_2 production in the atmosphere can be considered sustainable and crucial for environmental protection. In this regard, to put it simply, we can say that biochar extracts CO_2 from the atmosphere and traps this greenhouse gas in its tridimensional carbon net.

To understand the latter sentence, one must consider that carbon dioxide is directly involved in the photosynthesis summarized in eqn (12.1):

$$6CO_2 + 6H_2O \rightarrow C_6H_{12}O_6 + 6O_2 + \text{Energy} \tag{12.1}$$

In other words, plants capture CO_2 from the air to synthesize glucose ($C_6H_{12}O_6$), an essential brick to build up plant structures. Once plants term their life, they fall to the soil, where microorganisms and physical–chemical reactions mediated by wet–dry cycles and day–night and seasonal alternances decompose them. These transformations occur mainly under oxidative conditions that quickly convert most plant residues into CO_2, which is, in turn, returned to the atmosphere.

Being biochar chemically very stable,[1] collection and thermal conversion of plant biomasses to biochar followed by application of this material to soil is a valuable technique to reduce CO_2 atmospheric pollution. Figure 12.1 shows the biomass cycle upon biochar production. Once the plant life ends, the biomass can be converted to biochar, which can find application as a soil amendment.

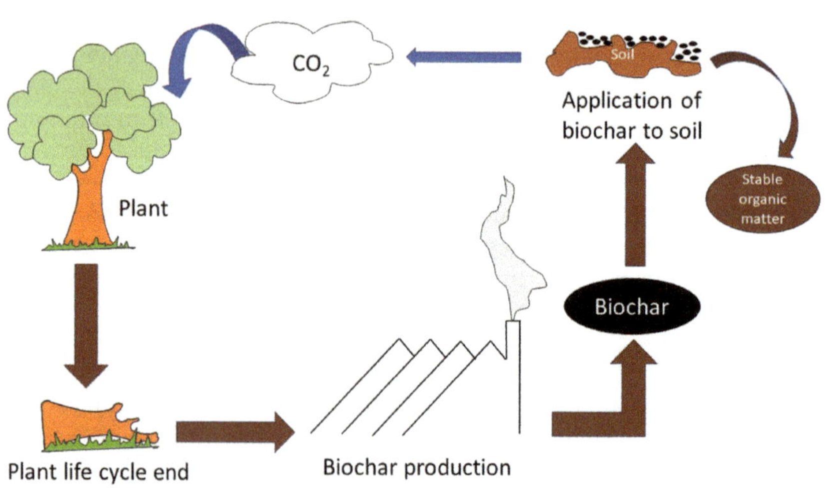

Figure 12.1 Biomass cycle when biochar is produced. Carbon dioxide is used in photosynthesis to produce plant structures. At the end of the plant life, the biomass can be converted to biochar which, in turn, can be applied to soil. Here, biochar is slowly and partly converted to atmospheric carbon dioxide, while the rest is transformed into stable organic matter. Biochar used as a soil amendment enhances soil fertility and crop production.

Due to the chemical recalcitrance of biochar, this material slowly converts into atmospheric CO_2, while producing stable soil organic matter. Thus, biochar application to soils allows the entrapment of carbon into such an environmental compartment, thereby preventing CO_2 pollution. Some studies report that biochar recalcitrance in soils goes from 400 years up to a millennium, supporting the idea that this material can be crucial in carbon sequestration into soils.[6] In addition, biochar ameliorates soil quality by changing its chemical–physical parameters, such as ionic exchange capacity, pH, and structure, and allows the reduction of nutrient leaching loss and the bioavailability of environmental contaminants. Indeed, biochar used as a soil amendment increases soil fertility and crop production even in highly weathered and degraded soils.[1,11–15]

Finally, it is worthy of note that the sorption properties of biochar make it a valuable filter for water and air remediation.[16–20]

12.3 Chemical–Physical Biochar Characteristics

Biochar is obtained by thermal degradation under oxygen-starved conditions (*i.e.*, pyrolysis) of the biomasses reported in Table 12.1. From a chemical point of view, water removal from the biomass cells is the first reaction occurring when pyrolysis temperature is applied (usually it is >300 °C).[21] This event allows the formation of a pore network, such as that shown in Figure 12.2, deriving from cell lysis due to rapid water evaporation.

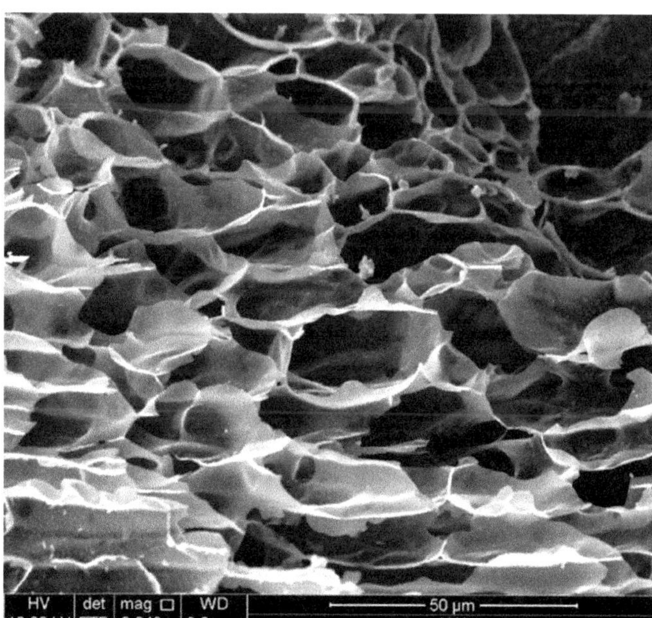

Figure 12.2 Pore system in a biochar obtained by pyrolyzing chicken manure at a temperature of 600 °C for 2 hours. This scanning electron micrograph comes from a collection of one of the authors (PC).

In plant systems, hemicellulose subsequently undergoes decomposition at a temperature ranging between 200 and 260 °C, while cellulose breaks down at 240–350 °C and lignin at 280–500 °C.[22] Thermal degradation of hemicellulose starts with depolymerization leading to oligosaccharides, which, in turn, are transformed into xylopyranose. The latter is further degraded to small-molecular-weight compounds such as furfural and fragments containing two or three carbon atoms. Similarly, the depolymerization of cellulose into oligosaccharides occurs. Then, the glycosidic bonds of the oligosaccharides are broken to produce D-glucopyranose, whose intramolecular rearrangement produces levoglucosan, which further decomposes to form small-molecular-weight compounds. In contrast, pyrolysis of lignin generates phenolic compounds such as guaiacol-, phenol-, syringol-, and catechol-like systems.[22]

The number and size of pores in biochar depend on many factors, among which the feedstock nature, pyrolysis temperature, rate of pyrolysis, and residence time of the biomass in the pyrolizer play the major role.[1]

Specifically, due to thermal degradation, a volume reduction follows the loss of volatile organic compounds, thereby allowing the retention of the rudimentary feedstock porosity and structure in the arising mineral and carbon skeleton. The residual cellular structures from the different feedstock are present and identifiable in biochars, and contribute to the majority of the biochar macro-porosity.[6] Figure 12.3 shows the pore structure of two biochars obtained by chicken manure (Figure 12.3(A)) and fir residues (Figure 12.3(B)) pyrolyzed at 500 °C for 45 and 30 minutes, respectively. The difference in the pore structures is evident. The ordered honeycomb-like structure deriving from the tridimensionality of the cell layout typical of plants (Figure 12.3(B)) is absent in the amorphous chicken manure alongside its

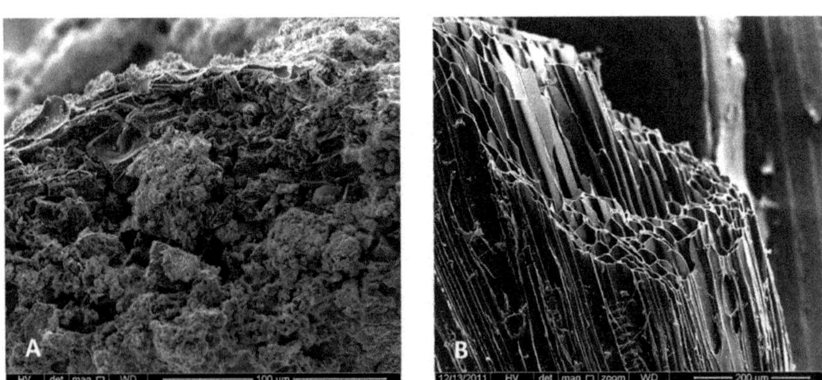

Figure 12.3 Pore structures of biochars from chicken manure (A) and fir residues (B). Both biochars were obtained by pyrolysis at 500 °C in a muffle stove for 45 and 30 minutes, respectively. Reproduced from ref. 1, https://doi.org/10.3390/agronomy11040615, under the terms of the CC BY 4.0 license, http://creativecommons.org/licenses/by/4.0/.

derived biochar (Figure 12.3(A)). Moreover, as the pyrolysis temperature increases (generally above 500 °C), the sizes of biochar pores change. Indeed, the high pressure generated by salt melting and the production of progressively larger amounts of volatile organic compounds (VOCs) determine the collapse of the carbonized cell walls and coalescence of contiguous pores, leading to larger-sized pores.[1]

The same effect on biochar pore size for temperature increase is obtained when raising the residence time in the muffle furnace.[1] Conversely, the pyrolysis rate modulates the amount of the different thermal degradation products. In particular, fast pyrolysis (*i.e.*, thermal degradation conducted in a time interval <2 s) allows the formation of 70–80% (w w^{-1}) tar (also referred to as bio-oil), 10–15% (w w^{-1}) syn-gas (*i.e.*, a mixture of H_2, CO, CH_4, and CO_2 at different ratios) and 10–15% (w w^{-1}) biochar; conventional pyrolysis (*i.e.*, thermal degradation conducted from 5 minutes up to a few hours) leads to the formation of 40–50% (w w^{-1}) tar, 20–30% (w w^{-1}) biochar, and 20–30% (w w^{-1}) syn-gas; slow pyrolysis (that is, a thermal degradation conducted for days) results in ~30% (w w^{-1}) tar, ~35% (w w^{-1}) biochar, and ~35% (w w^{-1}) syn-gas.[1]

The feedstock pretreatment, pyrolysis atmosphere, and biochar post-pyrolysis treatment can also modify the surface physical chemistry of biochar.[1] The former is usually needed to facilitate thermal degradation of biomass. The intricate structure of biomasses, where cellulose, hemicellulose, and lignin are rigidly connected through non-covalent bonds and covalent cross-linkages, is responsible for the recalcitrance to thermal degradation, which causes undesirable energy expenditure when this process is used for energy production (*e.g.*, gasification).[23–28] Biomass pretreatment also allows the formation of porous systems having specific properties functional for wastewater treatment, air cleaning, soil remediation, and catalytic purposes to facilitate energy production.[23–28] For instance, washing biomasses before pyrolysis by either acidic or basic solutions leads to the removal of mineral ashes up to 70% (w w^{-1}) and hemicellulose decomposition (under acidic conditions) or lignin disruption (under alkali conditions), thereby leading to – as pyrolysis is applied – the formation of higher amounts of tar and gases at the expense of the solid residue.[1] The latter, however, when obtained after acidic washing, shows a better capacity to adsorb heavy metals such as cadmium because of the development of mesopores from the acid treatment.[29,30] Recently, ionic liquids (ILs) have been also used to modify the biomass structure before pyrolysis.[24] ILs disrupt the complex H-bond net inside the lignocellulosic biomass system (Figure 12.4), facilitating thermal decomposition during pyrolysis.[24]

The biomass pretreatment by using metal-containing solutions (*e.g.*, salts of Mg, Fe, Ca, and Al) enables obtaining biochars with higher adsorption capacities,[25] as adsorbed metals act as Lewis acids and likely interact with Lewis bases such as electron-rich organic pollutants.[1,25]

Other possible biomass pretreatments are the electrospinning of cellulose acetate solution on the biomass to generate carbon fibers suitable for

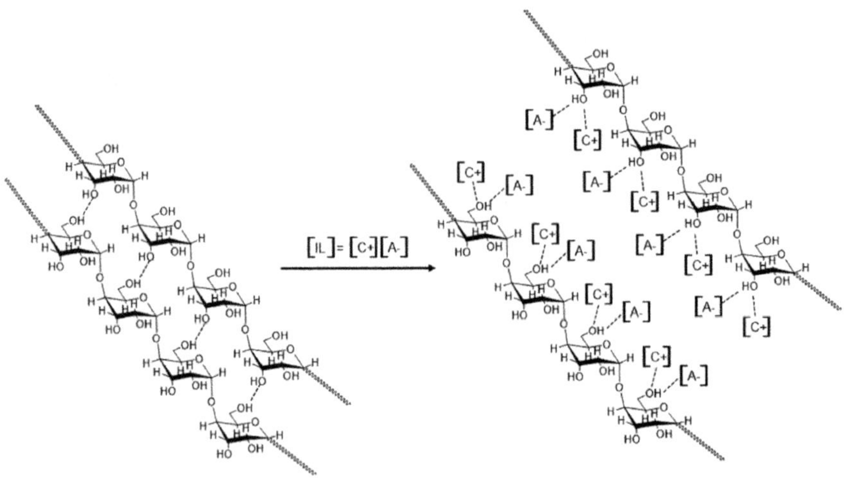

Figure 12.4 H-bond rupture mechanism by ionic liquids. IL stands for ionic liquid; C+ is the cation moiety, while A− is the anion one. Reproduced from ref. 1, https://doi.org/10.3390/agronomy11040615, under the terms of the CC BY 4.0 license, http://creativecommons.org/licenses/by/4.0/.

electrode production,[31] the low-temperature pre-carbonization combined with KOH chemical activation to obtain tubular nanomaterials for supercapacitors,[32] and $(NH_4)_2HPO_4$ addition before carbonization to increase biochar porosity, cross-linking, and availability of nitrogen and phosphorous.[32] In this regard, a more detailed review of biomass treatments before pyrolysis is provided by Bi *et al.*[33]

The surficial physical chemistry of biochar is also affected by pyrolysis atmosphere.[1] For example, conducting pyrolysis under CO_2 causes a series of reactions, among which the Boudouard one plays an important role ($CO_2 + C \rightarrow 2CO$).[34] These reactions lower the biochar yield and allow high surface areas and low H/C and O/C ratios over the systems obtained in a N_2 atmosphere.[34] An alternative atmosphere made by NH_3 is usually applied to produce N-enriched biochars suitable as plant fertilizers or useful activated carbon in materials science (*e.g.*, electronic, optic, or transmission applications).[35–37] Combining CO_2 and NH_3 as gases to control pyrolysis determines the production of N-enriched and highly adsorptive biochars.[38,39] Indeed, "*while carbon dioxide allows pore volume and surface area improvement, ammonia reacts with the pore walls, thereby forming a kind of glue-like layer which improves adsorption properties of the biochar as compared to the materials obtained either in pure CO_2 or NH_3 atmospheres. As the temperature is raised above the value of around 800 °C, pore collapse and degradation of N-containing groups occurs*".[1]

Applying post-pyrolysis treatments can change biochar properties to obtain new materials for very different and often not even connected fields.[1] For instance, using persulfate salts generates radical sulfate and active oxygen on the biochar surface.[40,41] The latter chemical species are responsible for degrading organic pollutants in biochar-treated soils and water.[1] Hydrodynamic cavitation is an innovative and costless way to produce biochars with a surface area larger (120%) than similar non-functionalized carbonaceous systems.[42] Yet, biochars soaked with metals feature better adsorption properties. Indeed, trapping zero-valent iron in the porous biochar system generates hydrophilic biochars useful to decontaminate soils and water either from organic[43–45] or inorganic[46] pollutants.

As this chapter aims to understand how NMR techniques can aid biochar characterization, the list of feedstock pretreatments and biochar post-treatments is deliberately incomplete. However, Conte *et al.*[1] provide more detailed information on this topic.

12.4 Biochar and Nuclear Magnetic Resonance Techniques: High-resolution Spectroscopy

Different NMR technologies are suitable for analyzing biochar characteristics. Considering the insolubility of biochars, their chemical nature can be investigated only by solid-state NMR spectroscopy[47,48] (see Appendix A for the NMR spectroscopy theory). In particular, cross-polarization (CP) magic angle spinning (MAS) NMR spectroscopy appears to be the most used technique due to its suitability in quickly acquiring structural information about complex and poorly soluble systems.[49] The CPMAS ^{13}C NMR pulse sequence is very simple (Figure 12.5). It consists of applying a ^{1}H 90° pulse needed to flip the proton magnetization along the *x*-axis. Then, a spin lock pulse is applied on both the ^{1}H and ^{13}C channels in order to obtain the

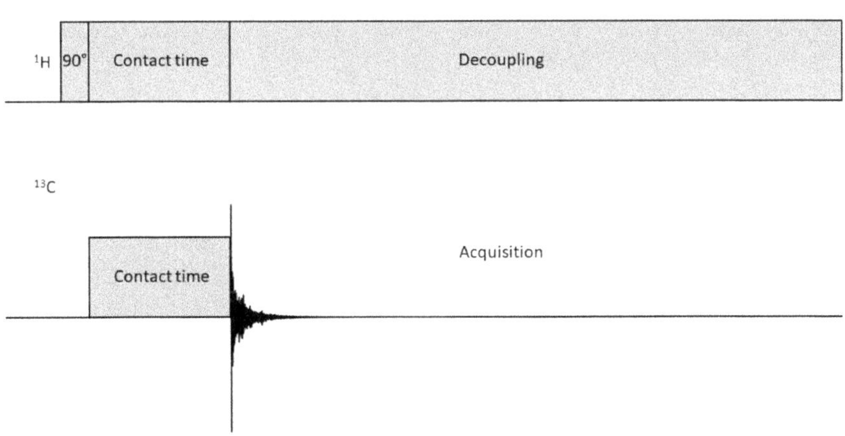

Figure 12.5 Scheme of the CPMAS ^{13}C NMR pulse sequence. A 90° pulse on protons is applied in order to excite those nuclei. Then, a contact time on both ^{1}H and ^{13}C NMR channels is used. It is needed to satisfy the Hartmann–Hahn condition, according to which the ^{1}H nuclei relax to the ground state, while the ^{13}C ones are excited. Finally, a ^{13}C FID is acquired while protons are decoupled.

matching of the energy levels of the two nuclei. The energy level matching is known as the Hartmann–Hahn condition, and it is maintained for a specific duration called the contact time (CT) during which a polarization transfer occurs from abundant nuclei (^{1}H in this case) to dilute nuclei (^{13}C in this case), thus allowing a maximization of the ^{13}C signals. It is important to note that the CT value cannot exceed a specific value typical for the analyzed sample, after which relaxation processes occur, causing a loss of signals. Finally, a ^{13}C NMR signal is acquired at the end of the contact time alongside the proton decoupling.

CPMAS ^{13}C NMR spectroscopy reveals that the main structural biochar feature is aromatic.[21,50–55] In some cases, when biochars are obtained at relatively low temperatures (*e.g.*, up to 500 °C), CPMAS ^{13}C NMR spectra may also show signals attributable to oxygenated and nitrogenated carbons (due, as an example, to the presence of residual oligosaccharides and oligopeptides), as well as alkyl systems.[21,51,52] However, for clarity, this type of spectroscopy presents many limitations for biochar analysis.

Among the different technical issues to account for, one of the most important is biochar softness, which varies according to the temperature used for its production. In particular, as stated in the literature,[56] the lower the temperature, the softer the biochar. Conversely, increasing the temperature causes a decreased biochar softness, making it harder and more grindable.[56] Moreover, biochar softness prevents good packing into the zirconia rotors used to run the experiments. Therefore, when the rotor is spun at a very high rate (*i.e.*, >5 kHz), a vortex effect occurs due to the centrifugal force leading the packed material to be flatted towards the rotor walls.[57] The physical changes occurring during the fast rotation affect spectral resolution, thereby allowing the formation of artifacts.[57] The vortex effect can also be

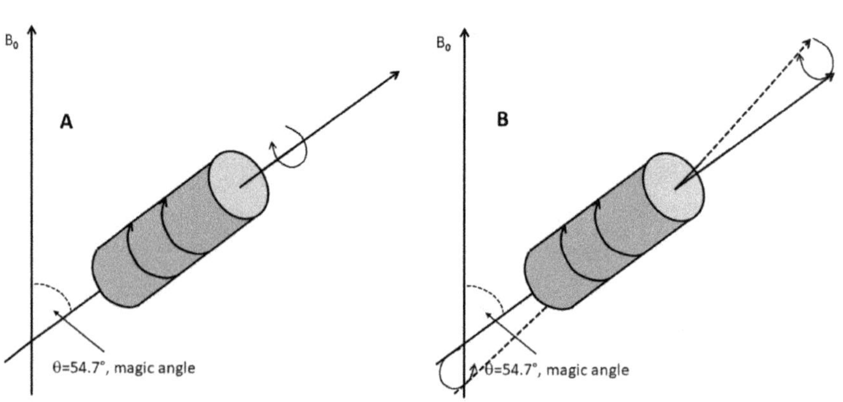

Figure 12.6 Representation of a zirconia rotor spun at the magic angle during a CPMAS ^{13}C NMR experiment (A). Due to the vortex effect described in the text, an asymmetrical rotation can occur (B). This may lead to the collision between the rotor and the walls of the coil, thereby producing probe damage.

responsible for the physical asymmetry in the fast rotation (Figure 12.6), which can lead to a collision between the rotor and the walls of the coil where the NMR experiment is physically performed. Hence, the NMR probe can sustain damages.

Reducing the size of the spinning rotors where the solid samples are packed can limit the vortex effect, as the smaller the rotor, the lower the amount of the packed sample.[57]

When CPMAS ^{13}C NMR experiments must be run by applying slow rotor spin rates (*e.g.*, ≤5 kHz), spinning side bands (SSBs) appear in the spectra (Figure 12.7).[52,53,55,58] Figure 12.7 shows the typical CPMAS ^{13}C NMR spectrum of biochar produced by pyrolysis at *ca.* 500 °C of forest residues. This spectrum was acquired with a Bruker Advance 400, operating at a ^{1}H Larmor frequency of 400 MHz and a ^{13}C Larmor frequency of 100 MHz, by applying a rotor spin rate of 5 kHz, 4000 scans, a contact time of 1 ms, a recycle delay of 3 s, and a spectral width of 29 762 Hz, being the spectral center at 100 ppm. A ramped sequence was also applied to account for magnetic field inhomogeneity due to the 5 kHz rotor spinning speed.[59]

As reported above, the sole significant band is visible at 128 ppm and derives from the resonance of the aromatic carbons. However, the chemical shift anisotropy generated by the aromatic planes produced spinning side bands distanced from the main one by 5000 × *N* Hz, where *N* is an integer non-null number (the reader is referred to Appendix A for the meaning of chemical shift anisotropy). Due to this, the right-hand side part of the spectrum in Figure 12.6 shows three spinning side bands: the first order SSB at a distance of 5 kHz (*i.e.*, $N = 1$) from the aromatic band, the second order

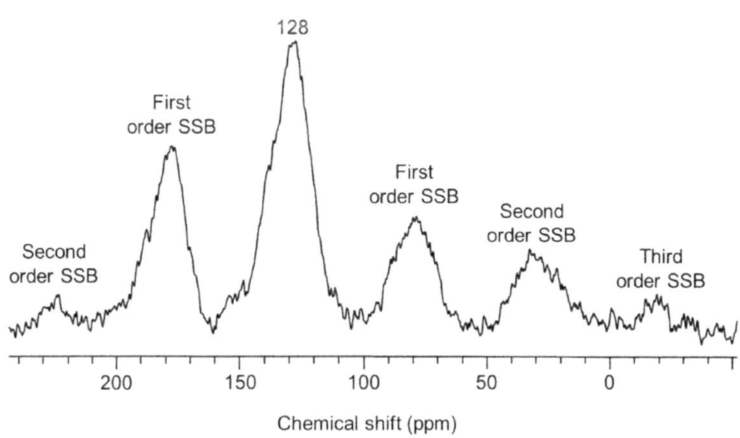

Figure 12.7 Typical CPMAS ^{13}C NMR spectrum of a biochar obtained at a temperature of *ca.* 500 °C. The spectrum has been acquired by applying a rotor spin rate of 5 kHz, 4000 scans, a contact time of 1 ms, and a recycle delay of 1 min. A RAMP sequence was used to account for magnetic field inhomogeneity during the rotor spinning. SSB stands for "spinning side band". This spectrum has never been published and it belongs to the collection of one of the authors (PC).

SSB at 10 kHz (*i.e.*, $N=2$) from the main aromatic signal, and the third order SSB at 15 kHz (*i.e.*, $N=3$), far away from the band at 128 ppm.

Spinning side bands are symmetrically displaced around the main signals from which they originate. Therefore, the left-hand side part of the spectrum in Figure 12.7 shows the same SSB pattern as that previously described. However, the third-order SSB is not visible to the left because of the insufficient spectral width.

The presence of spinning side bands strongly affects spectral interpretation.[53] The two first-order and the second-order (right-hand side part of the spectrum) SSBs fall in the regions where oxygenated/nitrogenated and alkyl carbons may appear, respectively. Thus, retrieving information on the nature of the carbon systems resonating under the spinning side bands is unlikely. In some studies, however, a mathematical operation is performed to obtain information on the chemical nature of the nuclei resonating below the SSBs.[55,60] In this regard, the intensity of the SSBs on the left- and right-hand sides of the main NMR signal should theoretically be the same. Therefore, some researchers published results based on the areas of the down-field spinning side bands where no NMR signals should appear (*i.e.*, the SSB on the left-hand side of the main signal). Let us indicate these areas with A_{SSB}. The latter can be subtracted by those of the up-field bands where both side bands and signals appear overlaid (*i.e.*, $A2$, where $A2=A_{SSB}+A_{Signal}$). The difference between $A2$ and A_{SSB} gives the signal area (*i.e.*, $A2 - A_{SSB} = A_{SSB}+A_{Signal}-A_{SSB}=A_{Signal}$), providing both nature and quantitative assessment of the nuclei whose resonance peak is hidden by the SSBs.

This mathematical elaboration contains two errors. On the one hand, spinning side bands are symmetrical around the signal from which they are generated but not for what concerns the areas. This phenomenon is not only clearly stated in the literature,[49] but it is evident by the visual inspection of Figure 12.7, where, for example, the two first-order SSBs show different areas, with the left-hand side one larger than the right-hand side SSB. The difference between the two areas produces a negative value, which is an NMR non-sense. On the other hand, the second source of error relates to the cross-polarization itself. CP is based on transferring the population from protons to carbons, making the latter nuclei as sensitive as protons. The reliability of the cross-polarization technique depends on the homogeneous ^{1}H distribution in the investigated molecular system. As the protons are not homogeneously distributed, there is a lack of population transfer and a reduction in NMR sensitivity.[61] This issue is significant when dealing with highly aromatic carbon systems such as biochars. For this reason, all the quantitative information about biochars from CPMAS ^{13}C NMR spectra cannot be considered reliable.

Concerning the problems above, one may argue that applying special suppression pulse sequences can help remove SSBs.[48,62] However, this operation also removes the signals overlaid by the SSBs. Therefore, incomplete information about biochar's chemical nature is obtained. Moreover, to overcome the problem of the inhomogeneous proton distribution preventing any suitable CPMAS ^{13}C NMR quantitative assessment, the direct polarization (DP) MAS ^{13}C NMR technique can be applied. This technique suffers two main disadvantages.[48] Firstly, it is difficult to set up the correct recycle delay. Therefore, if the recycle delay is too short, the signal becomes saturated and spectral sensitivity decreases. Conversely, if the recycle delay is too long, excessive machine time is required, increasing the costs of NMR analyses. Secondly, direct polarization pulse sequences lack Hahn echo, thereby providing baseline distortions because of dead-time effects.[48] To overcome all these problems, Mao *et al.*[48] suggested the application of a DPMAS ^{13}C NMR with a recoupling dipolar dephasing technique (Figure 12.8). This technique allows a quantitative assessment of biochar spectra. Namely, while the proton channel is continuously decoupled, a first 90° pulse on the carbon nuclei is applied to let the ^{13}C magnetization flip toward the *x*-axis. Subsequently, a ^{13}C 180° pulse is applied after a rotation period (t_R; referred to as the rotor spinning speed applied during NMR analyses) to avoid distortions due to deadtime effects. After a $2t_R$ period, a refocusing of the ^{13}C magnetization is obtained, and acquisition starts. This sequence works well when the rotor spin rate is >14 kHz, ensuring the disappearance of the spinning side bands, which may affect both qualitative and quantitative evaluation of the NMR spectra.

The limitation of the DPMAS ^{13}C NMR with dipolar dephasing technique is related to the high rotor spinning speed which is not always achievable as highlighted in Section 12.5.

12.5 Biochar Characteristics *Via* 2D Hetero-correlated Solid-state Spectroscopy

A suitable alternative to CPMAS ^{13}C NMR spectroscopy to study biochar nature without the reported limitations is the application of 2D ^{1}H–^{13}C solid-state NMR

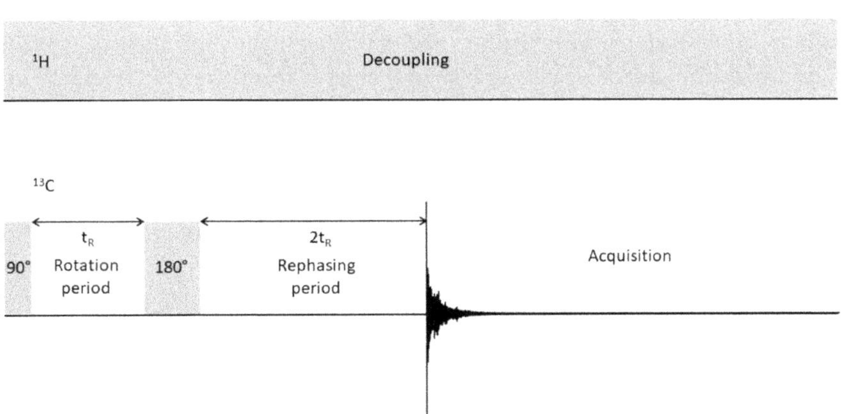

Figure 12.8 DPMAS ^{13}C NMR with recoupling dipolar dephasing. The reader is referred to the main text for the detailed explanation of this pulse sequence.

spectroscopy.[62–64] 2D experiments were applied to analyze a biochar produced at 350 °C from an α-cellulose powder and allowed understanding that:

> *"Carbohydrates did not survive the used pyrolysis conditions. Therefore, O-alkyl C was assigned to ethers. Phenols were not identified to a higher extent suggesting that furan and benzofuran-type units determine the O-functionality of the aromatic domains. The latter are directly connected to alkyl chains".*[64]

These results followed comparable findings from Le Brech *et al.*[63] on biochars obtained by pyrolyzing *Miscanthus* residues at 300 °C.

Notwithstanding the promising results from the application of 2D $^{1}H–^{13}C$ solid-state NMR spectroscopy, few studies deal with this subject, likely due to the long durations and high costs of experiments.

12.6 Other Physical Limitations in the Application of High-resolution Spectroscopy for Biochar Analyses

The above paragraph reports the advantages and limitations of high-resolution solid-state NMR spectroscopy for biochar characterization. Nevertheless, dealing with highly aromatic carbonaceous systems may cause additional problems in preparing the NMR instrument for the analyses.

One of the first settings to perform on an NMR instrument is probe *tuning* and *matching*.[65] In this regard, an NMR probe can be considered as a radio that receives the "*sound*" from a radio source, the nucleus. The radio is physically a coil, also referred to as a receiver. *Tuning* a probe means that one must tune the coil to the radiofrequency of the relevant nucleus, which "*transmits*" its own "*voice*". Whenever the receiver coil is improperly tuned (*i.e.*, mistuning), the mismatch between the receiver and transmitter frequencies (Figure 12.9(A)) causes the reception of noise and a poor (or even null) signal, similar to a radio broadcast that becomes swamped with hissing noise. When there is a good matching between receiver and transmitter frequencies (Figure 12.9(B)), good frequency tuning reduces the level of the noise produced by the electrical current in the circuits, allowing the probe to "*hear*" nuclear "*voices*".

The second step in preparing the NMR instrument for the analyses is "*matching*" – *i.e.*, maximizing the power transfer from the transmitter to the sample and from the latter to the receiver. This operation requires equalizing the probe impedance (the resistance to alternating current) with that of the receiver and the transmitter. Even when the tuning is matched, the impedance level is still not optimized to enable the maximum signal transfer from the transmitter, the sample, and the receiver (Figure 12.9(C)). Then, one needs to regulate the impedance to make the wobbling curve reach the minimum, as depicted in Figure 12.9(D).

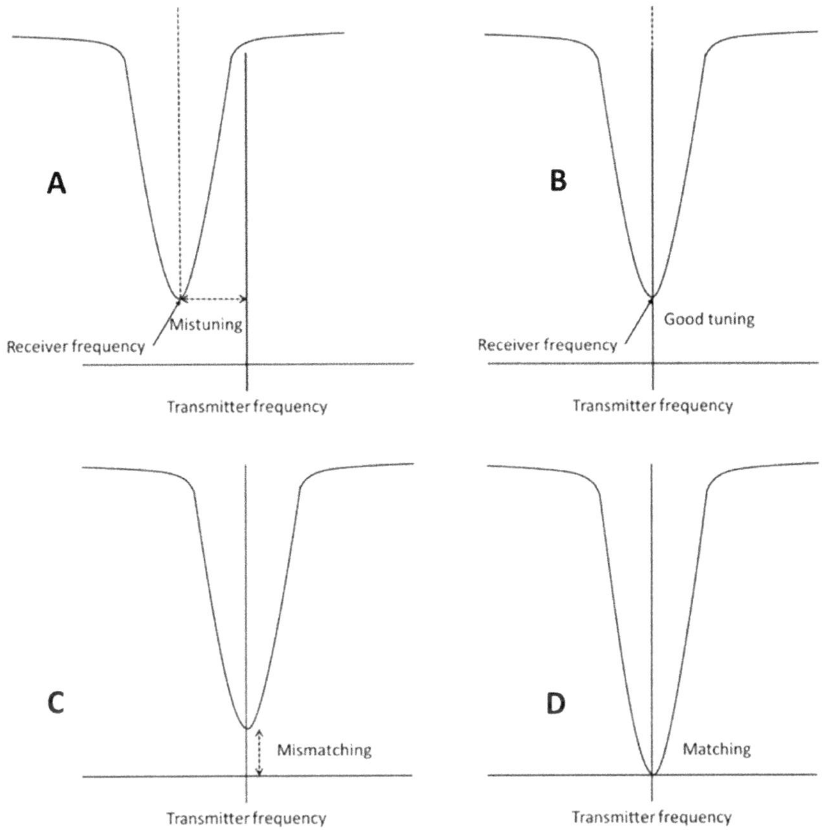

Figure 12.9 Wobbling curves showing (A) the mistuning of the probe; (B) the good tuning where the frequencies of the receiver and the transmitter coincide with each other; (C) the mismatching preventing the maximum power transfer between the probe and the receiver/transmitter; and (D) the good matching where the minimum impedance is reached.

The presence of the sample in the probe significantly affects the electrical currents in charge for NMR probe tuning and matching. Dielectric or highly conductive samples may cause eddy currents, which make tuning and matching difficult.[58,66,67] Moreover, when dealing with highly conductive samples, such as high-temperature biochars containing broad π-electron clouds, additional problems may arise. Solid-state NMR spectroscopy experiments require the use of high-power equipment. When tuning and matching are achieved, the high power needed to run the pulse sequences on highly conductive samples (*e.g.*, high temperature biochars) may generate a voltaic arch due to the electron transfer from the sample to the micro-imperfections present in the coil. This arching may generate temperatures as high as 600 °C, resulting in probe burning.

All the limitations discussed here make high-resolution solid-state NMR far from being the ideal solution for probing biochar characteristics.

Low-resolution NMR techniques seem to fit better the aims of understanding biochar properties.

12.7 Low-resolution NMR

In the NMR language, the locution "*low-resolution NMR*" or "*LR-NMR*" refers to a set of techniques providing spectra with poor or no resolution. Unresolved spectra are due to wide and overlaid signals where no multiplets are distinguishable, and the convolution of different signals, deriving from the resonances of distinct nuclei, produces only one, thereby making correct assignments difficult (or even impossible).

NMR resolution increases when special technical care is used. For example, increasing the magnetic field strength may help to improve NMR resolution. Therefore, what is indicated as high-resolution NMR nowadays can become low-resolution in the future due to technical improvements. Certainly, technical improvements must not overcome the radiofrequency limits (<3 Hz–3 THz) within which NMR operates. However, conventionally, LR-NMR is intended as a technique for which the magnetic field strength ranges from 4×10^{-5} T (*i.e.*, Earth's magnetic field) to 2.5 T.

The spectral resolution does not affect nuclear relaxation. For this reason, LR-NMR techniques may find application in measuring the relaxation characteristics of complex systems for which resolution is not needed. An additional advantage is that LR-NMR allows monitoring of slow motions that are not "*seen*" when strong magnetic fields are applied.[68]

12.8 The Basics of the NMR Relaxation

Although relaxation is deeply described in all the appendices of this book, a brief report about its meaning and its relationship with molecular motion is illustrated here.

Once an ensemble of nuclei are placed in a stationary magnetic field ($\overrightarrow{B_0}$), a magnetization ($\overrightarrow{M}$) is generated. After the application of a magnetic field, $\overrightarrow{B_1}$, perpendicular to the direction of $\overrightarrow{B_0}$, the magnetization rotates by a tilt angle, producing a component along the *z*-axis ($\overrightarrow{M_z}$) and one in the *x*–*y* plane ($\overrightarrow{M_{xy}}$). Relaxation is a phenomenon consisting of the loss of magnetization coherence in the *x*–*y* plane and the recovery of the magnetization along the *z*-axis. The time needed for the magnetization to totally lose coherence in the *x*–*y* plane is called the spin–spin (or transversal) relaxation time. It is conventionally indicated as T_2. The time needed for the complete recovery of the magnetization along the *z*-axis is referred to as the spin–lattice (or longitudinal) relaxation time, and it is conventionally indicated as T_1. The inverse of the relaxation time is indicated as the relaxation rate. Therefore, $R_1 = 1/T_1$ and $R_2 = 1/T_2$ represent the longitudinal and transversal relaxation rates, respectively.

The relationships between relaxation times (or relaxation rates), the correlation time (τ_C), and the intensity of the magnetic field expressed as the Larmor frequency (ν_L) are as follows:[69]

$$R_1 = \frac{1}{T_1} = \frac{3}{10}\frac{\hbar^2 {\mu_0}^2 \gamma^4}{r^6}\left(\frac{\tau_C}{1+(\nu_L\tau_C)^2} + \frac{4\tau_C}{1+4(\nu_L\tau_C)^2}\right) \tag{12.2}$$

$$R_2 = \frac{1}{T_2} = \frac{3}{20}\frac{\hbar^2 {\mu_0}^2 \gamma^4}{r^6}\left(3\tau_C + \frac{5\tau_C}{1+(\nu_L\tau_C)^2} + \frac{2\tau_C}{1+4(\nu_L\tau_C)^2}\right) \tag{12.3}$$

where r is the average internuclear distance, μ_0 is the nuclear magnetic moment, $\hbar$ is the reduced Planck constant $\left(i.e., \frac{h}{2\pi}\right)$, and γ is the magnetogyric ratio. The correlation time is the time needed by a molecule either to rotate by 1 rad or to move a distance equal to its gyration radius when the molecule is approximatively spheric or features other shapes, respectively.

According to the relationships in eqn (12.2) and (12.3), the transversal relaxation time monotonically decreases with the correlation time, irrespective of the applied magnetic field strength (Figure 12.10). Conversely, the longitudinal relaxation time shows a minimum that allows differentiation among three motional regimes (Figure 12.10). Namely, the condition $(\nu_L\tau_C)^2 \ll 1$ indicates a fast-motion or an extremely-narrowing regime. The latter is associated with very short correlation times (*i.e.*, $\leq 10^{-10}$ s) occurring when small molecules (such as bulk water) are present in a system. The condition $(\nu_L\tau_C)^2 \gg 1$ refers to a slow-motion or spin-diffusion regime, which occurs when slowly tumbling molecules (*e.g.*, polymers) with correlation times $>10^{-7}$ s are present. Finally, the condition $(\nu_L\tau_C)^2 \approx 1$ indicates an intermediate-motion regime noticeable when correlation times are in the range of $\sim 10^{-9}$ s to $\sim 10^{-6}$ s, such as those for some nanosponges[70] and ionic liquids.[71]

Following these outlines, T_1 and T_2 values vary according to the small magnetic field fluctuations generated by the molecular motions affecting dipolar interactions among molecules in the same phase or in different phases, such as at the liquid–solid interfaces.[72,73] Therefore, monitoring both relaxation times is useful to unveil the physical–chemical characteristics of complex systems.

12.9 T_2 Monitoring for Biochar Characterization

Although the evaluation of T_2 is widely used to monitor agro-food systems,[68,74–77] there are not too many studies dealing with T_2 and biochar. Interestingly, Wong *et al.*[78] studied biochar porosity using an innovative technique that combines cryoscopy and T_2 monitoring. The technique, called nuclear magnetic resonance cryoporometry (NMRC),[79] is based on the addition of a liquid (such as water) to the biochar porous system. Then, the

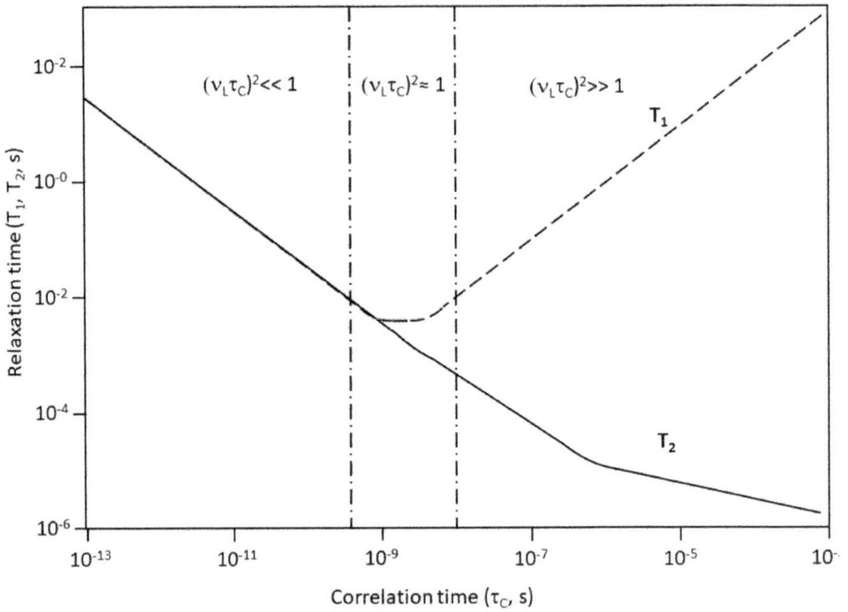

Figure 12.10 Variation of longitudinal (T_1) and transversal (T_2) relaxation times with the correlation time (τ_C). The latter, whose meaning is provided in the main text, is directly proportional to the strength of the applied magnetic field $\overrightarrow{B_0}$. The conditions $(\nu_L\tau_C)^2 \ll 1$, $(\nu_L\tau_C)^2 \sim 1$, and $(\nu_L\tau_C)^2 \gg 1$ refer to the fast-, intermediate-, and slow-motion regimes explained in the main text.

whole system is frozen. Finally, a T_2 relaxometry sequence is applied to measure the melting point of the liquid. In particular, the melting point increases as the pore size decreases due to stronger interactions between the liquid and the pores' boundary wall. Conversely, as the pore size increases, the liquid liquefies at lower temperatures because of the weaker interactions with the pore walls. Noticeably, the pore volumes studied by NMRC appear to be affected by the pyrolysis temperature and feedstock nature,[78] thereby confirming the findings from studies using other analytical techniques.[1]

Goñi-Urtiaga *et al.*[80] used T_2 measurements at 20 MHz "*to estimate the distribution of water on different biochar porosities for unaged and aged samples in contact with adsorbed and liquid water*". The authors showed that the bulk water amount decreases in time, while that trapped in the largest biochar pores increases, thereby revealing that water mainly lies in the biochar meso- and macropores. Conversely, the amount of water in the smallest pores remains constant over time, thus revealing "*that the number of small and accessible pores does not increase with the exposure time*".

Recently, Chen *et al.*[81] studied the effect of different amounts of biochar to monitor the effects on some physical–chemical soil properties. Specifically, they combined T_2 measurements with the filter paper method to obtain the pore size distributions and the soil–water retention curves of different soils with the addition of various amounts of biochar. The authors stated that:

> *The impact of biochar on soil PSD* [pore size distribution] *and SWRC* [soil–water-retention-curve] *varied depending on the soil type and biochar content. In fine-grained soils, the addition of 8% biochar resulted in the greatest decrease in mesopore size and percentage, while in medium-grained soils, 6% biochar had the most significant effect, and in coarse-grained soils, the optimum biochar content was 2%. At low suction states, the addition of 8% and 2% biochar improved the water retention of fine-grained and medium-grained soils, respectively, but had little effect on coarse-grained soils. Overall, the effect of biochar on soil–water characteristics was non-monotonic and depended on the relative sizes of biochar and soil particles. Therefore, selecting the optimum biochar-to-soil ratio requires careful consideration of these factors.*[81]

All the methods described above assume that relaxivity is homogeneous all over the porous surface, which may not be true in natural or environmentally relevant systems.[82] Therefore, porosity determination and porosity-based evaluations can be affected by approximations due to the above assumption.

Song[83] designed a DDIF (decay due to diffusion in internal field) sequence by which the problem of homogeneous relaxivity in natural systems has been overcome. According to Song:[83]

> *The essence of* [the] *method is to first establish a nonuniform magnetization distribution within individual pores. Such modulation is created by allowing spins to precess in the nonuniform internal magnetic field. Then, the time-evolution (mostly decay) of such modulation is monitored. The time constant of this decay* [namely, T_2] *is a direct measure of the pore size.*

De Pierri *et al.*[84] were the first to use the DDIF sequence to monitor the porosity of biochars obtained under different pyrolysis conditions in order to acknowledge their ability in the sorption of agro-chemicals. In particular, the authors stated that:

> *the specific surface area (SSA) of pores with diameters >5.0 μm increased with pyrolysis temperatures and seemed to dictate biochar's retention, which was >70% of the applied amounts at 650 °C. These macropores appear to act as main arteries for herbicide intra-particle diffusion into smaller pores, thus enhancing herbicides retention. Biochar granulometry had little, but herbicide aging had a significant effect on sorption, mainly of Tebuthiuron. However, soils amended with 10 000 kg ha^{-1} of the biochars showed low sorption potential. Therefore, higher than usual biochar rates or proper incorporation strategies, i.e., surface incorporation, will be needed to remediate areas contaminated with these highly mobile herbicides, thus precluding their leaching to groundwaters.*

12.10 Application of FFC NMR Relaxometry to Unveil Biochar Properties

In the last decade, an innovative LR-NMR technique has emerged as a valid tool to evaluate biochar characteristics. This technique is named fast field cycling (FFC) NMR relaxometry, and it relies on the rapid exchange of the intensity of the applied magnetic field within which magnetization evolves under the effect of small magnetic field fluctuations due to molecular dynamics.[85] The fundamentals of FFC NMR relaxometry are reported in Appendix B of this book. Therefore, in the present paragraph, only details about the use of the technique for biochar evaluation are reported.

Very likely, the first paper dealing with FFC NMR relaxometry and biochar is from De Pasquale *et al.*[50] who evaluated the surface characteristics of biochars obtained by using the same conditions as those in an industrial thermochemical process. In particular, having ruled out the chemical-based differences among the biochars, FFC NMR relaxometry revealed that the different behaviors of water molecules on the surface of a poplar and a conifer biochar were due to their different pore sizes. Namely, poplar biochar appeared richer in small-sized pores, while larger pores were characteristic of conifer biochar.

In a later study, Conte *et al.*[73] used variable temperature FFC NMR relaxometry to understand the mechanisms involved in the water–biochar interface interactions. Results revealed that the activation energy (E_a) related to the interactions between the water and biochar surface ranged between 2 and 6 kJ mol^{-1}. These E_a values are typical for weak hydrogen bonds such as those involved in O–H···O=C, C–H···O, O–H···π systems.[86] Consequently, the authors determined that water molecules can penetrate into biochar pores and are trapped therein due to the formation of the aforementioned weak hydrogen bonds. In other studies, where biochar surface transformations were investigated after their addition to soils,[87] the effect of biochar particle size on water dynamics was analyzed,[88] the role of water in biochar nitrate uptake and release was investigated,[15] and the mechanisms of organic coating of biochar were evaluated,[89,90] which supported the findings by Conte *et al.*[73] All these studies are of paramount importance in understanding the role of biochar in improving soil fertility and the environmental remediation when applied as an amendment. Indeed, biochar can be used to remove organic and inorganic contaminants[16,17,27,91–94] from soils and water and can increase soil fertility *via* changes in pH,[95] structure,[12] and microbial activity.[96]

12.11 Conclusions and Perspectives

The present chapter summarizes the different NMR techniques used for biochar characterization. Being insoluble, most researchers tend to apply high-resolution CPMAS ^{13}C NMR spectroscopy, commonly used to evaluate the chemical nature of many different environmentally relevant systems.[51,62,64,97,98]

Less used solid-state techniques explore the correlation between different NMR observable nuclei.[62–64] However, when biochar is under investigation, many problems may arise. In particular, the presence of multi-ordered spinning side bands, difficulties in tuning and matching, and possible arching effects are critical, and justify the preference for low-resolution NMR approaches based on T_1 and T_2 measurements, as these drawbacks affect neither nuclear relaxation nor the setting of the LR-NMR machines. Specifically, fast field cycling NMR relaxometry appears to be very promising to unveil the physical chemistry of the biochar surface. Its sole limitation is related to a lack of mathematical models that are appliable to data interpretation. In fact, it seems that a specific mathematical model is developed for each kind of sample (see Appendix B for details). Therefore, FFC NMR relaxometry research is nowadays going towards a kind of "unified model" appliable for the understanding of the physical chemistry of a wide variety of materials. To this aim, an automatic L1-based regularization method for the analysis of FFC dispersion profiles (described in Appendix B) has been recently developed.[99] This method provides correlation time distributions whose interpretation is far yet from being well understood when biochar is investigated. This is due to a lack of molecular models to compare with biochar samples. Therefore, more investigations on biochar should be conducted *via* FFC NMR relaxometry to validate the features and strength of this automatic L1-based regularization method.

References

1. P. Conte, R. Bertani, P. Sgarbossa, P. Bambina, H. P. Schmidt and R. Raga, *et al.*, Recent Developments in Understanding Biochar's Physical–Chemistry, *Agronomy*, 2021, **11**(4), 615. Available from: https://www.mdpi.com/2073-4395/11/4/615.
2. K. Popper, *The logic of scientific discovery*, Verlag von Julius Springer, Vienna, Austria, 1959, p. 545.
3. C. Bao, Y. Guo, B. Yuan, Y. Hu and L. Song, Functionalized graphene oxide for fire safety applications of polymers: A combination of condensed phase flame retardant strategies, *J. Mater. Chem.*, 2012, **22**(43), 23057–23063.
4. Z. He, S. M. Uchimiya and M. Guo, Production and Characterization of Biochar from Agricultural By-Products: Overview and Use of Cotton Biomass Residues, in *Agricultural and Environmental Applications of Biochar: Advances and Barriers*, ed. M. Guo, Z. He and S. M. Uchimiya, Soil Science Society of America, Inc., Madison, USA, 1st edn, 2015, pp. 63–86. Available from: DOI: 10.2136/sssaspecpub63.2014.0037.5.
5. H. Knicker, How does fire affect the nature and stability of soil organic nitrogen and carbon? A review, *Biogeochemistry*, 2007, **85**(1), 91–118.
6. J. Lehmann and S. Jospeh, *Biochar for Environmental Management: Science and Technology*, ed. J. Lehmann and S. Joseph, Earthscan, London, 1st edn, 2009.

7. European Biochar Foundation, Guidelines for a Sustainable Production of Biochar, Vol. version 9, European Biochar Foundation (EBC), 2020, pp. 1–22. Available from: https://www.european-biochar.org/biochar/media/doc/ebc-guidelines.pdf.
8. B. J. Brown, M. E. Hanson, D. M. Liverman and R. W. Merideth, Global sustainability: Toward definition, *Environ. Manage.*, 1987, **11**(6), 713–719.
9. P. Ronan, What are the properties of a greenhouse gas?, ACS Climate Science Toolkit. Available from: https://www.acs.org/climatescience/greenhousegases/properties.html.
10. S. Solomon, J. S. Daniel, T. J. Sanford, D. M. Murphy, G. K. Plattner and R. Knutti, *et al.*, Persistence of climate changes due to a range of greenhouse gases, *Proc. Natl. Acad. Sci. U. S. A.*, 2010, **107**(43), 18354–18359.
11. H. Schmidt, B. Pandit, V. Martinsen, G. Cornelissen, P. Conte and C. Kammann, Fourfold Increase in Pumpkin Yield in Response to Low-Dosage Root Zone Application of Urine-Enhanced Biochar to a Fertile Tropical Soil, *Agriculture*, 2015, **5**(3), 723–741.
12. G. Baiamonte, C. De Pasquale, V. Marsala, G. Cimò, G. Alonzo and G. Crescimanno, *et al.*, Structure alteration of a sandy-clay soil by biochar amendments, *J. Soils Sediments*, 2015, **15**(4), 816–824.
13. P. Conte, H. P. Schmidt and G. Cimò, Research and Application of Biochar in Europe, in *Agricultural and Environmental Applications of Biochar: Advances and Barriers*, ed. M. Guo, Z. He and S. M. Uchimiya, Soil Science Society of America, Inc., Madison, USA, 2015, pp. 409–422.
14. P. Conte and V. Ferro, Measuring hydrological connectivity inside soils with different texture by fast field cycling nuclear magnetic resonance relaxometry, *Catena*, 2022, **209**(P2), 105848.
15. C. I. Kammann, H. P. Schmidt, N. Messerschmidt, S. Linsel, D. Steffens and C. Müller, *et al.*, Plant growth improvement mediated by nitrate capture in co-composted biochar, *Sci. Rep.*, 2015, **5**, 11080.
16. A. G. Caporale, M. Pigna, A. Sommella and P. Conte, Effect of pruning-derived biochar on heavy metals removal and water dynamics, *Biol. Fertil. Soils*, 2014, **50**(8), 1211–1222.
17. M. A. Rao, G. Di Rauso Simeone, R. Scelza and P. Conte, Biochar based remediation of water and soil contaminated by phenanthrene and pentachlorophenol, *Chemosphere*, 2017, **186**, 193–201.
18. M. Kamali, L. Appels, E. E. Kwon, T. M. Aminabhavi and R. Dewil, Biochar in water and wastewater treatment – a sustainability assessment, *Chem. Eng. J.*, 2021, **420**(P1), 129946.
19. W. Gwenzi, N. Chaukura, C. Noubactep and F. N. D. Mukome, Biochar-based water treatment systems as a potential low-cost and sustainable technology for clean water provision, *J. Environ. Manage.*, 2017, **197**, 732–749.
20. W. Gwenzi, N. Chaukura, T. Wenga and M. Mtisi, Biochars as media for air pollution control systems: Contaminant removal, applications and future research directions, *Sci. Total Environ.*, 2021, **753**, 142249.

21. G. Cimò, J. Kucerik, A. E. Berns, G. E. Schaumann, G. Alonzo and P. Conte, Effect of heating time and temperature on the chemical characteristics of biochar from poultry manure, *J. Agric. Food Chem.*, 2014, **62**(8), 1912–1918.
22. W. J. Liu, W. W. Li, H. Jiang and H. Q. Yu, Fates of Chemical Elements in Biomass during Its Pyrolysis, *Chem. Rev.*, 2017, **117**(9), 6367–6398.
23. O. D. Nartey and B. Zhao, Biochar preparation, characterization, and adsorptive capacity and its effect on bioavailability of contaminants: An overview, *Adv. Mater. Sci. Eng.*, 2014, **2014**, 715398.
24. J. Baruah, B. K. Nath, R. Sharma, S. Kumar, R. C. Deka and D. C. Baruah, *et al.*, Recent trends in the pretreatment of lignocellulosic biomass for value-added products, *Front. Energy Res.*, 2018, **6**, 1–19.
25. O. M. Rodriguez-Narvaez, J. M. Peralta-Hernandez, A. Goonetilleke and E. R. Bandala, Biochar-supported nanomaterials for environmental applications, *J. Ind. Eng. Chem.*, 2019, **78**, 21–33.
26. E. I. P. Pereira, E. C. Suddick, I. Mansour, F. N. D. Mukome, S. J. Parikh and K. Scow, *et al.*, Biochar alters nitrogen transformations but has minimal effects on nitrous oxide emissions in an organically managed lettuce mesocosm, *Biol. Fertil. Soils*, 2015, **51**(5), 573–582.
27. T. Xie, K. R. Reddy, C. Wang, E. Yargicoglu and K. Spokas, Characteristics and applications of biochar for environmental remediation: A review, *Crit. Rev. Environ. Sci. Technol.*, 2015, **45**(9), 939–969.
28. J. Qian, B. Dai, B. Wang, Y. Zha and Q. Song, Traceability in food processing: problems, methods, and performance evaluations—a review, *Crit. Rev. Food Sci. Nutr.*, 2020, 1–14.
29. Q. Li, L. Tang, J. Hu, M. Jiang, X. Shi and T. Zhang, *et al.*, Removal of toxic metals from aqueous solution by biochars derived from long-root Eichhornia crassipes, *R. Soc. Open Sci.*, 2018, **5**(10), 180966.
30. H. Zhang, X. Yue, F. Li, R. Xiao, Y. Zhang and D. Gu, Preparation of rice straw-derived biochar for efficient cadmium removal by modification of oxygen-containing functional groups, *Sci. Total Environ.*, 2018, **631–632**, 795–802.
31. J. Cai, H. Niu, H. Wang, H. Shao, J. Fang and J. He, *et al.*, High-performance supercapacitor electrode from cellulose-derived, inter-bonded carbon nanofibers, *J. Power Sources*, 2016, **324**, 302–308.
32. Y. Cao, L. Xie, G. Sun, F. Su, Q. Q. Kong and F. Li, *et al.*, Hollow carbon microtubes from kapok fiber: Structural evolution and energy storage performance, *Sustainable Energy Fuels*, 2018, **2**(2), 455–465.
33. Z. Bi, Q. Kong, Y. Cao, G. Sun, F. Su and X. Wei, *et al.*, Biomass-derived porous carbon materials with different dimensions for supercapacitor electrodes: A review, *J. Mater. Chem. A*, 2019, 7(27), 16028–16045.
34. C. Guizani, F. J. Escudero Sanz and S. Salvador, Effects of CO_2 on biomass fast pyrolysis: Reaction rate, gas yields and char reactive properties, *Fuel*, 2014, **116**, 310–320.
35. L. Xu, Q. Yao, J. Deng, Z. Han, Y. Zhang and Y. Fu, *et al.*, Renewable N-Heterocycles Production by Thermocatalytic Conversion

and Ammonization of Biomass over ZSM-5, *ACS Sustainable Chem. Eng.*, 2015, **3**(11), 2890–2899.
36. N. Indayaningsih, F. Destyorini, R. I. Purawiardi, D. R. Insiyanda and H. Widodo, Production of activated carbon by using pyrolysis process in an ammonia atmosphere, *J. Phys.: Conf. Ser.*, 2017, **817**, 012006.
37. W. Chen, Y. Chen, H. Yang, K. Li, X. Chen and H. Chen, Investigation on biomass nitrogen-enriched pyrolysis: Influence of temperature, *Bioresour. Technol.*, 2018, **249**, 247–253.
38. Z. Xiong, Z. Shihong, Y. Haiping, S. Tao, C. Yingquan and C. Hanping, Influence of NH_3/CO_2 Modification on the Characteristic of Biochar and the CO_2 Capture, *BioEnergy Res.*, 2013, **6**(4), 1147–1153.
39. X. Zhang, S. Zhang, H. Yang, Y. Feng, Y. Chen and X. Wang, *et al.*, Nitrogen enriched biochar modified by high temperature CO_2-ammonia treatment: Characterization and adsorption of CO_2, *Chem. Eng. J.*, 2014, **257**, 20–27.
40. J. He, Y. Xiao, J. Tang, H. Chen and H. Sun, Persulfate activation with sawdust biochar in aqueous solution by enhanced electron donor-transfer effect, *Sci. Total Environ.*, 2019, **690**, 768–777.
41. J. Liu, S. Jiang, D. Chen, G. Dai, D. Wei and Y. Shu, Activation of persulfate with biochar for degradation of bisphenol A in soil, *Chem. Eng. J.*, 2020, **381**, 122637.
42. L. Albanese, S. Baronti, F. Liguori, F. Meneguzzo, P. Barbaro and F. P. Vaccari, Hydrodynamic cavitation as an energy efficient process to increase biochar surface area and porosity: A case study, *J. Cleaner Prod.*, 2019, **210**, 159–169.
43. H. Dong, C. Zhang, K. Hou, Y. Cheng, J. Deng and Z. Jiang, *et al.*, Removal of trichloroethylene by biochar supported nanoscale zero-valent iron in aqueous solution, *Sep. Purif. Technol.*, 2017, **188**, 188–196.
44. M. Lawrinenko and D. A. Laird, Anion exchange capacity of biochar, *Green Chem.*, 2015, **17**(9), 4628–4636.
45. S. Mortazavian, T. Jones-Lepp, J. H. Bae, D. Chun, E. R. Bandala and J. Moon, Heat-treated biochar impregnated with zero-valent iron nanoparticles for organic contaminants removal from aqueous phase: Material characterizations and kinetic studies, *J. Ind. Eng. Chem.*, 2019, **76**, 197–214.
46. F. Yang, S. Zhang, Y. Sun, K. Cheng, J. Li and D. C. W. Tsang, Fabrication and characterization of hydrophilic corn stalk biochar-supported nanoscale zero-valent iron composites for efficient metal removal, *Bioresour. Technol.*, 2018, **265**, 490–497.
47. P. Conte and A. Spinella, 2021-Conte_Spinella_BOOK.pdf, in *Spectroscopy for Materials Characterization*, 1st edn, 2021, pp. 281–317.
48. J. Mao, X. Cao and N. Chen, Characterization of Biochars Using Advanced Solid-State ^{13}C Nuclear Magnetic Resonance Spectroscopy, in *Advanced Biofuels and Bioproducts*, ed. J. W. Lee, New York, USA, 1st edn, 2013, pp. 47–55.

49. P. Conte, R. Spaccini and A. Piccolo, State of the art of CPMAS ^{13}C-NMR spectroscopy applied to natural organic matter, *Prog. Nucl. Magn. Reson. Spectrosc.*, 2004, **44**, 215–223.
50. C. de Pasquale, V. Marsala, A. E. Berns, M. Valagussa, A. Pozzi and G. Alonzo, *et al.*, Fast field cycling NMR relaxometry characterization of biochars obtained from an industrial thermochemical process, *J. Soils Sediments*, 2012, **12**(8), 1211–1221.
51. G. Bonanomi, F. Ippolito, G. Cesarano, F. Vinale, N. Lombardi and A. Crasto, *et al.*, Biochar chemistry defined by ^{13}C-CPMAS NMR explains opposite effects on soilborne microbes and crop plants, *Appl. Soil Ecol.*, 2018, **124**, 351–361.
52. K. Jindo, H. Mizumoto, Y. Sawada, M. A. Sanchez-Monedero and T. Sonoki, Physical and chemical characterization of biochars derived from different agricultural residues, *Biogeosciences*, 2014, **11**(23), 6613–6621.
53. H. Ben, N. Hao, Q. Liu and A. J. Ragauskas, Solid-State NMR Investigation of Bio-chars Produced from Biomass Components and Whole Biomasses, *Bioenergy Resour.*, 2017, 1036–1044.
54. X. Li, Q. Shen, D. Zhang, X. Mei, W. Ran and Y. Xu, *et al.*, Functional Groups Determine Biochar Properties (pH and EC) as Studied by Two-Dimensional ^{13}C NMR Correlation Spectroscopy, *PLoS One*, 2013, **8**(6), e65949.
55. S. Munira, J. J. Dynes, M. Islam, F. Khan, T. Adesanya and T. Z. Regier, *et al.*, Relative proportions of organic carbon functional groups in biochars as influenced by spectral data collection and processing, *Chemosphere*, 2021, **283**, 131023.
56. M. H. Mohd Hasan, R. T. Bachmann, S. K. Loh, S. Manroshan and S. K. Ong, Effect of Pyrolysis Temperature and Time on Properties of Palm Kernel Shell-Based Biochar, *IOP Conf. Ser.: Mater. Sci. Eng.*, 2019, **548**(1), 012020.
57. P. Conte, R. Spaccini and A. Piccolo, State of the art of CPMAS ^{13}C-NMR spectroscopy applied to natural organic matter, *Prog. Nucl. Magn. Reson. Spectrosc.*, 2004, **44**, 215–223.
58. P. Srinivasan, A. K. Sarmah, R. Smernik, O. Das, M. Farid and W. Gao, A feasibility study of agricultural and sewage biomass as biochar, bioenergy and biocomposite feedstock: Production, characterization and potential applications, *Sci. Total Environ.*, 2015, **512–513**, 495–505.
59. A. E. Berns and P. Conte, Effect of ramp size and sample spinning speed on CPMAS ^{13}C NMR spectra of soil organic matter, *Org. Geochem.*, 2011, **42**(8), 926–935.
60. A. Piccolo, P. Conte and A. Cozzolino, Effects of mineral and monocarboxylic acids on the molecular association of dissolved humic substances, *Eur. J. Soil Sci.*, 1999, **50**(4), 687–694.
61. P. Conte, A. Piccolo, B. Van Lagen, P. Buurman and P. A. De Jager, Quantitive differences in evaluating soil humic substances by liquid- and solid-state ^{13}C-NMR spectroscopy, *Geoderma*, 1997, **80**(3–4), 339–352.

62. N. Baccile, C. Falco and M. M. Titirici, Characterization of biomass and its derived char using ^{13}C-solid state nuclear magnetic resonance, *Green Chem.*, 2014, **16**(12), 4839–4869.
63. Y. Le Brech, L. Delmotte, J. Raya, N. Brosse, R. Gadiou and A. Dufour, High Resolution Solid State 2D NMR Analysis of Biomass and Biochar, *Anal. Chem.*, 2015, **87**, 843–847.
64. H. Knicker, M. Velasco-Molina and M. Knicker, 2D Solid-State Hetcor ^{1}H–^{13}C NMR Experiments With Variable Cross Polarization Times As a Tool for a Better Understanding of the Chemistry of Cellulose-Based Pyrochars—a Tutorial, *Appl. Sci.*, 2021, **11**(18), 8569.
65. J. Keeler, *Understanding NMR Spectroscopy*, John Wiley & Sons, Ltd, Chichester, England, 1st edn, 2005, p. 459.
66. A. Ross, G. Schlotterbeck, F. Dieterle and H. Senn, NMR Spectroscopy Techniques for Application to Metabonomics, in *The Handbook of Metabonomics and Metabolomics*, ed. J. C. Lindon, J. K. Nicholson and E. Holmes, Elsevier B.V., Amsterdam, The Netherlands, 2007, pp. 55–112.
67. J. A. Stringer, C. E. Bronnimann, C. G. Mullen, D. H. Zhou, S. A. Stellfox and Y. Li, *et al.*, Reduction of RF-induced sample heating with a scroll coil resonator structure for solid-state NMR probes, *J. Magn. Reson.*, 2005, **173**(1), 40–48.
68. P. Conte, Applications of Fast Field Cycling NMR Relaxometry, in *Annual Reports on NMR Spectroscopy*, ed. G. A. Webb, Academic Press, Elsevier, London, UK, 1st edn, 2021, pp. 141–188.
69. M. H. Levitt, *Spin Dynamics-Basics of NMR*, Wiley, 2013, vol. 53, pp. 1689–1699.
70. P. Lo Meo, F. Mundo, S. Terranova, P. Conte and D. Chillura Martino, Water Dynamics at the Solid-Liquid Interface to Unveil the Textural Features of Synthetic Nanosponges,*J. Phys. Chem. B*, 2020, **124**(9), 1847–1857.
71. D. Kruk, R. Meier, A. Rachocki, A. Korpała, R. K. Singh and E. A. Rössler, Determining diffusion coefficients of ionic liquids by means of field cycling nuclear magnetic resonance relaxometry, *J. Chem. Phys.*, 2014, **140**(24), 244509.
72. V. I. Bakhmutov, *Practical NMR Relaxation for Chemists*, John Wiley & Sons, Inc., Chichester, England, 1st edn, 2004.
73. P. Conte, V. Marsala, C. De Pasquale, S. Bubici, M. Valagussa and A. Pozzi, *et al.*, Nature of water-biochar interface interactions, *GCB Bioenergy*, 2013, **5**(2), 116–121.
74. J. V. Bayer, F. Jaeger and G. E. Schaumann, Proton nuclear magnetic resonance (NMR) relaxometry in soil science applications, *Open Magn. Reson. J.*, 2010, **3**(SPEC. ISS. 1), 15–26.
75. C. Buchmann, M. Meyer and G. E. Schaumann, Characterization of wet aggregate stability of soils by H-NMR relaxometry, *Magn. Reson. Chem.*, 2015, **53**(9), 694–703.
76. P. Conte, G. Cuccurullo, A. Metallo, A. Micalizzi, L. Cinquanta and O. Corona, Comparing different processing methods in apple slice drying. Part 2 solid-state Fast Field Cycling 1H-NMR relaxation

properties, shrinkage and changes in volatile compounds, *Biosyst. Eng.*, 2019, **188**, 345–354.
77. E. Piacenza, D. F. Chillura Martino, L. Cinquanta, P. Conte and P. Lo Meo, Differentiation among dairy products by combination of fast field cycling NMR relaxometry data and chemometrics, *Magn. Reson. Chem.*, 2022, **60**(3), 369–385.
78. J. W. C. Wong, J. B. W. Webber and U. O. Ogbonnaya, Characteristics of biochar porosity by NMR and study of ammonium ion adsorption, *J. Anal. Appl. Pyrolysis*, 2019, **143**, 104687.
79. J. B. W. Webber, A review of the use of simple time-domain NMR/MRI for material-science: Time-domain benchtop and mobile NMR/MRI for liquid and solid material science, and nano- to micro-pore-size measurement, *SN Appl. Sci.*, 2021, **3**(10), DOI: 10.1007/s42452-021-04791-2.
80. A. Goñi-Urtiaga, D. Courtier-Murias, G. Picca, J. L. Valentín, C. Plaza and M. Panettieri, Response of water-biochar interactions to physical and biochemical aging, *Chemosphere*, 2022, **307**, 136071.
81. X. Chen, L. Li, X. Li, J. Kang, X. Xiang and H. Shi, *et al.*, Effect of Biochar on Soil–Water Characteristics of Soils: A Pore-Scale Study, *Water*, 2023, **15**(10), 1–16.
82. H. Liu, M. Nogueira D'Eurydice, S. Obruchkov and P. Galvosas, Determining pore length scales and pore surface relaxivity of rock cores by internal magnetic fields modulation at 2 MHz NMR, *J. Magn. Reson.*, 2014, **246**, 110–118.
83. Y. Q. Song, Using internal magnetic fields to obtain pore size distributions of porous media, *Concepts Magn. Reson., Part A: Bridging Educ. Res.*, 2003, **18**(2), 97–110.
84. L. de Pierri, E. H. Novotny, C. E. Pellegrino Cerri, A. José de Souza, B. B. Mattos and V. L. Tornisielo, *et al.*, Accessing biochar's porosity using a new low field NMR approach and its impacts on the retention of highly mobile herbicides, *Chemosphere*, 2022, **287**, 132237.
85. R. Kimmich and E. Anoardo, Field-cycling NMR relaxometry, *Prog. Nucl. Magn. Reson. Spectrosc.*, 2004, **44**(3–4), 257–320.
86. G. Gilli and P. Gilli, *The Nature of the Hydrogen Bond: Outline of a Comprehensive Hydrogen Bond Theory*, Oxford University Press, Oxford, UK, 1st edn, 2009, p. 337.
87. G. Fellet, P. Conte, V. Bortolotti, F. Zama, G. Landi and D. F. Chillura Martino, *et al.*, Changes in Physicochemical Properties of Biochar after Addition to Soil, *Agriculture*, 2022, **12**(3), 320.
88. P. Conte and N. Nestle, Water dynamics in different biochar fractions, *Magn. Reson. Chem.*, 2015, **53**(9), 726–734.
89. P. Conte and V. A. Laudicina, Mechanisms of organic coating on the surface of a poplar biochar, *Curr. Org. Chem.*, 2017, **21**(6), 559–565.
90. N. Hagemann, S. Joseph, H. P. H. P. Schmidt, C. I. C. I. Kammann, J. Harter and T. Borch, *et al.*, Organic coating on biochar explains its nutrient retention and stimulation of soil fertility, *Nat. Commun.*, 2017, **8**(1), 1–11.

91. M. Pipíška, E. K. Krajčíková, M. Hvostik, V. Frišták, L. Ďuriška and I. Černičková, *et al.*, Biochar from Wood Chips and Corn Cobs for Adsorption of Thioflavin T and Erythrosine B, *Materials*, 2022, **15**(4), 1492.
92. G. Fellet, P. Conte and L. Marchiol, Biochar Effects on Ce Leaching and Plant Uptake in Lepidium sativum L. Grown on a Ceria Nanoparticle Spiked Soil, *Appl. Sci.*, 2023, **13**(11), 6846.
93. J. Paz-Ferreiro, H. Lu, S. Fu, A. Méndez and G. Gascó, Use of phytoremediation and biochar to remediate heavy metal polluted soils: A review, *Solid Earth*, 2014, **5**(1), 65–75.
94. X. Yang, S. Zhang, M. Ju and L. Liu, Preparation and modification of biochar materials and their application in soil remediation, *Appl. Sci.*, 2019, **9**(7), 1365.
95. R. Chintala, J. Mollinedo, T. E. Schumacher, D. D. Malo and J. L. Julson, Effect of biochar on chemical properties of acidic soil, *Arch. Agron. Soil Sci.*, 2014, **60**(3), 393–404.
96. Y. Sheng and L. Zhu, Biochar alters microbial community and carbon sequestration potential across different soil pH, *Sci. Total Environ.*, 2018, **622–623**, 1391–1399.
97. R. Angelico, C. Colombo, E. D. Iorio, M. Brtnický, J. Fojt and P. Conte, Humic Substances: From Supramolecular Aggregation to Fractal Conformation — Is There Time for a New Paradigm?, *Appl. Sci.*, 2023, **13**, 1–33. Available from: https://www.mdpi.com/2076-3417/13/4/2236.
98. P. Conte and A. E. Berns, Dynamics of cross polarization in solid state nuclear magnetic resonance experiments of amorphous and heterogeneous natural organic substances, *Anal. Sci.*, 2008, **24**(9), 1183–1188.
99. G. Landi, G. V. Spinelli, F. Zama, D. C. Martino, P. Conte and P. Lo Meo, *et al.*, An automatic L1-based regularization method for the analysis of FFC dispersion profiles with quadrupolar peaks, *Appl. Math. Comput.*, 2023, **444**, 1–16.

CHAPTER 13

The Use of NMR Techniques in Cultural Heritage

A. SPINELLA,*[a] D. F. CHILLURA MARTINO[b] AND P. LO MEO*[b]

[a] Advanced Technologies Network Center (ATeN Center), v.le delle Scienze ed. 18, 90128 Palermo, Italy; [b] Department of Biological, Chemical and Pharmaceutical Sciences and Techniques (STEBICEF) – University of Palermo, v.le delle Scienze ed. 17 "S. Cannizzaro", 90128 Palermo, Italy
*Emails: alberto.spinella@unipa.it; paolo.lomeo@unipa.it

13.1 Introduction

The preservation of cultural heritage, from architecture and monuments to figurative artwork, is largely dependent on environmental conditions. Air pollution from sulphur oxides, nitrogen oxides or PM is largely responsible for the deterioration of artefacts, in particular as long as acid rain corrosion, soiling and black crust formation are concerned.[1–4] Therefore, monitoring the state of cultural heritage is strictly related to environmental problems. It is important to stress here that the study of an artwork always requires a multidisciplinary approach,[5] with a careful consideration of diverse aspects (chemical, physical, biological, ecological, engineering, but also historical, artistic, economic and even legal). Therefore, the use of diverse, complementary investigation techniques is generally needed.[6] In the last few decades, the interest in the characterization of artefacts and materials in the field of cultural heritage has progressively grown. This is due to the fact that the determination of the state of conservation of artistic handworks, and the study of the materials used for making them, has become an increasingly important topic in order to obtain a deeper knowledge of the causes of deterioration and the materials that one needs to use for their consolidation.

New Developments in NMR No. 32
The Environment in a Magnet: Applications of NMR Techniques to Environmental Problems
Edited by Pellegrino Conte, Delia Francesca Chillura Martino and Paolo Lo Meo

Published by the Royal Society of Chemistry, www.rsc.org

These studies have been mainly aimed at the development of new consolidating materials and more effective diagnostic methods. In this regard, it is mandatory to use non-destructive, micro-invasive and (when possible) non-invasive diagnostic approaches.[7]

Although NMR is a powerful technique in many research fields,[8] its application to cultural heritage is relatively recent.[5] This is because for a long time NMR has been almost exclusively used as a technique for the study of samples in solution and thus not applicable to insoluble solids such as glass, paper, stone materials and so on. Even if the characterization of organic substances in solution, such as residues found in ancient artefacts, can be carried out by NMR spectroscopy in the liquid phase, the analysis of these samples is often difficult because, in most cases, they are made by complex mixtures of compounds. Nowadays, through the use of NMR techniques such as measurement of molecular self-diffusion coefficients,[9] complex mixtures of organic compounds can be successfully analysed. At the same time, 2D NMR techniques may be applied to identify the structure of each compound. Furthermore, cryogenically cooled NMR liquid phase probes (cryoprobes), which can greatly enhance the signal-to-noise ratio of the spectra, have been recently introduced. Using these probes, the problems related to the low sensitivity of NMR spectroscopy have been overcome, thus making the analysis of sample quantities down to the microgram scale possible.[10] Moreover, with the development of multinuclear probes that are able to investigate many nuclei of the element table and of solid-state 1D and 2D NMR techniques, NMR has become a powerful technique in analytical and structural determinations for both organic and inorganic substances in the cultural heritage field. Finally, the development of mobile single-sided NMR sensors[11–15] has made it possible to provide a precise relaxometric profiling of large surfaces *in situ*.

In the diagnostics for cultural heritage, sampling constitutes a critical point. On the one hand, the big and heavy cryomagnets used in ordinary NMR spectroscopy cannot be moved. Therefore, an ordinary spectroscopic investigation would require the possibility of extracting a sample (or a set of samples) from an artwork, provided that such a sampling does not cause any unacceptable damage to the artefact itself. This requirement can be occasionally satisfied. Nevertheless, with the introduction of high resolution solid-state NMR techniques[16,17] such as magic angle spinning (MAS), cross-polarization (CP), and high power dipolar decoupling (HPDD), which can greatly enhance the sensitivity of low abundant nuclei, recently NMR has become a powerful technique for the characterization of materials in the field of cultural heritage. On the other hand, in many cases the artefact itself cannot be moved because of its intrinsic nature, its location, or any other reason. In these cases, "*the mountain must come to Mohammed*", *i.e.* NMR must somehow transform into a "portable" technique, specifically designed for carrying out *in situ* studies of arbitrarily sized objects. This can be achieved by the use of unilateral NMR sensors, constituted by an open magnet and surface rf coils, which generate a sensitive volume out of the sensor and inside the artefact under study. From this standpoint, the introduction of the NMR-MOUSE® (MObile Universal Surface Explorer) instrumentation[13–15] has provided researchers with a powerful and versatile tool for the non-invasive investigation of paintings (frescoes or canvas) and of wood artefacts or human remains[5] as well.

13.2 Theory of Solid-state NMR

Although a deep evaluation of the NMR theory is presented in Appendices 1 and 2, here a brief overview of the technique is reported. Namely, in the liquid state the dipolar interactions among nuclei are averaged to zero by the Brownian motion of molecules, and thus the signal's position in the spectrum does not depend on the orientation of the molecules with respect to the applied magnetic field ($\overrightarrow{B_0}$). Therefore, the spectra of liquid samples are composed of sharp and well resolved signals. Conversely, in the solid state all the interactions among nuclei are not averaged to zero and this causes signal enlargement. Furthermore, signal positions depend on the orientation of the molecules with respect to $\overrightarrow{B_0}$. As a consequence, a distribution of lines (powder spectrum) which are very close to each other is generated by each single crystal, causing a further line width enlargement (chemical shift anisotropy, CSA). Finally, if a certain nucleus has a spin $>\frac{1}{2}$, an interaction between the nuclear quadrupolar electric momentum and the electric field gradient takes place, and this is another source of line broadening. However, the case of quadrupolar nuclei will not be discussed. As the nuclear relaxation times of solid samples are very long, spectral acquisition may be very time-consuming. As already reported, in order to overcome these problems, three techniques of acquisition have been developed, namely MAS, HPDD and CP. Dipolar interactions among nuclei and the chemical shift anisotropy effects are averaged by the first two techniques, while the third one allows one to obtain spectra of low abundant (dilute) nuclei in a shorter time.

Many terms of the Hamiltonian that describe spin interactions, which cause line broadening in solid samples, contain the geometric factor $(3\cos^2\theta - 1)$, which accounts for the orientation of the shielding/dipolar tensor with respect to the applied magnetic field B_0. It can be shown that, by spinning the sample at the magic angle, *i.e.* 54.74°, the geometric factor $(3\cos^2\theta - 1)$ averages to zero on the NMR timescale. Therefore, this technique is routinely used to remove the effects of chemical shift anisotropy and to help in the removal of dipolar coupling. The MAS technique consists in the high speed spinning of a rotor filled with the ground solid sample at the magic angle with respect to the direction of the static magnetic field (Figure 13.1).

The dipolar decoupling technique consists in the irradiation of abundant nuclei (typically protons) having dipolar interaction with dilute nuclei (such as carbons) that one wants to observe. The high power dipolar decoupling thus minimizes the dipolar broadening in order to obtain a spectrum with sharper lines.

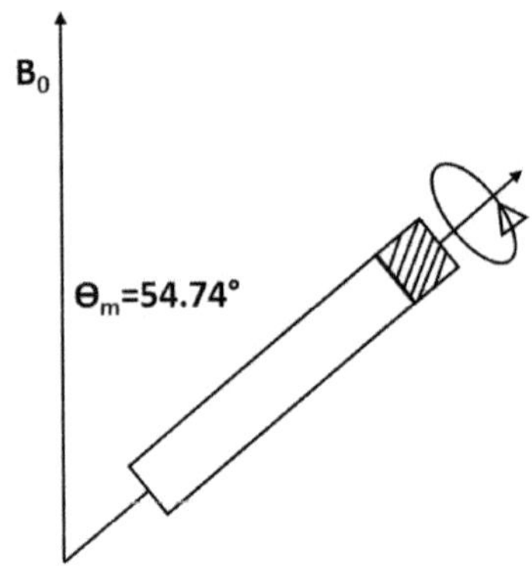

Figure 13.1 The magic angle spinning of a solid sample.

The cross-polarization technique is applied to observe dilute nuclei such as ^{13}C or ^{29}Si having a low natural abundance (1.1% for ^{13}C and 4.7% for ^{29}Si) and a low gyromagnetic ratio, and thus possessing a very weak observable net magnetization.

As already indicated, the acquisition of spectra of dilute nuclei is very time-consuming and requires a large number of scans to obtain a reasonable S/N ratio because of the low nuclear sensitivity. The application of the CP technique by which the high polarization of abundant nuclei is transferred to dilute nuclei causes the enhancement of the S/N ratio. This transfer of polarization between different nuclei is achieved by modulating the radio-frequency pulses on the receiving channels of the two nuclei allowing the matching of their energy levels in the rotating frame. This energy level matching is known as the Hartman–Hahn condition. For example, in the case of systems containing protons and carbons, the Hartman–Hahn condition is as follows: $\gamma_C B_{1C} = \gamma_H B_{1H}$, where γ_C and γ_H are the carbon and proton gyromagnetic ratios, respectively. Since the proton gyromagnetic ratio is four times the carbon gyromagnetic ratio, the Hartman–Hahn condition will be satisfied if the power of the applied radio-frequency field B_{1C} is four times the power of the applied radio-frequency field B_{1H}. As a consequence, there is an enhancement of the dilute spin signal intensity by as much as the ratio of the gyromagnetic ratios of the abundant and dilute spins. In the case of proton and carbon nuclei, because $\gamma_H/\gamma_C = 4$, the enhancement factor is 4. It is important to note that the Hartman–Hahn condition (*i.e.* the contact between the two nuclear species) must be maintained for a time (contact time τ_C) that allows the maximization of the NMR signal. On the other hand, τ_C cannot exceed a certain value, depending on the analysed sample, above which a loss of signal intensity occurs because of the relaxation processes. A representation of the CP pulse sequence is shown in Figure 13.2.

In the presence of cross-polarization, since polarization is being transferred from protons to carbons, the recycle delay for signal averaging is dictated by the shorter T_1 relaxation time of protons. As a result, the acquisition time of

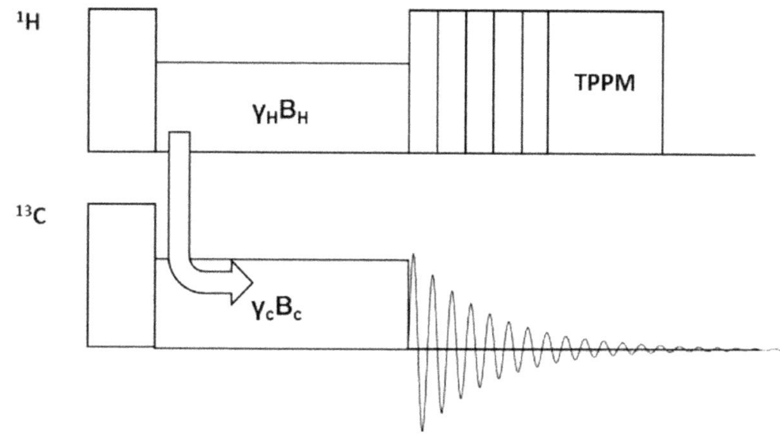

Figure 13.2 Cross-polarization pulse sequence.

carbon spectra is noticeably shortened with respect to carbon spectra acquired by direct polarization.

Even though this technique requires 50–80 mg of material to perform the analysis, the method is absolutely non-destructive, and the sample can be recovered after the NMR experiment and studied with other techniques. This is possible because the sample does not require any treatment prior to acquisition. Furthermore, the amount of information obtained from a single sample by means of an NMR experiment is often reason enough to resort to this type of analysis. The use of sample holders, equipped with inserts which reduce their volume, allows reduction of the quantity of sample required for the analysis (10–12 mg).

13.3 Theory of Single-sided NMR

In ordinary NMR techniques, a sample is deeply inserted into a magnetic field and this poses severe limitations to the size of the sample itself. Conversely, the "mobile" NMR techniques rely on a simple but not trivial idea to exploit the inhomogeneous stray magnetic field any magnet produces in its external surroundings,[11] in such a way to enable sampling of a volume up to a certain depth onto an extended surface. This led to the development of portable instrumentations, such as the MOUSE (MObile Universal Surface Explorer)[15] and the GARfield (Gradients At Right angles to field),[18] which have been proven outstandingly useful in cultural heritage investigations. Indeed, the geometry of this sensor generates a magnetic field with a uniform gradient to resolve the near-surface structure of large samples.[19]

The theory and the main practical aspects of single-sided NMR techniques have been summarized in an excellent review by Blümich *et al.*[11] (interested readers are strongly advised to consult it for more exhaustive explanations and insights). In brief, for the very exemplificative purposes of this

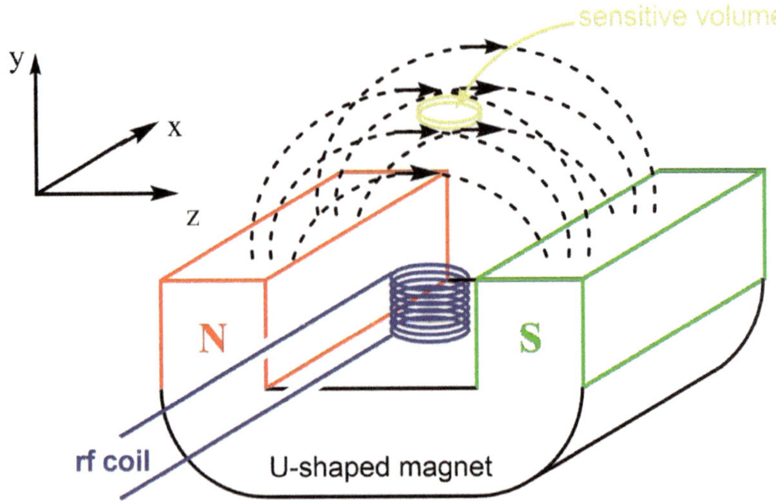

Figure 13.3 Simplified scheme of a MOUSE-NMR apparatus.

presentation, let us consider a U-shaped magnet that generates a magnetic field with arch-shaped force lines (Figure 13.3), as an ideal model arrangement.

In a tiny central slice region, the line forces (along the *z*-axis, according to the depiction in Figure 13.3) will be almost perfectly parallel to the ideal face of the magnet (the *xz*-plane), with a significant gradient (even in the order of tens of $T\,m^{-1}$) along the orthogonal direction (*y*-axis). This slice constitutes the sensitive sampling region.[11,12] Then, a radio-frequency (rf) coil, placed between the magnet poles orthogonal to the magnet face, provides both the rf pulses and the receiver to generate and detect the NMR signal. One can exploit the magnetic field gradient on *y* to explore different depths of the sampling region by changing the radio-frequency emitted by the rf coil. Alternatively, different depths can be sampled with a fixed radio-frequency by simply moving the magnet arrangement towards or away from the analysed surface. The effects from the magnetic field inhomogeneities can be overcome by the use of a classical spin-echo pulse sequence or, better, of a CPMG (Carr–Purcell–Meiboom–Gill) multiple echo sequence,[12,20,21] which generates a train of echoes of decaying amplitude. Other arrangements, such as the simple bar magnet equipped with a figure-eight rf coil, which are not described herein for brevity, work in a conceptually similar way. See the review by Blümich *et al.*[11] for details. By the MOUSE technique, it is possible to sample the volume under a surface up to 2.5 cm depth or so, with a millimeter depth resolution. Finally, it is worth mentioning that it is even possible to obtain localized proton spectra with a portable open magnet with a fair spectral resolution (0.25 ppm) by shimming the field of an open magnet to high homogeneity over a large volume external to the sensor.[22]

The main application of MOUSE-NMR is undoubtedly in the achievement of the relaxometric profiles of an artefact by a suitable scan of its surface (for the general theory of NMR relaxometry the reader is referred to Appendix B of this book). Detection of both the longitudinal (T_1) and the transversal (T_2) relaxation of the ^{1}H nuclei can be exploited, in particular whenever the presence of moisture within a porous matrix can be investigated. In general, for fluids in a pore network, the longitudinal (R_1) and transversal (R_2) relaxation rates can be, respectively, expressed as

$$R_1 = \frac{1}{T_1} = \frac{1}{T_{1,\mathrm{bulk}}} + \rho_1\left(\frac{S}{V}\right) \tag{13.1}$$

$$R_2 = \frac{1}{T_2} = \frac{1}{T_{2,\mathrm{bulk}}} + \rho_2\left(\frac{S}{V}\right) + \frac{D(\gamma_\mathrm{H} G t_\mathrm{E})^2}{12} \tag{13.2}$$

where $T_{1,\mathrm{bulk}}$ and $T_{2,\mathrm{bulk}}$ are the longitudinal and transverse relaxation times of bulk water, S and V are the specific pore volume and the specific area of the material pore network, and ρ_1 and ρ_2 are the longitudinal and transversal specific relaxivities, respectively. Notably, the transverse relaxation is also affected by signal attenuation due to molecular diffusion through the magnetic field gradient. Hence, by application of the CPMG sequence, eqn (13.2) must take into account this effect with an additional term, where D is the diffusion coefficient, γ_H is the ^{1}H gyromagnetic ratio, G is the magnetic field gradient and t_E is the echo time. Unfortunately, eqn (13.1) and (13.2) work in the case of a homogeneous pore size distribution. In real cases where non-homogeneous distributions occur, both longitudinal and transverse relaxation can be expressed as a superimposition of different contributions, *i.e.* the relaxation kinetics can be expressed as a summation of elementary exponential contributions:

$$M(t) = M_\infty + \sum_i M_i \cdot \exp\left(-\frac{t}{T_i}\right) \tag{13.3}$$

where $M(t)$ is the magnetization of the sample as a function of time and M_∞ is the magnetization at equilibrium (more complex models are even possible[23]). Of course, the amplitude of the magnetization signal will be a function of the hydrogen atom density throughout the sample. Eqn (13.3), of course, can be adapted to both longitudinal and transverse relaxation. Even better, the real system can be viewed as a continuum of microdomains. Hence, the relaxation kinetics can be expressed as an integral form:

$$M(t) = M_\infty + M_0 \int_{T=0}^{\infty} D(T) \cdot \exp\left(-\frac{t}{T}\right) \mathrm{d}T \tag{13.4}$$

where $D(T)$ is a suitable distribution function, which can be retrieved as the inverse Laplace transform[24] of the relaxation kinetics data. The use of the UPEN algorithm[25] provides an excellent tool for the task. The latter type of analysis is particularly useful because it enables individuating at a glance the presence of different dynamic domains in the sample. Consequently, the presence of differently sized pore populations in a wet mortar for instance, or the presence of different molecular components in an organic material, having a different diffusivity, can be immediately envisaged.

13.4 Applications of NMR Techniques to Cultural Heritage: Selected Examples

As has been already mentioned, application of NMR in cultural heritage has experienced a large increase in popularity during the last few decades, and some very good and exhaustive reviews are available in recent literature.[12,26–28] Therefore, for the aims of this presentation, only some examples have been selected. They are relevant to the application of both CP-MAS and MOUSE, for their particularly significant and illustrative character.

13.4.1 CP-MAS Solid-state NMR

A first very intriguing aspect of solid-state NMR is the possibility to exploit it for the investigation of amorphous materials such as glass and amorphous polymers used as conservation materials.[29] The structure of the aluminosilicate glass network can be studied by the quantitative determination of the coordination sites of aluminum using ^{27}Al MAS NMR. In fact, in most cases, the Al spectrum of such materials consists mainly of two signals, of which one around 0 ppm is due to the six-coordinated aluminum species (*i.e.*, octahedral Al) and the other around 60 ppm is due to the four-coordinated species (*i.e.*, tetrahedral Al).[30] Through the observation of the ^{29}Si e ^{27}Al nuclides, the characterization of stone materials coming from archaeological sites can be performed in order to determine their state of conservation and to plan restoring interventions.[31] Nowadays, the isotopic abundance and the sensitivity are the only limitations of an NMR experiment. Most of the elements of the periodic table may be analysed in the liquid as well as in the solid state (Figure 13.4).

As long as other, less conventional, elements are concerned, the use of ^{119}Sn and ^{207}Pb NMR for studying the formation of lead soaps and the deterioration of oil paintings containing heavy metal-based pigments has been recently reviewed.[28]

An interesting application of NMR is represented by the analysis of archaeological rubber.[32] This natural polymeric material is slightly soluble and volatile, and techniques such as gas chromatography (GC–MS) or liquid chromatography (LC–MS) cannot be used. In the cited article, the authors demonstrated that the rubber obtained from the latex extracted from *Castilla elastica* possessed enhanced elastic properties due to a particular purification process. Recently, ^{13}C CP-MAS NMR and ^{13}C DP-MAS NMR techniques have been used to assess the state of degradation of archaeological leather by the analysis of the glycine, hydroxyproline and proline fractions of the collagen.[33] The molecular dynamics changes of the collagen were

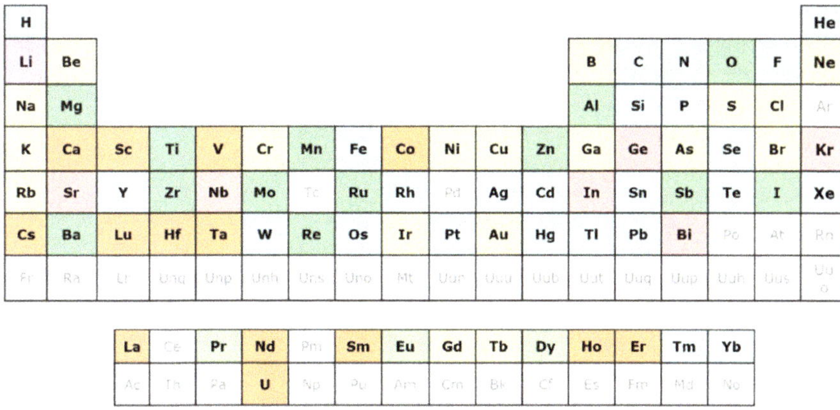

Figure 13.4 A periodic table where nuclei potentially detectable with NMR are evidenced. Different colours indicate nuclei having different spin quantum numbers *I*. Blue ($I = 1/2$), pink ($I = 1$), yellow ($I = 3/2$), green ($I = 5/2$), and orange ($I = 7/2$).

detected by the determination of the T_{1C} spin–lattice relaxation times and their values were correlated with the state of conservation of the samples. An interesting example of a novel diagnostic approach to paper is the one involving the analysis of compounds arising from the paper degradation. This kind of study was carried out by the application of HR-MAS NMR techniques, both 1D and 2D.[34] In particular, in all paper samples, carboxylic, dicarboxylic and hydroxycarboxylic acids were found which can trigger further degradation processes. Furthermore, NMR spectra can be used as a fingerprint of samples consisting of a complex mixture of analytes through the use of standards and references. This is the case of the evaluation of the geographical origin of archaeological amber samples.[35] NMR can also be applied in the diagnosis of the state of degradation of paper and parchment. Paper can in fact be characterized by the crystallinity degree determination of the cellulose and, if degradation caused by hydrolysis has occurred, by quantitative determination of the oligomers. Parchment degradation can be assessed by studying the shape of the signals of the amino acids forming the collagen and by evaluating the presence of fatty acids.[36]

A typical application field of solid-state NMR is undoubtedly constituted by cellulose-based materials, such as paper and wood. In particular, it is possible to study the interactions between ancient wood and varnishes or conservation materials and to evaluate the state of conservation even after the treatment. Few illustrative cases of study have been selected and are presented hereinafter.

In the first one, a set of varnishes were prepared in laboratory using different linseed oil/colophony ratios to reproduce the ones used by the great Italian violin maker A. Stradivari.[37] Indeed, varnishes used by Stradivari for his instruments are generally made by a mix of linseed oil and colophony[6] or metal rosinates in different ratios ranging between 75/25 (oil/resin) and

60/40 (oil/rosinate). Nonetheless, the influence of the mixture composition on the structural and/or functional features of the resulting varnish is still under study. Several solid- and liquid-state NMR techniques were applied to obtain information about the chemical–physical characteristics of the varnishes. Firstly, the authors found that the diffusion coefficient D_0 values obtained by diffusion ordered spectroscopy (DOSY) for linseed oil and colophony were clearly different for two components. This is because, before the varnish drying process, probably due to scarce interactions at the microscopic scale, linseed oil and colophony have different motional behaviours. This was confirmed by 2D-NOESY spectra. The cross peaks in the 2D-NOESY spectrum revealed only intramolecular dipolar interactions related to fatty acids composing the linseed oil. No cross peaks between the signals of linseed oil and colophony were observed in any of the analysed mixtures, and this indicates the absence of strong intermolecular interactions between linseed oil and colophony. From the analysis of the $T_{1\rho}$H relaxation times obtained by solid-state NMR, it is worth noting that linseed oil and colophony have shown different values in all samples, except for the 75/25 mixture. This indicates that, as $T_{1\rho}$H values determined for the resin and oil components are equivalent within the experimental error, the latter sample can be considered as a homogeneous material in a scale of tens of angstroms.

The second case reported here is the follow-up of a previous work dealing with the study of the conservation state of the wooden part of a roman rostrum (Figure 13.5) soon after its finding in the seabed of Acqualadroni

Figure 13.5 The roman Acqualadroni rostrum.

(Messina, Italy). The archaeological survey and recovery of this artefact were particularly relevant since it is at present one of the two rostrums ever found together with its wooden part.

In previous work,[38] the ^{13}C CP-MAS NMR spectral analysis was performed in order to obtain qualitative information on the presence of the components of the wood. For this purpose, a sample of fresh wood of the same species (*Pinus pinaster*) was analysed as a reference. Furthermore, by variable contact time (VCT) experiments, quantitative parameters such as the holocellulose/lignin ratio and the cellulose crystallinity index were determined and the state of conservation of the lignin was assessed by non-quaternary suppression (NQS) spectral analysis. Finally, information on the cellulose and lignin polymeric dynamics and on the interactions between them was obtained through $T_{1\rho}$H and T_1H relaxation time determination. In fact, it is possible to determine if wood behaves as a blend on a molecular scale (dynamically homogeneous) or as a mixture of non-interacting polymers, leading to distinct spin systems. Following these analyses, the wood of the rostrum was consolidated by immersion in a melamine-formaldehyde resin (Kauramin) aqueous solution for eight months at the "Centro di Restauro del Legno Bagnato" (Pisa, Italy).

A most recent paper[39] deals with the study of the state of conservation of the wood after ten years of the consolidation treatment. In particular, solid-state NMR spectroscopy was used to determine at a microscopic scale the wood state of conservation and to highlight the interactions of Kauramin with wood components, ten years after the consolidation treatment. A coring sampling of the wood was exceptionally authorized, and the obtained core was divided into seven aliquots as a function of depth.

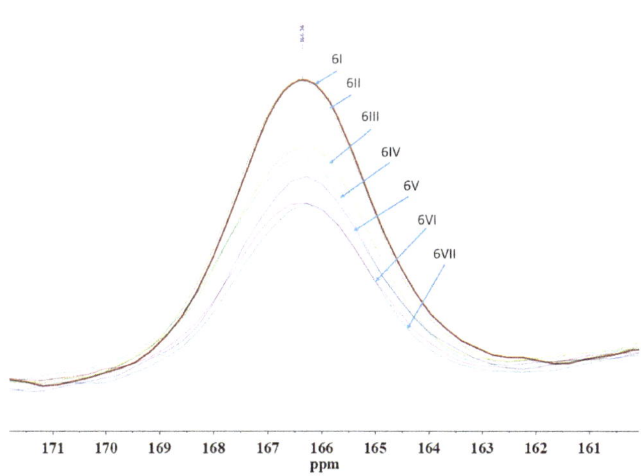

Figure 13.6 The signal at 167 ppm related to the quaternary carbons of the triazine rings of Kauramin. This indicates the presence of the consolidant up to at least a depth of 14 cm.

In order to determine the cellulose crystallinity index, the lignin condensation degree and the holocellulose–lignin ratio ^{13}C cross-polarization magic angle spinning (^{13}C{^{1}H} CP-MAS NMR) spectra and variable contact time (VCT) experiments have been carried out. These experiments revealed that the polysaccharide components were degraded mainly in the surface layers, while they were better preserved at greater depths. The surface samples showed high crystallinity index values, and the middle samples showed lower values of the crystallinity index, which remain constant up to a depth of 14 cm. This trend was an evidence that amorphous cellulose has largely been conserved in the deeper samples, while in the surface areas it has been degraded. In all the investigated samples, lignin remained almost unaffected. Solid-state ^{13}C spectra showed the presence of Kauramin in all the analysed samples (Figure 13.6), thus indicating the penetration properties of this consolidant. $T_{1\rho}$H relaxation time determination and two-dimensional ^{13}C–^{1}H correlation (FSLG CP HETCOR) NMR experiments were used to highlight possible interactions between the Kauramin and the wood components. By these experiments, the authors demonstrated the existence of interactions between the aromatic rings of lignin and the 1,3,5-triazine ring of Kauramin and of a network of hydrogen bonds between the consolidant and both components of the wood. These interactions can finally explain at a microscopic level the irreversibility of the treatment with Kauramin reported in the literature.

In a third example, an analytical protocol was set up by combining ^{13}C CP-MAS solid-state NMR with portable unilateral NMR and other techniques, and was successfully applied for studying an Egyptian wooden sarcophagus.[40] In this way, information was obtained on the organic substances as well as on the paramagnetic elements inside and on the surface of the wooden matrix. This study revealed also the presence of compounds such as polyvinyl acetate used for a former restoring intervention and a rigorous assessment of the conservation state of the artefact.

Many methods to identify drying oils, binders and natural resins coming from paintings after extraction with deuterated acetone involving NMR techniques are reported in the literature.[41] The hydrolysis and oxidation degree assessment of organic compounds such as linseed oil, egg tempera and an acrylic copolymer used as binders can be carried out by 1D and 2D NMR experiments. This analysis was carried out focusing on the detection of specific markers which are typical of oxidative and hydrolytic degradation phenomena. Di Tullio and Proietti recently reported on the use of ^{1}H HR-MAS-NMR for the characterization of varnishes and binders, such as linseed oil, shellac, sandarac and colophony resins on paintings.[42]

Finally, as long as conservative agents for porous stone materials are concerned, an interesting example concerns the behaviour of two different fluorinated polymers.[43] Two commercial materials (Fluo AQ and Fluoline HY) based on a fluorinated elastomer consisting of vinylidene fluoride and hexafluoropropylene copolymers were applied on samples of Cusa's stone to give it water repellence and vapour permeability. Solid-state NMR was used to investigate the interactions between the stone samples and the polymeric material and these chemical–physical interactions were correlated with the observed protection action. Scanning electron microscopy and X-ray microscopy were applied to investigate the distribution of the polymers inside the stone pores. The ^{19}F MAS NMR, $^{13}C\{^{19}F\}$ CP-MAS NMR and $^{13}C\{^{1}H\}$ CP-MAS NMR spectra were acquired to characterize all samples. The comparison between the NMR spectra of pure polymers and those of polymers after their application revealed no significant changes in the chemical environment of the observed nuclei and evidenced modifications in the organization and dynamics of the two protectives, as well as in the mobility of polymer chains (Figure 13.7).

These modifications were confirmed by the analysis of the T_1F and T_2F relaxation times. From the analysis of these values, the two protectives in contact with the walls of the pores were found to be less mobile probably due to a confinement effect also confirmed by the increase in the intensity of the spinning sidebands in the spectra of the polymers applied to the stone. In addition, the greater water repellence given to the stone by the polymer Fluo AQ can be attributed to its lower mobility associated with its higher degree of cross-linking. Indeed, a more homogeneous surface film can be the result of the cross-linking of the polymeric system.

13.4.2 MOUSE-NMR

NMR methods based on the measurement of the relaxation times and diffusion coefficients of water are particularly suitable for studying the

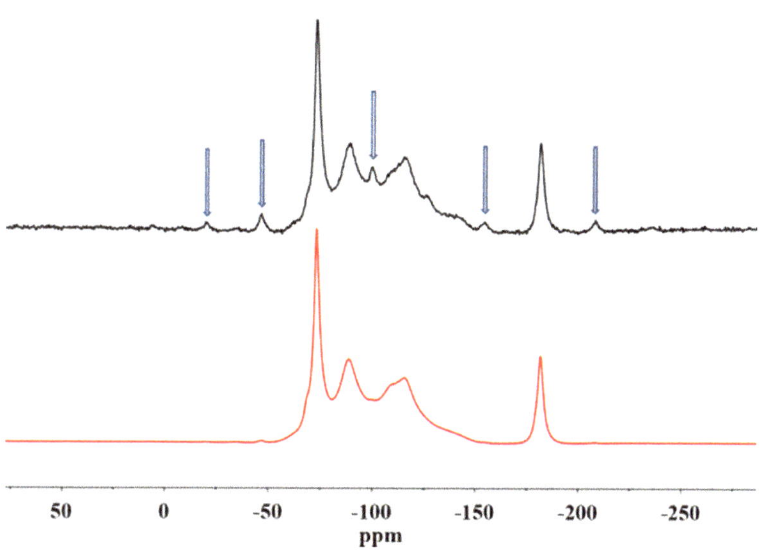

Figure 13.7 Solid-state 19F MAS NMR spectra of pure Fluorine HY (lower spectrum) and after the application on stone (upper spectrum).

structures of porous materials, which are largely present in cultural heritage. In fact, when a fluid is present in a confined environment, its nuclear relaxation parameters are strictly related to the pore size. In particular, the smaller the pores of a certain material, the faster the relaxation process of a fluid inside them. Examples of pore size distribution measurements of water-saturated natural rocks obtained by NMR compared with the results obtained with other techniques such as mercury intrusion porosimetry (MIP) are available in the literature.[44]

Analysis of ancient frescoes is undoubtedly a field in which the MOUSE technique can exploit its full potential. A particularly good example was reported in 2010 by Di Tullio *et al.*,[45] who examined a degraded wall painting of St. Clement Basilica, Rome. By mapping the moisture distribution at different depths from the surface, analysis of the moisture gradients enabled localization of the origin of a water infiltration. The same authors recently reported on the conservation risk assessment, which was similarly accomplished on the vault of the Turkish room in Villa Medici, Rome.[42] Another similar very recent example by Blümich *et al.* regards the study of wall paintings in Ostia Antica,[46] which allowed recognizing in various sites a discontinuity in the relaxation times and in the moisture content depth profile, revealing the presence of different superimposed painted mortar layers.

As long as easel paintings are concerned, MOUSE-NMR can be particularly useful in monitoring the state of varnishes[42] and paint binders,[47] and in monitoring the effect of aqueous based systems or organic solvents used for cleaning.[48,49] This is particularly important because paintings cleaning may constitute a particularly delicate and potentially harmful operation owing to its irreversibility.[42] In this context, time-dependent relaxometric profiling enabled collection of valuable information on the effect of the cleaning treatment on the paint layer of acrylic emulsion or vinyl-based model paintings.[48] Attempts were even made to assess paintings' ageing by relaxometric determinations, in order to assess a possible dating method.[47]

Additionally, in the case of frescoes and, more generally, of mortar substrates, MOUSE-NMR can be helpful in monitoring the state of stone artworks and building materials. The topic has been recently reviewed by Nagel and co-workers.[50] Keeping in mind that the signal intensity, *i.e.* the magnetization amplitude is proportional to the amount of H nuclei present in the sample,[51] moisture distribution profiles, capillary adsorption dynamics and time-dependent hydration of porous systems can be viably assessed. Studies of this type have been carried out on materials such as plaster, gypsum, limestone or cements.[52–55] Recently, Brown and Orlowsky[56] and Kunz *et al.*[57] reported on the long-term application of hydrophobic silanes, siloxanes and silicone resins for the preservation of natural stones (namely, Baumberger and Obernkirchener sandstone) from weathering. Hence, NMR in-depth profiling of moisture adsorption of differently exposure aged (up to 24 years) samples enabled determination of the efficacy and durability of the treatment.

MOUSE-NMR can be profitably exploited, as well as the CP-MAS techniques discussed in the previous section, also for studying ancient woods. In particular, Stagno and Capuani reported[58] on the characterization of a waterlogged Roman age pine wood (from the Roman harbor of Naples, 5th century AD). Analysis of the T_2^*-weighed images of the ancient wood in a suitable section (compared with a modern one) enabled evidencing very well physiological structures such as tracheids and resin canals, as well as the most degraded zones characterized by bacterial erosion and deposition of inorganic paramagnetic impurities. At the same time, relaxometric data enabled assessment of the presence of two different diffusion compartments, reasonably corresponding to two different pore size distributions, as well as the average diameter and the tortuosity of the channels. MOUSE-NMR has been profitably exploited in conjunction with other non-invasive techniques (optical coherence tomography, FT-IR, X-ray fluorescence) for the stratigraphic study of ancient Cremonese violins.[6]

13.5 Summary and Perspectives

Modern NMR offers a set of powerful and versatile tools for studying and monitoring the conservation status of diverse artworks and is able to provide valuable complementary information when used in conjunction with other spectroscopic or spectrometric techniques (FT-IR, HR-MS, XRD, OCT and so on). In particular, solid-state CP-MAS-NMR and portable MOUSE-NMR can be profitably used in this context. Of course, along with the physico-chemical nature and the features of the artefact under examination, and with the particular kind of investigation or monitoring action needed, the choice of the most suitable technique depends on various factors, including in particular the availability of samples obtained with negligible damage to the artefact itself or, otherwise, the impossibility to sample or move it.

From a conceptual standpoint, solid-state NMR focuses on the molecular/microscopic structure of the sample, the same way as in materials chemistry. Conversely, portable NMR exploits relaxometry (monitoring both longitudinal and transverse relaxation). It is important to stress that, unlike solution NMR (that is well and widely known as the main tool for retrieving the structure of molecular compounds), and despite the fact that the relevant instrumentation has already been commercialized for decades, CP-MAS and MOUSE are still viewed as niche techniques. Their potentialities are still underrated not only among non-specialist people (art historians, archaeologists, museum curators, art restorers, and so on) involved in cultural heritage preservation, but even in technicians with a solid chemistry or physics cultural background. A possible reason for this unfortunate fault might be envisaged in the very limited space left to non-conventional NMR techniques in undergraduate teaching courses. Therefore, it is the authors' prediction that these interesting techniques may benefit from increased attention in the future, so that NMR-based tools may at least gain the importance they actually deserve.

References

1. K. Vidović, S. Hočevar, E. Menart, I. Drventić, I. Grgić and A. Kroflič, *Environ. Sci. Pollut. Res.*, 2022, **29**, 46405–46437.
2. R. Hamilton and H. Crabbe, in *The Effects of Air Pollution on Cultural Heritage*, ed. R. Hamilton, V. Kucera, J. Tidblad and J. Watt, Springer US, Boston, MA, 2009, pp. 1–27.
3. R. A. Livingston, *Atmos. Environ.*, 2016, **146**, 332–345.
4. V. Comite, A. Miani, M. Ricca, M. La Russa, M. Pulimeno and P. Fermo, *Environ. Res.*, 2021, **201**, 111565.
5. D. Capitani, V. Di Tullio and N. Proietti, *Prog. Nucl. Magn. Reson. Spectrosc.*, 2012, **64**, 29–69.
6. C. Invernizzi, G. Fiocco, M. Iwanicka, M. Kowalska, P. Targowski, B. Blümich, C. Rehorn, V. Gabrielli, D. Bersani, M. Licchelli and M. Malagodi, *Microchem. J.*, 2020, **155**, 104754.
7. P. Baglioni and D. Chelazzi, *Chem. – Eur. J.*, 2021, **27**, 10798–10806.
8. E. D. Becker, *High Resolution NMR: Theory and Chemical Applications*, Elsevier, 1999.
9. G. Pagès, V. Gilard, R. Martino and M. Malet-Martino, *Analyst*, 2017, **142**, 3771–3796.
10. M. Spraul, A. Freund, R. Nast, R. Withers, W. Maas and O. Corcoran, *Anal. Chem.*, 2003, **75**, 1536–1541.
11. B. Blümich, J. Perlo and F. Casanova, *Prog. Nucl. Magn. Reson. Spectrosc.*, 2008, **52**, 197–269.
12. C. Rehorn and B. Blümich, *Angew. Chem., Int. Ed.*, 2018, **57**, 7304–7312.
13. M. Baias, *Magn. Reson. Chem.*, 2017, **55**, 33–37.
14. W. M. Awad and M. Baias, *Magn. Reson. Chem.*, 2020, **58**, 792–797.
15. B. Blümich and J. Anders, *Magn. Reson.*, 2021, **2**, 149–160.
16. K. J. MacKenzie and M. E. Smith, *Multinuclear Solid-state Nuclear Magnetic Resonance of Inorganic Materials*, Elsevier, 2002.
17. K. Schmidt-Rohr, K. Schmidt-Rohr and H. W. Spiess, *Multidimensional Solid-state NMR and Polymers*, Academic Press, 1994.
18. P. S. Aptaker, P. J. McDonald and J. Mitchell, *Magn. Reson. Imaging*, 2007, **25**, 548.
19. J. Perlo, F. Casanova and B. Blümich, *J. Magn. Reson.*, 2005, **176**, 64–70.
20. H. Y. Carr and E. M. Purcell, *Phys. Rev.*, 1954, **94**, 630–638.
21. S. Meiboom and D. Gill, *Rev. Sci. Instrum.*, 2004, **29**, 688–691.
22. J. Perlo, F. Casanova and B. Blümich, *Science*, 2007, **315**, 1110–1112.
23. M. Peyron, G. K. Pierens, A. J. Lucas, L. D. Hall and R. C. Stewart, *J. Magn. Reson., Ser. A*, 1996, **118**, 214–220.
24. V. Bortolotti, R. Brown, P. Fantazzini, G. Landi and F. Zama, *Inverse Probl.*, 2016, **33**, 015003.
25. G. C. Borgia, R. J. S. Brown and P. Fantazzini, *J. Magn. Reson.*, 2000, **147**, 273–285.
26. M. Wagner, J. Catalano, V. D. Tullio, R. Pigliapochi, N. Zumbulyadis, S. A. Centeno and C. Dybowski, in *Comprehensive Inorganic Chemistry III*

(Third Edition), ed. J. Reedijk and K. R. Poeppelmeier, Elsevier, Oxford, 2023, pp. 788–836.
27. N. Proietti, D. Capitani and V. D. Tullio, *Magnetochemistry*, 2018, **4**, 11.
28. J. Catalano, V. Di Tullio, M. Wagner, N. Zumbulyadis, S. A. Centeno and C. Dybowski, *Magn. Reson. Chem.*, 2020, **58**, 798–811.
29. R. Youngman, *Materials*, 2018, **11**, 476.
30. V. Alzari, E. Kamseu, C. Leonelli, A. Spinella, G. Malucelli, G. Bianco, D. Nuvoli, D. Sanna, F. Armetta and M. Rassu, *Mater. Lett.*, 2021, **295**, 129808.
31. D. Capitani, P. Ciccioli, V. Di Tullio, N. Proietti and N. Proietti, *Chem. Biol. Technol. Agric.*, 2017, **4**, 1–37.
32. D. Hosler, S. L. Burkett and M. J. Tarkanian, *Science*, 1999, **284**, 1988–1991.
33. Y. Zhang, Y. Li, X. Liu, Y. Wang, H. Tang, L. Qu, Y. Shang and W. Chen, *J. Cult. Herit.*, 2022, **58**, 179–185.
34. C. Corsaro, D. Mallamace, J. Łojewska, F. Mallamace, L. Pietronero and M. Missori, *Sci. Rep.*, 2013, **3**, 1–10.
35. J. B. Lambert, Y. Wu and J. A. Santiago-Blay, *J. Nat. Prod.*, 2005, **68**, 635–648.
36. A. E. Aliev, *Biopolymers*, 2005, 77, 230–245.
37. A. Spinella, M. Malagodi, M. L. Saladino, M. L. Weththimuni, E. Caponetti and M. Licchelli, *J. Polym. Sci., Part A: Polym. Chem.*, 2017, **55**, 3949–3954.
38. S. Bastone, A. Spinella, D. F. C. Martino, S. Tusa and E. Caponetti, *Microchem. J.*, 2016, **124**, 831–836.
39. A. Spinella, D. F. Chillura Martino, M. L. Saladino, F. Sammartino, F. Caruso and E. Caponetti, *Cellulose*, 2021, **28**, 1025–1038.
40. N. Proietti, F. Presciutti, V. Di Tullio, B. Doherty, A. M. Marinelli, B. Provinciali, N. Macchioni, D. Capitani and C. Miliani, *Anal. Bioanal. Chem.*, 2011, **399**, 3117–3131.
41. A. Spyros and D. Anglos, *Appl. Phys. A: Mater. Sci. Process.*, 2006, **83**, 705–708.
42. V. D. Tullio and N. Proietti, *Magnetochemistry*, 2020, **6**, 21.
43. V. Ciaramitaro, A. Spinella, F. Armetta, R. Scaffaro, E. F. Gulino, G. Kourousias, A. Gianoncelli, E. Caponetti and M. L. Saladino, *Appl. Sci.*, 2021, **11**, 5767.
44. R. L. Kleinberg, *Magn. Reson. Imaging*, 1996, **14**, 761–767.
45. V. Di Tullio, N. Proietti, M. Gobbino, D. Capitani, R. Olmi, S. Priori, C. Riminesi and E. Giani, *Anal. Bioanal. Chem.*, 2010, **396**, 1885–1896.
46. B. Blümich, E. Del Federico, D. Jaschtschuk, M. Küppers, K. Fallon, A. Steinfeld and P. Tomassini, *Heritage*, 2021, **4**, 4421–4438.
47. F. Busse, C. Rehorn, M. Küppers, N. Ruiz, H. Stege and B. Blümich, *Magn. Reson. Chem.*, 2020, **58**, 830–839.
48. P. Moretti, L. Cartechini and C. Miliani, *Anal. Bioanal. Chem.*, 2020, **412**, 1063–1075.
49. V. Di Tullio, R. Pigliapochi, N. Zumbulyadis, S. A. Centeno, J. Catalano, M. Wagner and C. Dybowski, *Microchem. J.*, 2023, **190**, 108582.

50. S. M. Nagel, C. Strangfeld and S. Kruschwitz, *J. Magn. Reson. Open*, 2021, **6-7**, 100012.
51. K.-J. Dunn, D. J. Bergman and G. A. LaTorraca, *Nuclear Magnetic Resonance: Petrophysical and Logging Applications*, Elsevier, 2002.
52. R. J. Gummerson, C. Hall, W. D. Hoff, R. Hawkes, G. N. Holland and W. S. Moore, *Nature*, 1979, **281**, 56–57.
53. C. Nunes, L. Pel, J. Kunecký and Z. Slížková, *Constr. Build. Mater.*, 2017, **142**, 395–409.
54. M. Fleury, G. Berthe and T. Chevalier, *Magn. Reson. Imaging*, 2019, **56**, 32–36.
55. T. Colinart and P. Glouannec, *Energy Procedia*, 2015, **78**, 1484–1489.
56. F. Braun and J. Orlowsky, *J. Cult. Herit.*, 2020, **41**, 51–60.
57. A. Kunz, M. Groh, F. Braun, S. Brüggerhoff and J. Orlowsky, *J. Cult. Herit.*, 2022, **53**, 176–183.
58. V. Stagno and S. Capuani, *Acta IMEKO*, 2022, **11**, 10.

CHAPTER 14

Detection of Adulterations and Contaminations in Food Products by NMR

V. DOMENICI

Dipartimento di Chimica e Chimica Industriale, University of Pisa, *via* Moruzzi 13, 56124 Pisa, Italy
Email: valentina.domenici@unipi.it

14.1 Introduction

All food can be in principle the object of adulterations; however, the most targeted food and agricultural products for adulterations are those produced in high quantities around the world and those of high quality and commercial value.[1] Several types of fraud in food are known, such as dilution (*i.e.* dilution of wine with water or dilution of extra-virgin olive oil with low quality olive oils), addition (*i.e.* the addition of sugar or syrup to honey or juices), mislabelling (*i.e.* false labels stating the botanic or geographic origin of food) and substitution (*i.e.* substitution of herbs and spices in food products). Several strategies can be adopted to search for specific frauds. As a general approach, it is possible to detect adulterations by the identification and quantification of specific markers, such as chemical compounds present in a particular agricultural product (molecules associated to a plant, such as, for instance, polyphenols, organic acids, amino acid derivatives and so on) or biological and morphological markers, such as pollens in honey and similar products.[1–3] Another possibility is to search for a group of chemical compounds (such as fatty acids in olive oils) or chemical–physical properties (interval of pH, water content or conductivity), which are specific

New Developments in NMR No. 32
The Environment in a Magnet: Applications of NMR Techniques to Environmental Problems
Edited by Pellegrino Conte, Delia Francesca Chillura Martino and Paolo Lo Meo

Published by the Royal Society of Chemistry, www.rsc.org

to a type of food: if a product has a different combination of chemical compounds, or an anomalous value in the set of measured chemical–physical properties, this can be a sign of adulteration.[3,4] A third approach consists in taking into account the whole metabolomic profile or, more realistically, a portion of the metabolomic profiles, namely a large group of metabolites, which are specific to a product in terms of chemical composition and relative concentrations. This last approach is nowadays the most common in food science, since the complexity of food matrices and the extremely high number of chemical compounds that can be detected by means of different analytical techniques allow a detailed characterization of a food product, with the possibility of classifying food samples based on biological, genetic or geographic provenance.

From the experimental point of view, each approach has been explored by almost all techniques: chromatography, such as high-performance liquid chromatography (HPLC) and gas chromatography (GC) coupled with mass spectroscopy (MS), atomic absorption (AAS) or atomic emission (AES) spectroscopy, calorimetry, such as differential scanning calorimetry (DSC), molecular spectroscopies, such as Fourier transform infrared (FT-IR), UV-vis absorbance spectroscopy and nuclear magnetic resonance (NMR).[4–9] As it will be shown throughout this chapter, in many cases the detection of adulterations in food by means of NMR spectroscopic techniques is associated to a massive use of chemometric methods.[10] These mathematical and statistical methods are necessary to deal with complex systems, such as food and agricultural products, where many chemical and physical variables concur to define and characterize them.[7] Several methods can be adopted starting from the simplest one, called principal components analysis (PCA), aimed to find analogies and differences among samples which can then be organized into groups. The classification and discrimination of samples is usually performed also with other methods, such as linear discriminant analysis (LDA) or soft independent modelling of class analogy (SIMCA). These methods are used to build 'fingerprint' models and to predict the classification of a new sample. The presence of a sample which is placed outside of a group with known classification parameters may reveal an adulterated sample. Principal component regression (PCR) and partial least squares (PLS) regression are the additional methods applied to the multivariate analysis of large and complex data sets such as whole NMR spectra.[7]

14.2 NMR Methodologies to Detect Adulterations in Food Products

Similarly to other analytical methods, NMR spectroscopy can be used to characterize food quality and authenticity, as well as to detect adulterations. A first classification of NMR methods can be done between targeted and untargeted approaches.[4,6] Targeted metabolite analysis aims for the identification and quantification of specific molecular markers (*i.e.* a single marker or a restricted group of metabolites), mainly by means of high resolution NMR methods. More often, untargeted approaches are used to identify and characterize the entire metabolite profile, also referred to as 'fingerprinting' methodologies. Several NMR techniques, often coupled with multivariate statistical analysis, were developed giving a very important contribution to metabolomic science.[6]

As shown later in this chapter, NMR methods used to detect food adulterations can also be classified into high resolution and low resolution, and/or low field, NMR techniques. The principles of these methods and the range of applicability can be rather different among them; however, in all cases, some general aspects need to be taken into consideration in order to get reliable, reproducible and efficient NMR methodologies:[4] sample collection and sample size, sample preparation and eventual sample treatment, data acquisition and experimental parameters, data analysis including eventual pre- and post-data treatment, interpretation of data and eventual multivariate statistical analyses. These aspects are part of the so-called 'experimental design' and they need to be optimized to produce a 'protocol' which should help for inter-laboratory comparison and for the standardization of NMR procedures.[5,6]

Depending on the specific NMR technique adopted to characterize food and identify possible adulterations, NMR spectroscopy has several advantages with respect to other analytical methods.[9] For instance, high resolution NMR methods (reported in Section 14.2.1), such as ^{1}H NMR spectroscopy, are characterized by high selectivity, the capability to detect simultaneously a great number of chemical constituents even at low concentrations, simple sample preparation and very high reproducibility. Low resolution (sometimes referred to as low field) NMR techniques (discussed in Section 14.2.2), such as low field NMR relaxometry, are usually fast methods, do not require any sample treatment or a minimal sample preparation, and can be performed on portable instruments. Magnetic resonance imaging (MRI) techniques (see Section 14.2.3) are non-destructive methods, suitable for on-site and in-line screening of food and agricultural products.

14.2.1 High Resolution NMR

High resolution NMR techniques are usually exploited for quantitative and structural investigations of any molecular system containing magnetic nuclei, namely nuclei with a non-zero intrinsic magnetic moment (I), such as ^{1}H, ^{13}C, ^{31}P, ^{15}N and so on. The number of studies based on high resolution NMR spectroscopy in foodomics, defined as the field of science that studies food and nutrition by applying the typical 'omics' technologies with the aim to analyse the effects of food on human health, integrating genomics, transcriptomics, proteomics and metabolomics, is increasing fast in recent years.[4] The reasons of the progressive success of high resolution NMR spectroscopy in foodomics is mainly related to the very high reproducibility of NMR spectra, which can be achieved by controlling a relatively low number of parameters (*i.e.* magnetic field strength, magnetic homogeneity,

NMR sequence's parameters and post-processing parameters). Moreover, high resolution NMR spectra can be easily organized in a database and they can in principle be used directly for classification of food, without analysing them, in the untargeted approaches. On the other hand, several applications are reported where high resolution NMR spectroscopy is applied for the identification and quantification of single molecular markers, present at low concentrations (from a few to tens of ppm). The quantification of the absolute concentration of food constituents also refers to the so-called 'quantitative NMR' (qNMR) techniques, and it is used to detect adulterations in targeted approaches. In the following paragraphs, significant examples of both methodologies are reported and discussed based on ^{1}H NMR, ^{13}C NMR and other nuclei NMR.

14.2.1.1 ^{1}H NMR

High resolution NMR techniques gave a great contribution to food safety, authenticity and fraud detection. Most of these studies focus on the ^{1}H nucleus, due to its highest sensitivity and to the fact that all organic molecules present in food contain protons. When dealing with high resolution NMR spectroscopy, the typical operating magnetic field corresponds to ^{1}H Larmor frequencies between 400 MHz and 600 MHz, even though in recent years, thanks to technological improvements, benchtop NMR machines working at 40–90 MHz can be used to acquire high resolution NMR spectra, too. ^{1}H NMR techniques are usually very fast, reproducible and robust. The disadvantages are related to the cost of NMR instrumentations and to the problem of ^{1}H signal overlapping, which can occur when analysing complex molecular systems due to the relatively small spectral window of ^{1}H signals (maximum of 10–12 ppm). As shown in the following, depending on the physical properties of food and on the typical chemical composition, the acquisition of NMR spectra may require a physical and/or chemical sample treatment, such as solvent extraction or pH stabilization. In some cases, however, ^{1}H NMR spectra are acquired without any sample treatment. In the following, high resolution ^{1}H NMR studies are overviewed focusing on the possibility of identifying and detecting adulterations in different food matrices.

Among food products, virgin (VOO) and extra-virgin olive oils (EVOO) have been the object of intense NMR studies and many different techniques have been developed in order to evaluate olive oil quality as well as to identify frauds.[11–15] Common adulterations of olive oil involve the use of unqualified oils, such as fried recycled waste cooking oils, refined or treated oils, such as decolorated or deodorized oils, and cheaper edible oils mixed with higher quality olive oil in different percentages. Despite the large number of applications of NMR spectroscopy combined with chemometric techniques to the analysis of EVOO and VOO, olive oil remains one of the most fraudulent food products. This aspect is one of the main motivations for research into new reliable methods based on NMR spectroscopy. Most of the high resolution ^{1}H NMR techniques applied to the identification of olive oil adulterations focus their attention on the saponifiable part of olive oils, which represent 98–99% of the whole mass. This part is mainly composed of triglycerides, formed by glycerol units and fatty acids. From the ^{1}H NMR point of view, all edible oils can be indeed characterized by signals ascribable to saturated fatty acids, such as palmitic and stearic acids, unsaturated fatty acids, such as oleic, linoleic and α-linolenic acids, triglycerides and 1,3′-diglycerides. In minor amounts, compounds such as polyphenols and their derivatives have been identified by ^{1}H NMR spectroscopy in high quality EVOOs, too. As reported in ref. 11 and shown in Figure 14.1(a), some regions of the proton NMR spectrum are very sensitive to the amount of linolenic acid (triplet centred at 0.98 ppm), which is detectable in soybean oil (SO) and not in other oils, such as peanut oil (PE).

The spectral region between 1.98 and 2.08 ppm (see Figure 14.1(b)) is quite different in peanut (PE), sesame (SE) and rapeseed (RA) oils with respect to corn (CO), soybean (SO) and other seed (SS) oils, due to the different contents in linolenic and linoleic to oleic acids.[11] The total amount of saturated fatty acids can be derived from the integral of peaks centred at 1.255 and 1.300 ppm, which is indeed higher in peanut (PE) oil (see Figure 14.1(c)). Based on these considerations, ^{1}H NMR spectroscopy is used to determine the relative content of fatty acids in edible oils. In particular, the contents of linolenic, linoleic, oleic acids and the total amount of saturated fatty acids can be easily calculated from the integrals of specific proton signals in the high resolution ^{1}H NMR spectrum. In addition, the characteristic ^{1}H NMR spectra of unqualified oils, such as fried oils and waste cooked oils, have been investigated[11] and some significant differences with edible oils are shown in Figure 14.2.

It is interesting to note that waste cooking oils present an intense peak due to water (see the region between 3.0 and 3.5 ppm), which is related to the hydrolytic rancidity. The aromatic region, between 5.8 and 7.3 ppm, is within the reach of signals in fried and waste cooking oils due to the production of

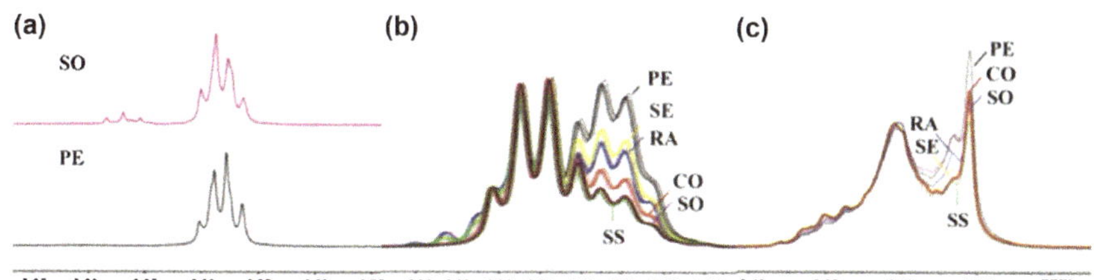

Figure 14.1 Characteristic ^{1}H NMR peaks in the different species of edible oils: (a) linolenic acid in rapeseed oil and soybean oil; (b) the ratio of linolenic acid to oleic acid and linoleic acid in the different species of edible oils; and (c) saturated fatty acids in the different species of edible oils. Labels: CO, corn oil; PE, peanut oil; RA, rapeseed oil; SE, sesame oil; SO, soybean oil; SS, seed oil. Reproduced from ref. 11 with permission from Elsevier, Copyright 2022.

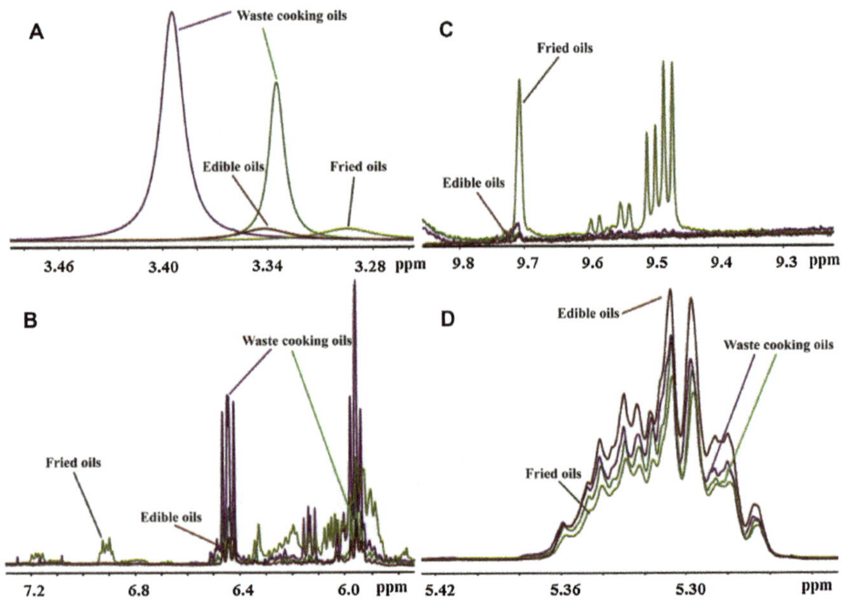

Figure 14.2 ^{1}H NMR spectral characteristics of waste cooking oils, fried oils and qualified edible oils: (A) spectral region of water; (B) spectral region of aromatic hydrocarbon; (C) spectral region of aldehyde; and (D) spectral region of methine with double bonds. Reproduced from ref. 11 with permission from Elsevier, Copyright 2022.

polycyclic aromatic hydrocarbons occurring during the refining and recycling processes. These molecules, which usually are not present in edible oils, are harmful to humans. The region typical of aldehydes, from 9.2 to 9.8 ppm, is very different between fried oils and all other oils, thus indicating the damage of the pristine edible oils due to high-temperature treatment. Analogous effects can be observed in the region between 5.20 and 5.40 ppm typical of the methine peaks of double bonds. These spectral differences are at the basis of several protocols that use ^{1}H NMR spectroscopy as a fundamental tool in detecting frauds in olive oils.

The complexity and variety of information that can be achieved from the analysis (qualitative and quantitative ones) of the ^{1}H NMR spectra of oils stimulated the development of expert systems, such as the one reported in ref. 11 based on a software used for the automatic recognition of possible adulterations, which combine the NMR data with quality parameters, such as the acid, iodine, carbon and hydroxyl values. Results reported in the literature concerning the potentialities of ^{1}H NMR spectroscopy as an analytical tool to detect adulteration of olive oils imply the use of chemometric analysis as well as the evaluation of analytical protocols tested by different laboratories, as reported by Mannina *et al.*[12] In this work, the ability of a ^{1}H NMR-based analytical protocol in detecting the presence of low levels (10%) of refined hazelnut oil in olive oils was demonstrated. Similarly, other works[13–15] showed how high resolution ^{1}H NMR methodologies combined with chemometrics were successfully applied to the detection of frauds in the field of olive oils.

Another example of detection of contaminants in food matrices by using high resolution ^{1}H NMR spectroscopy is that reported in ref. 16 dealing with carbonated soft drinks. The analysis by means of ^{1}H NMR spectroscopy of beverages, such as beer and wine, fruit juices and soft drinks,[8] has the advantage of a minimum treatment of the sample, which usually consists in adding a small amount of a solution with an internal standard, for quantitative analysis, or a reference sample, for the chemical shift scale, and a minimum amount of deuterated solvent needed for NMR measurements. In the case of carbonated soft drinks, the knowledge of characteristic ^{1}H NMR signals, ascribable, for instance, to glucose, sucrose and citric acid, and the construction of a spectral database obtained from a high number of spectra of soft drinks, are at the basis of the methodology proposed by Charlton *et al.*[16] The authors have registered the ^{1}H NMR spectra of three contaminants, namely, *p*-cresol, paraquat and glyphosate, deliberately added in different amounts to several soft drink samples. Paraquat and glyphosate can be present in commercial drinks, but they are classified as toxic, while *p*-cresol is responsible for a particular off-taste. In this work,[16] a supervised multivariate statistical method, namely the SIMCA model, has been used to discriminate contaminated from not contaminated soft drinks, showing that the combination between high resolution ^{1}H NMR spectroscopy and chemometrics is successful in detecting the presence of these contaminants at a level much below the level that can pose serious risks to human health.

Other food samples in the liquid state, for instance, obtained by diluting food powders, as in the case of milk powders,[17] or obtained from fermentation processes, such as the sourdough fermentation of *Triticum dicoccum*,[18] are good examples of food matrices which can be analysed without complex and time-consuming chemical treatments of the samples before acquiring high resolution ^{1}H NMR spectra. In the case of milk powders, the combination of liquid state ^{1}H NMR and non-targeted methods, based on the conformity index analysis on a large NMR spectral data set was able to detect the presence of several contaminants, such as melamine, at low concentrations.[17]

In solid food matrices, such as fruits, meat and non-water-soluble powders, and in soft matrices, such as cheese and honey, a physical/chemical treatment of the sample is needed before acquiring any high resolution ^{1}H NMR spectra. It is the case of another study[19] reporting a NMR-based method to detect adulterations in *Dendrobium officinale*, which is used as herbal medicine or functional food, and it is subject to adulterations with low-cost herbs or *Dendrobium officinale* powder obtained from leaves instead of stems. Here, the identification of specific spectral features on a large number of ^{1}H NMR spectra obtained on microwave-assisted extracts of pure *Dendrobium officinale* stem powder samples was followed by the development of a machine-learning kernel based on a least-squares support vector machine model.[19] This model was used for classification of unknown powders

Phloridzine (P) Epicatechin (E) Catechin (C) Chlorogenic Acid (A)
P
E
C
A
ppm
7
6
Ruggina
Nesta
Sassola
Mantovana

and for prediction purposes, and it was able to distinguish adulterated *Dendrobium officinale* powders with an overall accuracy of 98%. Usually, when the analysis of samples requires the preparation of extracts from the pristine food sample, special care has to be taken to the extraction procedure and its reproducibility and effectiveness in separating different classes of chemical compounds. As reported in ref. 20 for the detection of peanut adulteration in food samples, several extraction procedures can be examined and the identification of chemical markers in the high resolution ^{1}H NMR spectrum is a fundamental step toward the development of a reliable method. The identification and quantification of specific markers, which is typical of targeted NMR methods, is done both to identify adulterations, as in the case of peanuts,[20] and to assess the botanic or geographic origin of food samples, as in the case of apples produced in a specific region from different apple plants.[21] In this last case, apple extracts were prepared with an optimized procedure[21] and high resolution ^{1}H NMR spectra were analysed to identify and quantify specific compounds, namely phloridzin, epicatechin, catechin and chlorogenic acid. The results were then analysed by means of an unsupervised multivariate method, namely the PCA method, which is widely used to distinguish samples on the basis of botanic and/or geographic origins (see Figure 14.3).[21]

Unsupervised multivariate statistical methods, such as PCA or pattern recognition, and supervised ones, such as orthogonal partial least squares discriminant analysis (OPLS-DA), are also used to analyse the whole ^{1}H NMR spectra, in the so-called ^{1}H NMR metabolomic fingerprinting approach. For instance, several prediction models were built based on ^{1}H NMR metabolite fingerprint analysis to discriminate adulterated stingless bee honey from non-adulterated honey.[22] In another recent work,[23] the authentication and the prediction of authentic and adulterated saffron (*Crocus sativus* L.) was carried out by developing a ^{1}H NMR metabolite fingerprinting approach. Other untargeted ^{1}H NMR methods were successfully applied to the identification of adulterations in coffee[24] and paprika powders.[25]

14.2.1.2 ^{13}C and Other Nuclei

Proton NMR spectroscopy is characterized by a very high sensitivity, but a relatively small spectral window; in contrast, carbon-13 NMR suffers from a much lower sensitivity, due to the low natural abundance of the carbon-13

Figure 14.3 Centre: example of the ^{1}H NMR spectrum of an extract sample obtained from apple dried pulp. The region of interest (5–8 ppm) is shown with the assignment of the proton signals of the polyphenolic moieties of the different analytes present in the extract: phloridzin (P), epicatechin (E), catechin (C) and chlorogenic acid (A). The molecular structure of the four analytes and their NMR spectra are also shown. Examples of ^{1}H NMR spectra and characteristic signals of extracts obtained from four types of apples (Mantovana, Sassola, Nesta and Ruggina) are reported. Reproduced from ref. 21 with permission from Elsevier, Copyright 2022.

isotope, but it is characterized by a large spectral window (up to 170 ppm). This last feature is at the basis of the high potentiality of ^{13}C NMR spectroscopy in detecting specific chemical compounds even in very complex systems, such as food, since the problem of signal overlapping is generally not present. ^{13}C NMR spectroscopy has been successfully used to detect adulterations in several food samples, such as olive oil[12,26] and wine.[27,28] A combination of 1D, 2D ^{1}H and ^{13}C NMR methods have been applied to the identification and quantification of amino acids, such as alanine, valine, leucine, isoleucine and proline, and organic acids, such as acetic, tartaric, lactic, succinic, and malic acids, in wines.[27,28] This method was at the basis of a reliable distinction among wines of different varieties, vintage and geographical areas. In the case of olive oils, ^{13}C NMR spectroscopy is able to fully characterize the saponifiable fraction[26] as well as some minor components, such as squalene,[12,29] which is present in extra-virgin olive oils at higher concentrations with respect to lower grade olive oils. Recently, ^{13}C NMR spectroscopy was used to recognize adulterant vegetable oils in essential oils commonly used in food industries[30] and to distinguish milk produced by different animals.[31] The refocused adiabatic ^{13}C INEPT (insensitive nuclei enhanced by polarization transfer) sequence was used to build a solid and relatively fast method[32] to discriminate food containing different edible oils and fats, such as butters, margarines and animal fats (see Figure 14.4).

As a first step, Guyader *et al.*[32] have recorded and compared the ^{13}C NMR spectra of several classes of oils and fats used in the food industry, such as sunflower, canola and palm oils, and butter, showing that in the typical ^{13}C NMR spectral region of olefins (from 126 to 132 ppm), these spectra have different features due to the different chemical compositions of the fatty acids and triglycerides' components (see Figure 14.4). The obtained spectral database was then used to create two models based on linear discriminant analysis (LDA):[32] the first one was used to quantify the discrimination between samples of butter and plant oils, while the second one was used to detect the presence of palm oil in food products, such as spreads and biscuits (see Figure 14.5).

The first step was the fat extraction from the finished food products, which was necessary due to the low and variable percentages of fat, generally not higher than 20% of the whole food matrix. A selection of extracts of biscuits, spreads and cakes having a single type of fats or oils were compared to the spectral database previously built (see Figure 14.5). The second step was the development of a non-targeted chemometric approach applied to the whole NMR spectral fingerprints of commercial food products. Several tests were carried out to detect the presence of palm oil in spreads and biscuits, showing a precision of 100% on the set of samples analysed.[32]

Among other magnetic nuclei, few works have been published to identify adulterations in food, based on ^{19}F and ^{31}P high resolution NMR techniques. ^{19}F NMR spectroscopy has been applied to the trace analytical detection of fluorinated pesticides in vegetable oils and wines with high sensitivity,[33] while ^{31}P NMR spectroscopy has been used to analyse the contents of

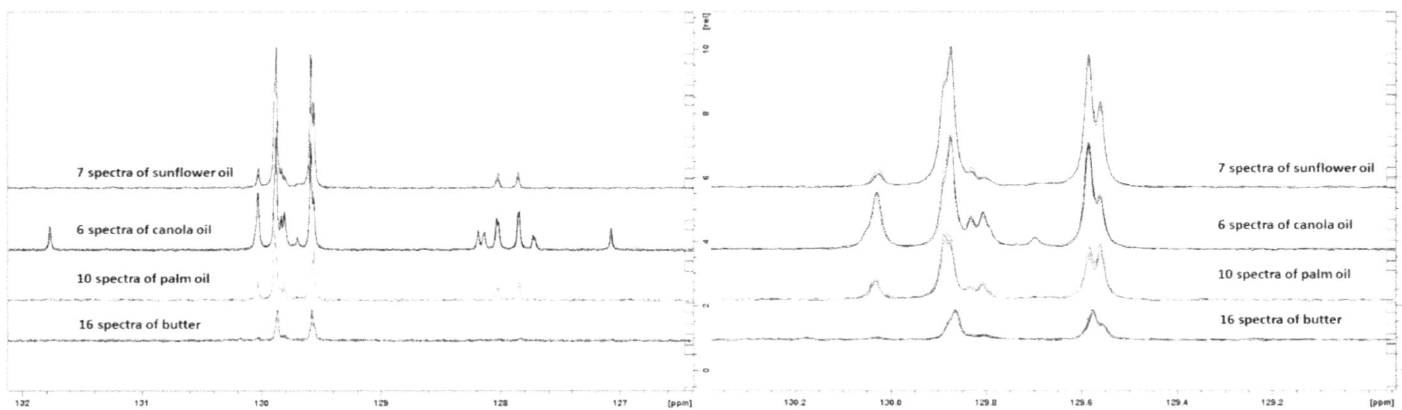

Figure 14.4 Superposition of the ^{13}C spectral database for 4 different origins of raw fats (on the left) between 126.5 ppm and 132.0 ppm. On the right, a typical spectral region between 28.6 ppm and 30.0 ppm is reported. Reproduced from ref. 32 with permission from Elsevier, Copyright 2022.

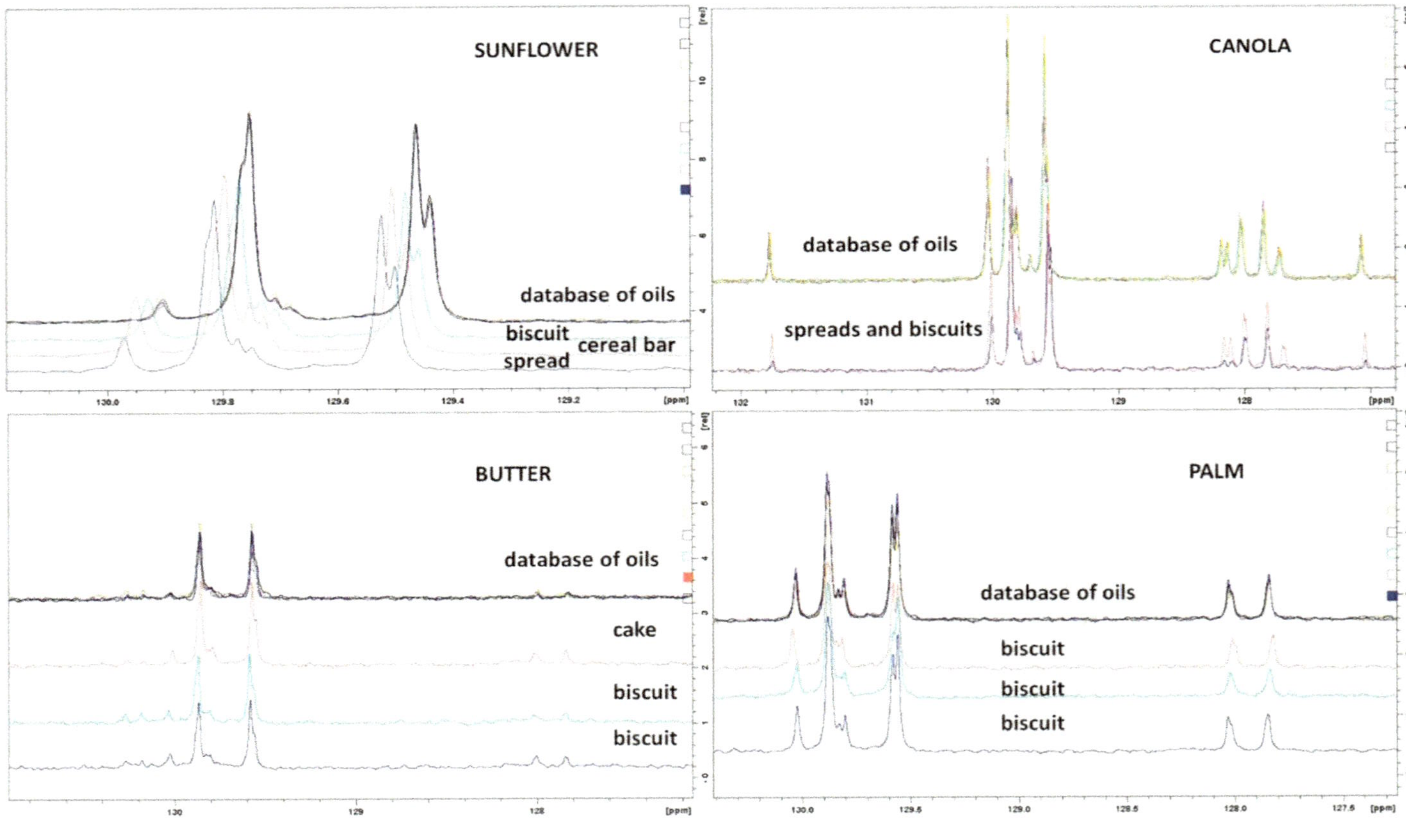

Figure 14.5 Superposition of ^{13}C NMR spectra acquired for some finished product extracts (spectra at the bottom of the figure) taken as examples with the corresponding database (shown on the top of the figure) for the origin of [top left] sunflower, [top right] canola, [bottom left] butter and [bottom right] palm. Reproduced from ref. 32 with permission from Elsevier, Copyright 2022.

phosphates and their derivatives in different types of milk.[34] A method was optimized[35] to detect the presence of compounds that contain free –OH groups in olive oils; the protocol implies the reaction with a phosphorus derivatizing agent and the acquisition of ^{31}P NMR spectra. This method has been applied to the study of thermal degradation of olive oils and to identify the presence of products of degradation in commercial oils.[36]

14.2.2 Low Resolution and Low Field NMR Techniques

High field NMR is nowadays considered the 'standard' NMR, even though, from the historical point of view, low field NMR instruments were the first ones, fundamental for many discoveries from the chemical shift to the 2D correlation experiments.[37] Technically speaking, NMR working at ^{1}H Larmor frequencies below 100 MHz is called low field NMR. From the very beginning, low field NMR was characterized by permanent magnets or electromagnets, which are now becoming quite attractive, thanks to the miniaturization and electronics advancements. These technological improvements led to the development of cheap commercial portable and tabletop NMR instruments, opening many applications also in food analysis. As shown in the following paragraphs, some of these recent compact NMR machines have a very high sensitivity and they can be used to acquire high resolution NMR spectra, even though the ^{1}H Larmor frequency is less than 100 MHz.[37]

Low resolution ^{1}H NMR techniques generally refer to several different techniques, such as ^{1}H NMR fast field cycling relaxometry, time-domain (TD) and low field (LF) ^{1}H NMR relaxometry, as well as ^{1}H NMR diffusometry and magnetic resonance imaging (MRI). Many applications of both low resolution and low field NMR techniques are known in the field of quality control of agricultural products and foods and comprehensive reviews can be found in ref. 38–41. In the following paragraphs, we will focus on applications of these techniques specifically to detect food adulterations and frauds.

14.2.2.1 Time-domain NMR and Low Field NMR Relaxometry

Non-invasive food analysis is based on NMR instruments with permanent magnets having a wide bore of gap where food and other products can be directly inserted. Several configurations and set-ups exist in low field NMR which are also used to perform in-line screening and on-site fast analysis. The magnetic field strength in typical low field NMR is below 0.7 T and the field inhomogeneity of the field, which is usually higher than 100 ppm, does not allow recording of NMR spectra, while it is perfect to measure relaxation times and diffusion coefficients.[38] Most of the applications related to low field NMR relaxometry imply the acquisition of relaxation times, longitudinal (T_1) and transverse (T_2), by means of different NMR sequences, such as inversion recovery (IR) and Carr–Parcell–Meiboom–Gill (CPMG) spin-echo train sequences, to measure T_1 and T_2, respectively. Relaxation times, T_1 and T_2, can be correlated to the dynamic properties of protonated molecules, which can be very different depending on the local environments, on the physical properties of the materials, such as viscosity, and on the presence of different phases (liquid–liquid, liquid–solid, liquid–gas, ...) typical of heterogeneous systems. For these reasons, low field NMR techniques have been largely adopted in quality analysis of foods, such as meat, cheese, and several types of fruits, as reported in a recent review by Moraes and Colnago.[38] Several applications are known concerning the possibility of using non-invasive time-domain NMR methods to detect adulterations in food matrices. An example of a non-invasive study of kiwifruits, based on a portable NMR instrument working at a low magnetic field, is reported in ref. 42 and shown in Figure 14.6.

In this work,[42] the authors analyse the CPMG decays (magnetization *versus* time) in kiwifruits at different stages of ripening to check their water content, since T_2 is generally more sensitive than T_1. As shown in Figure 14.6, portable low field NMR is unable to measure T_2 relaxation times at a depth of about 0.5 cm from the peel surface. The trends of CPMG decays of nine kiwi samples during two ripening campaigns (*i.e.* August and December) are presented in Figure 14.7(A) and (C), showing that the decays measured in

Figure 14.6 (A) Measurement of intact kiwifruit with a portable unilateral NMR instrument. (B) Section of kiwifruit showing the depth of measurement with a portable NMR instrument. Reproduced from ref. 42 with permission from Elsevier, Copyright 2022.

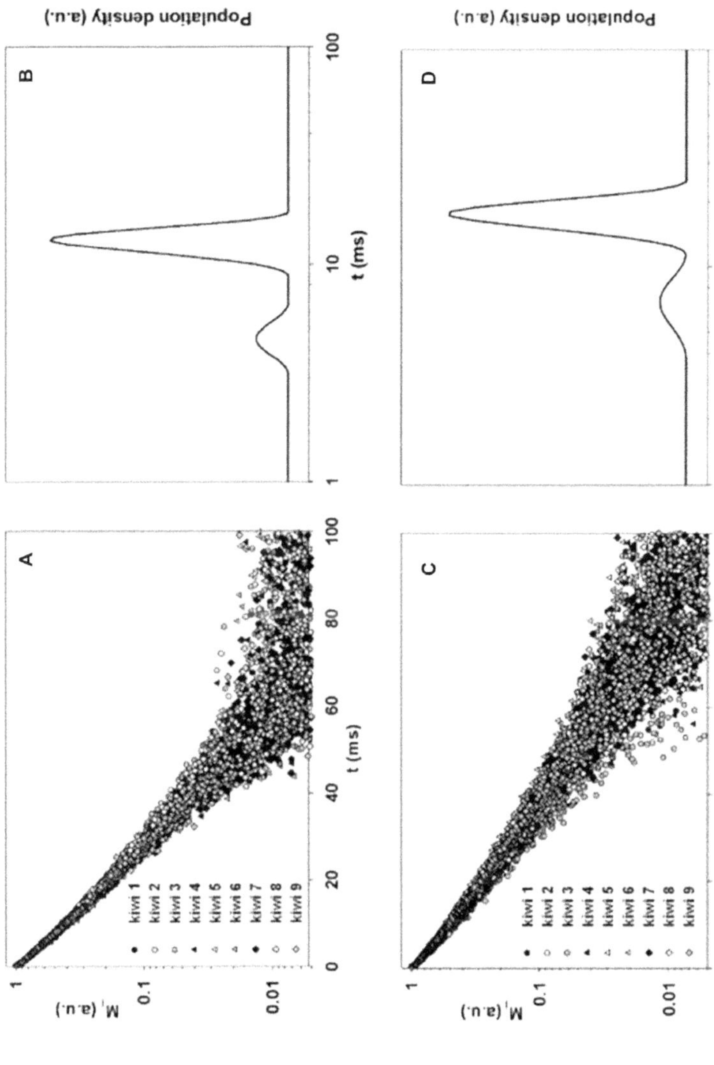

Figure 14.7 CPMG trends (*i.e.* decay of the magnetization M_i versus time, after applying the CPMG pulse sequence) measured on nine kiwifruits in August (A) and in December (C) (a.u. = arbitrary units). As an example, a T_2 distribution measured in August (B) and one measured in December (D) are reported: the maxima of the distributions are centred at the corresponding most probable T_2 values, whereas peak areas correspond to the relative population of the two T_2 values. Reproduced from ref. 42 with permission from Elsevier, Copyright 2022.

December are all much longer than those measured in August. The spread of the decays observed in both CPMG decays reflects small differences in the state of ripening among the investigated samples. Usually, to better visualize the results, it is preferred to apply the inverse Laplace transform (ILT) to the relaxation decays and to see how the relaxation times (in this case, T_2) are distributed in the food samples.

As shown in Figure 14.7(B) and (D), ILT plots clearly show the presence of two maxima (peaks) of the distribution, thus indicating two most probable T_2 values, which are probably associated to different types of 'water' in the fruits, which is typical of heterogeneous systems with different compartments (*i.e.* bulk water, water in intercellular and intracellular spaces). The position of the maxima of T_2 reported for kiwifruits studied in August (B) and in December (D) is indicative of the different ripening stage. TD NMR relaxometry is indeed a very sensitive technique to detect water distribution in food, which can be associated to several physical and chemical properties of food, as well as food storage conditions, heating or cooling treatments and physical damage.

Analogous studies have been performed on mayonnaise and salad dressing, analysed without removing their packages,[43] to get information about the fat content. In this work, the CPMG trend obtained on a large set of salad dressing samples was correlated with fat content and the method was used to detect eventual adulterations in commercial samples. Similar methodologies based on low field NMR relaxometry have been reported to detect adulterations in soft cheeses[44] and meat.[45]

Not all low field permanent magnets or electromagnets have bores wide enough to get the food inside; in this case, normal NMR tubes are used to perform the analysis of food preparations or food extracts. By using this type of NMR equipment, TD NMR methods were developed to detect adulterations of vegetable oils,[46] honey samples adulterated with syrups,[47] and soft candies adulterated with not declared additives.[48] Recently, a methodology based on ^{1}H NMR relaxation times, T_1 and T_2, measured on a 0.5 T benchtop permanent magnet was optimized to detect adulterations in dyes, such as malachite green, copper sulphate and Sudan red, on vegetables.[49] Some of these methods based on low field NMR relaxometry became standard procedures, recognized at the international level, to determine the fat or oil content in food.[50,51] Among the main advantages of low field TD NMR methods with respect to high resolution NMR spectroscopy, it is worth noting the possibility of having portable instruments for on-site and on-line measurements.[52–54]

14.2.2.2 ^{1}H NMR Diffusometry

Low field NMR portable instruments, such as MOUSE NMR and benchtop NMR, which do not have enough spectroscopic resolution for signal detection, have been widely used for industrial applications based on the measurement of both relaxation times and diffusion coefficients. Self-diffusion of molecules, such as water or fat, inside food or agricultural products, can give useful information to check the quality and eventual damage due to heating and cooling processes.[39] Diffusion coefficients measured with low field NMR were of help in detecting olive oil adulterations[55] and the addition of corn-syrup to soft candies, such as Turkish delights.[56]

The diffusion coefficients of complex systems can also be measured by using standard high resolution NMR instruments. For instance, the so-called diffusion ordered spectroscopy (DOSY) ^{1}H NMR experiment is performed to measure and better visualize the differences among different components of a mixture depending on self-diffusion. As an example, DOSY ^{1}H NMR was used to identify several adulterations in herbal medicines and dietary supplements in different market formulations and the same methodology was used to detect undeclared ingredients.[57]

A recent study on olive oils produced in two areas of Italy, Tuscany and Apulia was performed based on both high resolution and low resolution NMR methods.[58,59] In these works, ^{1}H NMR relaxation times (T_1 and T_2) as well as diffusion coefficients (D), through the DOSY ^{1}H NMR technique, were measured to determine the differences among different vegetable oils. In particular, the diffusion coefficients of linseed oil (LO), sunflower oil (SuO) and soybean oil (SoO) were found to be sensibly higher than those of extra-virgin olive oils (EVOOs), as shown in Figure 14.8.

In contrast, macadamia oil (MO) is characterized by similar values of the self-diffusion coefficients of EVOOs. Based on these results, several tests were performed to explore the potentialities of self-diffusion measurements to detect specific adulterations in EVOOs.[59]

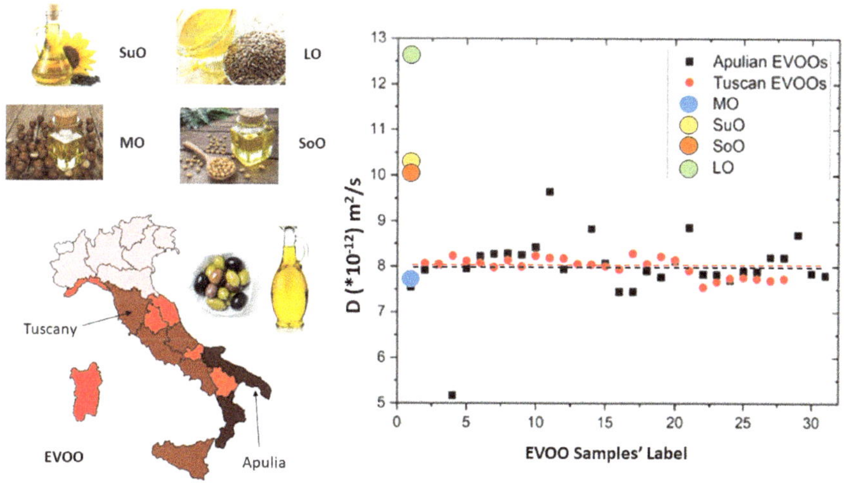

Figure 14.8 Average diffusion coefficients (D) measured by the ^{1}H NMR DOSY technique on a set of extra-virgin olive oils, namely, Apulian and Tuscan EVOOs, and vegetable oils from macadamia (MO), sunflower (SuO), soybean (SoO), and linseed (LO) oils. Reproduced from ref. 59, https://doi.org/10.1021/acs.jafc.1c00914, under the terms of the CC BY 4.0 license, https://creativecommons.org/licenses/by/4.0/.

14.2.2.3 Fast Field Cycling ^{1}H NMR Relaxometry

A very well-known technique to investigate food quality is fast field cycling (FFC) ^{1}H NMR relaxometry.[40,60] Specially designed electromagnetic NMR instruments are needed and specific pulse sequences, with or without prepolarizing the samples, can be used to measure ^{1}H NMR relaxation times, mainly T_1, in a wide Larmor frequency range, from a few KHz to hundreds of MHz. Spin–lattice relaxation times, T_1, are sensitive to spin fluctuations and, as a consequence, to dynamic motions, such as exchange processes, molecular and internal reorientations, conformational motions, self-diffusion and cross-relaxations, as well as collective motions. Each motion is generally associated to a characteristic correlation time, τ_c. For instance, the shorter the τ_c value, the faster the molecular motions. Spin–lattice relaxation becomes usually effective when the correlation rate (*i.e.* inverse of the correlation time τ_c) of the dynamic process is of the same order of magnitude as the Larmor frequency.[40] For this reason, the possibility of investigating the frequency dependence of T_1 is fundamental in order to obtain a complete description of the sample in terms of dynamic processes. The analysis of the relaxation rates, R_1 (*i.e.* inverse of the relaxation times, T_1), in food and agricultural products is usually done to obtain quantitative information about the main motional processes, such as correlation times or diffusion coefficients, as in the case reported in Figure 14.9(a) for extra-virgin olive oil.[58] Here, the frequency dependence of R_1 is analysed in terms of three dynamic processes ascribable to two reorientation processes (*Rot 1* and *Rot 2*) of fatty acids and triglycerides, plus the average self-diffusion (SD) motion.

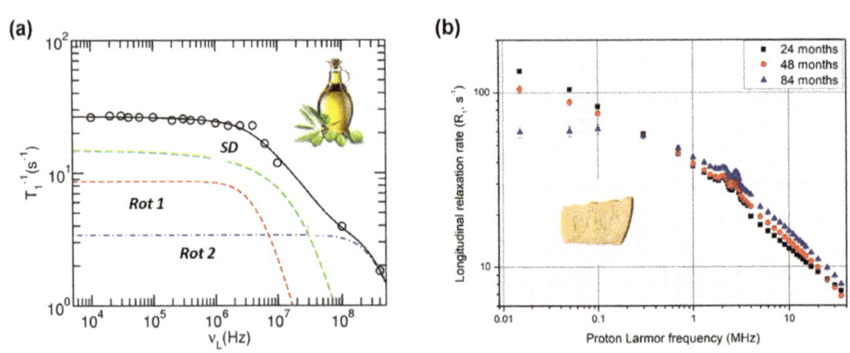

Figure 14.9 (a) Spin–lattice relaxation rate, R_1, (circled) of a sample of extra-virgin olive oil as a function of the ^{1}H Larmor frequency, ν_L. Contributions of individual mechanisms (*Rot 1*, *Rot 2* and SD) are shown together with the sum of the contributions (solid black line). Adapted from ref. 58, https://doi.org/10.1021/acs.jafc.1c00622, under the terms of the CC BY 4.0 license, https://creativecommons.org/licenses/by/4.0/. (b) ^{1}H dispersion NMR profiles for the Parmigiano Reggiano cheese aged for 24 months (squares), 48 months (circles), and 84 months (triangles). Reproduced from ref. 64 with permission from Elsevier, Copyright 2022.

The proton relaxation rate dispersion curves, as the ones shown in Figure 14.9, are very sensitive to different factors (chemical composition and physical properties) which influence dynamic motions in complex systems, such as food. In recent years, the number of applications of FFC ^{1}H NMR relaxation in the field of food and manufactured derived products has increased significantly.[40,58,60–64] Several works show that T_1 frequency dispersions of olive oils are different from those of other vegetable edible oils and this can be at the basis of oil discrimination and eventual adulterations, such as the dilution of olive oils with other vegetable oils.[58,61] As shown in Figure 14.9(b), FFC ^{1}H NMR relaxometry can be applied to differentiate the changes at the molecular level occurring in Parmigiano Reggiano cheese during ageing.[64] Recently, the combination between ^{1}H NMR dispersions and chemometrics was helpful in investigating adulterations in dairy products, such as several types of cheese, on the basis of the nature of milk used in the preparation.[62] Adulteration in meat is also another important issue in several countries and the possibility of using FFC ^{1}H NMR relaxometry in discriminating sausages produced from different kinds of meat, such as beef, chicken, turkey, and pork, was investigated.[63]

14.2.2.4 MRI

A widely adopted non-destructive method to study food is magnetic resonance imaging (MRI), which was used since the eighteen to analyse the internal morphology of horticultural products, such as fruits and vegetables.[41] MRI is a typical diagnostic method, where 2D or 3D images are reconstructed based on the average values of relaxation times (T_2 and T_1 weighted images) or diffusion coefficients (*D*-weighted images) measured within a volume of the analysed object. Usually, the ^{1}H signals monitored in MRI methods are those of water or fat molecules which represent the major constituents depending on the type of food or agricultural product. Water or fat molecules have different mobility depending on the chemical and physical properties of their local environment. For this reason, MRI sensors can be used in on-line detection of eventual internal defects or to evaluate the freshness of the products.[65] Analogously to non-destructive TD NMR relaxometry, MRI has also been applied to check eventual changes in fat content due to heating/cooling treatments or imperfect storage conditions, as reported in a recent paper studying the effects of frying on potatoes.[66] Several examples of applications of MRI to detect adulterations were published, such as the case of water-injected pork meat, which is a serious problem in meat production,[67] and to identify adulterated prawns injected with different hydrocolloids, such as gelatine, carrageenan and agar.[68] The results obtained by MRI are indeed very promising in achieving better food production processes as well as in the detection of food adulterations. Like low field NMR relaxometry, MRI is a robust, non-destructive and relatively rapid method, which does not require sample pre-treatment.[69]

14.2.2.5 Low Field Benchtop NMR

Improvements in magnets' technology and design gave rise to low field benchtop NMR suitable for high resolution NMR spectroscopy.[37] In fact, portable NMR spectrometers providing ^{1}H NMR chemical shift resolution became commercially available since 2010. A recent review has been published by van Beek[70] showing many different applications of low field portable NMR instruments, from fraud detection[4] to process control and phytochemical analysis.[70] The possibility of acquiring ^{1}H NMR spectra with these low field NMR instruments allowed determination, for instance, of the ratio between polyunsaturated fatty acids (PUFAs) and monounsaturated fatty acids (MUFAs) in different oils, in a very fast and robust way. Adulterations of special oils used as fragrances in food[71] and the presence of pesticides and not declared additives were also possible.[72,73]

14.2.3 Other Magnetic Resonance Techniques

14.2.3.1 SNIF-NMR

SNIF-NMR ('site-specific natural isotopic fractionation nuclear magnetic resonance') is a technique based on the measurement of the stable isotope content of a product and it has been applied to detect adulterations in food and agricultural products, such as wines, oils and fruit juices.[8] The ratio of isotopes, such as carbon (^{12}C, ^{13}C), nitrogen (^{14}N, ^{15}N), hydrogen (^{1}H, ^{2}H) and oxygen (^{16}O, ^{18}O), is unique for each plant, since it depends on many factors, such as the biochemical properties of the plant, and the physical and chemical properties of the specific soil, so it can be correlated with the geographic origin and so on. In principle, if the isotopic ratio is altered this can be an indication of adulteration. For instance, SNIF-NMR was used to determine the ratio between deuterium and hydrogen isotopes (D/H) at a specific site of ethanol and this approach has been applied to wine,[74] to check if sugar or other syrups were added as adulterants. A similar approach was applied to maple syrup and a specific protocol was developed to identify the adulteration of maple syrup with beet and cane sugar, which are commonly used adulterants.[75] In this procedure, sugar is fermented in ethanol and the isotopic D/H ratios are measured by recording the intensities of the ^{2}H NMR peaks corresponding to the three isotopomers of ethanol, compared with that of a known internal standard (TMU: *N,N*-tetramethylurea), as shown in Figure 14.10(a). The intensity of signals shown in Figure 14.10(a) is proportional to the deuterium/proton isotopic abundance, which is specific of a particular type of sugar or syrup.

A useful tool of the SNIF-NMR technique is the 'adulteration triangle', such as the one shown in Figure 14.10(b). In this case, a plot is built reporting the typical isotopic abundance of ^{13}C (δ^{3}C) *versus* the D/H ratio of maple syrup, cane sugar and beet sugar. The eclipse shown in Figure 14.10(b) indicates the 95% confidence for authentic maple syrup from a particular

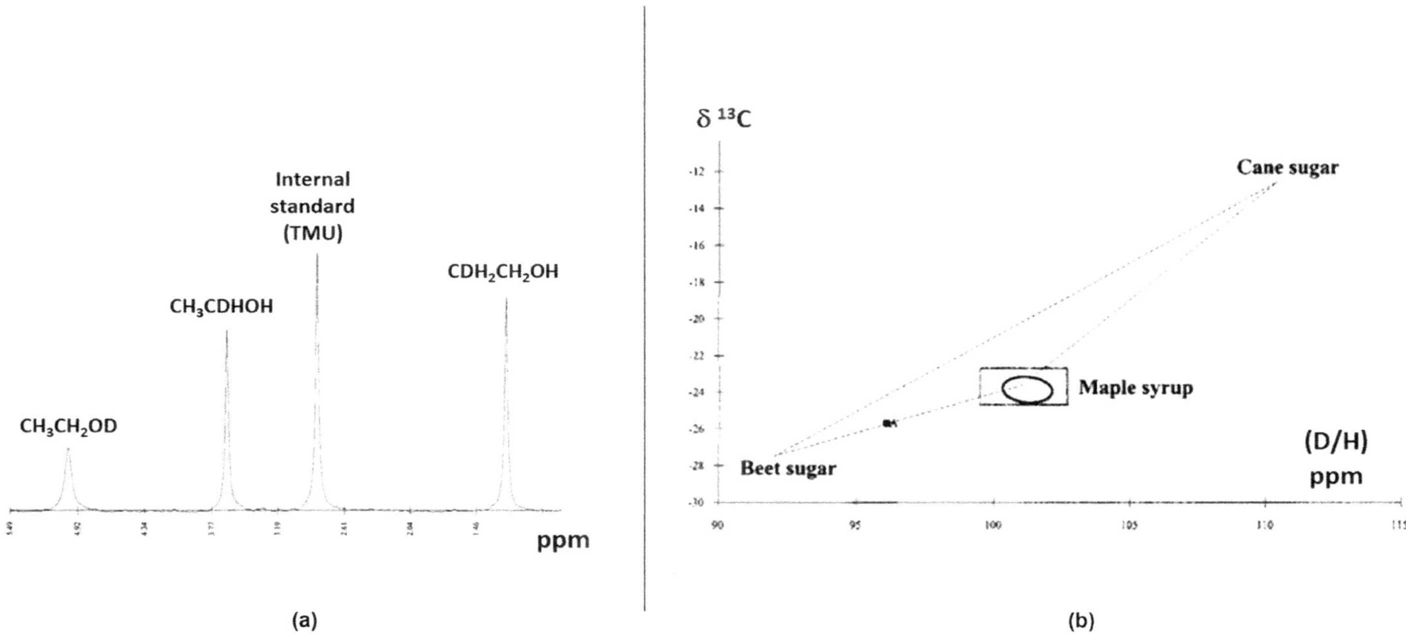

Figure 14.10 (a) Deuterium spectrum of ethanol fermented from maple syrup, showing the three signals due to the three isotopomers of ethanol and the internal standard (TMU). (b) Adulteration triangle showing an adulterated "market sample" (see the black arrow), the rectangle defined by the cutoff points used to detect samples, and the 95% confidence eclipse for authentic maple syrup samples from a particular geographic area (*i.e.* Vermont). Reproduced from ref. 75 with permission from American Chemical Society, Copyright 1996.

geographic area. These plots are indeed useful not only to identify adulterated samples, but also to quantify the percentage of adulterants and to discriminate among different geographic origins. SNIF-NMR techniques combined with multivariate statistical analysis, such as PCA and LDA, were successfully applied to identify adulterated wines and juices.[76] In recent years, more robust approaches were developed based on SNIF-NMR in combination with isotope ratio mass spectroscopy (IRMS),[8] as reported in a paper concerning the detection of undeclared acetic acid added to vinegar and other products.[77]

14.2.3.2 Ultra-low-field NMR with SQUID

Ultra-low-field (ULF) nuclear magnetic resonance (NMR) has attracted attention because it is a quite compact, easy-to-handle, and relatively low-cost technology. Typically, ULF NMR instruments are characterized by magnetic fields of about a few μT, with ^{1}H Larmor frequencies of a few kHz, and the NMR sensor is a superconducting quantum interference device (SQUID).[78] With these devices, it is possible to acquire both ^{1}H NMR signals and 1D MRI signals. For instance, several applications utilizing a high-temperature superconductor (HTS) SQUID coupled with a permanent magnet were reported for contaminant detection in food and drinks. In particular, it was shown that these instruments can be used to detect both magnetic and nonmagnetic contaminants, such as glass, ceramics, stainless steel, aluminium and polymers, in several drink samples.[79,80]

14.3 Conclusions and Future Perspectives

The problem of food adulterations and contaminations has several implications for human health and the economy, and it concerns both food consumers and producers. In the last twenty years, nuclear magnetic resonance techniques had a tremendous expansion giving rise to many applications in the fields of food safety, authenticity and fraud detection. In this brief review, the state of the art of the actual methodologies based on NMR is reported focusing on the capability of NMR in detecting adulterations in food and agricultural products. The variability of NMR techniques is such that many different types of food matrix can be analysed, from liquids to solids, and many kinds of problems can be solved, such as the detection of adulterants and contaminants at low concentrations and eventual damage due to bad storage conditions. Starting from high resolution ^{1}H NMR spectroscopy to low resolution and time-domain NMR relaxometry, with targeted or untargeted methodologies, this review aims to represent the richness of NMR techniques in this hot field. In the future, further developments are expected related to the technological improvements in ultra-low-field NMR, thanks to the sensitivity enhancement with hyperpolarization methods, and in the field of portable benchtop NMR for on-line and on-site food screening.

References

1. C. Cordella, I. Moussa, A.-C. Martel, N. Sbirrazzuoli and L. Lizzani-Cuvelier, *J. Agric. Food Chem.*, 2002, **50**, 1751.
2. A. Bordoni and F. Capozzi, *Curr. Opin. Food Sci.*, 2015, **4**, 124.
3. G. P. Danezis, A. S. Tsagkaris, F. Camin, V. Brusic and C. A. Georgiou, *Trends Anal. Chem.*, 2016, **85**, 123.
4. L. Laghi, G. Picone and F. Capozzi, *Trends Anal. Chem.*, 2014, **59**, 93.
5. A. P. Sobolev, S. Circi, D. Capitani, C. Ingallina and L. Mannina, *Curr. Opin. Food Sci.*, 2017, **16**, 59.
6. A. P. Sobolev, F. Thomas, J. Donarski, C. Ingallina, S. Circi, F. C. Marincola, D. Capitani and L. Mannina, *Trends Food Sci. Technol.*, 2019, **91**, 347.
7. Y. B. Monakhova, T. Kuballa and D. W. Lachenmeier, *J. Anal. Chem.*, 2013, **68**, 755.
8. N. Ogrinc, I. J. Kosir, J. E. Spangernberg and J. Kidric, *Anal. Bioanal. Chem.*, 2003, **376**, 424.
9. T. Kuballa, T. S. Brunner, T. Thongpanchang, S. G. Walch and D. W. Lachenmeier, *Curr. Opin. Food Sci.*, 2018, **19**, 57.
10. D. L. Massart, *Chemometrics: A Textbook*, Elsevier, New York, 1988.
11. S. Cai, Y. Zhang, F. Xia, G. Shen and J. Feng, *J. Food Compos. Anal.*, 2019, **84**, 103316.
12. L. Mannina, M. D'Imperio, D. Capitani, S. Rezzi, C. Guillou, T. Mavromoustakos, M. D. M. Vilchez, A. H. Fernández, F. Thomas and R. Aparicio, *J. Agric. Food Chem.*, 2009, **57**, 11550.
13. C. L. Ray, J. A. Gawenis and C. M. Greenlief, *Molecules*, 2022, **27**, 213.
14. G. Vlahov, *Prog. Nucl. Magn. Reson. Spectrosc.*, 1999, **35**, 341.
15. R. M. Alonso-Salces, L. A. Berrueta, B. Quintanilla-Casas, S. Vichi, A. Tres, M. I. Collado, C. Asensio-Regalado, G. E. Viacava, A. A. Poliero, E. Valli, A. Bendini, T. Gallina Toschi, J. M. Martínez-Rivas, W. Moreda and B. Gallo, *Food Chem.*, 2022, **366**, 130588.
16. A. J. Charlton, P. Robb, J. A. Donarski and J. Godward, *Anal. Chim. Acta*, 2008, **618**, 196.
17. M. M. Bergana, K. M. Adams, J. Harnly and J. C. Moore, *J. Food Compos. Anal.*, 2019, **78**, 49.
18. R. Colosimo, M. Gabriele, M. Cifelli, V. Longo, V. Domenici and L. Pucci, *Food Chem.*, 2019, **305**, 125510.
19. L. Jiang, L. Pan, H. Gao and H. Zheng, *J. Food Meas. Charact.*, 2020, **14**, 1427.
20. C. Schmitt, T. Bastek, A. Stelzer, T. Schneider, M. Fischer and T. Hackl, *J. Agric. Food Chem.*, 2020, **68**, 14364.
21. A. Francini, S. Romeo, M. Cifelli, D. Gori, V. Domenici and L. Sebastiani, *Food Chem.*, 2017, **221**, 1206.
22. C.-H. Yong, S. A. Muhammad, F. A. Aziz, F. I. Nasir, M. Z. Mustafa, B. Ibrahim, S. D. Kelly, A. Cannavan and E.-K. Seow, *Food Chem.*, 2022, **368**, 130808.

23. E. A. Petrakis, L. R. Cagliani, M. G. Polissiou and R. Consonni, *Food Chem.*, 2015, **173**, 890.
24. M. V. de Moura Ribeiro, N. Boralle, H. Redigolo Pezza, L. Pezza and A. T. Toci, *J. Food Compos. Anal.*, 2017, **57**, 24.
25. B. Horn, S. Esslinger, C. Fauhl-Hassek and J. Riedl, *Food Control*, 2021, **128**, 108205.
26. R. Sacchi, F. Addeo and L. Paolillo, *Magn. Reson. Chem.*, 1997, **35**, S133.
27. I. J. Košir and J. Kidric, *Anal. Chim. Acta*, 2002, **458**, 77.
28. I. J. Košir, M. Kocjancic and J. Kidric, *Analusis*, 1998, **26**, 97.
29. G. Vicario, A. Francini, M. Cifelli, V. Domenici and L. Sebastiani, *Antioxidants*, 2020, **9**, 1245.
30. E. Truzzi, L. Marchetti, S. Benvenuti, A. Ferroni, M. C. Rossi and D. Bertelli, *J. Agric. Food Chem.*, 2021, **69**, 8276.
31. G. Andreotti, R. Lamanna, E. Trivellone and A. Motta, *J. Am. Oil Chem. Soc.*, 2002, **79**, 123.
32. S. Guyader, F. Thomas, V. Portaluri, E. Jamin, S. Akola and V. Silvestre, *et al.*, *Food Control*, 2018, **91**, 216.
33. R. D. Mortimer and B. A. Dawson, *J. Agric. Food Chem.*, 1991, **39**, 1781.
34. P. S. Belton and R. L. J. Lyster, *J. Dairy Res.*, 2009, **58**, 443.
35. P. Dais and A. Spyros, *Magn. Reson. Chem.*, 2007, **45**, 367.
36. C. Lucas-Torres, Á. Pérez, B. Cabañas and A. Moreno, *Food Chem.*, 2014, **165**, 21.
37. B. Blümich, *J. Magn. Reson.*, 2019, **306**, 27.
38. T. B. Moraes and L. A. Colnago, *Braz. J. Phys.*, 2022, **52**, 43.
39. G. Guthausen, *TrAC, Trends Anal. Chem.*, 2016, **83**, 103.
40. E. G. Ates, V. Domenici, M. Florek-Wojciechowska, A. Gradisek, D. Kruk, N. Maltar-Strmecki, M. Oztop, E. B. Ozvural and A. L. Rollet, *Trends Food Sci. Technol.*, 2021, **110**, 513.
41. E. Kirtil, S. Cikrikc, M. J. McCarthy and M. H. Oztop, *Curr. Opin. Food Sci.*, 2017, **17**, 9.
42. D. Capitani, L. Mannina, N. Proietti, A. P. Sobolev, A. Tomassini, A. Miccheli, M. E. Di Cocco, G. Capuani, R. De Salvador and M. Delfini, *Talanta*, 2010, **82**, 1826.
43. F. M. V. Pereira, A. P. Rebellato, J. A. L. Pallone and L. A. Colnago, *Food Control*, 2015, **48**, 62.
44. G. de Oliveira Machado, G. G. Teixeira, R. H. D. S. Garcia, T. B. Moraes, E. Bona, P. M. Santos and L. A. Colnago, *Molecules*, 2022, **27**, 4434.
45. T. Monaretto, A. Souza, T. B. Moraes, V. Bertucci-Neto, C. Rondeau-Mouro and L. A. Colnago, *Magn. Reson. Chem.*, 2019, **57**, 616.
46. Q. Zhang, A. S. M. Saleh and Q. Shen, *Food Bioprocess Technol.*, 2013, **6**, 2562.
47. R. O. R. Ribeiro, E. T. Marsico, C. S. Carneiro, M. L. G. Monteiro, C. C. Junior and E. F. O. Jesus, *J. Food Eng.*, 2014, **135**, 39.
48. P. Pocan, E. Ilhan and M. H. Oztop, *Magn. Reson. Chem.*, 2019, **57**, 661.
49. S. Shomaji, N. V. R. Masna, D. Ariando, S. D. Paul, K. Horace-Herron, D. Forte, S. Mandal and S. Bhunia, *Foods*, 2021, **10**, 2232.

50. I.O.F. Standardization, Animal and vegetable fats and oils – Determination of solid fat content – Pulsed nuclear magnetic resonance method, ISO 8292, 1991.
51. I.O.F. Standardization, Oilseeds – Simultaneous Determination of Oil and Water Contents – Method Using Pulsed Nuclear Magnetic Resonance Spectrometry, ISO 10565, 1995.
52. M. C. Ezeanaka, J. Nsor-Atindana and M. Zhang, *Food Bioprocess Technol.*, 2019, **12**, 1435.
53. D. Capitani, A. P. Sobolev, V. Di Tullio, L. Mannina and N. Proietti, *Chem. Biol. Technol. Agric.*, 2017, **4**, 17.
54. M. Lia, B. Lib and W. Zhang, *Food Chem.*, 2018, **242**, 16.
55. Z. Xu, R. H. Morris, M. Bencsik and M. I. Newton, *Sensors*, 2014, **14**, 2028.
56. P. Pocan, M. Knapkiewicz, A. Rachocki and M. H. Oztop, *J. Agric. Food Chem.*, 2021, **69**, 12089.
57. J. Vaysse, S. Balayssac, V. Gilard, D. Desoubdzanne, M. Malet-Martino and R. Martino, *Food Addit. Contam.: Part A*, 2010, **27**, 903.
58. A. Gradisek, M. Cifelli, D. Ancora, A. Sepe, B. Zalar, T. Apih and V. Domenici, *J. Agric. Food Chem.*, 2021, **69**, 12073.
59. D. Ancora, J. Milavec, A. Gradisek, M. Cifelli, A. Sepe, T. Apih, B. Zalar and V. Domenici, *J. Agric. Food Chem.*, 2021, **69**, 12081.
60. P. Conte, *Annual Reports on NMR Spectroscopy*, 2021, vol. 104, p. 141.
61. P. Conte, V. Mineo, S. Bubici, C. De Pasquale, F. Aboud, A. Maccotta, D. Planeta and G. Alonzo, *Anal. Bioanal. Chem.*, 2011, **400**, 1443.
62. E. Piacenza, D. F. C. Martino, L. Cinquanta, P. Conte and P. L. Meo, *Magn. Reson. Chem.*, 2022, **60**, 369.
63. S. S. Uguz, E. B. Ozvural, M. J. Beira, M. H. Oztop and P. J. Sebastiao, *Mol. Phys.*, 2019, **117**, 1015.
64. P. Conte, L. Cinquanta, P. Lo Meo, F. Mazza, A. Micalizzi and O. Corona, *Food Res. Int.*, 2021, **139**, 109845.
65. M. Mazhar, D. Joyce, G. Cowin, I. Brereton, P. Hofman, R. Collins and M. Gupta, *Biol. Technol.*, 2015, **100**, 33.
66. D. Yang, G. C. Wu, P. Y. Li, X. G. Qi, H. Zhang, X. G. Wang and Q. Z. Jin, *Food Chem.: X*, 2020, 7, 100095.
67. S. M. Gai, Z. H. Zhang, Y. F. Zou and D. Y. Liu, *Int. J. Food Eng.*, 2019, **15**, 20180313.
68. M. Li, B. Li and W. J. Zhang, *Food Chem.*, 2018, **42**, 16.
69. E. Kirtil and M. H. Oztop, *Food Eng. Rev.*, 2016, **8**, 1.
70. T. A. van Beek, *Phytochem. Anal.*, 2021, **32**, 24.
71. A. Krause, Y. Wu, R. Tian and T. A. van Beek, *Planta Med.*, 2018, **84**, 953.
72. D. McDowell, M. Defernez, E. K. Kemsley, C. T. Elliott and A. Koidis, *LWT–Food Sci. Technol.*, 2019, **111**, 490.
73. M. P. Choules, J. Bisson and C. Simmler, *et al.*, *J. Pharm. Biomed. Anal.*, 2020, **178**, 112915.
74. G. J. Martin and M. L. Martin, *Models Methods Plant Anal.*, 1888, **6**, 258.
75. G. G. Martin, Y. L. Martin, N. Naulet and H. J. D. McManus, *J. Agric. Food Chem.*, 1996, **44**, 3206.

76. G. G. Martin, V. Hanote, M. Lees and Y. L. Martin, *J. AOAC Int.*, 1996, **79**, 62.
77. A. Gregova, E. Neradova, V. Kruzik, J. Mazac, P. Havelec and H. Cizkova, *Eur. Food Res. Technol.*, 2014, **239**, 169.
78. P. J. Ganssle, H. D. Shin, S. J. Seltzer, V. S. Bajaj, M. P. Ledbetter, D. Budker, S. Knappe, J. Kitching and A. Pines, *Angew. Chem., Int. Ed.*, 2014, **53**, 9766.
79. Y. Hatsukade, T. Abe, S. Tsunaki, M. Yamamoto, H. Murata and S. Tanaka, *IEEE Trans. Appl. Supercond.*, 2013, **23**, 1602204.
80. Y. Hatsukade, S. Tsunaki, M. Yamamoto, T. Abe, J. Hatta and S. Tanaka, *Phys. C*, 2013, **494**, 199.

CHAPTER 15

MRI, Pollution, and Human Health: Recent Advances in Clinical Practice

G. RIVA* AND A. SPADARO

Casa di Cura Clinica del Mediterraneo, *via* Ettore Fieramosca, 100, Ragusa, Italy
*Email: Drgaetanoriva@gmail.com

15.1 Introduction

Pollution refers to the presence of harmful substances and products in the environment. As already stated in different chapters of this book, pollution may affect all the environmental compartments in the way that when one of them is polluted, all the others are affected by the health status of the former. Grave consequences to the planet and its inhabitants, including human health problems and damages to natural ecosystems, occur due to pollution. Therefore, it is mandatory to account for all the available measures to reduce or, even better, prevent environmental contamination in order to protect the environment and preserve natural resources for the future generations.

Atmospheric pollution can seriously affect human health. Indeed, it can cause respiratory problems such as asthma and bronchitis, cardiovascular diseases, and even premature death. The most affected people are the most vulnerable ones, such as children, elderly persons, and those with pre-existing health conditions (such as asthma). There are many sources of air pollution, including vehicle emissions, industrial activities, and the burning of fossil fuels such as coal and oil.

New Developments in NMR No. 32
The Environment in a Magnet: Applications of NMR Techniques to Environmental Problems
Edited by Pellegrino Conte, Delia Francesca Chillura Martino and Paolo Lo Meo

Published by the Royal Society of Chemistry, www.rsc.org

To reduce the adverse health effects of air pollution, it is important to reduce the amount of pollutants released into the air. This goal can be achieved through a variety of methods, such as stricter emission standards for vehicles and industries, transitioning to cleaner energy sources, and promoting the use of public transportation.

The International Agency for Research on Cancer (IARC)[1] classified air pollution, particularly the fine particulate matter known as $PM_{2.5}$, as a major cause of cancer. $PM_{2.5}$ is a particulate matter having a size less than or equal to 2.5 microns. For the most, it originates from different human activities such as agriculture and industry. A simple example to explain the $PM_{2.5}$ origin is car brakes. Indeed, every time brakes are used, $PM_{2.5}$ emission occurs. In general, the $PM_{2.5}$ origin can be either primary or secondary. The latter is a consequence of the chemical–physical transformations of many materials that can give rise to formation of small dust particles, as well as many other solid compounds containing nitrates, sulphates, ammonia, and a large variety of salts.

$PM_{2.5}$ is a type of air contaminant that can be inhaled deeply into the lungs and be the cause for respiratory problems, including coughing, wheezing and difficulty in breathing. Moreover, it appears that increasing risk for heart attack and strokes can be related to the atmospheric $PM_{2.5}$ systems.

Ozone, nitrogen dioxide, polycyclic aromatic hydrocarbons, sulphur-based compounds and many others can be considered as air pollutants that can not only affect the respiratory system causing symptoms such as coughing and breath shortness, but also cause many other illnesses affecting many organs in the body, thereby leading to complications or exacerbating existing health conditions.

The Report No. 10/2019 from the European Environment Agency (EEA)[2] highlighted the causes related to air pollution of the premature deaths occurred in 2019 in the 27 EU member states. In particular, about 307 000 premature deaths are attributable to $PM_{2.5}$, 40 400 premature deaths to nitrogen dioxide (NO_2), and 16 800 premature deaths to ground-level ozone. Based on this, in 2021, the World Health Organization (WHO) updated the 2005 guidelines for the air quality to protect human health.

The Lancet Commission on Pollution and Health found in 2022[3] that 16 percent of all deaths globally in 2019, *i.e.*, about 9 million deaths, were caused by atmospheric pollution, thereby making it the world's largest environmental risk factor for disease and premature death. This also caused economic losses of about US$4.6 trillion (6.2 percent of global economic output) that year. The Lancet noticed that most of the deaths, about 92 percent, related to pollution and the resulting economic losses occurred in low- and middle-income countries (LMICs). However, comparisons of the chemical nature of pollutants from the recent past up to now showed that pollutants are changing. Indeed, since 2000, the number of deaths caused by domestic air and water pollution (*i.e.*, due to unsafe water drinking and inadequate sanitation) has declined, while deaths attributable to more modern forms of pollutants, *e.g.*, lead and other different chemicals due to increasing industrialization and ever-increasing urbanization, have increased.

Among the most frequent pathologies that can be directly imputed to pollution are diseases of the respiratory system, which, due to physiological characteristics along with the digestive system, appears most exposed and therefore vulnerable. Some randomized studies have ascertained a direct correlation between air pollution and cancers of the digestive system related to the transport of pollutants by the ciliary apparatus present in the nose into the pharynx through which it reaches both the respiratory and gastroenteric systems by swallowing (we swallow pollutants every 15 minutes).

Although CT (computed tomography) is the most widely used method for studies in the field of oncology, there is no doubt that magnetic resonance imaging (MRI) is starting to play an increasing role for human health monitoring due to modern innovations that have broadened its uses. Moreover, the absence of ionizing radiations makes MRI preferable in the investigations on young subjects and pregnant women.

15.2 Present and Future Applications of Pulmonary MRI

MRI is an important and comprehensive clinical examination and has been widely used, to date, for the study of the digestive tract and abdomen. There are various clinical protocols now standardized in this regard. Triphasic studies of the liver and the use of hepatobiliary contrast agents have consolidated the success of MRI in the oncologic study of the liver, thereby allowing more accurate characterization of dubious focal lesions on CT.

The role of MRI in the study of chronic inflammatory bowel diseases (Crohn's, rectal ulcerative colitis, *etc.*) seems equally undisputed. Many of the sequences in use today arose precisely to facilitate differential diagnosis in the context of abdominal organ pathologies. The real revolution in the clinical field, however, is represented by the recent use of MRI to study the respiratory system. Lung MRI, in fact, provides structural and quantitative functional images of the lungs without ionizing radiation.

The lack of ionizing radiation makes lung MRI an ideal technology for paediatric examinations, pregnant women, and patients requiring serial and longitudinal follow-up. So far, however, this method has limited clinical use because of the low signal intensity from the lung parenchyma. However, recent technical advances have helped MRI to overcome its well-known limitations related to low proton density in the lungs and the rapid signal decay due to susceptibility artifacts at air–tissue interfaces.

Luckily, recent MRI techniques, including ultra-short echo time and zero echo time (the reader is referred to Appendix C for the meaning of the terms and the understanding of MRI theory), are expanding the clinical opportunities for lung MRI, thereby making this examination less and less niche and more routine. In addition to chest X-ray, which is the most commonly used first-line test for chest disorders, and computed tomography (CT), up to now the most comprehensive and detailed modality for cross-sectional and three-dimensional (3D) imaging of the lungs, MRI is becoming the third important and alternative method for evaluating lung disease.

More than any other technique, MRI combines excellent soft tissue contrast and functional information. It allows multiple and repeated measurements and can be used to assess thoracic organ motion and perfusion. These advantages are already well appreciated in MRI of the heart, where morphologic information (wall thickness, edema, myocardial scarring) with functional assessment (wall motion, contractility, ejection fraction) and viability/perfusion are combined. The key technique for lung morphology MRI is based on the high-frequency resonant signal of protons in tissues and fluids. This is also referred to as proton-MRI or ^{1}H-MRI. Particularly with the use of parallel multi-coil acquisitions and acceleration methods, these techniques make lung MRI very practical for the evaluation of vascular diseases of the lung parenchyma and lungs. Especially useful are SSH (SINGLE SHOT) sequences using SENSE (the reader is referred to Appendix C for details about SSH and SENSE).

Pathological conditions in the lungs related to changes in the lung structure result in increased lung density (collapse of the lungs with atelectasis, accumulation of fluid within the alveolar space and/or lung interstitium, accumulation of fluid and cells in infiltrates or soft tissue growth resulting in nodules and masses) or decreased lung density (overload due to air entrapment, enlargement of air spaces in emphysema).

Given the well-known limitations of MRI in visualizing low proton density structures with high susceptibility artifacts at air–tissue interfaces, any pathology resulting in increased lung density (and consequently proton density) should be easy to detect, whereas decreased lung density will remain more difficult to visualize with MRI. However, most conditions that lead to destruction and overinflation of the lung parenchyma, as in emphysema, are associated with characteristic changes in the remaining lung parenchyma (fibrosis, septa, parenchymal distortion, bronchial wall thickening, *etc.*). This means that morphologic MRI can still provide a correct diagnosis based on enlarged spaces of missing signals due to overinflation and results associated with increased signal intensity. Direct visualization of the airspace remains the domain of sophisticated imaging technology in specialized centers.

In addition, changes in lung perfusion (hypoxic vasoconstriction, pulmonary embolism, congestion due to heart failure) cause changes in the lung signal on MRI. MRI not enhanced by mdc (contrast medium) can directly visualize the vessel tree to detect macrovascular pathology. Much experience exists for the detection of pathological conditions that cause increased lung proton density. Recent studies have demonstrated the high sensitivity of MRI with T_2-weighted sequences and proton density for the study of the lungs. It can be concluded that the sensitivity of MRI for lung infiltrates is at least as high as that of chest X-ray and CT.

Most of the applied protocols are based on T_2-weighted fast spin echo sequences or proton density, with respiratory gating or triggering or in the apnea acquisition mode, taking advantage not only of ultrafast sequences but also of the recent opportunities offered by artificial intelligence to overcome the technical limitations dictated by acquisition times as discussed in Appendix C. Another key clinical requirement is the detection of small solid or soft tissue lesions ("nodules"). The sensitivity of MRI for lung nodules greater than 4 mm varies between 80 and 90% and reaches 100% for lesions greater than 8 mm. Depending on the sequencing technique and signal intensity of the lesions and given optimal conditions (*e.g.*, the patient can hold his or her breath for 20 s or perfect gating/triggering), one can take a threshold size of 3–4 mm for lung nodule detection with MRI. Compared to CT, it may be even faster and more efficient to read lung MRI for lung nodules, as they appear with a bright signal against the dark background of healthy lung tissue.

Calcified nodules tend to disappear in the background because they have no intrinsic signal, whereas contrast-filled vascular lesions will be highly visible on T_1-weighted images. So far, a variety of sequence types have been evaluated for lung nodule detection with MRI. The spectrum includes T_2-weighted fast spin echo imaging with and without fat saturation, inversion recovery techniques, T_1-spin echo and gradient echo sequences. Implications for protocol design should include at least one T_2-weighted or proton-weighted or short τ inversion recovery (STIR) sequence to cover infiltrates and nodular lesions with high fluid content. This should be combined with a second sequence to also cover nodules with a high signal on T_1-weighted images. In particular, for detection of malignant lesions with hyperperfusion and intense enhancement, i.v. administration of the paramagnetic contrast material could be useful and even increase detection rates.

15.3 Artificial Intelligence (AI): Compressed Sensing, Machine Learning and Deep Learning

As already anticipated, one of the main limitations that, in part, has hindered the widespread use of MRI in lung disease is the difficulty of acquiring images in apnea to avoid motion artifacts, and, therefore, ultrafast sequences are required, which only technological innovation in recent years has made possible. In addition, the increasing use of AI (artificial intelligence) algorithms could further facilitate the resolution of these technical problems, on the one hand, and automate the analysis of the results, on the other hand, through the automatic detection of pathology by exploiting the main feature of lung diseases that always involve an increase in proton density.

Practically speaking, it is possible, for example, to take advantage of compressed sensing to speed up acquisition due to the reduced sampling operated by this algorithm. In practice, compressed sensing (also known as compressive sensing, compressive sampling, and sparse sampling) is a signal processing technique for efficiently acquiring and reconstructing an electrical signal by finding solutions to linear systems under certain conditions.

Any MRI acquisition needs to analyze the signal from the relaxation of each individual atom and locate it precisely in the three dimensions of space. Acquiring this multitude of data implies the need for a time that might be excessively long to perform the apnea investigation. Therefore, thanks to this particular AI algorithm, it has been possible to reduce the sampling, acquiring only some data and not all and then reconstructing the missing data to obtain a homogeneous set realized through automatic and mathematical filling of the missing data rather than their direct acquisition. In practice, one optimizes the sparsity of a signal by recovering it from far fewer samples than required by the Nyquist–Shannon sampling theorem, the treatment of which is beyond the scope of this presentation.

Candès *et al.*[4] showed that, given the knowledge of the sparsity of a signal, the signal itself can be reconstructed with even fewer samples than required by the sampling theorem, and this idea is the basis of compressed sensing: complete sample obtained from partial data. To this type of technical solution for achieving diagnostic acquisition of lung images, as mentioned above, it is also possible to associate other AI protocols that exploit, for example, the machine learning or the self-learning ability of equipment that over time can improve the recognition of pathological *versus* non-pathological areas, made possible due in part to the peculiar characteristic of the lung parenchyma to undergo an increase in proton density under all conditions other than normal.

Machine learning is currently used everywhere. When we interact with banks, shop online or use social media, machine learning algorithms are used to make our experience efficient, easy, and safe. Machine learning and associated technology are developing rapidly, and we are just beginning to discover their capabilities. Algorithms are the engines that power machine learning. The two main types of machine learning algorithms currently used are supervised machine learning and unsupervised learning. The difference between these two types is defined by the way each algorithm learns data to make predictions.

Supervised machine learning algorithms are the most widely used. With this model, a data set acts as a guide and teaches the algorithm the results to generate. Exactly as a child learns how to identify fruits by memorizing them from a picture book, in supervised machine learning the algorithm learns from an already labelled dataset with a predefined output. *Unsupervised machine learning* uses a more independent approach, in which a computer learns to identify complex processes and patterns without the careful and constant guidance of a person. Unsupervised machine learning involves training based on data that are unlabelled and for which no specific output has been defined. To continue using the previous analogy, unsupervised machine learning is similar to a child learning to identify fruits by observing colours and patterns, rather than by memorizing names with the help of a teacher. The child will look for similarities between images and divide them into groups, assigning each group the corresponding new label. The *k*-means clustering algorithms, principal and independent component analysis, and association rules are examples of unsupervised machine learning.

15.4 Role of AI in MRI

While accuracy is still the main goal, there is also a shift towards automation that aims to make radiologists' lives easier. Since radiologists have to review many images every day, it is unavoidable that this part of their work might be automated. Algorithms can be trained to find and detect various types of abnormalities based on images, so why should not we allow them to do this work so that radiologists can spend their time on the more difficult issues? With the possibilities of deep learning, algorithms are able to learn on their own while radiologists monitor their effectiveness. The longer it is used, the more effective it will be, and it is too good an opportunity to miss. Radiology could quickly become one of the most creative specialties where problem-solving and a holistic approach are key.

The applications of artificial intelligence that are imagined to bring about the most change in radiology practice are those related to the new potential of automatic quantification. In particular, the most promising ones are as follows:

- Radiomics: this discipline is related to the extraction, from images, of large quantities of quantitative indicators (features), thanks to which it is possible to perform in-depth analyses and build predictive models. Machine learning techniques are needed to be able to correlate them with clinical data (clinical outcome, case stratification, *etc.*). This new branch can complement and find support in the information provided by other disciplines using molecular biology (genomics) or pathological anatomy (pathomics) data. Some companies selling software packages for radiomic image analyses are: Aquilab, TexRAD, HealthMyne, Olea Medical, Quantib, Sophia Genetics, Quibim, and many others that also offer various different commercial products in the field.
- The introduction of new biomarkers: medical images contain information that is complementary to and correlated with other clinical, genomic, and laboratory indicators, from the integration of which and with appropriate data-mining we obtain imaging-based biomarkers and/or biomarkers. Through the identification and application of these biomarkers, hitherto little used because they are dependent on the availability of advanced analysis and quantification tools, one is able to improve the diagnosis and prognosis of a given patient's pathology (precision medicine).

- The implementation of virtual biopsies: solid tumors are heterogeneous phenotypically, physiologically, and genomically and evolve over time. The use of medical imaging both as a guidance tool and also as a tool for non-invasive *in vivo* analysis (biomarkers) should improve the outcome of biopsies (combination and/or integration of real and virtual biopsies). By being able to repeat images multiple times without invasiveness to the patient, it would be possible to improve their monitoring over time and ensure earlier characterization and optimization of therapy. In addition to these applications, there are other potentials peculiar to artificial intelligence that are geared toward the improvement of acquired images and that are mainly applied in the image reconstruction and processing stages. Among them, it is useful to mention these soon-to-be-commercialized applications:
- Thanks to the new deep learning algorithms, the analysis (outcomes) produced by CADe and CADx devices supporting diagnosis is improved, enhancing their prediction and differentiation capabilities: differentiation of benign from malignant calcifications.
- Thanks to the new deep learning networks, it is possible to reduce the number of projections required for good image reconstruction in three dimensions, partially overcome the critical issues related to the presence of noise and/or the presence of artifacts, and improve the low quality of a given image. In particular, it seems possible to
    - enable ultrafast scans (CT, MR, PET) resulting in reduced motion-related effects (the heart and lungs);
    - limit the hardware demands of iterative algorithms;
    - avoid data loss in image reconstruction by enabling automatic feature extraction and classification, directly using raw data acquired from equipment (tomographs);
    - reconstruct an image with incomplete data; and
    - build a tomographic image from previous data already available (CT, MR), improving registration (deformable and non-deformable) between images and enhancing automatic segmentation techniques.

To conclude, MRI is becoming an undisputed protagonist in the study of pulmonary pathology in recent years, offering advantages increasingly appreciated by radiologists over methods currently considered gold standard, even though with fewer advantages over MRI which allows morphological, functional, and perfusion studies without the use of ionizing radiation:

- The future is now.

References

1. International Agency for Research on Cancer (IARC), Air pollution and cancer, *Clinical Handbook of Air Pollution-Related Diseases*, ed. K. Straif, A. Cohen and J. Samet, International Agency for Research on Cancer, France, 2018, pp. 445–457.

2. European Environment Agency, Air quality in Europe report, EEA. Luxembourg: Publications Office of the European Union: European Environment Agency, 2016, p. 88.
3. R. Fuller, P. J. Landrigan, K. Balakrishnan, G. Bathan, S. Bose-O'Reilly and M. Brauer, *et al.*, Pollution and health: a progress update, *Lancet Planet Health*, 2022, **6**(6), e535–47.
4. E. J. Candès, J. K. Romberg and T. Tao, Stable signal recovery from incomplete and inaccurate measurements, *Commun. Pure Appl. Math.*, 2006, **59**(8), 1207–1223.

APPENDIX A

Basics of NMR Spectroscopy (for Novices)

P. LO MEO

Department of Biological, Chemical and Pharmaceutical Sciences and Techniques (STEBICEF) – University of Palermo, V.le delle Scienze ed. 17 "S. Cannizzaro", 90128 Palermo, Italy
Email: paolo.lomeo@unipa.it

A.1 Introduction

The NMR phenomenon is based on the detection of how an atomic nucleus with a non-null spin interacts with a magnetic field. Thus, NMR spectroscopy can be used to study any chemical element having a spinning nucleus. For organic chemists, this is a technique *par excellence* employed to retrieve the structure of "small" molecular compounds, by detecting ^{1}H and ^{13}C nuclei; however, with a few adaptations, it constitutes a very useful tool of general applicability for the detection of other elements and for the characterization of complex systems such as supramolecular aggregates, biomolecules and organic or inorganic materials. This, in turn, opens the way to diverse applications in industrial, biomedical and environmental contexts. NMR is a very versatile technique: indeed, it should be rather considered as a constellation of different interrelated techniques, *e.g.* 1D and 2D (correlation spectroscopy) spectroscopy in the liquid phase, NOE-based, TOCSY and solid-state techniques, relaxometry, and so on. Any chemist is perfectly aware of the topics covered in this chapter; therefore, the following presentation aims at outlining the fundamental concepts and aspects of modern NMR for those people (engineers, biologists, environmentalists, physicians, *etc.*) who are not familiar with the subject. For a more

New Developments in NMR No. 32
The Environment in a Magnet: Applications of NMR Techniques to Environmental Problems
Edited by Pellegrino Conte, Delia Francesca Chillura Martino and Paolo Lo Meo

Published by the Royal Society of Chemistry, www.rsc.org

detailed and exhaustive discussion, interested readers are strongly advised to consult the reference textbooks listed at the end of this presentation.

A.2 Nuclear Spin and Magnetization

Spin is a quantum property of subatomic particles, and atomic nuclei as well. It is an *intrinsic* property of matter, such as mass or electric charge. Despite widespread opinion, spin is *not* a rotation around the particle axis; nevertheless, it gives rise to a phenomenology, the classical counterpart of which can be envisaged in the rotation around an axis.

Spin is a *quantized* property; hence, any subatomic particle or atomic nucleus is characterized by a *spin quantum number* (S) that quantifies it; S can assume any integer or semi-integer value from 0 up (of course, if $S=0$ one simply states that the particle or nucleus is spinless). Protons, electrons and neutrons all have $S=\frac{1}{2}$; conversely, for atomic nuclei, (i) if both the atomic number and the mass number are even, then S is null (the nucleus has no spin, *e.g.* ^{12}C, ^{16}O, ^{32}S, *etc.*); (ii) if the atomic number is odd and the mass number is even, then S is an integer (*e.g.* ^{2}H, ^{10}B, and ^{14}N have $S=1$); (iii) if the mass number is odd (irrespective of the atomic number), then S is a semi-integer (*e.g.* ^{1}H, ^{13}C, ^{15}N, ^{19}F, and ^{31}P have $S=\frac{1}{2}$; ^{11}B has $S=3/2$, *etc.*).

To any spinning particle or nucleus, a *spin angular momentum* vector $\boldsymbol{s}$ is associated, the module of which is given by the relationship:

$$|\boldsymbol{s}| = \hbar\sqrt{S(S+1)} \tag{A.1}$$

where $\hbar = h/2\pi$ is the reduced Plank constant. Hence, for $S=\frac{1}{2}$, 1 or 3/2, $|\boldsymbol{s}|$ is equal to $0.866\hbar$, $1.414\hbar$ or $1.936\hbar$, respectively, and so on. Moreover, an electrically charged spinning particle possesses also a *spin magnetic dipole moment*, $\boldsymbol{\mu}$, according to the relation:

$$\boldsymbol{\mu} = \gamma_x \boldsymbol{s} \tag{A.2}$$

where γ_x is the *gyromagnetic* ratio of the particular particle considered (from this fact, the idea that spin is equivalent to a rotation stems by analogy with electrically charged rotating macroscopic objects). Thus, nuclear spin can be in principle evidenced by making the nucleus interact with an external magnetic field ($\mathbf{B}_0$). However, because the nucleus is a microscopic object, its behavior is governed by quantum mechanics. In particular, the spinning particle *cannot* align its spin vector $\boldsymbol{s}$ along the direction of $\mathbf{B}_0$ but *must* assume a tilted orientation in such a way that the projection of $\boldsymbol{s}$ with respect to $\mathbf{B}_0$ is an integer (for S integer) or a semi-integer (for S semi-integer) multiple of $\hbar$. In the most common case of $S=\frac{1}{2}$ (*i.e.* for ^{1}H, ^{13}C, *etc.*), this implies that the particle has allowed two orientations only, characterized by a tilt angle (ϑ) with respect to $\mathbf{B}_0$ as large as (Figure A.1 (left))

$$\vartheta = \arccos\left[(\hbar/2)/|\boldsymbol{s}|\right] = \arccos(1/\sqrt{3}) \approx \pm 54°44' \tag{A.3}$$

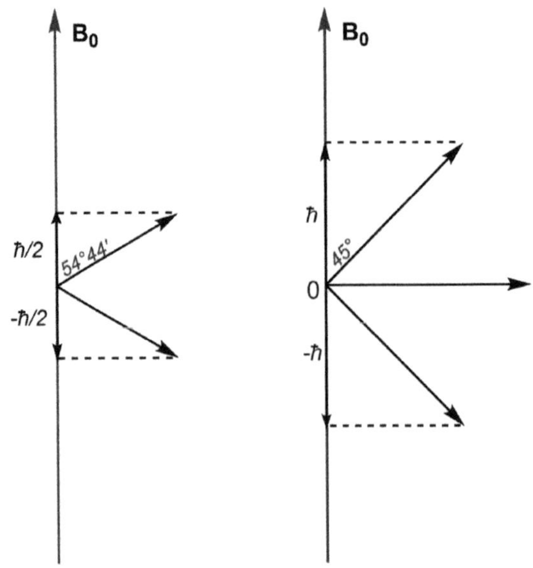

Figure A.1 Allowed orientations of a spinning and charged particle/nucleus in a magnetic field: $S = \frac{1}{2}$ (left) and $S = 1$ (right).

Analogously, it can be easily shown that for $S = 1$ there are three orientations allowed at tilt angle values as large as $\vartheta = \arccos 1/\sqrt{2} = \pm\ 45°$ or $\vartheta = 90°$ (Figure A.1 (right)), and so on (in general, the number of allowed orientations for any particle will be $2S + 1$).

Before proceeding, it is important to stress that the basic mathematical formalisms of NMR spectroscopy apply to any spinning nucleus, irrespective of its chemical nature, with only minor adaptations. Therefore, for the sake of simplicity, the discussion hereinafter and in the following sections will mainly focus on the proton (*i.e.* the ^{1}H nucleus) as a model spinning particle, for the simple reason that historically ^{1}H NMR spectroscopy has been the most important one. Applications to the study of other nuclei will be addressed subsequently.

The fact that the ^{1}H nucleus, after having been placed into a magnetic field, has two orientations allowed only implies that it can access two distinct energy states, characterized by a potential energy value (ε) as large as

$$\varepsilon = -\boldsymbol{\mu} \cdot \mathbf{B}_0 = -\gamma_{\mathrm{H}} \cdot |\mathbf{s}| \cdot |\mathbf{B}_0| \cdot \cos \vartheta = \mp \frac{1}{2} \cdot \hbar \cdot \gamma_{\mathrm{H}} \cdot \mathbf{B_0} \tag{A.4}$$

where $\mathbf{B}_0$ is the strength of the magnetic field and γ_{H} is the proton gyromagnetic ratio. The two energy states differ for the parallel or antiparallel orientation of vector $\boldsymbol{s}$ projection along $\mathbf{B}_0$. Hence, we can define a *spin state quantum number* (m_s) that identifies the spin state of the proton as

$$m_{\mathrm{s}} = -\varepsilon/(\hbar \cdot \gamma_{\mathrm{H}} \cdot \mathbf{B_0}) = \pm \frac{1}{2} \tag{A.5}$$

In this way, the energy of the state can be suitably written as $\varepsilon(m_s) = -m_s \cdot \hbar \cdot \gamma_H \cdot \mathbf{B}_0$.

In principle, it is possible to promote a proton from its *fundamental* (α) spin state ($m_s = +\frac{1}{2}$) to its *excited* (β) spin state ($m_s = -\frac{1}{2}$) by covering the energy gap ($\Delta\varepsilon$) with a photon of the right frequency (according to Planck's law), *i.e.*

$$\Delta\varepsilon = \varepsilon_{(\beta)} - \varepsilon_{(\alpha)} = \hbar \cdot \gamma_H \cdot \mathbf{B_0} = h \cdot \nu_0; \quad \text{thus: } \nu_0 = (\gamma_H / 2\pi) \cdot \mathbf{B_0} \tag{A.6}$$

where ν_0 is the frequency of the electromagnetic radiation able to promote the proton. It is worth stressing that $\Delta\varepsilon$ and ν_0 linearly depend on $\mathbf{B}_0$. For example, at a magnetic field strength of *ca.* 11.74 T, we have $\Delta\varepsilon = 0.20 \text{ J mol}^{-1}$ and $\nu_0 = 500$ MHz. The latter frequency falls in the region of radio waves.

For a macroscopic sample placed in a magnetic field, under thermal equilibrium conditions with its environment, protons are distributed between the fundamental and the excited state according to Boltzmann's distribution law, *i.e.*

$$P_\beta / P_\alpha = \exp[-\hbar \cdot \gamma_H \cdot \mathbf{B_0} / k_B T] \tag{A.7}$$

where P_α and P_β are the populations of the states, k_B is Boltzmann's constant, and T is the absolute temperature. The difference between the populations will then be

$$\Delta P = P_\alpha - P_\beta = P_0 \cdot \frac{1 - \exp[-\hbar \cdot \gamma_H \cdot \mathbf{B_0} / k_B T]}{2 + \exp[-\hbar \cdot \gamma_H \cdot \mathbf{B_0} / k_B T]} \approx (P_0/2) \cdot \hbar \cdot \gamma_H \cdot \mathbf{B_0} / k_B T \tag{A.8}$$

where P_0 is the overall spin population (ΔP is usually very small. For instance, in the previous example for $\mathbf{B}_0 = 11.74$ T one can easily calculate $\Delta P \approx 0.00004 \cdot P_0$).

In order to understand what happens when a real macroscopic sample of hydrogen-containing molecules is placed into a magnetic field $\mathbf{B}_0$, it is convenient to consider a Cartesian reference system having the *z-axis* coincident with the direction of $\mathbf{B}_0$. First, because of the tilt angle ϑ, proton magnetic momenta $\boldsymbol{\mu}$ show a *precession* motion (Larmor's precession; its classical counterpart can be envisaged in the precession motion for the rotation axis of a macroscopic spinner under the action of gravity). The angular frequency of Larmor's precession (ν_L) *must* coincide with ν_0. This match is essential to ensure the *resonance* between $\boldsymbol{\mu}$ and the magnetic component of the radiofrequency, and the consequent transfer of the energy quantum. It is important to recall here that the magnetic moment of any proton in the system is forced to a $\pm 54°44'$ tilt with respect to the *z-axis*; conversely, its orientation onto the *xy* plane continuously changes, as the projection of $\boldsymbol{\mu}$ on *xy* rotates at an angular frequency as large as ν_L (or, which is the same, ν_0) because of Larmor's precession. Ideally, one can imagine collecting all the $\boldsymbol{\mu}$ vectors in the sample, rigidly translating them, and bringing their origins to coincide with the origin of the reference axis system. Then, all the $\boldsymbol{\mu}$ vectors will be placed onto two ideal cone surfaces (Figure A.2). Noticeably, because

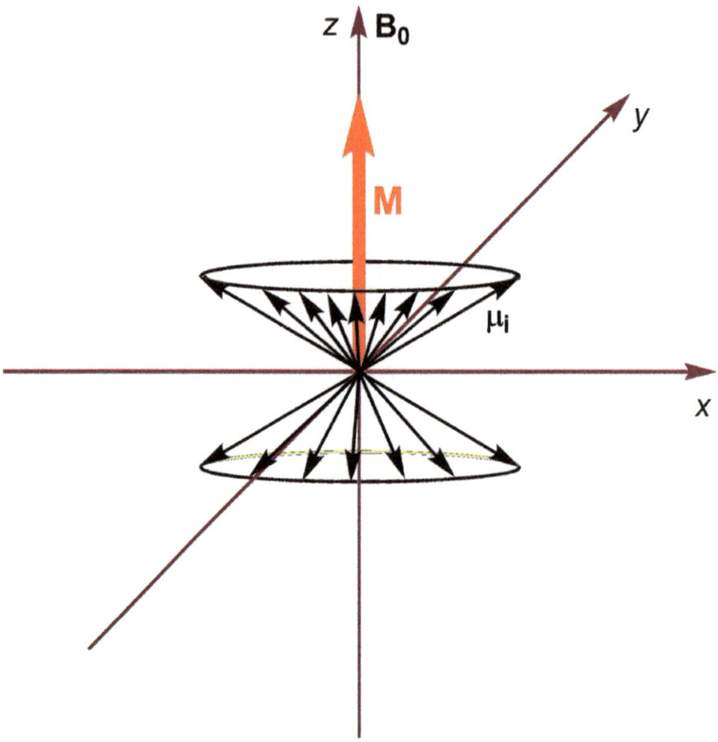

Figure A.2 Construction of the magnetization vector **M**.

there is no constraint along x and y, the two surfaces will be uniformly covered by the $\boldsymbol{\mu}$ vectors, forming a symmetric arrangement around the *z-axis*. Hence, the *magnetization* vector (**M**) of the sample is defined as the vector sum of the $\boldsymbol{\mu}$ spin magnetic momenta of the protons, *i.e.*, $\mathbf{M} = \sum_i (\boldsymbol{\mu}_i)$.

Trivial geometry considerations immediately lead to conclude that the direction and verse of **M** must be those of the *z-axis*, whereas its module $|\mathbf{M}|$ under thermal equilibrium conditions will be given by the expression:

$$|\mathbf{M}| = |\boldsymbol{\mu}| \cdot \cos\vartheta \cdot \Delta P = P_0 \cdot \hbar^2 \cdot {\gamma_H}^2 \cdot \mathbf{B_0}/2k_B T \tag{A.9}$$

A.3 Chemical Shift

In the previous discussion, we have been ideally referring to *protons*; however, in real contexts, we never have *naked* protons, but rather ^{1}H nuclei bound in a molecule, and macroscopic samples of molecules as well. Bound hydrogen nuclei are covered in electron density (the chemical bond *is* shared electron density, indeed), and electrons are sensitive to the presence of a magnetic field because of their own charge and spin. Hence, once the sample has been introduced into the main magnetic field $\mathbf{B}_0$, the electron density creates around the ^{1}H nucleus a weak local induced magnetic field ($\mathbf{B}_s$) opposing $\mathbf{B}_0$, which exerts onto the nucleus a *shielding effect*. This effect is proportional to the strength of the inducing field, *i.e.*, $\mathbf{B}_s = -\sigma \cdot \mathbf{B}_0$, where σ is the *shielding constant* for the nucleus considered. Consequently, the nucleus actually perceives an *effective* magnetic field ($\mathbf{B}_{\mathrm{eff}}$) resulting from the composition of $\mathbf{B}_s$ and $\mathbf{B}_0$, *i.e.*, $\mathbf{B}_{\mathrm{eff}} = (1 - \sigma) \cdot \mathbf{B}_0$. Then, the actual Larmor frequency of the nucleus will be as large as

$$\nu_{\mathrm{L}} = (\gamma_{\mathrm{H}}/2\pi) \cdot \mathbf{B}_{\mathrm{eff}} = (\gamma_{\mathrm{H}}/2\pi) \cdot (1 - \sigma) \cdot \mathbf{B_0} = (1 - \sigma) \cdot \nu_0 \tag{A.10}$$

where ν_0 is the ideal Larmor frequency of the naked nucleus. In general, the value of σ for a given hydrogen nucleus in a given molecule cannot be directly accessed. Rather, we can much more easily evaluate the difference between the σ values for the investigated nucleus and a ^{1}H nucleus in a suitable reference molecule added to the sample. Organic chemists have chosen tetramethylsilane (TMS, $Si(CH_3)_4$) as the reference because it is chemically inert and volatile (and thus it can be easily removed from the sample) and because its ^{1}H nuclei experience a very large shielding due to the inductive effect of silicon. Then, the *chemical shift* (δ_{X}) of the investigated ^{1}H nucleus (X) is defined as

$$\delta_{\mathrm{X}} = 10^6 \cdot (\sigma_{\mathrm{TMS}} - \sigma_{\mathrm{X}}) \tag{A.11}$$

where σ_{TMS} and σ_{X} are the shielding constants for TMS and X, respectively, and the coefficient 10^6 is a scaling factor introduced for the mere opportunity. Notably, δ_{X} (and σ as well) is an adimensional parameter; nevertheless, it is customary referring to it in terms of ppm (parts per million) as a sort of unit. Of course, its value is not dependent on the intensity $\mathbf{B}_0$ of the main magnetic field. In general, hydrogen atoms in "ordinary" organic molecules experience a weaker shielding effect in comparison to TMS, which results in positive δ values (*i.e.* they are more or less *de*shielded with respect to TMS). The δ values for real molecules are usually in the range 0–10 ppm (the so-called *^{1}H spectral window*), with a few exceptions. (De)shielding results from three main contributions, namely:

(i) *inductive electron effects*, *i.e.* the balance of the molecular electron density due to the differences in electronegativity between the atoms constituting the molecule. Topological proximity (in terms of the number of chemical bonds) to strongly electronegative atoms (O, N, halogens) causes deshielding.

(ii) *magnetic anisotropy effects*, which are typical of unsaturated groups (double bonds, triple bonds, aromatic rings). In this case, the electron density of a π-conjugated system creates a sort of ring current causing the formation of an induced local magnetic field. Figure A.3 illustrates the classical case of a benzene ring. According to Lenz's law, the induced field opposes the inducing $\mathbf{B}_0$ field inside the ring but reinforces it outside. Field lines allow outlining an ideal conical surface

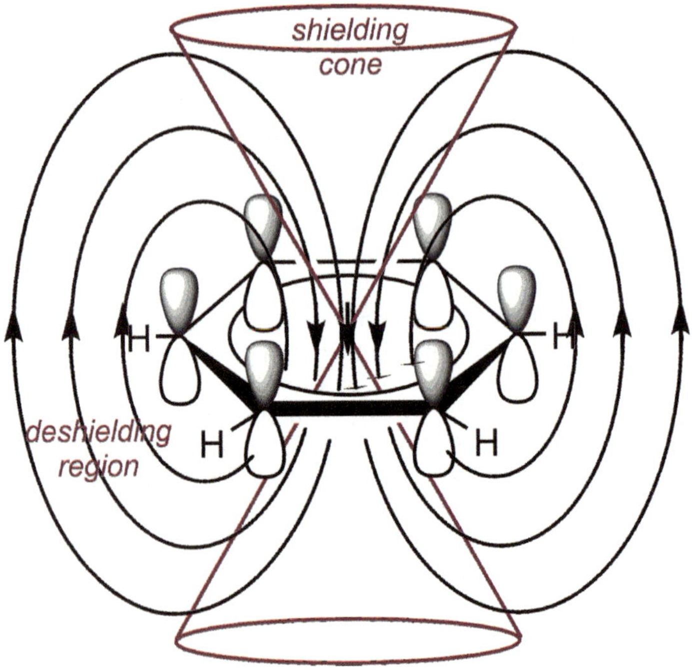

Figure A.3 Magnetic anisotropy of the benzene ring.

(shielding cone), in such a way that any ^{1}H nuclei inside the cone experience a shielding effect (δ values decrease), whereas ^{1}H nuclei outside the cone experience a deshielding effect (δ values increase).

(iii) *steric hindrance effects* tend to push away electron density from the nucleus, deshielding it.

Different ^{1}H nuclei in the same molecule experience different (de)shielding effects, and consequently precess and resonate at different frequencies. Moreover, (de)shielding effects combine together; therefore, the actual δ value of a given ^{1}H nucleus will result from the superimposition of different contributions. Typical δ range values for different typologies of hydrogen atoms in organic molecules are well-known and tabulated, and it is customary to illustrate them using the so-called *correlation tables* (a simplified example is depicted in Figure A.4).

A.4 NMR Spectrum and CW-NMR Instrumentation

Because different hydrogen atoms within a molecule perceive a different *chemical environment* and different (de)shielding effects, they precess and resonate at different frequencies. A *1D ^{1}H-NMR spectrum* is then a depiction of how the ^{1}H nuclei of a real sample placed into a magnetic field interact with a radiofrequency electromagnetic wave. More precisely, the spectrum

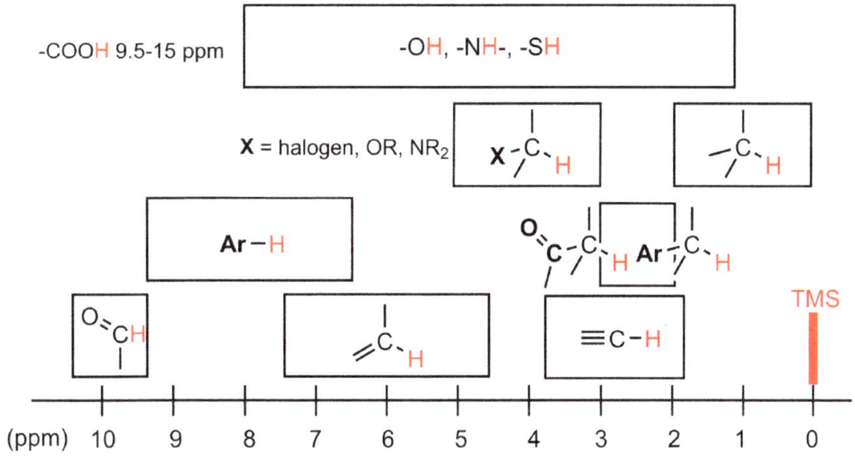

Figure A.4 Simplified correlation table for ^{1}H NMR spectroscopy.

represents the probability by which the sample absorbs every single frequency from the radio wave: the frequencies absorbed individuate different protons in the samples (experiencing different shielding effects), which resonate at the matching Larmor frequencies ν_L. Except for particular cases, signals in an NMR spectrum are narrow Lorentzian-shaped peaks centered at the various ν_L values.

Before proceeding, it is worth mentioning here that the spectra of organic substances are by rule recorded on liquid samples (we will further address the point in Section A.10). The substance is generally dissolved in a low-viscosity *deuterated* solvent (*i.e.* a solvent whose hydrogen atoms have been replaced by deuterium; otherwise, the signals of the substance would be buried under the signal(s) of the solvent) and placed in a tiny glass tube (ordinary NMR tubes are 0.5 cm wide and *ca.* 20 cm long). Deuterated chloroform ($CDCl_3$), deuterated dimethyl sulfoxide ($(CD_3)_2SO$) and deuterium oxide (D_2O) are the most common choices (deuterated methanol, dimethylformamide and acetone are also occasionally used).

In order to record a spectrum, suitable instrumentation is needed. Historically speaking, the first type of equipment employed for the task was the *continuous wave* (CW) apparatus. In its main lines, a CW instrument (Figure A.5) is constituted by (i) a static magnet or an electromagnet, providing the main magnetic field $\mathbf{B}_0$ (seldom larger than 2.35 T) placed horizontally; (ii) two small (*sweep*) coils able to provide a small adjustable additional magnetic field (B_{add}); and (iii) a *klystron*, inserted orthogonally into the magnetic field. The klystron is the "technological heart" of the instrument, comprising (i) the sample tube lodging, provided with a spinning system to make the sample tube rapidly spin around its axis; (ii) a *transmitter* coil providing a low power (*ca.* 0.2 W) continuous radiofrequency; and (iii) a *receiver* coil (*detector*), coaxially wrapped around the sample, producing a signal in correspondence with any absorbed frequency.

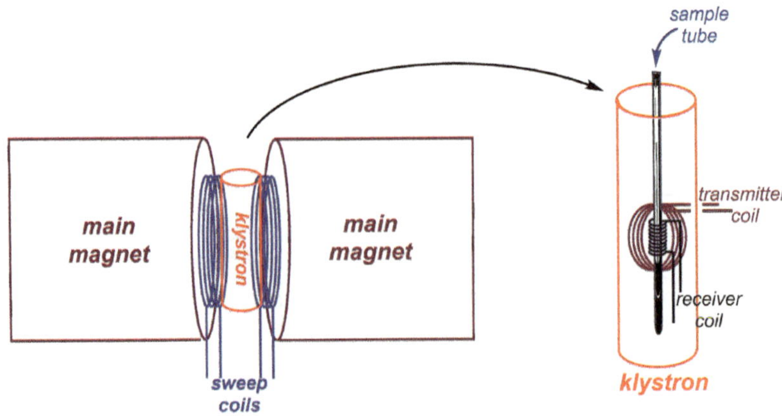

Figure A.5 Schematic representation of a CW-NMR instrument.

The instrument completes with a suitable arrangement of signal amplifiers and a recording system. It is interesting to notice that, in this kind of instrumentation, for technological opportunity reasons, the transmitter produces a single radiofrequency only. Hence, the resonance of the various protons in the samples is achieved by adjusting the main magnetic field with the additional variable magnetic field produced by the sweep coils. The larger the shielding effect for a given proton, the larger the additional field needed to make it resonate. Therefore, the chemical shift (δ_X) for a generic (X) proton is obtained as

$$\delta_X = 10^6 \cdot \left(\mathbf{B}_{add,TMS} - \mathbf{B}_{add,X}\right)/\mathbf{B}_0 \qquad (A.12)$$

where $\mathbf{B}_{add,TMS}$ and $\mathbf{B}_{add,X}$ are the additional fields needed to achieve resonance for the reference TMS and X, respectively. From the very fact that the transmitter coil in the klystron produces a single radiofrequency, it has become customary to qualify an NMR instrumentation by its operating frequency (60 MHz, 100 MHz, and so on), rather than by the strength of its main magnetic field.

From a physical standpoint, the interaction of the radiofrequency with the sample promotes both the quick excitation of the spins from the fundamental (α) to the excited (β) state and the quick reverse *relaxation* process. As a consequence, the difference ΔP between the two populations reduces to zero upon irradiation (as well as the magnetization vector **M**). A basic principle in NMR spectroscopy imposes that the intensity of a given signal recorded in the spectrum, more precisely the area enclosed by the signal peak, is proportional to the relevant population difference ΔP. This area can be easily obtained by integration of the peak, which is performed automatically by the instrument. For molecular compounds, this implies the well-known rule that *the area under a signal is proportional to the number of equivalent hydrogen atoms generating the relevant peak.* Equivalence between hydrogens may stem from rapid rotation around simple bonds (*e.g.* the H atoms in a methyl group) or molecular symmetry.

A.5 FT-NMR Instrumentation and Relaxation Processes

The CW technique has severe limitations and rapidly became obsolete, being superseded by the modern pulse Fourier transform (FT-NMR) one. Modern FT instrumentation benefits from the use of *cryomagnets*, *i.e.* liquid helium-cooled superconducting electromagnets, which allow obtaining much stronger magnetic fields than in CW instruments (up to 20 T and even beyond; at present, the most powerful instruments have crossed the threshold of 1 GHz, whereas routine instruments working at 400 MHz are quite common). This deeply changed the instrumental design (Figure A.6). In modern FT-NMR instruments, the magnetic field provided by the cryomagnet is placed vertically. The magnet is hosted in a suitably shaped thermal insulating vessel filled with liquid helium, which in turn is enclosed in a second vessel filled with liquid nitrogen (the *cryogenic liquids* must be periodically refilled). The cylindrical hollow space left by this arrangement hosts the *probe* (modern substitute of the old klystron), containing the sample lodging, the transmitter and receiver coils (see later), and the needed circuitry. However, the real strong point of the FT technique is constituted by the use of a *pulsed* radiofrequency, rather than a continuous one.

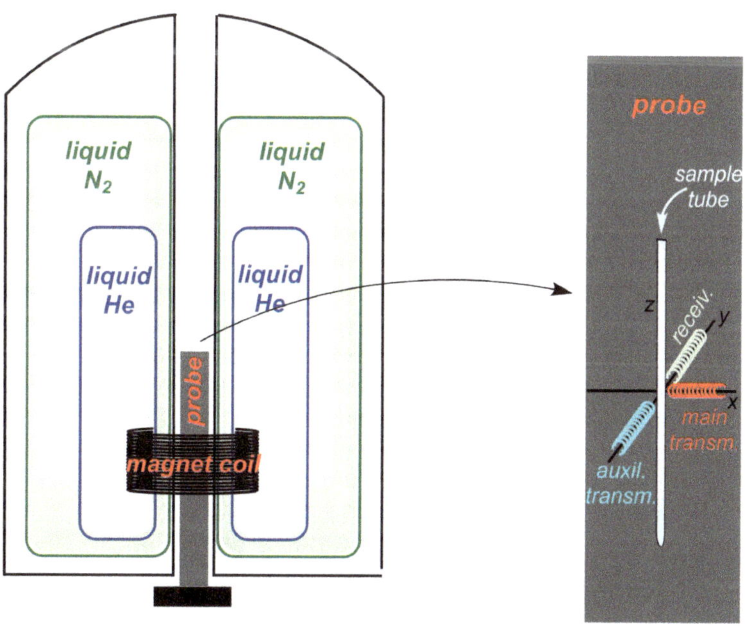

Figure A.6 Schematic representation of a modern FT-NMR instrument.

A *pulse* is a *short* and *intense* radiofrequency wave. In other words, the radio wave is released in a very short time (Δt, in general a few μs), but with a strong emission power (hundreds of W). Owing to the Heisenberg indetermination principle (in the form $\Delta E \cdot \Delta t > \hbar$), a short emission time implies that energy quanta are produced within a frequency range ($\Delta \nu$) inversely proportional to Δt. For emission times in the order of μs, this range is indeed some hundreds of ppm wide, which is large enough to cover the entire ^{1}H spectral window. Thus, a single pulse can simultaneously excite all the diverse protons in the sample. However, the most interesting aspect is the effect a pulse exerts on the magnetization vector **M**. Let us consider a Cartesian reference system with the *z-axis* coinciding with the direction of $\mathbf{B}_0$. Ideally, the main transmitter coil is placed in the probe along the *x-axis* and the receiver coil along the *y-axis* (see Figure A.6; a further auxiliary transmitter coil is placed on $-y$, but it is not necessary to consider it now). In general, it can be demonstrated that a pulse launched from a direction makes the magnetization vector **M** rotate clockwise by an angle Θ as large as

$$\Theta = \Delta t \cdot \gamma_{\mathrm{H}} \cdot \mathbf{B}_{\mathrm{I}} \tag{A.13}$$

where $\mathbf{B}_{\mathrm{I}}$ is the intensity of the oscillating magnetic field associated with the radio wave, which in turn is proportional to the emission power (expressed in W). It is worth stressing that, in the NMR language, usually, the duration of a pulse is not expressed in time units (μs), but rather in angle units (rad or deg), referring to the rotation produced onto **M**. Noticeably, the same angle can be obtained in principle with different emission times, by inversely varying the emission power.

From a physical standpoint, the rotation of **M** produced by the pulse implies significant consequences on the spin population. Two particular cases deserve to be considered:

(i) An 180° (or π rad) pulse produces the inversion of **M**, which rotates up to align along the *−z-axis*. This causes the inversion of the entire spin population, *i.e.* all protons in the α state are excited up to the β state, and all protons in the β state go back to the α state. Thus, the overall absorption of ΔP energy quanta by the sample occurs.

(ii) A 90° (or π/2 rad) pulse makes **M** rotate clockwise around the *x*-axis, and align along the *y*-axis. This causes a double effect on the spin population. First, the α and β populations become equal ($\Delta P/2$ energy quanta are absorbed by the sample); the second partial phase coherence is induced in the sample. *Phase coherence* is a delicate concept. It means that the ideal uniform distribution of $\boldsymbol{\mu}$ vectors on the conical surfaces mentioned in defining **M** (Figure A.2) is de-symmetrized by the action of the pulse.

It is immediately apparent that, after the sample has received the pulse, its spin population is largely perturbed from the thermal equilibrium distribution. Therefore, the system will subsequently undergo *relaxation* to recover thermal equilibrium.

Relaxation phenomena have paramount importance in FT-NMR and will be the object of exhaustive discussion in another presentation (Appendix B); nevertheless, a few basic concepts must be presented herein. In brief, relaxation involves two different simultaneous processes, referred to as *longitudinal* (or "spin–lattice") and *transverse* (or "spin–spin"). The former process relates to the loss of the excess energy quanta absorbed. The polar molecules of the solvent, which are subjected to continuous translational (Brownian) and rotational motions, each create a microscopic oscillating electromagnetic field in their surroundings. Therefore, if an excited (β) ^{1}H nucleus happens to get close to a micro-field oscillating at the right frequency (*i.e.* matching its Larmor frequency), it can give out its excess energy and return down to the fundamental (α) spin state (thus, an energy quantum is released to the *solvent lattice*). Conversely, transverse relaxation relates to the loss of phase coherence, *i.e.* the fact that protons' magnetic momenta *fan out* to recover the cone symmetric distribution with respect to the *z*-axis (as depicted in Figure A.2). This is achieved because of two different physical factors, namely: (i) the possible microscopic inhomogeneities of the main magnetic field, due to which protons with the same chemical shift placed in different points of the sample happen to precess at slightly different Larmor frequencies; (ii) exchange of an energy quantum between an excited nucleus and a nucleus with the same Larmor frequency in its fundamental state, which happens to be close in space but not perfectly in-phase (*spin–spin exchange*).

In principle, the two different relaxation processes are independent and should follow (in the simplest cases) a trivial first-order exponential kinetics. The relevant apparent kinetic constants are referred to as the *relaxation rates* (R, R_1 being the longitudinal one and R_2 the transverse one), whereas the inverses of the kinetic constants are referred to as the *relaxation times* (T_1 and T_2, respectively). By rule, transverse relaxation is much faster than the longitudinal one (see later). It is immediately apparent that relaxation phenomena, being connected to molecular motions, offer the opportunity to access information regarding molecular dynamics (this topic, as well as the methods to study relaxation kinetics, will be discussed in detail in Appendix B).

Now, the particular case of the 90° pulse must be analyzed in detail (Figure A.7). After the pulse has been absorbed by the sample, the ^{1}H population must keep on with the precession motion at the proper Larmor frequency ν_L, but starting from the new spin distribution (altered with respect to thermal equilibrium) produced by the pulse itself. This involves that the **M** vector, after having been rotated by the pulse and brought onto the *y-axis*, will start to rotate on the *xy* plane. Of course, if diverse hydrogen atoms are present in the sample, each with its own ν_L, **M** can be split into as many independently rotating components. As long as relaxation occurs, **M** rises (or, which is the same, its components do), following a sort of spiral-shaped

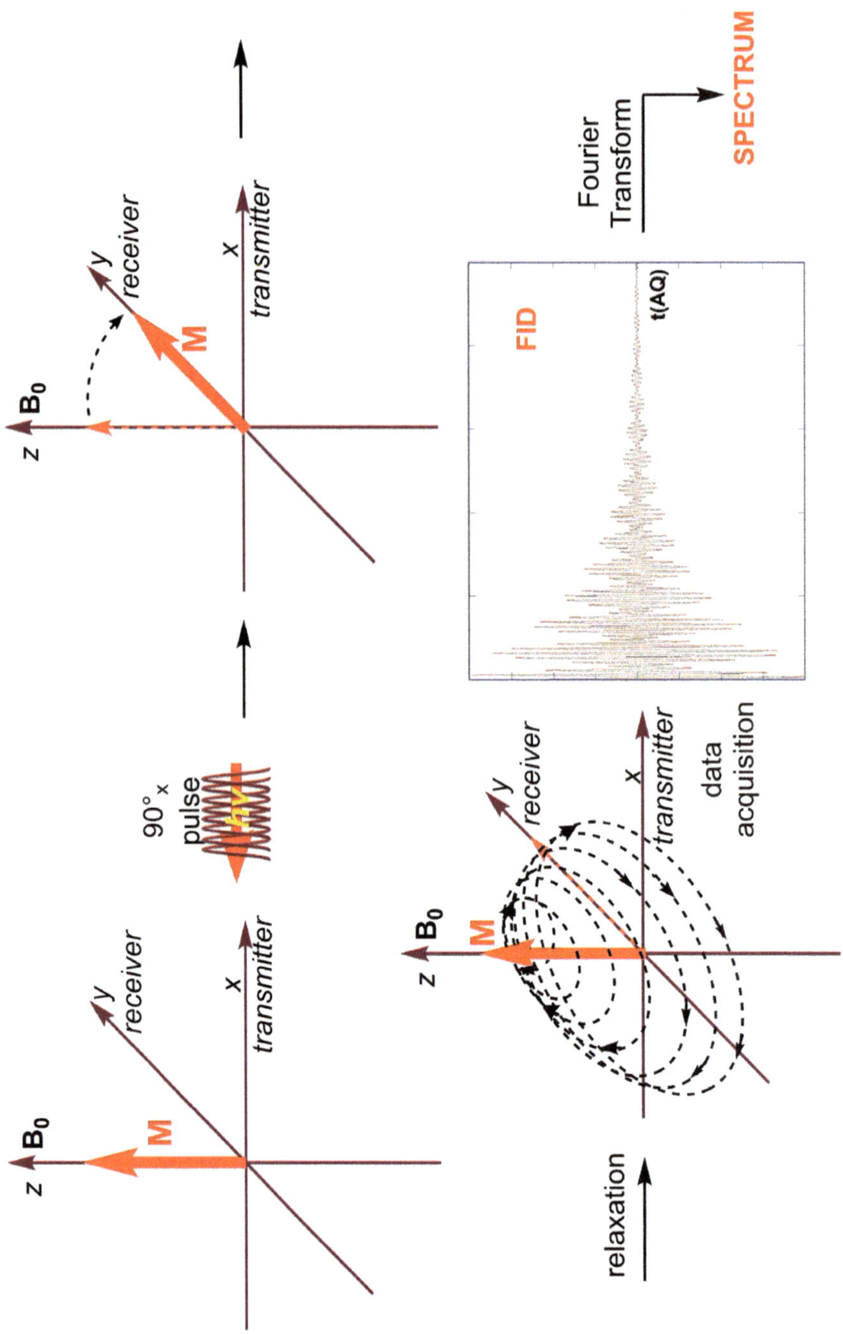

Figure A.7 FID acquisition in FT-NMR.

path, to reach again the *z-axis* direction, which corresponds with the complete loss of the perturbation induced by the pulse, and the full recovery of the thermal equilibrium distribution. During the process, the receiver coil placed on the *y-axis* perceives at any time the projection of **M** (or its components) along y, and its continuous variation, as a tiny induced current. The consequent electrical signal is digitally sampled, and its recording constitutes the so-called "*free induction decay*" (FID) trace. The FID is indeed an exponentially decaying sinusoid – or better, the superimposition of as many sinusoids as the number of components of **M** – collected in the time domain. Therefore, the FID is subjected to Fourier transform (FT) analysis to give ultimately back the spectrum (in the conjugated frequency domain) of the Larmor frequencies of the ^{1}H nuclei present in the sample. The chemical shift (δ_X) of a given X proton is then retrieved as

$$\delta_X = 10^6 \cdot (\nu_X - \nu_{TMS})/\nu_0$$

where ν_X, ν_{TMS} and ν_0 are, respectively, the resonance frequencies of the proton X and TMS, and the instrumental operating frequency.

Despite the much more complicated conceptual issues implied, the FT technique has overwhelming advantages as compared to the old CW one. As long as it is by far faster, the digitalized signal can be cumulated by performing subsequent pulse cycles, which enables improving the signal-to-noise ratio and the quality of the spectrum as well. This in turn allows us to record good quality spectra even if having available small amounts of the substance (less than 1 mg). Moreover, suitable *pulse sequences* can be set up, which makes information that cannot be gained otherwise accessible (a typical example of a pulse sequence will be discussed in Section A.8). As a final remark, because transverse relaxation is faster than longitudinal relaxation, FID decay is mainly governed by the former one. However, a too short T_2 value results in an enlargement of the signals, with a consequent loss in spectral resolution.

A.6 Spin Coupling

Spin coupling is a typical phenomenon in NMR spectroscopy, which provides very useful information in order to retrieve the structure of a molecule. In general, two or more different ^{1}H nuclei (but the same applies to any couple of spinning nuclei) that are topologically proximate, *i.e.* separated by not more than four chemical bonds (five in exceptional cases), *couple* together. The magnetic spin momenta of these nuclei interact, and this interaction provides an energy contribution that makes the signals of the two nuclei undergo *splitting* and assume a *fine structure*. In particular, in the simplest case of a 1-to-1 coupling, both the signals of the two hydrogen atoms split in two and assume the form of a *doublet* (Figure A.8(a)).

A rigorous and exhaustive explanation of the physics behind the coupling phenomenon is (of course) far beyond the scope and limits of this presentation. In a very simplified and naïve way, we can assume that the transition

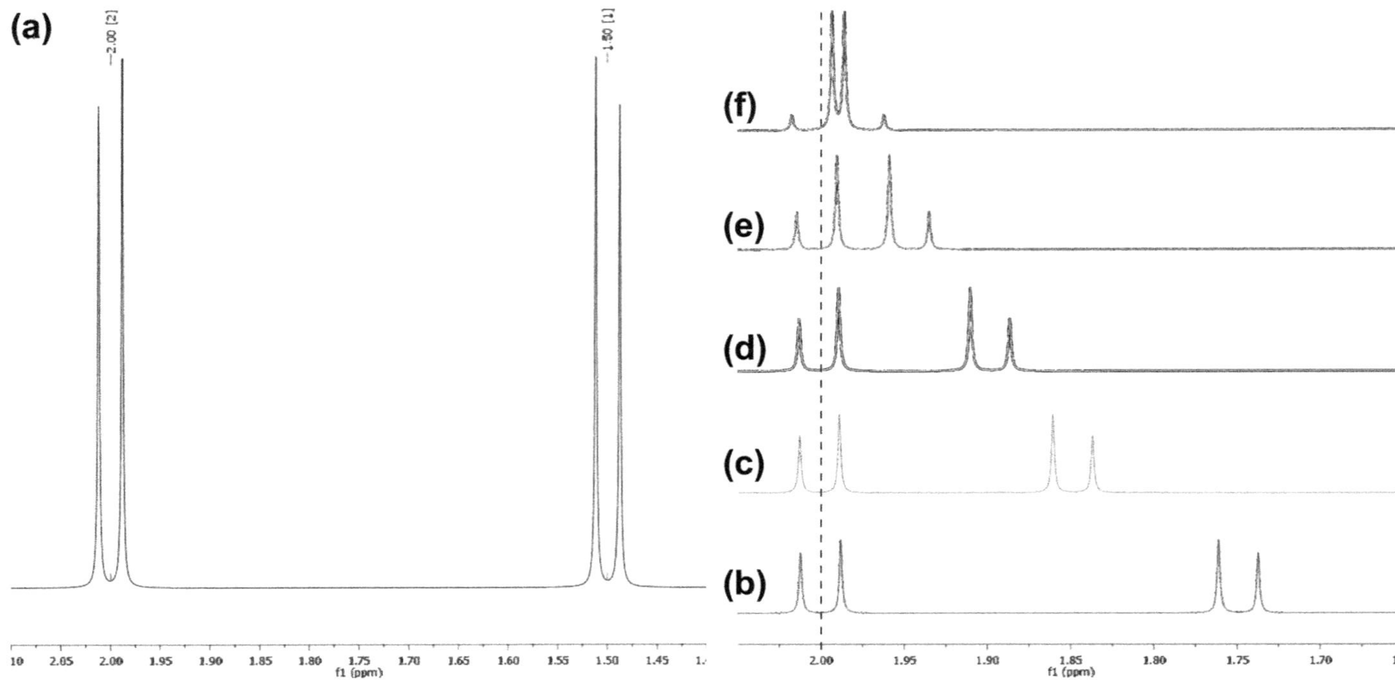

Figure A.8 Simulated spectra (500 MHz) for a ^{1}H–^{1}H coupling, $J = 12$ Hz; the first signal is centered at 2.0 ppm, whereas the second one is centered at (a) 1.5 ppm; (b) 1.75 ppm; (c) 1.85 ppm; (d) 1.9 ppm, (e) 1.95 ppm; and (f) 1.98 ppm. Desymmetrization of the doublet for the signal at 2.0 ppm on passing from a 1st-order to a 2nd-order spectrum is evidenced by the dashed line.

energy for a nucleus (A) interacting with another one (B) is different depending on the actual spin status of B (in a specular way, the transition energy for B is different depending on the spin status of A). Splitting is quantified by the difference in the frequencies (J, measured in Hz) between the spectral lines of the split signals; J is referred to as the *apparent coupling constants* between the interacting nuclei and depends on the energy (δE) associated with the coupling interaction, according to the relationship $\delta E = h \cdot J$. Noticeably, (i) line separation (measured in Hz) between the split signals of the two coupled nuclei *must* be the same (*i.e.* the two coupled nuclei show the same J value); (ii) the value of J *does not depend* on the intensity $\mathbf{B}_0$ of the main magnetic field. Therefore, the apparent line separation of each split signal in the spectrum, measured in terms of the chemical shift difference, is inversely proportional to $\mathbf{B}_0$.

In principle, the split lines should show the same intensity. This condition (referred to as a *first-order spectrum*) occurs whenever the difference between the resonance frequencies ($\delta\nu$) of the coupled nuclei is much larger than J (conventionally, for $\delta\nu/J \gg 10$). In this case, the frequency of the two spectral lines will be as large as $\nu_L \pm J/2$. Conversely, when the previous condition is not met, the internal lines of the coupled signals are more intense than the outer ones, and the signal system assumes a typical roof-like shape (Figure A.8(b)). This case is referred to as a *second-order* spectrum. This desymmetrization of the signal intensities becomes more and more apparent as $\delta\nu/J$ progressively decreases (noticeably, even the position of the two signals with respect to ν_L becomes nonsymmetric). At the limit, when $\delta\nu/J$ reduces to zero, *i.e.* when the coupled nuclei come to have the same chemical shift value, the external signals disappear and the internal ones merge into a unique signal. Hence, it is commonly said that nuclei with the same chemical shift do not show apparent coupling. Because $\delta\nu$ depends on $\mathbf{B}_0$, but J does not, a second-order spectrum progressively transforms into a first-order one on increasing the strength of the magnetic field.

Multiple couplings are very common even in the simplest organic molecules. In general, whenever a ^{1}H nucleus couples with n equivalent ^{1}H nuclei, its signal splits into $n+1$ spectral lines, the intensities of which are proportional to the coefficients of the well-known Pascal triangle (under first-order conditions), thus giving rise to a *multiplet* (*doublet, triplet, quartet, quintuplet, etc.*; the intensity pattern derives from combinatorial analysis). The lines, of course, must show the same separation, as large as J Hz. Furthermore, if a ^{1}H nucleus couples with two *non-equivalent* ^{1}H nuclei, with two different J values, its signal undergoes a double splitting (*i.e.* the proton signal independently splits for the two different couplings), giving rise to a *double doublet.* A combination of the two phenomena results in complex multiplicity patterns (for example, simultaneous coupling with three protons with a large J value, and with two further protons with a much smaller J, will give rise to a quartet of triplets). Indeed, possible combinations are virtually countless, and it is not uncommon that multiplicity patterns appear so complicated that they are barely interpretable (an illustrative example is depicted in Figure A.9).

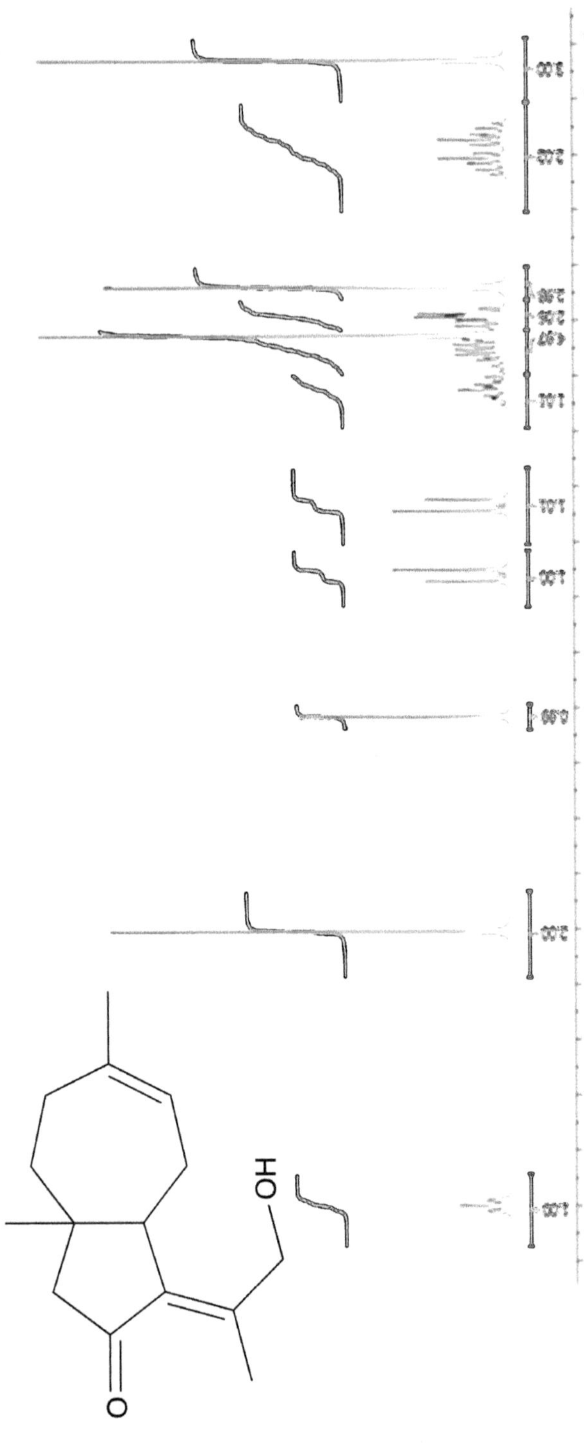

Figure A.9 Simulated ^{1}H NMR spectrum (300 MHz) for the molecule depicted.

As we mentioned previously, coupling constants J have a value (measured in Hz) that is characteristic of the coupling nuclei and does not depend on the intensity of the magnetic field. The actual value of J for a given coupling pattern is largely variable, depending on the system (in general, J values for ^1H–^1H coupling seldom exceed 20 Hz), and may provide useful additional information in order to retrieve the structure of an organic molecule. It is important to stress here that coupling is a *topological* phenomenon, *i.e.* it is related to the number of chemical bonds separating the nuclei, and has no relation with the actual distance in space. Conventionally, four different coupling typologies are distinguished:

(i) *direct* coupling (J_1) between directly bound nuclei (*i.e.* one bond of distance only). This situation does not occur in ^1H NMR spectroscopy, of course, but has great importance whenever other nuclei are considered.
(ii) *geminal* coupling (J_2), between nuclei bound to the same atom (*i.e.* two bonds of distance). In ^1H NMR this situation occurs in the $-CH_2-$ groups of a chiral molecule (diastereotopic hydrogens, which have different chemical shifts); geminal J_{H-H} values in the range of 10–20 Hz are typical.
(iii) vicinal coupling (J_3), between nuclei bound to adjacent atoms (*i.e.* three bonds of distance). It is the most common case in ^1H NMR spectroscopy, with J values largely varying, depending on the particular structural arrangement (for a detailed discussion the reader is advised to consult the reference textbooks).
(iv) long-range coupling (J_4) occurs between nuclei at more than three bonds of distance. By rule it is very weak (typical values seldom exceed 3 Hz) and can be actually observed only in particular cases (the *allylic* coupling and the *meta*-coupling in aromatics are the most common ones).

A.7 Double Resonance Techniques

Double resonance techniques require the simultaneous irradiation of a sample with two different radiofrequency waves. For this purpose, both the main (x) and the auxiliary ($-y$) transmitters are employed. Basically, the auxiliary transmitter emits a radio wave able to produce particular modifications in the sample, which can be detected once the main transmitter has launched the usual 90° pulse and the resulting FID has been collected and subjected to FT analysis. The main double resonance techniques are *selective decoupling* and the detection of the *nuclear Overhauser effect*.

Selective decoupling aims at simplifying complex coupling patterns and at demonstrating the actual coupling between two nuclei. For the sake of clarity, let us consider the simplest case of two coupled hydrogens A and B discussed previously. If the sample is irradiated with a continuous low-power radio wave having a frequency coinciding with the Larmor frequency of A (provided by the auxiliary transmitter), then its α–β transitions will be enormously speeded up, and the ΔP of its population will reduce to zero; in response, nucleus B loses its coupling with A. Hence, on recording the spectrum during irradiation, the signal of A disappears and the signal of B reduces to a *singlet*. Otherwise, if nucleus B is irradiated, its signal disappears and the signal of A becomes a singlet. Verification of selective decoupling provides the ultimate proof of reciprocal coupling. Noticeably, the decoupling immediately occurs upon irradiation, but the coupling is immediately recovered once the *decoupler* (*i.e.* the auxiliary transmitter) is turned off.

Unlike spin coupling, the *nuclear Overhauser effect* (NOE) is a spatial proximity phenomenon. When two nuclei are "close" in space (usually not more than 5 Å), on irradiating one of them, the signal relevant to the other one increases in intensity. The fact that the irradiated nucleus speeds up its spin state transitions stimulates the relaxation of the proximate non-irradiated nucleus, causing an increase in its ΔP. Of course, the extent of the effect strictly depends on the average internuclear distance. In ^{1}H NMR spectroscopy, this can lead to a signal intensification up to 15–20% for hydrogen atoms at the van der Waals distance; conversely, the effect rapidly fades away for a distance larger than *ca.* 4 Å. It is important to stress that the NOE effect has no relationship with spin coupling. Hydrogen atoms that are not topologically proximate (*i.e.* separated by 5 or more bonds), but happen to be close in space for mere structural reasons, show this effect. Therefore, the detection of NOE can be particularly useful in the resolution of stereochemistry problems. Moreover, its occurrence is bound to the kinetics of the relaxation processes. Therefore, unlike selective decoupling, NOE does not appear immediately after the auxiliary transmitter is turned on and does not disappear immediately after the transmitter is turned off.

A.8 Basics of ^{13}C NMR Spectroscopy

Carbon constitutes the backbone of all organic molecules; therefore, its NMR detection would provide a very important piece of information for retrieving molecular structure. However, there are various difficulties to cope with. As a matter of fact, C is constituted by two main natural isotopes: ^{12}C (98.84% abundance) and ^{13}C (1.16% abundance). Unfortunately, ^{12}C is spinless and cannot be detected by NMR. Conversely, ^{13}C has $S=\frac{1}{2}$ and can be detected, in principle, according to the same approach and mathematical formalism as for ^{1}H. However, the gyromagnetic ratio of ^{13}C (γ_C) is *ca.* $\frac{1}{4}$ as large as the one for ^{1}H. Therefore, at the same magnetic field strength, the energy separation between the α and β spin states, and the corresponding Larmor/resonance frequency as well, is much smaller in comparison to ^{1}H (for instance, at 11.74 T, as the proton resonates at 500 MHz, ^{13}C resonates at 124.7 MHz). For this reason, modern probes usually feature two distinct co-axial transmitters for irradiating separately the ^{1}H and ^{13}C nuclei. In general, the ensemble of instrumental and hardware resources used for detecting the

^{1}H nucleus is identified as the *^{1}H channel*, and, similarly, the *^{13}C channel* is the set of resources devoted to the ^{13}C nucleus.

Furthermore, a smaller gyromagnetic ratio as compared to proton implies also a much smaller population difference ΔP between the spin states. This, together with the low natural isotope abundance and other factors (in particular the quite slow longitudinal relaxation rate), results in the fact that carbon (as an element) has a *relative NMR sensitivity* as large as *ca.* 1/5700 with respect to hydrogen. In other words, as long as all other conditions are the same, the hydrogen signal is 5700 times more intense than the carbon one. Hence, the importance of the pulse technique is immediately evident: in fact, only massive accumulation of the signal enables obtaining ^{13}C NMR spectra of acceptable quality.

Another difficulty arises from the fact that ^{13}C couples with ^{1}H. For the ^{1}H–^{13}C coupling, direct J_1 values in the range 100–250 Hz and germinal or vicinal J_{2-3} values in the range 0–60 Hz are quite common. Therefore, fully coupled ^{13}C spectra show signals with very complex splitting patterns. In principle, it is possible to overcome this problem by *broadband decoupling*, *i.e.* by performing the simultaneous irradiation of the ^{1}H spectral region to eliminate ^{1}H–^{13}C coupling. Then, decoupled ^{13}C spectra appear enormously simplified because all signals are singlets. However, ^{1}H irradiation also causes a very strong NOE effect, the extent of which is a function of the number of H atoms directly bound to each C. In detail, the signals relevant to $-CH_3$ groups appear much more intense than the signal for $-CH_2-$ groups, which in turn appear stronger than the signals for $>$CH–. By contrast, quaternary carbons, receiving no NOE benefit, show small signals, often barely distinguishable from the baseline noise. Hence, the common rule derives that, unlike ^{1}H, the signals in a *^{13}C-NMR spectrum cannot be integrated, but only counted*. (It is worth mentioning, for the sake of completeness, that it is possible to overcome this difficulty and recover the possibility to integrate the ^{13}C signal by using the so-called *inverse-gated decoupling* technique. Interested readers are advised to consult reference textbooks for details.)

A further drawback of broadband decoupling is the fact that an important piece of information is lost, *i.e.* the number of H atoms directly bound to each C atom. It is possible to recover this information by using the *DEPT-135* and *DEPT-90* techniques. The DEPT spectra are obtained by applying suitable (complex) pulse sequences involving both the ^{1}H and ^{13}C channels. In detail, in the DEPT-135 spectrum methyl and methine groups appear as positive signals, methylene groups as negative signals and quaternary carbons disappear; in the DEPT-90 spectrum, only methine groups afford positive signals, whereas all other carbons disappear. Thus, a comparison of the DEPT spectra with the original one enables us to distinguish the four different types of C atoms (an illustrative example is shown in Figure A.10).

Besides, the concept of chemical shift applies to ^{13}C NMR spectroscopy in the very same way as to ^{1}H. TMS again provides the reference molecule. Noticeably, the *^{13}C spectral window*, conventionally ranging from 0 to 230 ppm, is much larger than the one for ^{1}H. Nevertheless, the general principles

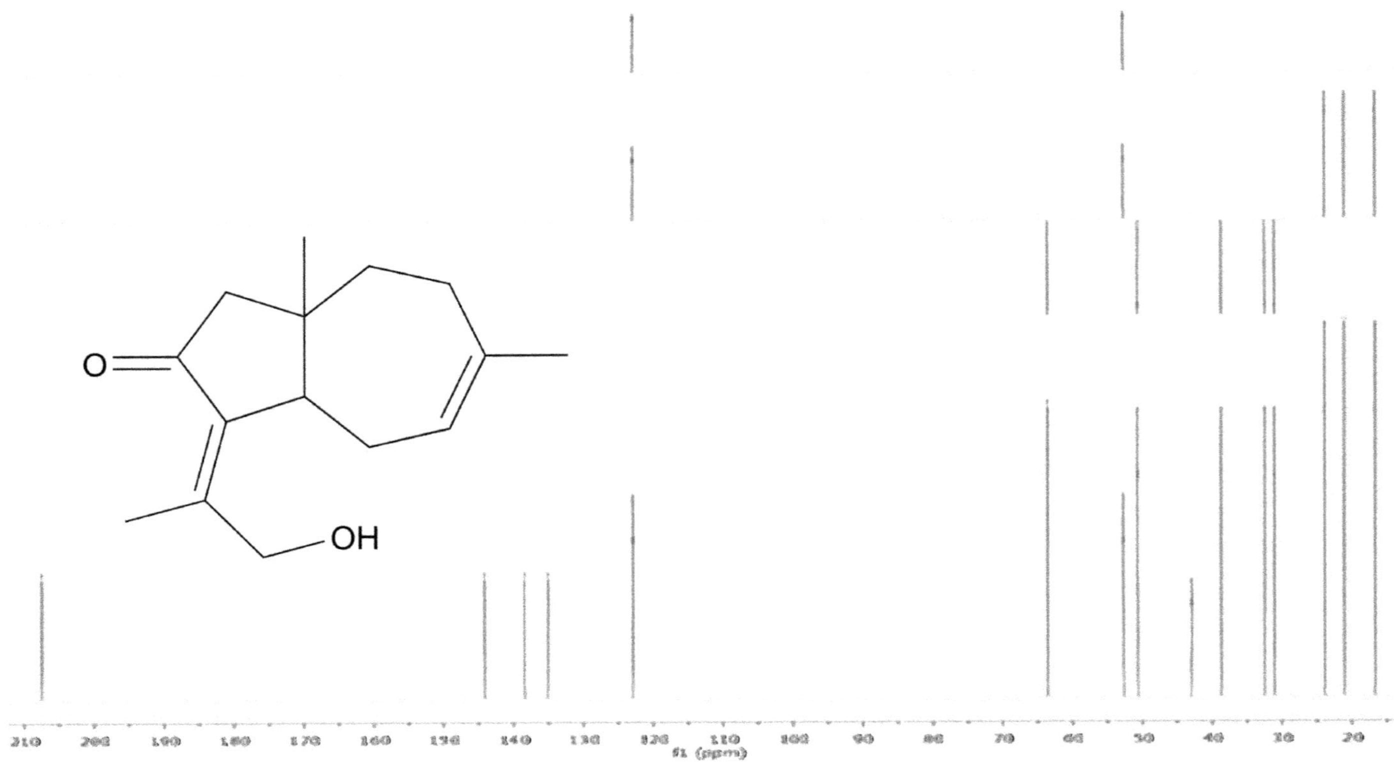

Figure A.10 Simulated ^{13}C NMR spectra (75 MHz) for the molecule depicted: broadband H-decoupled (bottom), DEPT-135 (middle), DEPT-90 (top).

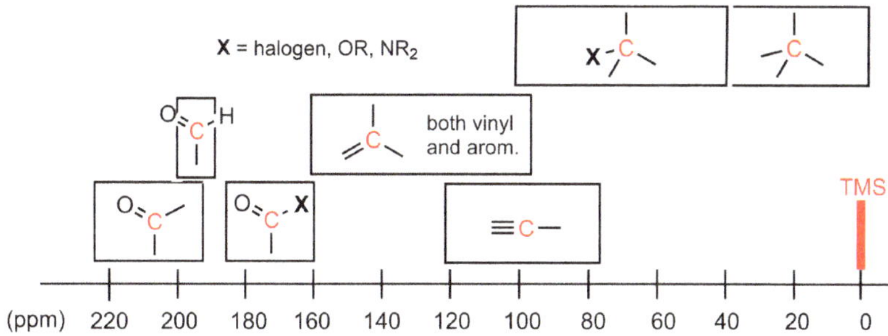

Figure A.11 Simplified correlation table for ^{13}C NMR spectroscopy.

governing ^{1}H chemical shifts apply nearly in the same way for ^{13}C. A simplified ^{13}C correlation table is depicted in Figure A.11.

As a final remark, unlike ^{1}H–^{13}C coupling, the ^{13}C–^{13}C direct coupling is not observed in spectra because, owing to the low isotope abundance, the probability that two particular carbon atoms in a molecule both have ^{13}C nuclei is only 1 out of 10 000.

A.9 Correlation (2D) Spectroscopy

In the widest sense, *correlation* NMR spectroscopy (or *2D-NMR spectroscopy*) aims at instantly evidencing the occurrence of a relationship between nuclei sharing a common feature, by correlating the relevant signals on a two-dimensional diagram. The most commonly investigated shared features are spin coupling and spatial proximity. It is customary to distinguish between a *homo-correlation* and a *hetero-correlation*. In the former case, the signals of the same spectrum are correlated; in the latter case, the signals in spectra relevant to different nuclei are correlated. Correlation spectroscopy requires the use of suitably structured pulse sequences, which can be even quite complicated. For explicative purposes, we will shortly discuss herein the *COSY* technique, which aims at individuating ^{1}H–^{1}H spin coupling (*i.e.* it is a homo-correlated ^{1}H–^{1}H spectrum).

The COSY pulse sequence can be symbolized as an instruction string:

$$[90^\circ_x] - t_\mathrm{D} - \left[90^\circ_{-y}\right] - \mathrm{AQ}$$

In detail, the sample is subjected in sequence to (i) a 90° pulse from the principal transmitter on x (*preparation* step), (ii) a wait for an arbitrarily chosen *delay* time t_D (*evolution* step), and (iii) a second 90° pulse from the auxiliary transmitter on $-y$, followed by the collection of the FID (*detection* step). The analysis of the effect of this pulse sequence on the magnetization vector of the sample would be a bit complicated. However, it is immediately apparent that the recorded FID and the relevant Fourier transform will depend on the length of t_D. Thus, a *collection* of FIDs is acquired, by continuously increasing t_D. Such a collection constitutes, as a whole, a two-variable function in the time domain (the independent variables being the FID acquisition time and the delay time), to which a double Fourier transform analysis can be applied. The outcome of the latter mathematical operation will be constituted by a two-variable function in the conjugated frequency domain, which features a set of peaks on a 3D surface. These peaks, in turn, can be conveniently represented on a 2D square diagram, in the same way as the mountains on a map (Figure A.12). Both the "coordinates" of such a map are indeed the original ^{1}H spectrum with its signals (identified by the relevant chemical shift values). The "map" will then show a characteristic butterfly-wings-like symmetrical arrangement of spots, with (i) a *main diagonal*, featuring as many "self-correlation" signals as the signals in the original 1D NMR spectrum, and (ii) some symmetrical "inter-correlation" signals, whose coordinates correspond to the signals of spin-coupled protons. Thus, the out-of-diagonal spots provide the requested information about spin coupling.

Other important 2D techniques aimed at detecting spin coupling are as follows:

- HMQC and HETCOR, which are hetero-correlated spectra retrieving direct (J_1) ^{1}H–^{13}C coupling (the two techniques use different pulse sequences and acquire the FID collection on either the ^{1}H or the ^{13}C channel);
- HMBC and COLOC, which are hetero-correlated spectra retrieving geminal (J_2) and vicinal (J_3) ^{1}H–^{13}C coupling (the differences are similar to those of the previous case);
- INADEQUATE is a homo-correlated spectrum retrieving direct (J_1) ^{13}C–^{13}C coupling.

All these techniques play a paramount role in the determination of the structure of organic molecules, both synthetic compounds and natural products.

A further important 2D technique based on spin coupling is TOCSY. This technique enables individuating *spin systems* at a glance, *i.e.* sequences of subsequently coupled nuclei. A TOCSY spectrum resembles a COSY one, but presents much more spots because the ^{1}H nuclei belonging to the same "coupling chain" appear as all coupled together (in other words, hydrogens belonging to the same coupling chain, but not reciprocally coupled, appear in the 2D TOCSY spectrum as if they were coupled). Therefore, this technique may become useful for the interpretation of the very complex spectra of bio-macromolecules such as polysaccharides and polypeptides, because it enables spotting out the signals relevant to hydrogen atoms belonging to the same sugar or amino acid subunit. The TOCSY technique can be also adapted for studying ^{1}H–^{13}C hetero-correlation.

Further important 2D techniques, devoted to the detection of the NOE effect, are NOESY and ROESY. Once again, these are ^{1}H–^{1}H homo-correlated spectra, whose appearance is similar to COSY. In particular, a NOESY or a

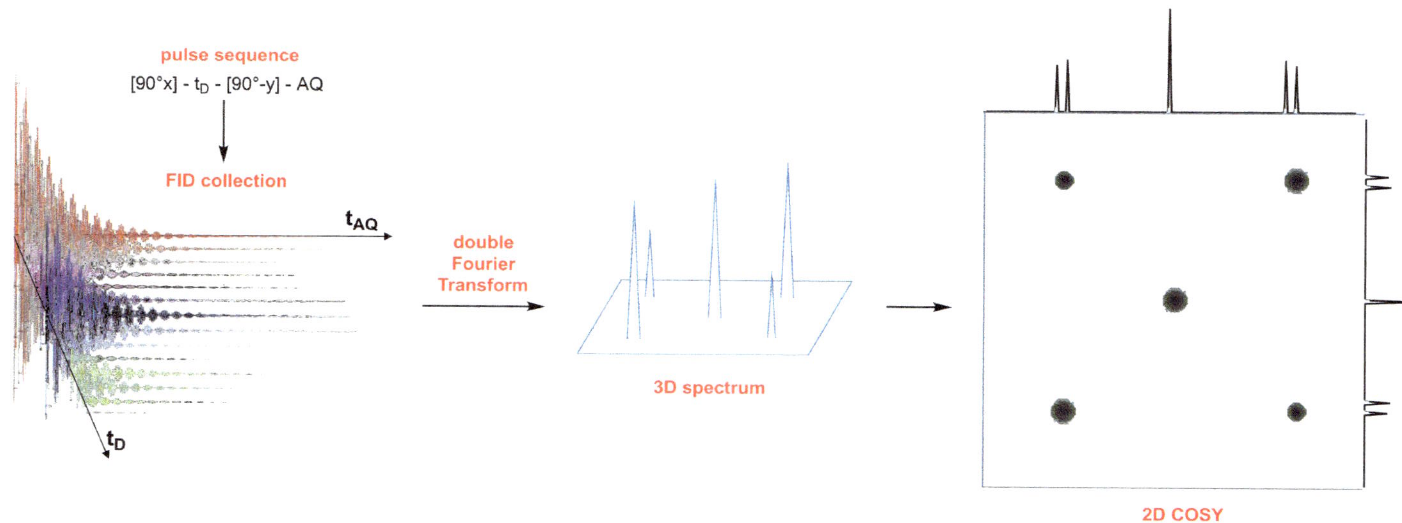

Figure A.12 Generation of the COSY spectra.

ROESY spectrum features all the spots already present in the relevant COSY due to spin coupling, and further additional spots due to the NOE effect, accounting for spatial proximity. Therefore, these techniques may be useful, for instance, for assessing the stereochemistry of natural products, for retrieving the secondary structure of proteins, and in particular for determining the structure of supramolecular systems such as host–guest complexes.

A.10 Solid-state NMR Techniques

As we discussed in Section A.3, shielding effects, mirrored by σ values, derive from the behavior of molecular electron density around the nuclei. In a similar way, theory provides that spin coupling, mirrored by J values, also depends on the interaction between nuclei and binding electrons. However, the electron density distribution is governed by orbital functions, whose mathematical form is strictly related to molecular symmetry and the directionality of chemical bonds. As a consequence, the values of both shielding and coupling constants derive from the superimposition of two distinct contributions, an *isotropic* and an *anisotropic* one. The former is an "intrinsic" contribution, not depending on the instant orientation of the molecule with respect to the main magnetic field $\mathbf{B}_0$. Conversely, the anisotropic contribution depends on the inclination angle (ϑ) of the ideal molecular/bond axis with respect to $\mathbf{B}_0$, by a proportionality factor as large as $[1 - 3(\cos\vartheta)^2]$. If the molecule is free to rapidly rotate, as occurs in low-viscosity media, then the anisotropic contributions mediate to zero, which results in well-resolved and easily readable spectra. For this very reason, as we mentioned above (Section A.4), NMR spectra are by choice recorded on liquid samples. By contrast, if the system is viscous, or if free rotation is slowed down (as in the case of polymers, or macromolecules in general), or worse if the sample is in the solid state, the anisotropy scaling factor cannot mediate to zero but assume different non-null values for every single molecule present. As a consequence, signals in the spectrum all undergo an enormous enlargement and loss of resolution, thus making the spectrum unreadable. There is a smart trick, however, which allows overcoming this difficulty.

It is immediately apparent that the anisotropy proportionality factor vanishes for an inclination angle as large as arccos $1/\sqrt{3} \approx \pm 54°44'$. This is referred to as the "*magic angle*". Therefore, if the sample is allowed to spin quickly enough around an axis tilted with respect to $\mathbf{B}_0$ by an angle as large as the magic angle, the time-averaged orientation of every molecule will equal the same angle; consequently, all anisotropy effects will "magically" fade away. From a practical standpoint, this involves a modification of the geometry of the probe. The sample is placed into a tiny tube (usually made in zirconia, a few cm long), which in turn is inserted into a *rotor*, ensuring the rapid spinning of the tube at the right angle. The spinning speed, usually in the range 3–15 kHz, is a critical parameter. Too low speed results in the appearance of intense sidebands for the signals. Sidebands can be easily recognized because the frequency difference with the main signal is a multiple of the spinning speed. Progressively increasing the latter one causes a decrease in the intensity of the sidebands, up to their almost complete disappearance.

Besides the differences in the preparation of the sample and in instrumental geometry, there is no significant conceptual difference, in principle, between a liquid state and a solid state NMR spectrum; therefore, from a practical standpoint interpretation involves no particular problem. For crystalline molecular compounds, spectral resolution is almost as good as that for liquid samples. Conversely, in amorphous samples signals undergo broadening.

A.11 Final Remarks and Selected Advanced Readings

The theoretical bases and the main aspects of modern NMR spectroscopy have been introduced in this presentation in the shortest and most simplified way possible. Some very peculiar NMR techniques, such as DOSY, dynamic NMR, relaxometry, MRI, *etc.*, have not been addressed herein for brevity. I deliberately did not provide any literature reference within the text, which indeed stems from my personal teaching experience. Again, it is important in my opinion to underline its merely introductive character, aimed at providing the reader with the ground level information to approach more advanced and specific bibliographic material. Interested readers can find a plethora of general textbooks on the subject and a large amont of advanced literature on particular topics.

A good historical presentation of NMR in the frame of the rise of organic spectroscopy has been drafted by N. C. Thomas in *J. Chem. Educ.*, 1991, **66**, 631–634.

An old, but everlasting classic of NMR spectroscopy undoubtedly is: A. Abragam, *The principles of nuclear magnetism*, Oxford University Press (1961) – a sort of "holy book" to which NMR-involved people pay veneration.

Good, quite accessible undergraduate textbooks are:

- M. R. Silverstein, F. X. Webster, D. J. Kiemle and D. L. Bryce, *Spectrometric Identification of Organic Compounds*, Wiley, 8th edn, 2019, ISBN: 978-0-470-61637-6.
- J. Keeler, *Understanding NMR Spectroscopy*, Wiley, 2nd edn, 2010, ISBN: 978-0-470-74608-0.
- H. Friebolin, *Basic One- and Two-Dimensional NMR Spectroscopy*, Wiley, 5th edn, 2010, ISBN: 978-3-527-32782-9.
- J. B. Lambert, E. P. Mazzola and C. D. Ridge, *Nuclear Magnetic Resonance Spectroscopy: An Introduction to Principles, Applications, and Experimental Methods*, Wiley, 2nd edn, 2019, ISBN: 978-1-119-29523-5.

APPENDIX B

Introduction to FFC NMR Theory and Models for Complex and Confined Fluids

V. BORTOLOTTI,*[a] L. BRIZI,[b] G. LANDI,[c] C. TESTA[b] AND F. ZAMA[c]

[a] Department of Civil, Chemical, Environmental, and Materials Engineering, University of Bologna, Bologna, Italy; [b] Department of Physics and Astronomy, University of Bologna, Bologna, Italy; [c] Department of Mathematics, University of Bologna, Bologna, Italy
*Email: villiam.bortolotti@unibo.it

B.1 General Introductory Remarks

This appendix intends to provide a very introductory guide to TD-NMR (time domain nuclear magnetic resonance) techniques, particularly to FFC (fast field cycling), to study fluids in complex systems, such as porous media or cement materials. We hope this simple guide will help researchers navigate the vastness of the specific literature and the complexity underlying FFC theories. Complexity and simplicity are opposite concepts. Therefore, simplification inevitably introduces approximations. In addition, the appendix must understandably remain limited to a reasonable number of pages. Over time, a wide variety of mathematical models have been developed to interpret the results of FFC measurements performed on samples of different types; but only a rough review of a short and arbitrary list of FFC models effective in practical applications is provided in the appendix. No application examples or comparisons among models are shown. Likewise, there is no

New Developments in NMR No. 32
The Environment in a Magnet: Applications of NMR Techniques to Environmental Problems
Edited by Pellegrino Conte, Delia Francesca Chillura Martino and Paolo Lo Meo

Published by the Royal Society of Chemistry, www.rsc.org

mention of the specific characteristics of FFC instruments, knowledge of which is essential for the effective use of the FFC technique. Therefore, the material reported on FFC theories does not claim completeness or total formal correctness. The reader is warmly invited to continue the study of the vast specialized literature, starting from the bibliography proposed at the end of the appendix (References denoted as superscripts in the main section or subsection headings) and the references quoted therein. A minimal knowledge of the concepts and terminology used in both quantum mechanics and NMR is required to efficiently and profitably use the material presented in this appendix.

B.2 Introduction[1–9]

As it has been shown in many publications (see References and the references therein for details), the investigation of the dependence of the NMR longitudinal relaxation rate (or spin–lattice, T_1), but also of the transverse relaxation rate, T_2, on the Zeeman field (that is, on the polarization external magnetic field $\boldsymbol{B}_0$) offers a rather direct and detailed insight into events at the molecular level that with other standard techniques cannot be obtained. The measurements of the nuclear magnetic relaxation rates as a function of the Larmor angular frequency ($\omega_0 = 2\pi\nu_0 = \gamma\boldsymbol{B}_0$, where γ is the gyromagnetic ratio of the nuclei and ν_0 the Larmor frequency), using the field cycling (FC) technique, provide a powerful tool for yielding rather directly the spectral energy distribution of all the molecular tumbling imposed on the magnetic moments. Although the origins of the FC methodology go back to the early 1950s, just a few years since the NMR discovery, its development has been relatively slow compared, for example, to the intense progress of NMR spectroscopy and MRI (NMR imaging). Many reasons can explain this slow evolution, but perhaps the most relevant are as follows: the complexity of NMR relaxation theories, with the consequent difficulty in developing molecular dynamics models to interpret the acquired data (the nuclear magnetic relaxation dispersion (NMRD) profiles), and the practical difficulties to create efficient and performant equipment to measure such NMRD curves. For example, many interesting molecular dynamics effects occur at low fields. However, the relaxation times tend to become shorter at lower fields, and the intensity of an NMR signal from a given population of spins exposed to a magnetic field, $\boldsymbol{B}_0$, approximately decreases to $\boldsymbol{B}_0^{3/2}$.

When the external magnetic field $\boldsymbol{B}_0$ is low (with the Larmor frequency below around a few kHz), the 'local field' may become comparable or higher, and secular approximation is not more valid (see the following sections). The low-field T_1 may be shorter than the longest correlation time (τ_c) so that the validity of the "weak collision" condition is violated (in the "weak collision" condition, the fluctuating perturbations are supposed to be of small intensity compared to the splitting energy due to $\boldsymbol{B}_0$, or in other terms $T_1 \gg \tau_c$).

Already from the pioneering Bloembergen–Purcell–Pound works (published in 1948), it appeared evident that the most pronounced field dependence relaxation phenomena occurred at relatively low fields, where low-frequency molecular motions can have a huge impact on the longitudinal relaxation times. Nevertheless, over time most efforts were primarily made to improve instruments and techniques operating at $\boldsymbol{B}_0$ fields as high as possible.

At low field for 'like' (nuclei with an equal gyromagnetic ratio) spins 1/2 the chemical shift, *j*-coupling, and rotational relaxation effects are negligible. The predominant spin–lattice relaxation mechanism is due to the fluctuating dipole–dipole couplings. Instead, spin-1 nuclei possess a finite electric quadrupole moment with a quadrupole coupling with local gradients of the molecular electric field, dominating the relaxation processes.

Each material or substance (at a specified temperature) presents a peculiar NMRD profile that fully characterizes it. Therefore, the NMRD approach is a "fingerprinting" method useful to characterize many substances.

The FFC technique applies to several fields, including, just to list a few, materials science (polymers, saturated porous media), food science (meat, milk-based products), pharmaceuticals, and much more.

This manuscript is mainly focused on giving some insight into the most important/used models now used to interpret the NMRD profiles to study molecular dynamics in complex systems deriving from dipole–dipole or quadrupolar interactions.

B.2.1 A Brief FC Historical Review

This section very synthetically tries to reconstruct the history of the evolution of field cycling relaxometry over time. The first NMR experiments using an FC approach were probably performed by E. Turner and A. Sachs, and by N. F. Ramsey and R. V. Pound at Harvard (1949–1951). The longitudinal relaxation time dependence on the magnetic field was soon named the T_1 *dispersion curve* or, later, the nuclear magnetic relaxation dispersion profile. The first NMRD profile was published by Ramsey and Pound (1950).

For almost three decades, most experiments were performed by moving the sample between two magnets (shuttling processes) either manually or mechanically with sophisticated mechanical devices. Because the shuttling process is cyclically repeated many times, this technique was called field-cycling NMR relaxometry. During this period, important improvements in FC instruments were possible, thanks to the work of many research groups, most of them in the United States. Between 1953 and 1960, the petroleum industries also developed technologies regarding FC NMR. For example, R. J. S. Brown and colleagues from Chevron (at that time Standard Oil Company of California), during the development of the nuclear magnetic logging (NML) tool, performed many studies and developed technologies for working with changing magnetic fields. IBM has also actively contributed with its Watson Laboratory research group (Columbia University, New York). During the 1960s, a few relaxometers were produced and installed worldwide: a couple in the USA, two in Italy, Florence and Turin, and one in Mons, Belgium. Since the early 1970s, a new approach was developed that allows handling samples characterized by faster relaxation times compared with the old shuttling method. The sample is held still while the field is switched among different values using an air core electromagnet. This new method was called fast field-cycling (FFC) NMR relaxometry and was initially investigated primarily by A. G. Redfield, F. Noack, S. H. Koenig, and R. Kimmich. Until the late 1990s, the FC technique was seldom applied, mainly because commercial instruments were unavailable. In 1996 the first commercial NMR-FFC relaxometer, based on the Noack–Schweikert technology, was produced by the Italian company Stelar (in Mede, Pavia). Since then, many hardware (*e.g.*, wide-bore magnets to perform rock core analysis) and software improvements have been introduced to extend the range of applications.

B.3 Theoretical Background[10–17]

As already stated in the introduction, what follows is a simplified, thus approximate, introductory version of the relevant theories concerning nuclear spin dynamics without a claim of completeness. For a complete and theoretically sound analysis and more details, the reader is referred to "References" at the end of the appendix.

When the main goal is to study the dynamics of nuclear spins, one can solve the time-dependent Schrödinger equation (see eqn (B.1)), the method used in this appendix. Alternatively, the Heisenberg approach can be used, where time dependence lies in the operators and not in the wavefunctions.

The quantum state of the whole sample, *i.e.*, information about the spin states, positions, and velocities of all electrons and nuclei, is entirely described by a *wavefunction* $|\Psi\rangle$.

The corresponding time-dependent Schrödinger equation is complete but useless and cannot be used in any realistic situation. For NMR purposes, one can work with a simplified approach, the *spin Hamiltonian hypothesis*, in which only the nuclear spin states appear, and using the corresponding time-dependent Schrödinger equation:

$$\frac{\mathrm{d}}{\mathrm{d}t}|\Psi_{\mathrm{spin}}(t)\rangle \approx -\frac{i}{\hbar}\mathscr{H}^{\mathrm{spin}}|\Psi_{\mathrm{spin}}(t)\rangle \tag{B.1}$$

where $|\Psi_{\mathrm{spin}}\rangle$ is the wavefunction of the spin state of the nuclei and $\mathscr{H}^{\mathrm{spin}}$ is the nuclear *spin Hamiltonian*.

Its validity is based on the separation of the timescales (*timescale separation*, which implies statistical independence) for nuclear and electronic motions. In fact, the electron motions are so rapid that nuclear spins are only affected by a time average of the fields generated by the electrons.

Moreover, the energies of nuclear spins are generally too small to influence both the motions of the molecules' electrons and the molecules themselves.

In practice, it can be assumed that the nuclear spin Hamiltonian hypothesis is a valid approximation for almost all systems at ordinary temperatures.

Below for simplicity of notation, the quantum states $|\Psi\rangle$ denote the nuclear spin states, and the operator $\mathscr{H}$ represents the nuclear spin Hamiltonian.

As it is well known, the $\mathscr{H}$ is a fundamental quantum mechanics operator with particular properties. Its *eigenvalues* are the energy values of the quantum system and its *eigenfunctions* form an orthonormal basis $|n\rangle$ (an *eigenbasis*):

$$\mathscr{H}|n\rangle = E_n|n\rangle,$$

where E_n is the energy associated with the *stationary* state $|n\rangle$. Therefore, the *wavefunction* of a spin system can be expressed as a linear combination of *eigenbasis*:

$$|\Psi\rangle = \sum_n c_n|n\rangle,$$

where the coefficients c_n are related to the probability that the system is in the state $|n\rangle$. It represents the *superposition of states*, a fundamental concept for the NMR theory.

B.3.1 Electric Spin Hamiltonian

Without any detailed calculations, using purely symmetry concepts, it is possible to show that, for spin $I = 1/2$ nuclei, the electric interactions vanish:

$$\mathscr{H}^{\text{ele}} = 0.$$

Namely, for spin 1/2, there are no electrical energy terms dependent on the internal structure or orientation of the nucleus. In the case of nuclei with spin $I > 1/2$, the distribution of the electric charge has no spherical symmetry, and the electric energy of the nucleus depends on its orientation with respect to the structure of the molecule. Therefore, the *electric quadrupole moment* of the nucleus interacts with the surrounding *electric field gradient*. The corresponding electric term of the spin Hamiltonian is called the electric quadrupole interaction:

$$\mathscr{H}^{\text{ele}} = \mathscr{H}^{Q}.$$

The quadrupole interaction depends on the nuclear quadrupole moment as well as on the electric field gradient, a property of the molecular environment. Therefore, the NMR of quadrupolar nuclei is a more complicated topic than that of spins 1/2 because there are both electric and magnetic influences on molecular dynamics.

B.3.2 Magnetic Spin Hamiltonian

From the basic NMR theory, it is well known that the *nuclear magnetic moment* $\boldsymbol{\mu}$ (or *magnetic dipole moment*, an intrinsic nuclear property intimately connected with its spin I) is equal to the spin angular momentum $\boldsymbol{I}$ multiplied by the gyromagnetic ratio γ:

$$\boldsymbol{\mu} = \gamma \boldsymbol{I},$$

$\boldsymbol{\mu}$ interacts with the surrounding magnetic field; classically the energy of a magnetic moment interacting with a magnetic field, $\boldsymbol{B}$, is given by the scalar product (or dot product) between them:

$$E = -\boldsymbol{\mu} \cdot \boldsymbol{B}.$$

So, the corresponding quantum expression is the Hamiltonian:

$$\mathscr{H}^{\text{mag}} = -\boldsymbol{\mu} \cdot \boldsymbol{B}. \tag{B.2}$$

Using an evocative visualization, just as a compass needle, which by aligning itself with the field reduces its magnetic energy to a minimum, the magnetic energy is at its minimum when $\boldsymbol{\mu}$ is parallel to $\boldsymbol{B}$, and it reaches its maximum when $\boldsymbol{\mu}$ and $\boldsymbol{B}$ are antiparallel. The magnetic and electric fields acting on a nuclear spin can come from an external apparatus (*external spin interactions*) and from the sample itself (*internal spin interactions*). External spin interactions are purely magnetic, and applied magnetics fields of various types are used to manipulate the nuclear spins. As mentioned before, the internal spin interactions for spin-1/2 are purely magnetic, while for spins $> \frac{1}{2}$ the electric quadrupolar interaction also acts. Nuclear spins are more strongly coupled with the "external apparatus" than with the molecular environment since external interactions are generally much more intense than internal ones.

For most forms of spectroscopy, the system acts the opposite of NMR, *i.e.*, the external perturbations are relatively weak, and the molecular structure governs the behavior of the spin system. The inverted situation in NMR leads to the extraordinary uniqueness of this technique because it means that the measuring instrument is, in a certain sense, part of the system under investigation. Figure B.1, which recalls the approach used in Levitt's book (listed in References, to which the reader is referred for a more enhanced description), summarizes the overall organization of the spin interactions.

B.3.3 External Magnetic Fields

During an NMR experiment, up to three kinds of external magnetic fields can be supplied:

- The instrumental magnet (superconducting, electromagnetic, or permanent magnet) provides a very strong, homogeneous, static magnetic field, $\boldsymbol{B}_0$.
- The radio frequency (RF) coil in the probe generates an RF oscillating field, $\boldsymbol{B}_{\text{RF}}(t)$, that is as spatially homogeneous as possible and, in

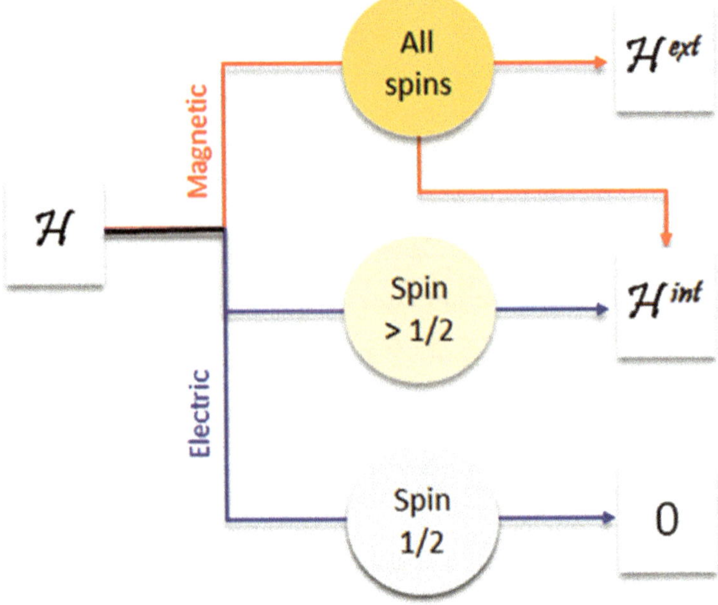

Figure B.1 Block diagram of the overall organization of the spin interactions.

normal circumstances (but not necessarily always), perpendicular to the static field $\boldsymbol{B}_0$.

- Relaxometers equipped with gradient coils can also provide a magnetic field gradient, $\boldsymbol{B}_{\text{grad}}(r, t)$, which is, therefore, position-dependent and can have a controlled time dependence. Usually, $\boldsymbol{B}_{\text{grad}}$ is much weaker than $\boldsymbol{B}_0$.

Therefore, the external part of the spin Hamiltonian can be expressed with a good approximation as

$$\mathscr{H}^{\text{ext}} = \mathscr{H}^{\text{static}} + \mathscr{H}^{\text{grad}}(r, t) + \mathscr{H}^{\text{RF}}(t),$$

and

$$\mathscr{H}^{\text{static}} = \sum_i \mathscr{H}_i^{\text{static}},$$

with $\mathscr{H}_i^{\text{static}}$ representing the interaction between the spin I_i and the static field $\boldsymbol{B}_0$. Analogously, the other terms of the Hamiltonian represent the interaction of all spins with the external magnetic fields $\boldsymbol{B}_{\text{RF}}$ and $\boldsymbol{B}_{\text{grad}}$.

B.3.4 External Spin Interactions

It is usual to choose the laboratory reference frame with an axis system in which $\boldsymbol{B}_0$ is along the z-axis and the oscillating field is in the xz-plane.

From eqn (B.2), the spin Hamiltonian related to the interaction of spins with the static external field $\boldsymbol{B}_0$ (Zeeman interaction) is given by

$$\mathscr{H}_i^{\text{static}} = -\gamma_i \boldsymbol{B}_0 I_{iz},$$

and

$$\omega_i = -\gamma_i \boldsymbol{B}_0$$

is the Larmor angular frequency of spin I_i. Note that since Planck's constant has disappeared from the resonance equation, then by Bohr's correspondence principle one can infer that this result is also valid in a classical (vectorial) framework.

It may be shown that, in ordinary situations, the longitudinal and non-resonant components of the oscillating RF field may safely be neglected (see References for details). Therefore, during the application of the RF field, $\mathscr{H}^{\text{RF}}(t)$ may be approximated as

$$\mathscr{H}_i^{\text{RF}}(t) \approx -\omega_i^{\text{nut}}\left[\cos(\omega_{\text{ref}}t+\varphi)I_{ix} + \sin(\omega_{\text{ref}}t+\varphi)I_{iy}\right],$$

where ω_i^{nut} is the *nutation frequency*, a measure of the resonant RF field strength experienced by spin expressed in angular frequency units, ω_{ref} is the spectrometer reference frequency, and φ is the initial phase of the RF field.

A spin I_i located in a spatial position $r = (x, y, z)$ can have interactions with possible external field gradients directed along the three laboratory frame axes (characterized by the gradients G_x, G_y, and G_z).

The transverse components of gradient fields are usually much weaker than $\boldsymbol{B}_0$ and can be ignored. Therefore, for any direction of the gradient, the relevant spin operator is always I_{iz}. So, for example, given a gradient along the y-axes of the laboratory frame, the corresponding Hamiltonian is

$$\mathscr{H}_i^{\text{grad}}(\boldsymbol{r}, t) = -\gamma_i G_y(t) y I_{iz}.$$

Analogous relations for the Hamiltonian can be written for the gradients G_x and G_z.

B.3.5 Internal Spin Hamiltonian

There are several important NMR experiments (such as in MRI) where one-spin dynamics can be used, *i.e.*, where nuclei can be assumed to be isolated and do not interact with each other. But, in general, the spin–spin couplings cannot be ignored for most samples. The internal Hamiltonian $\mathscr{H}^{\text{int}}$ represents the effect on the nuclei of magnetic and electric fields originating from the sample itself.

The NMR Hamiltonians for spin interactions can be expressed using a common general analytical form:

$$\mathscr{H}(t) = f\sum_k F^{(k)} O^{(k)}, \tag{B.3}$$

where f is a constant prefactor specific for the interaction, $O^{(k)}$ are the spin-operator functions acting on the spin variables, and $F^{(k)}$ are the functions related to the fluctuating structure (random functions of the relative positions of two spins). For example, the spatial functions $F^{(k)}$ for the intramolecular Hamiltonian dipolar coupling, in which the interdipole distance can be considered constant, can be described using second-order spherical harmonics $Y_{2,m}(\vartheta,\varphi)$, $m=0, \pm1, \pm2$, with $\vartheta(t)$ being the polar angle and $\varphi(t)$ the azimuthal angle.

The following discussion assumes that the substances are diamagnetic, *i.e.*, substances without unpaired electron spins (namely, without paramagnetic impurities). Fortunately, the list of spin interactions is short, synthetically they are as follows:

Direct dipole–dipole coupling (Figure B.2) represents the direct interactions of nuclear spins with each other, therefore not mediated by the cloud of molecular electrons. The magnetic dipole moment of a nucleus is also a source of a magnetic field that can be seen by other nuclear magnetic moments. This interaction may involve nuclei of the same molecule (intradipolar) or of different molecules (interdipolar). Furthermore, it may be homonuclear if the nuclei involved are of the same species or heteronuclear if not.

Quadrupolar coupling (Figure B.3) represents the electric-type interactions between nuclei with spin >1/2 and the surrounding electric fields.

A nucleus can have a structure with the electric charge distributed over a volume that can interact with the electric field gradient due to the cloud of electrons.

In *chemical shift* (Figure B.4), there is an indirect interaction of the external polarizing magnetic field with the nuclear spins through the involvement of the orbiting electrons that can cause a slight resonance frequency shift. It can have isotropic or anisotropic (*chemical shift anisotropy*, CSA) characteristics.

In the isotropic case, the orientation of the sample with respect to the direction of the magnetic field has no effect. It depends on the intensity of $\boldsymbol{B}_0$.

J-coupling (Figure B.5) represents the indirect interactions of nuclear spins with each other, mediated by the valence electrons. It does not depend on the intensity of $\boldsymbol{B}_0$.

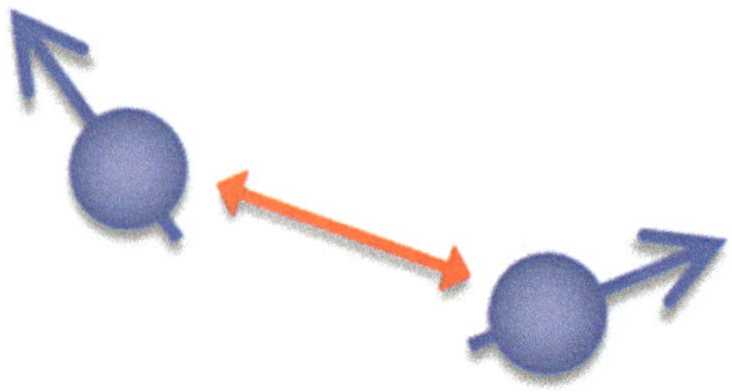

Figure B.2 Schematic picture of the *direct dipole–dipole coupling*.

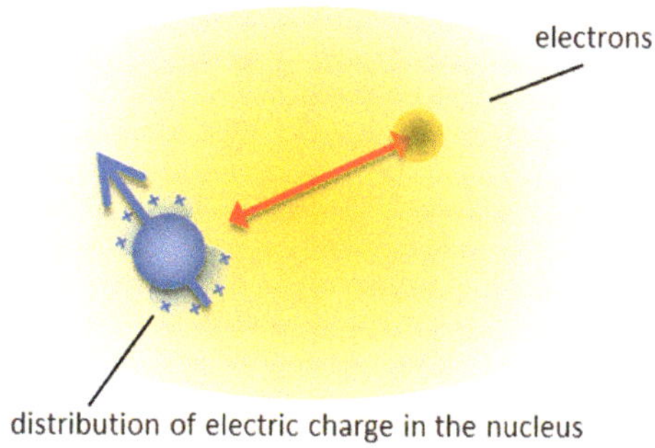

Figure B.3 Schematic picture of the *quadrupolar coupling*. The yellow spot represents the cloud of electrons.

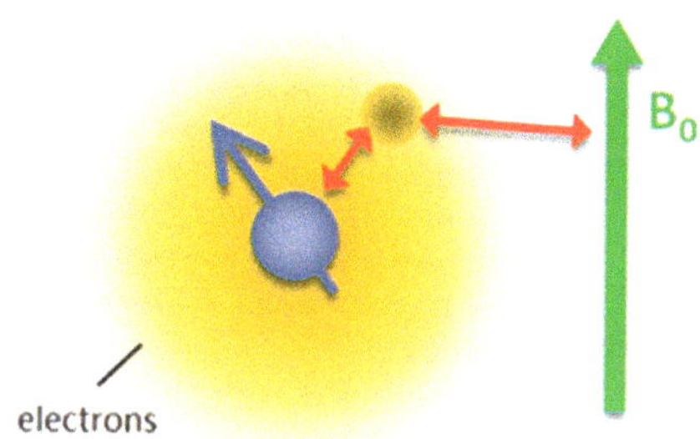

Figure B.4 Schematic picture of the *chemical shift interaction*. The yellow spot represents the cloud of electrons.

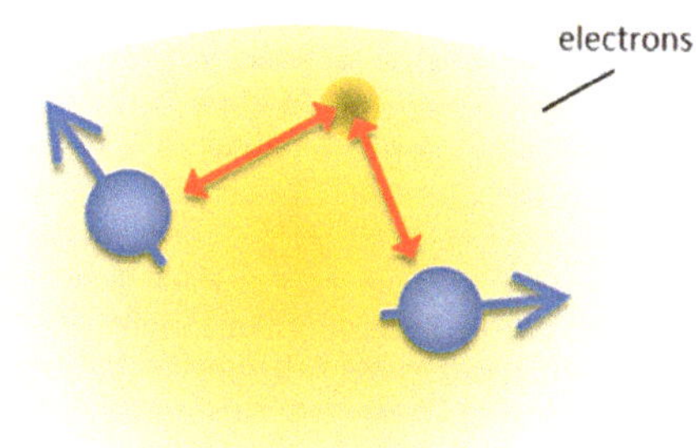

Figure B.5 Schematic picture of the *J-coupling interaction*.

Spin–rotation interaction represents the interactions between the nuclear magnetic moment and the magnetic fields generated by the electric current due to the rotational motion of the molecules (Figure B.6).

Figure B.7 shows a schematic summary representation of the organization of the internal spin interaction terms. The color-bar on the figure roughly displays the typical relative magnitudes of the interactions in the absence of molecular motion. The quadrupole coupling is usually the most prominent term but vanishes for spins 1/2.

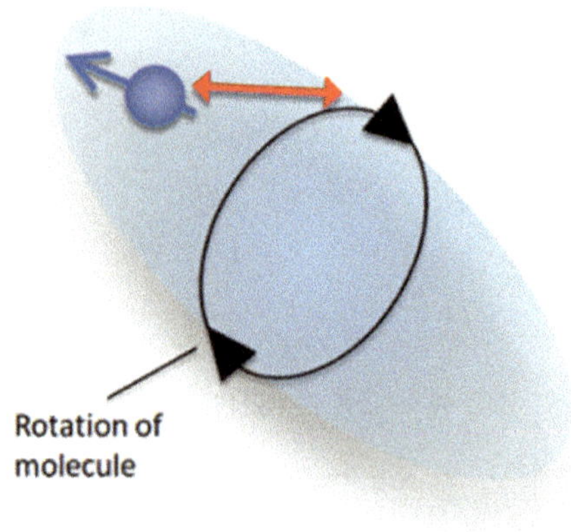

Figure B.6 Schematic picture of the *spin–rotation interaction.*

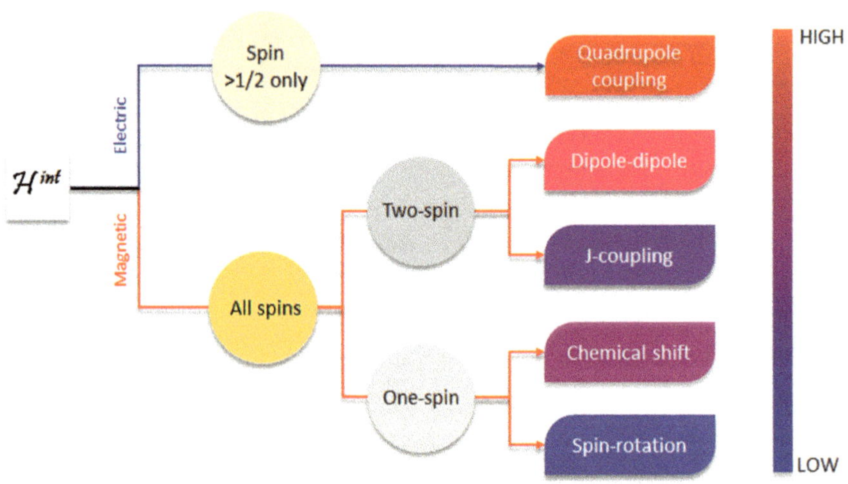

Figure B.7 Schematic picture of the organization of the internal spin interactions and their rough relative magnitudes (color bar).

The mathematical formulation of the nuclear spin interactions is somewhat complicated. However, a simplified form of the internal Hamiltonian can usually be used.

This occurs if it is possible to assume that the polarization external magnetic field is very strong (the *secular approximation*) and the molecular motion is rapid (the *motional averaging*).

The discarded parts of the Hamiltonian are responsible for the relaxation phenomena of the nuclear spin system, as described further on. These phenomena depend on the timescales involved and have rather complicated reciprocity. Figure B.8 summarizes the situation ignoring possible complications.

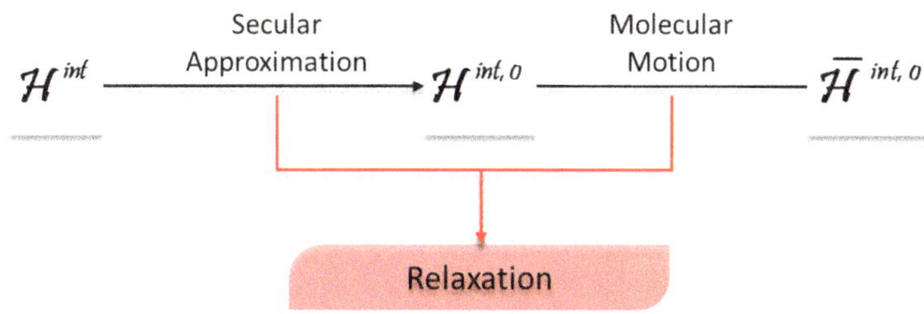

Figure B.8 Schematic picture of the process of simplification of the internal Hamiltonian.

The *secular approximation* can be used when the spin dynamics is dominated by the strong interaction with the external polarization magnetic field, which makes some internal spin interactions negligible.

This is usually a very good approximation, except in the case of quadrupolar spin interaction, where a complete description is often needed. The secular approximation then allows replacing $\mathscr{H}^{\text{int}}$ with a simplified form, called here $\mathscr{H}^{\text{int},0}$ (Figure B.8).

The *motional averaging* arises if the molecules undergo rapid motion, then the interaction terms fluctuate in time and $\mathscr{H}^{\text{int},0}$ may be replaced by an *averaged value*, $\bar{\mathscr{H}}^{\text{int},0}$ (see Figure B.8). In general, for gas and liquids, if the molecular motion is not too slow, the use of the averaged spin Hamiltonian is a good approximation.

There are three relevant modes of molecular motion, namely, *molecular translation*, *molecular rotations*, and *internal molecular motions.* Their effects on spin interactions depend on whether these interactions involve particles on the same molecule (intramolecular) or particles on different molecules (intermolecular). Figure B.9 schematically shows intramolecular and intermolecular couplings between nuclear spins.

In the case of some solids, and for gases and liquids, the *molecular orientation* depends on time, and often it is possible to invoke the ergodic hypothesis, namely, the average of the orientations of a molecule calculated over a sufficiently long period is equivalent to the average calculated on the ensemble of molecules at a given instant.

In *gases*, *rotational* and *translational* motions (diffusion) are extremely rapid. Therefore, the intramolecular spin interactions are averaged to their isotropic values, while all intermolecular interactions are averaged to a minimal value, essentially negligible.

In *liquids*, diffusion is much slower than in gas. It depends on the molecular size and fluid viscosity, and, for isotropic liquids, all *intramolecular* spin interactions are mediated to the isotropic values by the rotational motion of the molecules.

The *short-range intermolecular* interactions (which occur in the volume of space, domain, in which the molecule diffuses before NMR relaxation) are averaged to zero. Instead, the diffusional motion cannot average out the

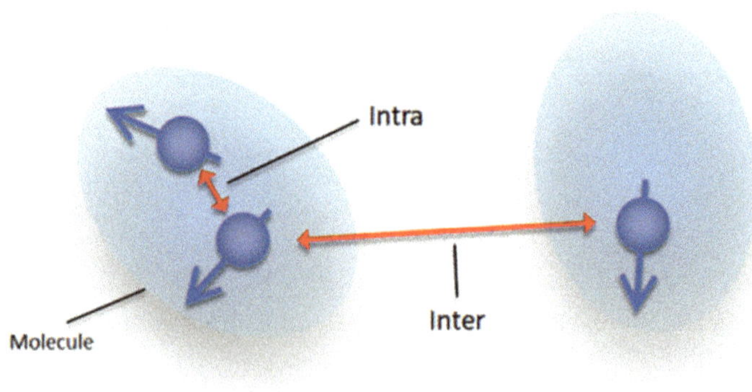

Figure B.9 Schematic illustration of intramolecular and intermolecular interactions.

long-range intermolecular interactions (occurring between molecules that are outside their respective diffusion domains).

In an anisotropic liquid, the motionally averaged spin interactions depend on the orientation of the liquid *director* (a direction along which the liquid crystal is oriented) with respect to the magnetic field.

The diffusive motion of molecules has an averaging effect on *short-range intermolecular* interactions by making them disappear, while *long-range intermolecular* interactions remain. The rotational motion of molecules tends to mediate intramolecular spin interactions at values different from isotropic ones. Atomic motion in a solid is typically highly constrained, and the internal spin Hamiltonian strongly depends on the solid's orientation concerning the magnetic field.

In conclusion, there is little averaging of the internal spin interactions, and both intramolecular and intermolecular interactions survive. The typical order of importance of relaxation mechanisms for spins 1/2 is as follows:

$$\text{dipole–dipole} > \text{CSA} > \text{spin–rotation}.$$

The CSA mechanism begins to be non-negligible compared with dipole–dipole at high magnetic fields. The spin–rotation mechanism can become significant for small molecules of non-viscous liquids or gases.

Instead, for spins $>\frac{1}{2}$, the usual order of importance of relaxation mechanisms is as follows:

$$\text{quadrupole} \gg \text{dipole–dipole} > \text{CSA} > \text{spin–rotation}.$$

In general, the interactions relevant to the relaxation process at low field are dipole–dipole and quadrupole interactions.

B.3.6 Random Field and Spin Relaxation

This section roughly illustrates the concepts of the *correlation function*, *correlation times*, and *spectral density function*. At thermal equilibrium, the distribution of the energy levels of populations of identical non-interacting spins in a magnetic field has a Boltzmann distribution. Immediately after a 90° pulse, the populations of the energy levels are equal, and a magnetization in the *xy*-plane is produced. Then the populations of the energy levels will return to the thermal equilibrium condition (longitudinal T_1 relaxation), and phase coherence between the spins is lost (transverse T_2 relaxation). The magnetization created in the *xy*-plane by a 90° pulse does not precess indefinitely around $\boldsymbol{B}_0$.

The thermal motion of the molecules causes the magnetic fields to fluctuate at the sites of the nuclear spins inducing NMR relaxation (both T_1 and T_2).

B.3.6.1 The Correlation Function, G(t)

Three terms are introduced to understand how the fluctuating magnetic field induces relaxation: the correlation time τ_c, the autocorrelation function $G(t)$ and its Fourier transform partner spectral density $J(\omega)$.

The autocorrelation function describes how rapidly the field B fluctuates, and it is generally of the form:

$$G(\tau) = \langle B(t) \cdot B(t+\tau) \rangle,$$

where $B(t)$ can be related to the molecular position and/or molecular orientation expressed in terms of polar coordinates and τ is a fixed time interval.

The triangular parentheses indicate either an average over the ensemble of many spins at a given time or an average over a long period of time for a single spin. The ergodic hypothesis is assumed; namely, these two types of averages are assumed to be equal. Stochastic fluctuations are subject to two fundamental properties:

- *stationarity*, namely, the correlation function $G(\tau)$ depends on the interval τ but is independent of the absolute time t; in other words, it is invariant with respect to a change of the origin of time. Therefore, it is often set equal to 0 for simplicity, so that $G(\tau) = \langle B(0) \cdot B(\tau) \rangle$;
- *time-reversal invariance*, namely, the correlation functions are *even* functions of τ, formally $G(\tau) = G(-\tau) = G(|\tau|)$.

If the interval τ is small compared with the timescale of the fluctuations, then the two values of the field at the two time points are very similar, and $G(\tau) \cong \langle B^2(0) \rangle$. Conversely, if the interval τ is long compared with the fluctuation timescale, the system loses the 'memory'. Consequently no consistent relationship exists between the values of the function and the ensemble average tends to zero. So, in general, $G(\tau)$ tends to be large for small values of τ and tends to zero for large values of τ. Although it is challenging to be theoretically justified, often it is assumed that the autocorrelation function has a simple exponential form:

$$G(\tau) = \langle B^2 \rangle e^{-\frac{\tau}{\tau_c}},$$

where the correlation time τ_c, a characteristic of the correlation decay, can be defined as

$$\tau_c = \frac{1}{G(0) - G(\infty)} \int_0^{\infty} [G(t) - G(\infty)] dt.$$

The correlation time is the timescale over which the field variations are observed. Using the exponential approximation for $G(\tau)$, it reads

$$\tau_c = \int_0^{\infty} e^{-\frac{t}{\tau_c}} dt,$$

and indicates qualitatively how long it takes before the stochastic field changes sign. Slow fluctuations involve a large value of τ_c, whereas rapid fluctuations involve a small value of τ_c.

The given definition of $G(\tau)$ is quite generic. But, in general, one is interested in describing the effects related to molecular motion. Let us consider a physical quantity function, $f(r(t))$, of a stochastic position, $r(t)$, where r is a random position vector. Being a function of r, it will also be a random function of t. It is also supposed that f is real ($f^* = f$, where * stands for the complex conjugate) and stationary, so it is invariant under the change of the origin of time. The autocorrelation function of $f(t)$ can therefore be defined as

$$G(\tau) = \langle f(0) \cdot f(\tau) \rangle = \int_{\mathbb{R}^3} \int_{\mathbb{R}_0{}^3} f(r) f(r_0) P(\boldsymbol{r}, \boldsymbol{r}_0, \tau) P_0(\boldsymbol{r}_0) d^3 r_0 d^3 r,$$

where $P(\boldsymbol{r},\boldsymbol{r}_0,\tau)$ is the conditional probability that a molecule is at the position $\boldsymbol{r}$ at time τ when it was at position $\boldsymbol{r}_0$ at time 0, and $P_0(\boldsymbol{r}_0)$ is the probability of finding the molecule at the position $\boldsymbol{r}_0$ at time 0. In the case of rotational and translational diffusion, $P(\boldsymbol{r},\boldsymbol{r}_0,\tau)$ can be computed by solving the classical diffusion equation. In a more general description, $G(t)$ is related to the terms $F^{(k)}$ of the sum of eqn (B.3). Therefore, the correlation function is related to the order k, $G_k(t)$. So, in the case of the spatial harmonic function $Y_{2,m}$, there are three correlation functions: $G_0(t)$, $G_1(t)$, and $G_2(t)$. Nevertheless, also for molecules in a confined environment, using a *reduced* (or normalized) *correlation function*:

$$\mathcal{G}(\tau) = \frac{\langle f(0) \cdot f(\tau) \rangle}{\langle |f|^2 \rangle},$$

it is practically not necessary in most cases to distinguish the individual reduced correlation functions:

$$G_0(t) = G_1(t) = G_2(t) = G(t).$$

Noticing that $G(0)=1$ and assuming that $G(\infty)=0$, then

$$\tau_c = \int_0^{\infty} \mathcal{G}(t)\mathrm{d}t.$$

B.3.6.2 The Spectral Density J(ω)

Spectral density measures the amount of molecular motion at the correct frequency to cause relaxation; in fact, the field will only affect the spins if it oscillates at the correct Larmor frequency.

It is defined as the Fourier transform of the autocorrelation function:

$$J(\omega) = \int_{-\infty}^{\infty} G(t)\mathrm{e}^{-i\omega t}\mathrm{d}t,$$

$J(\omega)$ describes the field fluctuations due to thermal molecular motions in the frequency domain; it is the conjugate counterpart of the correlation function in the time domain.

Since $G(\tau)$ is real and even then also $J(\omega)$ is real and even $(J(-\omega)=J(\omega))$:

$$J(\omega) = \int_{-\infty}^{\infty} G(t)\mathrm{e}^{-i\omega t}\mathrm{d}t = 2\int_0^{\infty} G(t)\cos(\omega t)\mathrm{d}t.$$

Using the *reduced* autocorrelation function, it is possible to define the *reduced (or normalized) spectral density*:

$$\mathcal{J}(\omega) = \int_{-\infty}^{\infty} \mathcal{G}(t)\mathrm{e}^{-i\omega t}\mathrm{d}t,$$

with the properties:

$$\mathcal{J}(0) = \int_{-\infty}^{\infty} \mathcal{G}(t)\mathrm{d}t = 2\tau_c,$$

and remembering the definition of the delta function $\delta(\tau) = \frac{1}{2\pi}\int_{-\infty}^{\infty} \mathrm{e}^{-i\omega\tau}\mathrm{d}\omega$, finally one obtains:

$$\frac{1}{2\pi}\int_{-\infty}^{\infty} \mathcal{J}(\omega)\mathrm{d}\omega = \mathcal{G}(0) = 1.$$

Using the previously defined quantities, it is possible to demonstrate that the frequency dependence of the spin–lattice relaxation rate $R_1(\omega)$ in the case of the intradipolar interaction of like spins follows a relation like the following (Bloch/Wangsness/Redfield theory):

$$R_1(\omega) = \frac{1}{T_1(\omega)} \propto \mathcal{J}(\omega) + 4\mathcal{J}(2\omega).$$

A similar relation can be developed for $R_2(\omega) = 1/T_2(\omega)$:

$$R_2(\omega) = \frac{1}{T_2(\omega)} \propto 3\mathcal{J}(0) + 5\mathcal{J}(\omega) + 2\mathcal{J}(2\omega).$$

B.4 FFC Models[2,13–15,18–21]

NMRD profiles are usually acquired using the FFC-NMR technique, which allows for extensive explorations of the magnetic field fluctuations, which are responsible for the nuclear spin relaxation.

In this way, exploring a very large range of Larmor frequencies ν_0 is possible (depending on instrumental limitations, roughly in the range $10\ \text{kHz} < \nu_0 < 40\ \text{MHz}$).

Many complex mathematical models have been proposed for the interpretation of NMRD curves characterizing the molecular dynamics of liquids in bulk and under confinement conditions.

Thus, this section only roughly reviews a short and arbitrary list of proven effective models in practical applications. It is intended as an introductory guide to the novice reader of these topics.

The discussion developed in the next section is done considering the following approximations.

The referred nuclei are mainly protons (spin $\frac{1}{2}$) and nuclei with spin 1 (for example, deuterons) in diamagnetic liquids. It is assumed that, for 'like' spins $\frac{1}{2}$ (as already mentioned, nuclei with the equal gyromagnetic ratio), the predominant T_1 relaxation mechanism is due to the dipole–dipole couplings and for nuclei with spin 1 due to the quadrupole couplings. Since quadrupole interaction is much more efficient than dipolar interaction, the influence of the dipolar relaxation mechanism can be neglected in the presence of quadrupolar interaction.

In condensed matter, dipolar coupled spins $\frac{1}{2}$ form multi-spin systems, whereas spins 1 may be considered isolated entities.

Since cross-correlation effects are normally of minor importance in the context of field-cycling NMR relaxometry, spin–lattice relaxation in multi-spin $\frac{1}{2}$ systems can be treated as a sum of two-spin $\frac{1}{2}$ relaxation rates of a reference spin interacting with all other spins in pairs.

Intermolecular dipolar interactions generally fluctuate significantly more slowly than *intramolecular* couplings and the spin–lattice relaxation rate resulting from both contributions may be added (separation of timescale approximation/hypothesis).

The '*weak collision*' (or "*high field limit*") is valid; that is, the amplitude of the fluctuating field (the 'local field') is much smaller than the quantizing Larmor field (therefore, before the spin relaxation to the thermal equilibration is appreciable, many fluctuation events occur. Namely, $\tau_c \ll T_1$).

The relaxation formalisms of dipolar-coupled homonuclear two-spin 1/2 systems and of quadrupolar-coupled spin 1 systems have much in common and lead to essentially equivalent analytical expressions.

B.4.1 Principles of NMR Dispersion in Bulk

The molecular dynamics of spin-bearing molecules in bulk liquids are characterized by intermolecular and intramolecular interactions, with the former tending to fluctuate much more slowly than the latter. In other words, they are fluctuations occurring on different timescales. The *intermolecular* couplings are governed by translational Brownian molecular motions over spatial distances that exceed the size of the molecule. In contrast, *intramolecular* couplings are governed by the molecular rotational diffusion, a much faster process. Dipole–dipole interactions can be both intramolecular and intermolecular, whereas quadrupolar interactions can only be intramolecular. The intermolecular dipole–dipole interactions have significant effects on the longitudinal relaxation only in the case of anisotropic molecular dynamics and occur at relatively low frequencies. Instead, their effects are negligible in the case of isotropic molecular dynamics. Therefore, assuming that rotational and translational molecular diffusion are stochastically independent, the overall longitudinal relaxation rate can be assumed to be the superposition of intermolecular and intramolecular dipole–dipole contributions among "like" or "unlike" spins. In diamagnetic samples, it may be written as

$$\frac{1}{T_1(\omega)} = \frac{1}{T_1^{\text{inter}}(\omega)} + \frac{1}{T_1^{\text{intra}}(\omega)}. \tag{B.4}$$

The strength with which intramolecular and intermolecular relaxation rates contribute to the sum in eqn (B.4) is related to the characteristics of the system under investigation (its composition, molecular dynamic characteristics, *etc.*) and to the frequency range probed in FFC-NMR experiments. In case quadrupolar couplings are present, intramolecular quadrupolar interactions occur at any frequency. Being much stronger than dipolar couplings, they usually dominate longitudinal relaxation, provided that there are no paramagnetic particles:

$$\frac{1}{T_1(\omega)} \simeq \frac{1}{T_1^{\text{intra}}(\omega)}.$$

In general, the spin dynamics must be described by a *conceptual model* representing the processes/interactions occurring in the spin population system from spins with spins and, in the case of confined fluids, from spins with the pore surface and by which the correlation function, $G(t)$, can be inferred and calculated. Therefore, the theoretical challenge lies in determining $G(t)$ from the conceived conceptual model. The Fourier transform of $G(t)$ then yields the spectral density function $J(\omega)$ which can be then related to the relaxation dispersion curves for both T_1 and T_2.

To this purpose, all these interactions can be described as fluctuating spherical coordinates representing the fluctuations of the dipole–dipole vectors or the fluctuations of the orientations of the electric field gradient.

So, for example, as specified before, dipolar interactions can be described using spherical coordinates, internuclear distance, azimuthal and polar angles $r = r(t)$, $\varphi = \varphi(t)$, and $\vartheta = \vartheta(t)$, respectively. While regarding the orientation of quadrupole couplings, azimuthal and polar angles are sufficient, *i.e.*, $\varphi = \varphi(t)$ and $\vartheta = \vartheta(t)$, and the variation of the internuclear distance can be supposed to be negligible.

B.4.2 BPP Theory

In 1948, Nicolaas Bloembergen, Edward Mills Purcell, and Robert Pound proposed the so-called BPP theory that introduced a standard formalism with which, in general, tentative discussions of relaxation processes start. It applies in case of the high-field limit (or 'weak collisions'), isotropic rotational diffusion of molecules and intramolecular dipole–dipole interaction, with fixed internuclear distances, of two-spin $\frac{1}{2}$ systems of "like spin" (namely, $I = S$ and $\gamma_I = \gamma_S$) in a bulk substance. The reader is referred to References at the end of the appendix for a detailed and analytical description.

Under the above assumptions, Bloembergen, Purcell, and Pound introduced the following transverse and longitudinal relaxation equations expressed in terms of reduced spectral density functions:

$$\frac{1}{T_2(\omega_0)} = \frac{1}{10}\left(\frac{\mu_0}{4\pi}\right)^2 \frac{\gamma^4\hbar^2}{r^6} I(I+1)[3\mathcal{J}(0) + 5\mathcal{J}(\omega_0) + 2\mathcal{J}(2\omega_0)],$$

$$\frac{1}{T_1(\omega_0)} = \frac{1}{5}\left(\frac{\mu_0}{4\pi}\right)^2 \frac{\gamma^4\hbar^2}{r^6} I(I+1)[\mathcal{J}(\omega_0) + 4\mathcal{J}(2\omega_0)].$$

Assuming that the molecular motion of a bulk liquid can be schematized simply by considering only isotropic rotational diffusions, then the reduced correlation function (in the reduced form, $G(0) = 1$) can be assumed to be a mono-exponential function:

$$\mathcal{G}(t) = \mathrm{e}^{-\frac{t}{\tau_c}}.$$

The corresponding reduced spectral density consequently has a Lorentzian (or Debye) form:

$$\mathcal{J}(\omega) = \int_{-\infty}^{\infty} \mathcal{G}(t)\mathrm{e}^{-i\omega t}\mathrm{d}t = \frac{2\tau_c}{1 + (\omega\tau_c)^2}.$$

So, for the dipolar coupling of two-spin $\frac{1}{2}$ systems, the longitudinal and transverse relaxation rates are, respectively:

$$\frac{1}{T_1(\omega_0)} = \frac{3}{10}\left(\frac{\mu_0}{4\pi}\right)^2 \frac{\gamma^4\hbar^2}{r^6}\left[\frac{\tau_c}{1 + {\omega_0}^2{\tau_c}^2} + \frac{4\tau_c}{1 + (2\omega_0\tau_c)^2}\right],$$

$$\frac{1}{T_2(\omega_0)} = \frac{3}{20}\left(\frac{\mu_0}{4\pi}\right)^2 \frac{\gamma^4\hbar^2}{r^6}\left[3\tau_c + \frac{5\tau_c}{1 + {\omega_0}^2{\tau_c}^2} + \frac{2\tau_c}{1 + (2\omega_0\tau_c)^2}\right],$$

where γ is the gyromagnetic ratio; μ_0 is the magnetic field constant; $\hbar$ is Planck's constant divided by 2π; r is the internuclear distance; and $\omega_0 = \gamma \boldsymbol{B}_0$

is the angular Larmor frequency (with $\boldsymbol{B}_0$ being the external magnetic field). In the case of non-viscous liquids, where the extreme narrowing condition ($\omega_0\tau_c \ll 1$) holds, relaxation rates R_1 ($= 1/T_1$) and R_2 ($= 1/T_2$) are equal:

$$\frac{1}{T_1} = \frac{1}{T_2} = \frac{3}{2}\left(\frac{\mu_0}{4\pi}\right)^2 \frac{\gamma^4\hbar^2}{r^6}\tau_c.$$

The assumption of mono-exponential correlation functions is an idealized situation that must be handled with care being scarcely relevant in many experimental systems.

The theory agrees well with experiments performed on pure substances but not in the case of complex environments.

B.4.3 Intermolecular Relaxation by Translational Diffusion

Even though the contribution of the intramolecular dipolar relaxation is generally much more relevant than intermolecular contribution, sometimes intermolecular relaxation becomes important. For example, in solutions containing paramagnetic compounds the electronic spin of the paramagnetic ions can interact with the nuclear spin of water protons that are not directly bound to them. Since the translational diffusion of molecules causes fluctuations in the dipole–dipole distance vector, then it contributes to spin relaxation.

The objective is to relate propagators of translational diffusion (the probability density for a displacement r of a molecule in a time interval τ) to dipolar autocorrelation functions of molecules assumed to form hard spheres, which are assumed to be "unaffected by any forces" until they touch.

The propagator can be obtained as a solution of *Fick's second law* (for example, using the usual Laplace transform method) for mutual self-diffusion with the Neumann boundary condition ('no-flux').

Synthetically the results for "like" spin $\frac{1}{2}$ are as follows:

$$\frac{1}{T_1(\omega)} = \frac{6}{5}\left(\frac{\mu_0}{4\pi}\right)^2 \frac{\gamma^4\hbar^2}{d^3}[J(\omega) + 4J(2\omega)],$$

$$\frac{1}{T_2(\omega)} = \frac{3}{5}\left(\frac{\mu_0}{4\pi}\right)^2 \frac{\gamma^4\hbar^2}{d^3}[3J(0) + 5J(\omega) + 2J(2\omega)],$$

$$J(\omega) = N\pi\tau_t \frac{3\omega\tau_t + 15/\sqrt{2}(\omega\tau_t)^{1/2} + 12}{(\omega\tau_t)^3 + 4\sqrt{2}(\omega\tau_t)^{5/2} + 16(\omega\tau_t)^2 + 27\sqrt{2}(\omega\tau_t)^{3/2} + 81\omega\tau_t + 81\sqrt{2}(\omega\tau_t)^{1/2} + 81},$$

where N is the number of spins per unit of volume and $\tau_t = \dfrac{d^2}{2D} = 9\tau_c$, with d being the diameter of a molecule and D the self-diffusion coefficient.

Being a long-distance interaction, it will be a slow phenomenon compared to rotational diffusion. Therefore, significant contributions to spin relaxation will only appear at relatively low frequencies.

B.4.4 Quadrupolar Relaxation in the High Field Limit

For spins larger than $\frac{1}{2}$, the coupling of the fluctuating molecular electric field gradient, due to the rotation of the non-uniform distribution of charges of the molecules, with the nuclear electric quadrupole moment produces the "quadrupolar relaxation" mechanism. It is, therefore, an inherently intramolecular interaction. If present, it is a very efficient source of relaxation because it is stronger than dipolar coupling.

Supposing to be in the *high-field limit*, the influence of Quadrupolar Hamiltonian is small compared to that of Zeeman Hamiltonian; therefore, quadrupolar couplings act as small perturbations of the Zeeman energy. In other words, the direction of quantization is governed by the external polarization field $\boldsymbol{B}_0$. Instead, in the so-called zero-field case or 'pure' nuclear quadrupole resonance (NQR), the energy levels are determined by the Quadrupolar Hamiltonian.

In general, similarly to the dipole–dipole case, it is possible to express relaxation times in terms of reduced spectral density:

$$\frac{1}{T_1(\omega_0)} = \frac{3}{80}\left(\frac{e^2qQ}{\hbar}\right)^2\left(1+\frac{\eta^2}{3}\right)[\mathcal{J}(\omega_0)+4\mathcal{J}(2\omega_0)],$$

where Q is the nuclear electric quadrupole moment, e is the electron charge, and q and η are parameters characterizing the electric field gradient (related to the electric field gradient tensor). The reduced spectral density $\mathcal{J}(\omega)$ refers to the reduced autocorrelation function computed for quadrupolar interaction.

For nuclei with spin quantum number S equal to 1 (for example, deuteron), it is possible to obtain for the transverse relaxation:

$$\frac{1}{T_2(\omega_0)} = \frac{3}{160}\left(\frac{e^2qQ}{\hbar}\right)^2\left(1+\frac{\eta^2}{3}\right)[3\mathcal{J}(0)+5\mathcal{J}(\omega_0)+2\mathcal{J}(2\omega_0)].$$

Several molecular dynamics NMR studies address deuterium, for example, when deuterated molecules are used to separate the intra- and intermolecular contributions in the relaxation of bulk liquids.

B.4.5 Cross Relaxation: NMRD Dips or Peaks

A special situation can occur when the I-spins are dipolar-coupled to quadrupole S-spins, and the nuclei S are also subjected to further quadrupole interaction with the electric-field gradients of the host molecules. If one of the nuclear quadrupolar energy levels matches the Larmor frequency of I-spins, then it produces the quadrupolar dips (or peaks) observed in the NMRD profile (*dips* on the $T_1(\omega)$ dispersion curve, *peaks* on the $R_1(\omega)$ dispersion curve).

For the Zeeman field approaching the low-field limit ($\boldsymbol{B}_0 \to 0$), the electric-field gradient acts as a quantization field. In the case of limited molecule motional averaging, quadrupole interaction yields the zero-field splitting of the S-spin energy levels. Conversely, in the high-field limit, in which $\boldsymbol{B}_0$ plays the role of the quantization field, the ordinary Zeeman splitting of the S-spin energies prevails. In both limits, the S-spins will permanently remain in thermal equilibrium independently of what happens to the I-spins.

The gyromagnetic ratio magnitude of the dipolar I-spins is usually much larger than that of the quadrupolar S-spins ($|\gamma_I| \gg |\gamma_S|$). In a low $\boldsymbol{B}_0$ field, if the resonance of the S-spins is still in the *low-field limit*, and thus the quadrupole interaction will dominate the splitting, the resonant frequency of the S-spins could be crossed by the resonance of the I-spin (determined by the Zeeman interaction).

At the crossings resonance $\omega_I = \omega_S$, where ω_S is the low-field resonance of the quadrupole nuclei and $\omega_I = \gamma_I \boldsymbol{B}_0$. In the case of the exponential correlation function, the corresponding zero-quantum ("*flip-flop*") reduced spectral density term, with a peak centered at $(\omega_I - \omega_S)$, has the form:

$$\mathcal{J}(\omega_I - \omega_S) \propto \frac{\tau_{\mathrm{dip}}}{1 + (\omega_I - \omega_S)^2 \tau_{\mathrm{dip}}^{\ 2}},$$

where the dipolar correlation time τ_{dip} depends on the different mechanisms causing fluctuations of the dipolar–dipolar interaction between I-spins and S-spins.

In other words, when the resonance frequencies of the two nuclear species approach each other, the magnetization of I-spin nuclei (usually protons) will relax particularly efficiently. As a result, the relaxation rates (times) as a function of the magnetic field will show the so-called quadrupole peaks (dips).

B.5 NMRD Modeling for Fluids in Complex Systems

The dependence on the magnetic field of the longitudinal relaxation rate provides a powerful tool to verify the theories on the behavior of the liquid under confinement in which the reduced dimensionality influences the interactions.

Therefore, having suitable theoretical models to link fluid molecular dynamics in the network of pores of a porous medium to the frequency-dependence of T_1, the results of NMRD relaxometry experiments can be used to understand the dynamics of fluids in confined geometries at the nanoscale dimension.

As mentioned previously, the interactions affecting the spin dynamics must be described by the correlation function $G(t)$.

As already discussed, the first successful dispersion theory was the BPP theory describing the bulk glycerin liquid by assuming that $G(t)$ decays exponentially. The complexity increases significantly for Magnetic Resonance studies for fluids in Porous Media (MRPM) or when high surface area particles are dissolved. In these cases, the motion of the fluid molecules is influenced by the pore surfaces or by the surface of the particles. Several MRPM or complex fluids' analytical mathematical models have been developed. These families of models simplify the complexity of the dynamics of the confined fluid, but at the same time capture the essential physical processes to describe NMRD data successfully. Among them, Levitz and co-workers have explored surface-mediated diffusion models chiefly for biological systems [around 2010]. Kimmich and co-workers [mid-1990s], prompted by the work of Bychuk and O'Shaughnessy [1995], developed a model of a fluid in which molecules desorb from a surface, diffuse in a bulk-like layer close to the surface, before returning to the surface. It is known as bulk-mediated surface diffusion (BMSD). Korb and co-workers [starting from the early 1990s] developed one of the most generally applicable theoretical descriptions for the relaxation time dispersion for fluids in porous media, in which the dominant relaxation mechanism for hydrogen-bearing molecules is due to the heteronuclear interaction with paramagnetic ion impurities (such as Fe^{3+} or Mn^{2+}) present on pore walls. Primarily these models work for low frequencies, therefore to long timescales, and can be thought of as variants of general modeling for a fluid-filled quasi-two-dimensional (Q2D) pore system characterized by three correlation times. Starting from early 2000s, D. Faux and co-workers developed a Q2D model, named the "3τ model", which considers the characteristic diffusion correlation times that characterize spins in a thin surface layer and in the bulk phase. Some quantities used in the "3τ modeling" are estimated using numerical methods.

B.5.1 Relaxation Model for Porous Media with and Without Surface Impurities[18,21,24]

J.-P. Korb and co-workers distinguished two classes of systems with high surface area: solid phases that are proton rich (*e.g.*, biological macromolecules and engineering polymers) and those that are proton poor (*e.g.*, microporous glasses, zeolites, plaster, cements or clay minerals).

In the first class, the magnetic field dependence of the observed relaxation rate is dominated by the cross-relaxation between the protons of the liquid and those of the solid, whereas, in the second class, the liquid spin relaxation is dominated by other effects, in particular those related to the translational diffusion of the liquid near the solid surfaces. The high surface area media that contain concentrations of paramagnetic centers (such as cements and rocks) may significantly affect the behavior of the NMR relaxation. The paramagnetic centers (such as manganese or iron) have a large magnetic moment and generate a local dipolar field interacting with the spins diffusing in the liquid. The effects of the electron magnetic moments largely dominate the ^{1}H longitudinal relaxation at low magnetic field strengths. The case in which no paramagnetic contaminants are present is also modeled.

The relaxation times of confined fluids are characterized by two limiting cases: slow-diffusion (diffusion-limited) relaxation, when bulk diffusion prevails, or fast-diffusion (surface-limited) relaxation, when surface processes prevail.

In the case of surface-limited relaxation, the overall proton relaxation rate $1/T_1$ is a linear combination of a surface relaxation rate, $1/T_{1,s}$, and a bulk relaxation rate, $1/T_{1,b}$. $1/T_{1,s}$ occurs in a thin layer of fluid close to the surface of thickness λ, in which the adsorbed proton can diffuse for a period of time before being desorbed and diffusing into the bulk liquid. In the typical frequency range used to study porous media with the FFC technique, $1/T_{1,b}$ has no dependence on the Larmor angular frequency.

Figure B.10 shows a schematic of a quasi-2D diffusion relaxation surface model of an aprotic (see below) liquid diffusing at the proximity of the pore surface.

For the evaluation of the influence of a surface distribution of paramagnetic sites on a proton located at a maximum distance from the interface of the grain equal to λ (having a size of several microns), the pore surface can be supposed to be a flat surface (its curvature is much larger than λ). ξ is the thickness of the relaxing surface distribution of paramagnetic impurities.

The term $1/T_{1,s}$ can be represented as a superposition of the contribution of the proton species diffusing near the fixed paramagnetic species, $1/T_{1,2D}$, and the contribution of the proton species trapped into the ligand field of the paramagnetic centers $1/T_{1,p}$:

$$\frac{1}{T_1(\omega)} = \frac{1}{T_{1,b}} + \frac{N_S}{N_b}\frac{1}{T_{1,2D}(\omega)} + \frac{N_P}{N_b}\frac{1}{T_{1,p}(\omega)},$$

where N_S represents the number of diffusing molecules within the thin transient layer λ close to the pore surface, N_P is the number of liquid molecules bonded to the paramagnetic impurities present at the surface, and N_b is the bulk liquid population.

The saturating fluid can be distinct between a *protic liquid* and an *aprotic liquid.* Protic liquids can exchange their protons; therefore, the moving

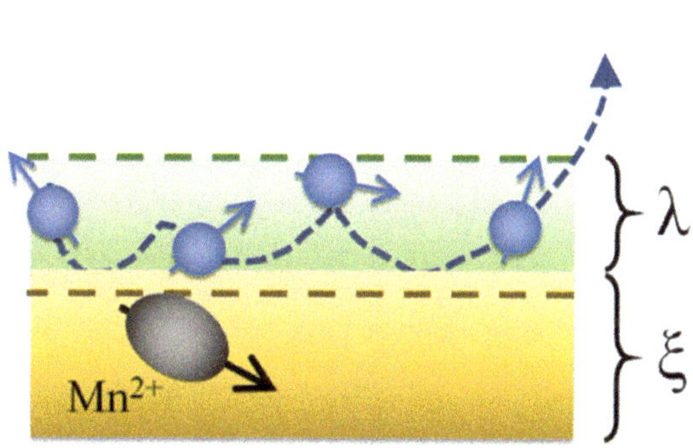

Figure B.10 Schematic illustration of a quasi-2D diffusion relaxation surface model.

protons can be trapped in the ligand field of the paramagnetic sites on the pore surface. Aprotic liquids cannot exchange their protons and moving protons cannot be trapped.

B.5.1.1 Aprotic Liquids

In this case, the protons do not tend to get trapped in the ligand field due to a paramagnetic site. This allows neglecting the $T_{1,\mathrm{p}}(\omega_I)$ contribution to the overall longitudinal relaxation.

The conceptual model of the whole diffusion process, which involves two correlation times, is resumed in Figure B.11.

Here τ_s, the *surface residence time*, is the finite time of residence of proton bearing molecules at a pore surface, a measure of how long the molecule diffuses in the thin surface layer λ and, therefore, a measure of how long the proton remains correlated to paramagnetic impurities, and τ_m, the *translational correlation time*, is associated with the individual molecular jumps in proximity to the surface.

It is possible to demonstrate that, at low frequencies, the overall longitudinal relaxation rate can be modeled by

$$\frac{1}{T_1(\omega_I)} = \frac{1}{T_{1,\mathrm{b}}} + \left(\frac{\mu_0}{4\pi}\right)^2 \frac{\pi S_\mathrm{P} \rho_\mathrm{liquid} \sigma_\mathrm{S} {\gamma_I}^2 {\gamma_S}^2 \hbar^2 S(S+1)}{\lambda \delta'^2 (1+x^2)} \tau_\mathrm{m} \times \left[3 \ln\left(\frac{1+{\omega_I}^2 {\tau_\mathrm{m}}^2}{\left(\frac{\tau_\mathrm{m}}{\tau_\mathrm{s}}\right)^2 + {\omega_I}^2 {\tau_\mathrm{c}}^2} \right) + 7 \ln\left(\frac{1+{\omega_S}^2 {\tau_\mathrm{m}}^2}{\left(\frac{\tau_\mathrm{m}}{\tau_\mathrm{s}}\right)^2 + {\omega_S}^2 {\tau_\mathrm{c}}^2} \right) \right],$$

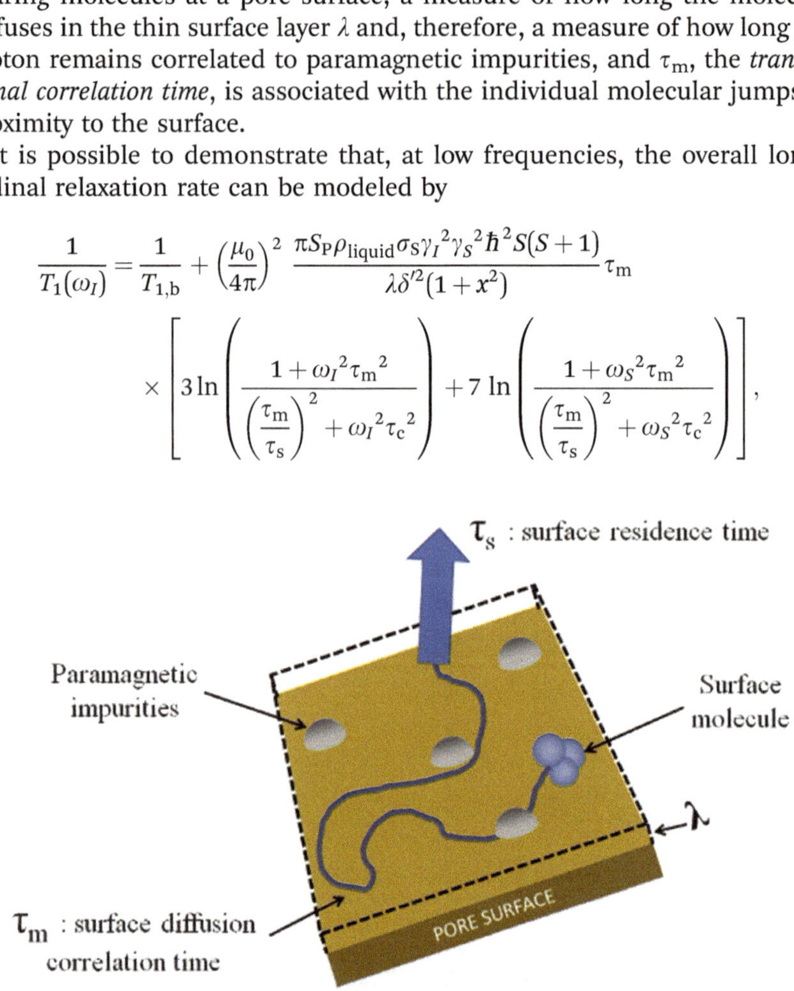

Figure B.11 Schematic illustration of the whole diffusion process of Korb's conceptual model.

where

- λ is the thickness of the surface layer (transient layer);
- x is a parameter to consider the variable distance δ' of the minimal approach between proton and impurity spins relative to the diameter δ of the molecules of the liquid.
- $\sigma_S = \eta_S \rho_S \xi$, the surface density of the (small) quantity of fixed paramagnetic species of S-spins uniformly distributed on pore surfaces. ρ_S is the density of the solid matrix, η_S is the volume concentration of paramagnetic ions, and ξ is the thin layer of paramagnetic ions.

B.5.1.2 Protic Liquids

In the case of protons (I), they might stay in the paramagnetic ion (S) ligand field for a time longer than the surface or bulk correlation times. In these cases, the $1/T_{1,\mathrm{p}}$ rate can be represented as

$$\frac{1}{T_{1,\mathrm{p}}(\omega_I)} = \frac{2}{15} \frac{\gamma_I^2 \gamma_S^2 \hbar^2 \, S(S+1)}{\delta'^6} \times T_{1,\mathrm{elec}}(\omega_S) \left[\frac{3}{1+\omega_I^2 T_{1,\mathrm{elec}}^2(\omega_S)} + \frac{7}{1+\omega_S^2 T_{1,\mathrm{elec}}^2(\omega_S)} \right],$$

where δ' is the distance of minimal approach between the proton and paramagnetic site. The electronic correlation time $T_{1,\mathrm{elec}}(\omega_S)$, where ω_s is the electronic angular frequency,

$$\frac{1}{T_{1,\mathrm{elec}}(\omega_S)} = H_S^2 \tau_\mathrm{v} \left[\frac{1}{1+\omega_S^2 \tau_\mathrm{v}^2} + \frac{4}{1+4\omega_S^2 \tau_\mathrm{v}^2} \right],$$

with τ_v being the correlation time for the electron–lattice fluctuating interaction and H_S^2 representing the intensity of the electron–spin fluctuations.

B.5.1.3 Pore Surface Without Paramagnetic Contaminants

To interpret the NMRD data from porous medium samples without paramagnetic contaminants, an analytical biphasic fast exchange model has been proposed, where the overall $1/T_1(\omega)$ is a superposition of surface and bulk contributions:

$$\frac{1}{T_1(\omega)} = \frac{1}{T_{1,\mathrm{b}}} + \frac{\lambda S}{V} \frac{1}{T_{1,\mathrm{S}}(\omega)}.$$

The surface contribution $1/T_{1,\mathrm{S}}(\omega)$ is weighed by the surface-to-volume ratio S/V. It depends on the proton Larmor angular frequency ω due to the numerous contacts within the thin surface relaxation layer λ between water protons and fixed surface hydrophilic protons. As previously highlighted, the term $1/T_{1,\mathrm{b}}$ is frequency independent in the studied range.

To express the frequency dependence of $1/T_{1,S}(\omega)$ as a linear combination of surface spectral density $J_S(\omega)$, the Levitz model was used. Therefore, for the surface contribution:

$$\frac{1}{T_{1,S}(\omega)} \propto J_S(\omega) + 4J_S(2\omega),$$

where surface spectral density is given by

$$J_S(\omega) \propto \tau_A \left[\sqrt{\frac{\omega}{\omega_A}} + \frac{\omega}{\omega_A} + \frac{1}{2}\left(\frac{\omega}{\omega_A}\right)^{3/2} \right]^{-1},$$

and $\omega_A = \dfrac{\delta^2}{2D{\tau_A}^2}$ is a characteristic angular frequency related to the water molecule size δ (assumed to be equal to 0.3 nm), τ_A, the average adsorption time of residence for water, and D, the translational bulk diffusion coefficient for water.

B.5.2 Anomalous Diffusion at the Interface[25,26]

The dynamics of adsorbed molecules, arising at unbounded interfaces between a liquid bulk phase and a second phase (for example, a solid phase), is a fundamental issue for confined fluids. If the adsorbed species are soluble in the liquid bulk, adsorption–desorption processes occur continuously, which can be a source of surface diffusion.

O. V. Bycbuk and B. O'Sbaugbnessy proposed that for a certain class of bulk-interface systems occurs a particular type of surface translational motion, an anomalous "superdiffusive" surface displacement, named Bulk Mediated Surface Diffusion (BMSD), in which the initially adsorbed molecules are retained for a much longer time (the surface retention time, t_h) than the desorption time (t_{des}), that is, $t_h \gg t_{des}$.

During t_h, molecules move on the surface through several desorption and readsorption loops. Therefore, they are not permanently lost to the bulk; instead, they are moving between surface locations. However, for times longer than t_h, the surface loses its initially absorbed molecules. During a loop, the interval of time between desorption and a successive readsorption, the molecules perform a Fickian diffusion in the bulk liquid, within the "capture range" distance b from the surface, of average duration τ (see the right part of Figure B.12).

The statistics of the BMSD phenomenon are non-Gaussian and Fick's law of diffusion is inapplicable. The molecules perform Lévy walks on the surface, a random walk in which the step-lengths follow a Lévy distribution (a heavy-tailed probability distribution for non-negative random variables), and for time $t < t_h$ the displacement r scales like t and not as $t^{1/2}$ as predicted by Fick's diffusion for Brownian walk/flight. Consequently, bulk-interface

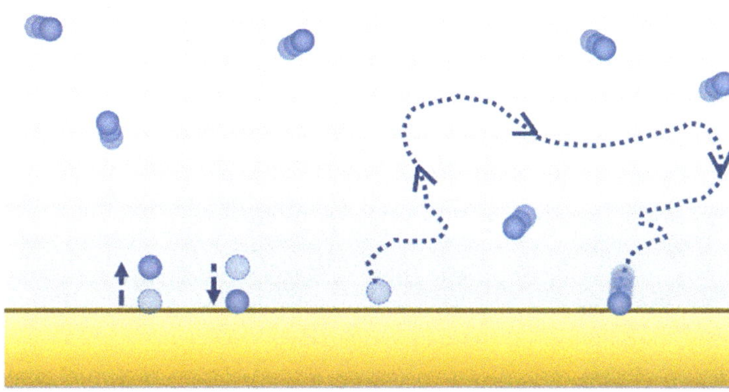

Figure B.12 Schematic illustration of the anomalous "superdiffusive" BMSD surface displacement.

systems can be classified as either "weakly" or "strongly" adsorbing. "Strong" systems exhibit anomalous diffusion, where for times $t \ll t_h$ a molecule spends almost all its time on the surface performing a sequence of desorption–readsorption events, while for $t \gg t_h$ it spends most of the time in the liquid bulk. In "weak" systems, the molecules can escape from the surface after just one desorption time (left part of Figure B.12). BMSD provides a remarkable mechanism for both surface self-diffusion and relaxation of surface density inhomogeneities.

B.5.3 Model for Molecular Flights in Large Surface Suspended Particles[27–30]

Confining systems characterized by large specific surface, such as porous materials or colloidal dispersions, strongly influence the diffusive motion of entrapped particles and the molecular dynamics of embedded fluids. The statistical properties of these random flights play a central role in nuclear magnetic relaxation of fluids confined in complex porous media.

To investigate the statistics of a random flight near an interface, one can study the dynamics nearby of a macroscopic plane surface. However, such a kind of experimental method can have sensitive issues. Otherwise, working with an aqueous colloidal suspension of flat and thin particles is possible.

Levitz and coworkers studied the case of a finite, bounded interface, in particular in the case of a suspension of anisotropic colloidal particles. They proposed a closed-form model that reproduces NMR relaxation features assessing the interaction between the Brownian dynamics of a fluid near a colloidal interface and the slowdown of the rotation of the particles.

Following Halle *et al.*, at low frequencies, one can consider a two-step model with statistically independent fast and slow motions occurring on different timescales: a faster local, slightly anisotropic reorientation superposed on a much slower motion, *e.g.*, translational diffusion of water molecules between different environments. Then the spin–lattice relaxation rate can be decomposed into two terms:

$$R_1(\omega) = R_1^{\text{slow}}(\omega) + R_1^{\text{fast}}.$$

At low frequencies, the fast term can be associated with the longitudinal relaxation rate of the fluid in the bulk phase, which, as is well known, can be assumed to be frequency independent, whereas the slow term can be associated with the interaction of the molecules with the surface and related to the spectral density using the usual relations:

$$R_1^{\text{slow}}(\omega) \propto J(\omega) + 4J(2\omega).$$

Following Bycbuk and O'Sbaugbnessy, when the surface is in the "strong adsorption regime", an individual trajectory near the solid–fluid flat interface can be described as an alternating succession of adsorption steps and Brownian flights in the bulk, an anomalous molecular diffusion on the surface characterized by a Lèvy flight (see Figure B.13(a)).

The local fluctuating magnetic interaction can be depicted using an indicator function, a "relaxation noise", $I(t)$, that is related to an alternate sum of Heaviside distributions $H(t)$.

In the simplest case, $I(t)$ represents a scalar interaction that takes a defined value during the adsorption step (A) (see Figure B.13(b)) and vanishes during the bulk loop (L). For quadrupolar or intradipolar magnetic interactions, $I(t)$ can also be a tensorial interaction in which each encounter with the surface takes a specific value directly related to the local surface orientation (Figure B.13(c)). For an unbounded surface,

$$I(t) = L(t) = \sum_{j=1}^{\infty} (-1)^j H(t - t_j).$$

In the case of a bounded surface, two "escaping tails" (T in Figure B.13(a)) delimit the overall exploration of the molecules on the surface. Therefore, the possibility that the molecule definitively leaves the finite flat interface can be

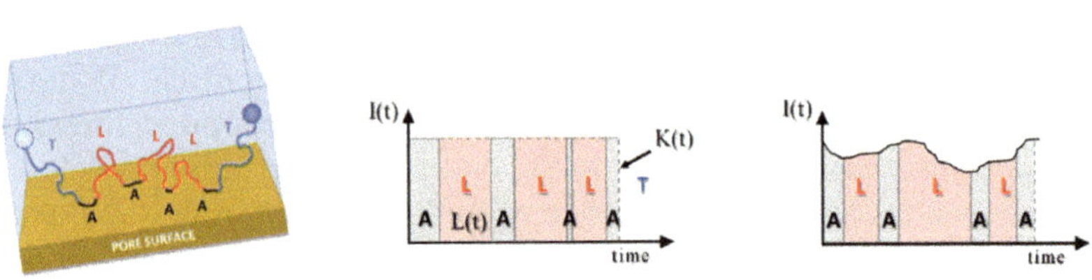

a) b) c)

Figure B.13 (a) Particle trajectory in the proximity of a finite flat surface. (b) Evolution of the indicator function $I(t)$, where A is the adsorption step and L is the bulk Brownian loop. (c) Molecular intermittent dynamic $I(t)$.

mathematically represented by a cut-off function, $K(t)$, that bounds $I(t)$, which is equal to 1 between two escaping tails and 0 elsewhere. Therefore formally

$$I(t) = K(t) \cdot L(t) = K(t) \cdot \sum_{j=1}^{\infty} (-1)^j H(t - t_j),$$

where the interaction of a molecule with the surface is in accord with the time series t_j.

As usual, the spectral density $\mathcal{J}(\omega)$ is the Fourier transform of the correlation function $G(t)$ that describes the rapidity of the fluctuations, $G(t) = \langle I(0) \cdot I(t) \rangle$. In the case of an unbounded flat interface, an explicit expression for the spectral density was obtained:

$$J(\omega) = \frac{\frac{\theta \tau_A}{2\omega_0}}{\left[\sqrt{\frac{\omega}{\omega_0}} + \frac{\omega}{\omega_0} + \frac{1}{2}\left(\frac{\omega}{\omega_0}\right)^{3/2}\right]},$$

where ω_0 is a characteristic angular frequency related to the water molecule size, the translational bulk diffusion coefficient of water molecules and τ_A, the average adsorption correlation time. θ is a constant related to the intensity of the magnetic interaction in the adsorption step.

For frequencies $\omega \ll \omega_0$, $J(\omega)$ evolves as $\frac{1}{\sqrt{\omega}}$; for $\omega \approx \omega_0$, $J(\omega)$ evolves as $\frac{1}{\omega}$; and finally for $\omega \gg \omega_0$, $J(\omega)$ evolves as $\frac{1}{\omega^{3/2}}$.

In the case of a bounded flat interface, assuming that the two random functions $L(t)$ and $K(t)$ are statistically independent, the spectral density is related to the following convolution:

$$J(\omega) = \tilde{C}_{esc}(\omega) \times \tilde{L}_2(\omega),$$

where $\tilde{L}_2(\omega)$ is the Fourier transform of $\langle L(0) \cdot L(t) \rangle$ and $\tilde{C}_{esc}(\omega)$ is the Fourier transform of

$$\tilde{C}_{esc}(t) = \int_t^{\infty} \langle K(\varsigma) \rangle d\varsigma,$$

with $\langle K(t) \rangle$ being the overall survival probability near a finite surface and characterized by an exponential behavior as demonstrated using Brownian dynamics simulations. In a liquid phase, during the timescale of the NMR experiment, the colloidal flat particles rotate, introducing a new fluctuating component participating to $I(t)$.

An angular director autocorrelation, $C_{ori}(t)$, can statistically define such dynamics, and $J(\omega)$ can be generalized as

$$J(\omega) = \tilde{C}_{ori}(\omega) \times \tilde{C}_{esc}(\omega) \times \tilde{L}_2(\omega).$$

Finally, in general, in porous media, the interfaces are very complex systems and exhibit a surface curvature map that cannot be easily straightforward-coupled with the intermittent dynamics approach; it therefore needs specific analysis.

B.5.4 Three Correlation Times Model[22,23,31–34]

Approximately around 2015, Faux and McDonald and co-workers proposed an alternative model, named the "3τ model", for the interpretation of NMRD curves acquired from porous media saturated with ^{1}H fluids.

Porous materials are characterized by a complex internal structure. Therefore, a confined fluid can exist in a wide range of environments with fluid molecules that rotate, vibrate and move in a very complex way. Concerning the relaxation processes, porous materials are classified into two categories: with and without paramagnetic ions. If paramagnetic contaminants are present, then the proton–proton interactions are normally negligible, and the relaxation rate is dominated by the dipolar interaction of moving proton spins with the paramagnetic ions (mobile if they exist as aqueous ions or fixed if they are embedded in the solid material). Although the study of NMR relaxation of solutions containing paramagnetic impurities is an important research area, this possibility will be ignored in the following description.

Instead, in porous materials without paramagnetic ions, the frequency-dependent relaxation rate is dominated by the combinations of interactions between pairs of protons in relative motion.

The challenge is, therefore, to develop a conceptual model able to capture the key physics of the dynamics of the molecule motion in the bulk and at the surfaces of pores for a variety of geometries and pore connectivity. In other words, in order to understand the information contained in the NMRD profiles, expressions must be derived for the various interactions between pairs of spins.

The 3τ model incorporates the key physics, grounded in molecular dynamics simulations, highlighted in the conceptual model sketched in Figure B.14. It approximates the complexity of the pore geometry as a quasi-two-dimensional (Q2D) system with locally-flat surfaces containing a bulk fluid of thickness h. It is characterized by three correlation times:

- τ_ℓ, representing the diffusion correlation time for the fluid at the pore surface moving slowly parallel to the pore surface. It is related to the surface water diffusion coefficient $D_\ell = \sigma^2/(6\tau_\ell)$, where σ is assumed to be equal to 0.27 nm, a distance roughly equal to the intermolecular distance between water molecules;
- τ_d, representing the desorption time, the time a water molecule remains on the surface before desorption, assuming that the rate of spin desorption follows an exponentially decreasing $\exp(-t/\tau_d)$;
- τ_b, the bulk fluid correlation diffusion time constant related to the water diffusion coefficient $D_b = \sigma^2/(6\tau_b)$.

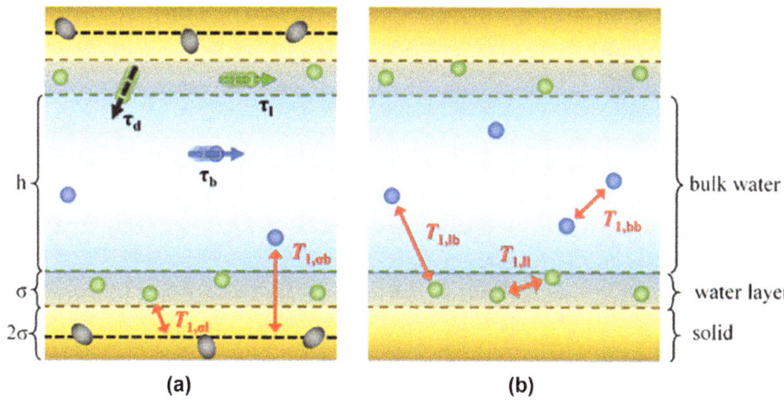

Figure B.14 Schematic illustration of the spin interactions in a planar (Q2D) pore. (a) Interactions dominated by paramagnetic impurities. (b) Interactions are dominated by proton–proton interactions.

Figure B.14 shows the conceptual model structure of a Q2D pore with the possible spin interactions, where the blue arrow represents the bulk diffusion, the green the in-layer spin diffusion, and the black the desorption. In Figure B.14(a), both the interactions of the protons in the surface layer (green) and in bulk (blue) are dominated by a layer (dashed line) of fixed paramagnetic impurities located 2σ below the surface. The effective layer of paramagnetic spins is supposed to be placed *below* the pore surface. Consequently, the long-time limit of $G(t)$ decays as t^{-2} in agreement with theoretical and computational demonstration (instead, when all spin species are contained in the 2D layer, then $G(t)$ decays as t^{-1}). These also have effects on the value of τ_ℓ computed by fitting the model to experimental data.

In Figure B.14(b), interactions are dominated by proton–proton interactions in layer–layer, bulk–bulk and bulk–layer. The bulk water spin density and the surface water density are assumed to be equal.

The 3τ model also takes into account the strong distance dependence of the dipolar interaction, enabling the direct estimate of τ_b (and hence the diffusion coefficient of the bulk fluid) from the high-frequency limit of the NMRD curve.

Each interaction is associated with a corresponding specific longitudinal relaxation rate. So, $T_{1,\sigma\mathrm{b}}^{-1}$ and $T_{1,\sigma\ell}^{-1}$ correspond to the interaction between the layer of spins of the electronic paramagnetic impurity (σ) in the crystal and the bulk (b) and surface (ℓ) fluids, respectively. $T_{1,\sigma\ell}^{-1}$ corresponds to the interaction between the spins in the bulk fluid and surface fluid, with the assumption that the surface water moves very slowly compared to the bulk (and can be assumed to be fixed), that is, $\tau_\ell \gg \tau_\mathrm{b}$.

$T_{1,\ell\ell}^{-1}$ is for the layer–layer spin interactions, at the pore surface layer, and $T_{1,\mathrm{bb}}^{-1}$ is for the bulk–bulk spin interactions. Finally, $T_{1,\mathrm{Mn}}^{-1}$ represents the relaxation rate due to paramagnetic ions, as, for example, manganese, dispersed in water (not used in the following description).

Finally, for systems with a negligible concentration of paramagnetic impurities or characterized by large pores, the measured NMRD curves $T_{1,\text{obs}}^{-1}$ can be modeled by

$$T_{1,\text{obs}}^{-1} = \frac{n_\ell}{n_\ell + n_\text{b}}\left(T_{1,\ell\text{b}}^{-1} + T_{1,\ell\ell}^{-1}\right) + \left(1 - \frac{n_\ell}{n_\ell + n_\text{b}}\right)\left(T_{1,\text{b}\ell}^{-1} + T_{1,\text{bb}}^{-1}\right), \tag{B.5}$$

where $\frac{n_\ell}{n_\ell + n_\text{b}}$ is the fraction of the total number of spins in the surface layer volume and $T_{1,\text{b}\ell}^{-1} = \frac{N_\text{b}}{N_\ell} T_{1,\ell\text{b}}^{-1}$, with N_ℓ and N_b being the spin volume density of the surface layer and bulk fluids, respectively.

For a porous system in which the relaxation is dominated by interactions with paramagnetic impurities at the pore surface, $T_{1,\text{obs}}^{-1}$ can be modeled as

$$T_{1,\text{obs}}^{-1} = \frac{n_\ell}{n_\ell + n_\text{b}} T_{1,\sigma\ell}^{-1} + \left(1 - \frac{n_\ell}{n_\ell + n_\text{b}}\right) T_{1,\sigma\text{b}}^{-1}$$

A further point that differentiates the 3τ model is that the analytical expressions developed for the dipolar correlation functions $G(t)$ are then numerically computed *via* Monte Carlo (MC) simulations to cover the full time domain, and consequently the corresponding spectral density functions $J(\omega)$ are generated *via* numerical integration. In addition, to parametrize the MC simulations, Molecular Dynamics (MD) simulations of bulk water and water in Q2D systems were performed.

Therefore, it is necessary to determine the specific time-dependent dipolar correlation function $G(t)$ for each of the interactions described by the above conceptual model (Figure B.14). To this end, Faux *et al.* derived the expression for $G_{\ell\text{b}}(t)$, $G_{\ell\ell}(t)$, $G_{\text{bb}}(t)$, $G_{\sigma\text{b}}(t)$, and $G_{\sigma\ell}(t)$.

The starting point was the expression:

$$G(t) = \frac{4\pi}{5}\int_{\mathbb{R}^3}\int_{\mathbb{R}_0{}^3}\left[\sum_{m=-2}^{2}\frac{Y_{2m}(x_0, y_0, z_0)Y_{2m}^*(x, y, z)}{r_0{}^3 r^3}\right] P(\boldsymbol{r}, t \cap \boldsymbol{r}_0) d^3 r_0 d^3 r.$$

This also entails devising an appropriate *probability density function*, $P(\boldsymbol{r}_0 \cap \boldsymbol{r}, t)$, dependent on a set of parameters related to the various interactions between pairs of spins (namely the self-diffusion coefficients for molecular species or, equivalently, the correlation times), which furnishes the probability that two spins are separated by the vector $\boldsymbol{r}_0$ at $t = 0$ and then by the vector $\boldsymbol{r}$ at time t. The Y functions are the spherical harmonic functions of degree 2 expressed in terms of the cylindrical coordinates (Y^* represents the complex conjugates).

Then, as is well known (see Section B.3.6.2 for details), the Fourier transform of $G(t)$ to obtain the spectral density function $J(\omega)$ may be written as

$$J(\omega) = 2\int_0^\infty G(t)\cos(\omega t)\text{d}t.$$

Finally, the usual relations are used to link relaxation rates with the spectral density functions. Specifically, for systems with paramagnetic ions:

$$\frac{1}{T_1(\omega)} = \frac{1}{3}\left(\frac{\mu_0}{4\pi}\right)^2 \gamma_I{}^2\gamma_S{}^2\hbar^2 S(S+1)[7J(\omega_S)+3J(\omega)],$$

$$\frac{1}{T_2(\omega)} = \frac{1}{6}\left(\frac{\mu_0}{4\pi}\right)^2 \gamma_I{}^2\gamma_S{}^2\hbar^2 S(S+1)[4J(0)+13J(\omega_S)+3J(\omega)],$$

and for systems without paramagnetic ions:

$$\frac{1}{T_1(\omega)} = \frac{1}{5}\left(\frac{\mu_0}{4\pi}\right)^2 \gamma_I{}^4\hbar^2 I(I+1)[J(\omega)+4J(\omega)],$$

$$\frac{1}{T_2(\omega)} = \frac{1}{10}\left(\frac{\mu_0}{4\pi}\right)^2 \gamma_I{}^4\hbar^2 I(I+1)[3J(0)+5J(\omega)+2J(2\omega)],$$

where ω $(= 2\pi f)$ is the proton Larmor angular frequency, ω_s $(= 658.2\omega)$ is the electron Larmor angular frequency of the impurity spins (typically Fe^{3+} and Mn^{2+}), γ_S is the gyromagnetic ratio of impurities, γ_I is the proton gyromagnetic ratio, and the spin numbers are $I = 1/2$ and typically $S = 5/2$.

Therefore, to implement the 3τ model, all the spectral density functions for the different interactions depicted in the conceptual model have to be numerically computed, namely $J_{\ell b}(\omega)$, $J_{\ell\ell}(\omega)$, $J_{bb}(\omega)$, $J_{\sigma b}(\omega)$, and $J_{\sigma\ell}(\omega)$.

The 3τ model has been designed for porous systems characterized by planar pores with a depth of h. However, the authors have shown that if the pore thickness is 5 nm or larger, the longitudinal relaxation time dispersion is independent of the pore thickness. Consequently, the 3τ model could apply to porous systems in which the fluid is confined to pores in which at least two dimensions exceed 5 nm.

However, from the practical point of view, it is difficult to fit the model to experimental data mainly because extremely complex calculations are required to derive the theoretical T_1^{-1} curve from the model. For this reason, a specific MATLAB script was developed, significantly reducing the operative difficulties. The code loads pre-calculated data sets based on the 3τ model (for a pre-defined grid of values for the τ_b, τ_ℓ, and τ_d parameters) and executes a least squares fitting, for example, of eqn (B.5). A version of the code with pre-calculated data sets is available for free download through GitHub.

B.5.5 Molecular Reorientation Mediated by the Translational Displacement Model[2,13,15,35,36]

In confined fluid systems, the adsorption of molecules at the pore walls of solid porous media or at the surfaces of a static or slowly moving particle in solution provides some preferential local molecular orientation. Diffusion in the bulk-like phase has no preferential orientations, but when molecules are adsorbed on the site of a surface, they adopt the intrinsic preferential orientation of that surface site. At first sight, only the molecular reorientation, that is, rotational diffusion, would seem to be affected by the fluid/surface interactions, while *translational diffusion*, which is sensitive to geometrical restrictions and obstacles (as, for example, adsorption at solid surfaces and exchange between different liquid phases), seems to be irrelevant.

Following the BMSD theory (see Section B.5.2), the dynamics of adsorbate molecules near the inner surfaces of a porous medium can be modeled as a process of diffusion/adsorption and desorption of molecules. This process is schematically depicted in Figure B.15, where a possible trajectory followed by a molecule is shown. The molecule can be adsorbed on the surface at site P_0 and then diffuse along the surface within the adsorbed phase (*solid black lines*) up to site P_1. Then the molecule can be desorbed and performs an excursion into the bulk-like phase (*red lines*). After a while, it can be re-adsorbed at site P_2 and diffuse again within the adsorbed phase up to site P_3 where it will again be desorbed, and so on until the molecule definitively leaves the surface and vanishes in the bulk-like phase (*arrow*). The set of *solid black* trajectories and *dashed straight lines* (representing a leap of the molecule between two points on the surface) forms the path of "surface diffusion", which represents the actual displacements of the molecule along the surface and which can be described as a Lévy walk.

Therefore, if the surface adsorption/desorption BMSD process is not negligible, there will be an interconnection between reorientations of the fluid molecules and their translations along surfaces. This mechanism is referred to as *reorientation mediated by translational displacements* (RMTD) and it is an important low-field diamagnetic relaxation mechanism for the molecules of the adsorbate phase.

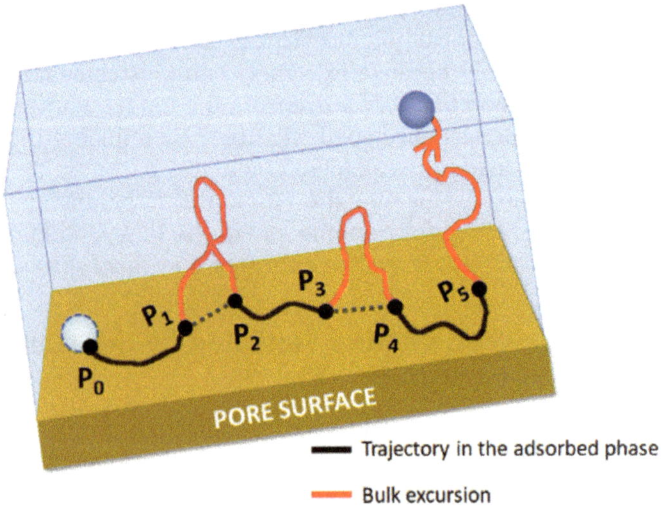

Figure B.15 Schematic illustration of surface diffusion of a particle of the adsorbate fluid at the solid surface of the adsorbent, as for example a pore surface.

Many elementary diffusion steps are required for the molecules to totally lose correlation with their initial orientation. The autocorrelation function can be many orders of magnitude slower than that of bulk liquid, consequently with evident effects on the low field spin–lattice relaxation times.

In the following, the interaction of the fluid with the pore surface will be described considering the common two-phase fast-exchange (TPFE) model and the spin–lattice relaxation with an RMTD approach.

As is well known, in a saturated solid matrix, the two-phase model can be represented as in the following schematic representation shown in Figure B.16. It is supposed that the molecules of the adsorbate liquid reside in two homogeneous phases called 'bulk-like' and 'adsorbate', respectively. In the 'bulk-like' phase, the molecular dynamics are practically the same as those in a bulk fluid, while in the 'adsorbate' phase, topological restrictions and molecule/surface interactions (possibly anisotropic) slow down the molecular motions with respect to the bulk-like phase.

Three types of characteristic times can be defined:

- τ_{ex} which characterizes the exchange kinetics, including transport to and from the interfaces, between *adsorbate* and *bulk-like* phases;
- τ_{rot} which is the correlation time for rotational diffusion in the *bulk-like* phase or, in a restricted form, in the *adsorbate* phase;
- NMR relaxation times, both T_1 longitudinal and T_2 transverse relaxation times, or in other words, the interval in which diffusive displacements are detected in experiments.

The TPFE model holds, *i.e.*, the system is in a fast diffusion regime, if "$\tau_{ex} \ll T_1$, T_2", or in other words, when the exchange rates are greater than relaxation rates in both phases. It is also always true that "$T_1 \gg \tau_{rot}$".

According to the definition of the correlation function, only the states of the spin system at time 0 and at a generic time t are important for its

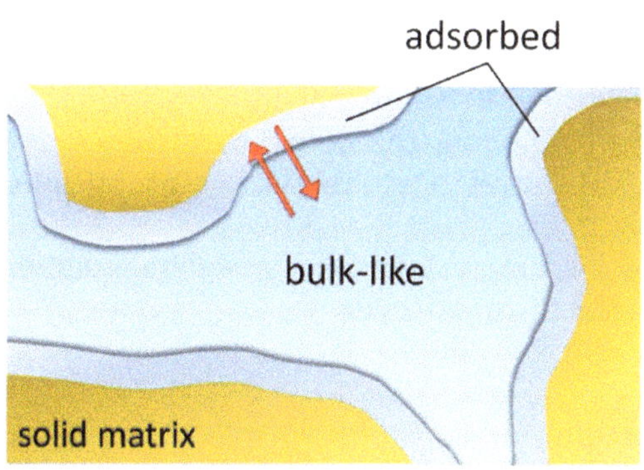

Figure B.16 Schematic illustration of the conceptual model for a two-phase fluid confined in a saturated porous medium.

computation (see Section B.3.6.1), without paying attention to what happens in between. Therefore, it is possible to subdivide the behavior of the populations of spins confined in a pore into four subensembles, identified by the following population fractions (where the subscripts 'a' and 'b' stand for "adsorbate" and "bulk-like" phases, respectively):

- $f_{a;a}(t)$, spins that initially and finally remained in the adsorbate phase;
- $f_{a;b}(t)$, spins that were initially in the adsorbate phase and at the end in the bulk-like phase;
- $f_{b;a}(t)$, spins that were initially in the bulk-like phase and at the end in the adsorbate phase;
- $f_{b;b}(t)$, spins that initially and finally remained in the bulk-like phase.

Of course, the following constraint holds:

$$f_{a;a}(t)+f_{a;b}(t)+f_{b;a}(t)+f_{b;b}(t)=1.$$

Furthermore, the correlation function can also be supposed to be a weighted average of the four independent correlation functions representing the behavior of the four spin families:

$$G(t)=f_{a;a}(t)G(t)_{a;a}+f_{a;b}(t)G(t)_{a;b}+f_{b;a}(t)G(t)_{b;a}+f_{b;b}(t)G(t)_{b;b}.$$

In the "long-time limit", $t \gg \tau_{rot}$, only $G(t)_{a;a}$ survives, whereas the other correlation functions will vanish, as the bulk-like phase molecules will be subject to rapid correlation losses. Therefore

$$G(t\gg\tau_{rot})\approx f_{a;a}(t)G(t)_{a;a}.$$

On the adsorbent surface, the adsorbed molecule may have a fast restricted rotational diffusion, *i.e.*, with a limited solid-angle range with respect to its preferential orientation. Therefore, $G(t)_{a;a}$ initially decays rapidly to a residual correlation value, which is then further reduced by the much slower RMTD process.

Owing to the "different timescale limit" between RMTD and the restricted rotational diffusion, the constant residual restricted rotational value will enter the correlation function as a multiplicative factor (stochastic independence).

With regard to the timescales of exchange and rotational diffusion, two further limits can be distinguished:

(1) $t \gg \tau_{ex} \gg \tau_{rot}$,
(2) $\tau_{rot} \ll t \ll \tau_{ex}$.

In case (1), on the timescale of the correlation decay, molecules will be exchanged frequently, and therefore the initial and final probabilities of finding a molecule in either phase can be established independently, so that:

- $f_{a;a}(t) \approx f_a^{\ 2}$
- $f_{a;b}(t) \approx f_{b;a}(t) \approx f_a(1 - f_a)$
- $f_{b;b}(t) \approx (1 - f_a)^2$

and consequently

$$G(t \gg \tau_{rot}) \approx f_a^{\ 2} G(t)_{a;a},$$

where f_a and $(1 - f_a)$ are the constant populations of adsorbate and bulk-like phases, respectively.

In case (2), the molecular exchange between adsorbate and bulk-like phases is therefore unlikely to occur:

- $f_{a;a}(t) \approx f_a,$
- $f_{a;b}(t) \approx f_{b;a}(t) \approx 0,$
- $f_{b;b}(t) \approx (1 - f_a)$

and consequently

$$G(t \gg \tau_{rot}) \approx f_a G(t)_{a;a}.$$

As usual, also for a TPFE system to obtain the low-frequency relaxation dispersion curve, $T_1 = T_1(\omega)$, it is necessary to obtain an explicit expression for $G(t)$, and its corresponding (reduced) spectral density function (see Section B.3.6.2 for details):

$$\mathcal{J}(\omega) = \int_{-\infty}^{\infty} G(t) \mathrm{e}^{-i\omega t} \mathrm{d}t.$$

Finally, as explained in Section B.3.6.2, the relaxation rate dispersion function is a linear combination of the spectral density function:

$$\frac{1}{T_1(\omega)} = C[\mathcal{J}(\omega) + 4\mathcal{J}(2\omega)],$$

where C is a constant related to the relevant specific couplings among spins.

As mentioned above, the BMSD process is assumed to govern the molecular dynamics of adsorbed molecules; see Section B.5.2 for details. In summary, BMSD is a process characterized by the 'strong adsorption limit', in which several cycles of desorption/unrestricted bulk excursion/re-adsorption occur before an adsorbed molecule definitively leaves the adsorbent surface to disappear in the bulk-like phase.

This "surface diffusion" implies that, initially and ultimately, the adsorbed molecule resides on the surface, irrespective of any bulk excursions in between and that it effectively diffuses following a process in which its surface displacements follow a non-Gaussian statistic (the surface propagator) in a topologically *quasi two-dimensional surface space* (*Q2DS*). BMSD is characterized by the 'retention time' t_h, indicating the time requested to replace the initial population of the adsorbate on a surface, and by the 'adsorption depth', $h = \sqrt{Dt_h}$, where D is the bulk diffusivity of the adsorbate.

For $s \ll \sqrt{Dt}$, where s is the displacement in Q2DS, the surface diffusion propagator is a 2-dimensional Cauchy distribution:

$$p_c(s,t) = \frac{1}{2\pi} \frac{\frac{D}{h}t}{\left[\left(\frac{D}{h}t\right)^2 + s^2\right]^{3/2}}.$$

In the long-time limit, $t > t_h$, the two-dimensional diffusion tends to follow the ordinary Gaussian propagator:

$$p_g(s,t) = \frac{1}{4\pi\sqrt{Dt}} e^{-\frac{s^2}{4Dt}}.$$

Following the RMTD approach, the correlation function, $G(t)_{a;a}$, is composed of a dynamic contribution, represented by the propagator $p(s, t)$ for effective displacements s in the *Q2DS*, and a geometric contribution, given by the surface correlation function $g(s)$, which characterizes the (micro)structural surface topology:

$$G_{a;a}(t) = \int g(s) p(s,t) 2\pi s \mathrm{d}s.$$

The decay of $g(s)$ reflects the roughness and the shape of the pore surfaces, thus also how the orientation of the surface changes.

Instead of using real space variables, it is often useful to employ a reciprocal space image based on the wavenumber k and the *spatial Hankel transform*:

$$G_{a;a}(t) = \frac{1}{2\pi} \int S(k) p(k,t) \mathrm{d}k,$$

where

$$S(k) = \frac{1}{2\pi} \int_0^\infty s g(s) J_0(ks) \mathrm{d}s$$

is the "spatial orientational structure factor", and

$$p(k,t) = 2\pi \int_0^\infty s P(s,t) J_0(ks) \mathrm{d}s$$

is the 'k spatial propagator', where $J_0(ks)$ is the Bessel function of order zero.

In the case of the Gaussian or Cauchy propagator, as in this case,

$$p(k,t) = e^{-\frac{t}{\tau_k}},$$

where the time constant can assume the values

$$\tau_k = \frac{1}{Dk^2}\ \text{(Gauss) and } \tau_k = \frac{h}{Dk}\ \text{(Cauchy)}$$

Finally, the RMTD correlation and the corresponding reduced spectral density read:

$$G_{\mathrm{a;a}}(t) = \frac{1}{(2\pi)^2}\int_0^{\infty} S(k)\mathrm{e}^{-\frac{t}{\tau_k}}\mathrm{d}k,$$

$$\mathcal{J}(\omega) = \frac{1}{(2\pi)^2}\int_0^{\infty} S(k)\frac{2\tau_k}{1+(\omega\tau_k)^2}\mathrm{d}k.$$

To apply this theory in practice, it is necessary to furnish an explicit form of the factor $S(k)$ related to the surface structure under examination and to evaluate the fraction $f_{\mathrm{a;a}}$.

So, as an illustrative example, noting that in a certain frequency range some samples exhibit $T_1(\omega)$ dispersion curves that follow a power law, it makes sense to assume power laws for the surface correlation function as well:

$$g(s) = as^{-\alpha},$$

where α (in the range $0\div1$) represents the surface roughness and a is a constant.

Then the orientational structure factor reads:

$$S(k) = bs^{-(1-\alpha)} = bs^{-\chi},$$

where b is a constant, finally obtaining,

$$T_1(\omega) \propto b^{-1}D^{(1-\chi)/2}\omega^{(1+\chi)/2}\ \text{(Gaussian propagator)},$$

$$T_1(\omega) \propto b^{-1}\left(\frac{h}{D}\right)^{(1-\chi)}\omega^{\chi}\ \text{(Cauchy propagator)}.$$

It is noteworthy that the same orientational structure factor yields different spin–lattice relaxation dispersion curves depending on the choice of the propagator.

B.5.6 Model-free Approach[37–44]

The idea behind the model-free approach for the analysis of NMRD curves from macromolecules or complex fluids was born around 1980 (Lipari & Szabo, King & Jardetzky) and then elaborated by B. Halle and coworkers, first from an implementation perspective and later from a theoretically more robust physical point of view.

If the experimental data are large enough to cover the low- and high-frequency plateaus of the NMRD curve reasonably well, then it is possible to extract some physical quantities related to molecular dynamics without invoking a specific model of interpretation, the model-free approach.

As is well known (see Section B.3.6.1 for more details), within the motional-narrowing regime, the relaxation behavior of the spin can be totally studied in terms of the time correlation function:

$$G(\tau) = \langle V(t) \cdot V(t+\tau) \rangle,$$

which describes the effects of a fluctuating lattice variable $V(t)$. Supposing $G(t)$ is real and even, then the spectral density function $J(\omega)$ is defined as

$$J(\omega) = \int_{-\infty}^{\infty} G(\tau) e^{-i\omega\tau} d\tau = 2\int_{0}^{\infty} G(\tau) e^{-i\omega\tau} d\tau = 2\int_{0}^{\infty} G(\tau)\cos(\omega\tau) d\tau.$$

It is noteworthy that

$$J(0) = 2\int_{0}^{\infty} G(\tau) d\tau. \tag{B.6}$$

So, $G(\tau)$ must be an integrable function, in other words $\underset{\tau\to\infty}{G(\tau)} \to 0$.

In the case of simple dynamics, such as spherical-top molecular rotational diffusion in an isotropic medium, then the time correlation function can have an exponential shape:

$$G(\tau) = G(0) e^{-|\tau|/\tau_c},$$

where the correlation time τ_c provides the timescale for the fluctuations of the lattice variable $V(t)$ and $G(0) = \langle V^2 \rangle$ is the mean-square amplitude of the fluctuations of the lattice variable. In this case, then, the spectral density function is of Lorentzian type:

$$J(\omega) = G(0) \frac{2\tau_c}{1 + (\omega\tau_c)^2},$$

a frequency function fully characterized by two parameters: $G(0)$ and τ_c.

From the quantum mechanical theory, in isotropic systems at the angular resonance frequency ω_0, the longitudinal relaxation dispersion rate ($R_{1\text{theory}}(\omega)$) for the "like" spin dominated by dipole–dipole or quadrupolar interactions depends on the spectral density function as (see Section B.3.6.2):

$$R_{1\text{theory}}(\omega_0) = K[J(\omega_0) + 4J(2\omega_0)],$$

where K is a constant related to the specific couplings among spins.

However, in general, NMRD profiles span a much wider frequency range, a "*stretched*" dispersion, than would result in the case of a pure Lorentzian spectral density.

While a pure Lorentzian dispersion is completely well characterized by two parameters, a stretched dispersion contains more information. The following describes how to compute two parameters using the model-free approach.

One such quantity is the integral of the dispersion curve, *i.e.*, the mean-square fluctuation of $V(t)$.

One such quantity is the integral of the dispersion profile, *i.e.* the mean-square fluctuation of the lattice variable. Considering the time reversal property of the correlation function $G(\tau)=G(-\tau)$, and assuming it is real, then it is possible (see Section B.3.6.2) to write:

$$\int_0^{\infty} J(\omega)\mathrm{d}\omega = \frac{1}{2}\int_{-\infty}^{\infty} J(\omega)\mathrm{d}\omega = \frac{1}{2}\int_{-\infty}^{\infty}\left(\int_{-\infty}^{\infty} G(\tau)\mathrm{e}^{-i\omega\tau}\mathrm{d}\tau\right)\mathrm{d}\omega$$
$$= \frac{1}{2}\int_{-\infty}^{\infty} G(\tau)\left(\int_{-\infty}^{\infty}\mathrm{e}^{-i\omega\tau}\mathrm{d}\omega\right)\mathrm{d}\tau = \frac{\cancel{2}\pi}{\cancel{2}}\int_{-\infty}^{\infty} G(\tau)\delta(\tau)\mathrm{d}\tau$$

where $\delta(\tau) = \frac{1}{2\pi}\int_{-\infty}^{\infty}\mathrm{e}^{-i\omega\tau}\mathrm{d}\omega$ is the Dirac function; therefore, then one obtains

$$\int_0^{\infty} J(\omega)\mathrm{d}\omega = \pi G(0).$$

Another parameter is the average motional time associated with a stretched dispersion, namely the time integral of the reduced time correlation function, which, recalling eqn (B.6), has the form:

$$\langle\tau_c\rangle = \frac{1}{G(0)}\int_0^{\infty} G(\tau)\mathrm{d}\tau = \frac{1}{2}\frac{J(0)}{G(0)}. \tag{B.7}$$

To determine the model-free parameters, a physically plausible mathematical model capable of representing most NMRD profiles, possibly with a small number of parameters, must be fitted to the measured raw data. As already proposed in previous works, a possible reasonable choice to represent the stretched spectral density is to use a multi-Lorentzian function, a sum of N Lorentzian terms:

$$J(\omega) = \sum_{n=1}^{N} c_n \frac{2\tau_n}{1+(\omega\tau_n)^2}. \tag{B.8}$$

Or in other words, this corresponds to assuming that the correlation function is a sum of exponentials:

$$G(\tau) = \sum_{n=1}^{N} c_n \mathrm{e}^{-|\tau|/\tau_n}.$$

It is easy to demonstrate that the molecular dynamics quantities $G(0)$ and $\langle\tau_c\rangle$ can be computed if it is possible to calculate the set of parameters

$\{\tau_n, c_n\}$. In fact, integrating $J(\omega)$ one can obtain

$$\begin{aligned}\int_0^{\infty} J(\omega)\mathrm{d}\omega &= \sum_{n=1}^{N} c_n\tau_n \int_0^{\infty} \frac{2}{1+(\omega\tau_n)^2}\mathrm{d}\omega \\ &= 2\sum_{n=1}^{N} c_n\tau_n \left[\frac{1}{\tau_n}\arctan(\omega\tau_n)\right]_0^{\infty} \\ &= 2\sum_{n=1}^{N} c_n\left[\frac{\pi}{2}-0\right] = \pi\sum_{n=1}^{N} c_n = \pi G(0),\end{aligned}$$

Hence

$$G(0) = \sum_{n=1}^{N} c_n,$$

and putting $\omega = 0$ in eqn (B.8)

$$J(0) = 2\sum_{n=1}^{N} c_n\tau_n,$$

then from eqn (B.7)

$$\langle\tau_c\rangle = \frac{1}{2}\frac{J(0)}{G(0)} = \frac{1}{2}\frac{2\sum_{n=1}^{N} c_n\tau_n}{\sum_{n=1}^{N} c_n}.$$

The next step is to select the fitting procedure to determine the regression parameters c_n and τ_n. Generally, it is possible to adopt two approaches: the non-linear discrete approach or the linear quasi-continuous approach.

B.5.6.1 Non-linear Discrete Model-free

Organizing the $2N$ unknown parameters in the vectors $\mathbf{c} \equiv (c_1,\ldots,c_N)^T$, and $\boldsymbol{\tau} \equiv (\tau_1,\ldots,\tau_N)^T$, they are computed by solving the classical non-linear least squares problem.

Defining the residual vector $\mathbf{r}(\mathbf{c}, \boldsymbol{\tau}) \in \mathbb{R}^M$ with components

$$r_i(\mathbf{c}, \boldsymbol{\tau}) \equiv R_1(\omega_i) - R_{1\text{theory}}(\omega_i), \quad i = 1, \ldots, M$$

where $R_1(\omega_i)$ are the NMRD measured data points and $R_{1\text{theory}}(\omega_i)$ are the *NMRD data points from the fitted model*, the values $\mathbf{c} \geq 0$ and $\boldsymbol{\tau} \geq 0$ can be computed by the following constrained non-linear weighted least squares problem:

$$\min_{(\mathbf{c},\boldsymbol{\tau})\geq 0} \sum_{i=1}^{M} w_i \mathbf{r}(\mathbf{c}, \boldsymbol{\tau})^2,$$

where $w_i \geq 0$ are the weights related to the data noise. If the data noise level is independent of i, then the unweighted least squares (*i.e.*, $w_i = 1$) can be safely applied. Otherwise, w_i should be given by means of the noise standard deviation σ_i at the point $R_1(\omega_i)$, $w_i = 1/\sigma_i^2$, or by applying strategies for estimating σ_i from data. A commonly used strategy consists of minimizing the linear least squares problem with $w_i = 1, \forall i$ and then using the fitted dispersion curve $R_{1\text{fit}}$ to obtain the weights as follows:

$$w_i = 1/R_{1\text{fit}}(i), \quad R_{1\text{fit}}(i) = \min(|R_{1i}|, \text{ErrFit}), \quad i = 1, \ldots, M,$$

where ErrFit is the threshold value computed as

$$\text{ErrFit} = \sqrt{\sum_{j=1}^{\bar{M}} \frac{(\mathbf{r}_{\bar{M}}(j))^2}{\bar{M}}},$$

where $\mathbf{r}_{\bar{M}}$ contains $\bar{M}$ smallest components of the residual vector $R_1(\omega_i) - R_{1i}$. Usually, $\bar{M} = 10$ is sufficient.

To minimize the least squares merit function, an accurate iterative solver such as the projected Newton method can be applied with the Hessian matrix modified as in the Levenberg–Marquardt method.

A cogent issue is the determination of the mathematical model and its validation, in other words, the number N of Lorentzian terms to be used in the fitting procedure. One possibility is to adapt their number by using the F test. Using the reduced χ_r^2 merit function:

$$\chi_r^2(N) = \frac{1}{M - 2N} \sum_{i=1}^{M} w_i r_i(\mathbf{c}, \boldsymbol{\tau})^2,$$

if the measurement errors follow a normal distribution, then the variable ratio $\chi_r^2(N)/\chi_r^2(N+1)$ follows an F distribution. Starting with $N = 1$, the number of terms is increased until the corresponding F probability satisfies the inequality: $P(N+1) < P_0$, where P_0 is a cut-off probability ($0.8 \leq P_0 \leq 0.9$ was recommended).

B.5.6.2 Model-free Quasi-continuous Approach

In this case, the merit function is linearized in the parameters by predetermining the values of the correlation time. Therefore, n correlation times $\tau_1, \ldots, \tau_N$ are assigned in a reference interval, *i.e.*, $[10^{-4}, 10^2]$ μs, according to a log-uniform distribution. The linear least squares problem assumes the form:

$$\min_{f \geq 0} \sum_{i=1}^{M} w_i \left(R_1(\omega_i) - \sum_{j=1}^{N} K_{i,j} f_j \right)^2,$$

where $\boldsymbol{f} = (f_1, \ldots, f_N)$ represents the unknown distribution intensity vector and

$$K_{i,j} = \frac{\tau_j}{1 + (\omega_i \tau_j)^2}, \quad j = 1, \ldots N; \quad i = 1, \ldots M.$$

The number of fixed correlation times N is greater than the number of observations M; the minimization is an undetermined problem. Moreover, the matrix K is ill-conditioned as the discretization of a Fredholm integral equation. Therefore, to compute stable and robust solutions, it is preferable to solve the following 1-norm penalized problem:

$$\min_{f\geq 0}\sum_{i=1}^{M} w_i\left(J(\omega_i)-\sum_{j=1}^{n}K_{i,j}f_j\right)^2+\alpha\sum_{j=1}^{n}|f_j|,$$

where $\alpha > 0$ is a regularization parameter.

These problems are generally difficult because the solution is extremely sensitive to the choice of the regularization parameter. However, balancing the residual norm with the 1-norm penalty term makes it possible to calculate the regularization parameter automatically and efficiently. This approach is implemented by the software *ModelFreeFFC Tool* available at "https://site.unibo.it/softwaredicam/en/modelfree".

B.6 Concluding Remarks

The study of the molecular dynamics of a variety of sample types can be effectively performed by the FFC technique and often also in a unique way. Nevertheless, its use is not widespread. Many reasons can explain this slow evolution, but perhaps the most relevant are the complexity of FFC theories, with the consequent difficulty in developing molecular dynamics models to interpret the acquired data (the NMRD profiles) and the practical difficulties to create efficient and performant equipment to measure such NMRD curves. The specific literature is highly specialized and difficult to access for those who want to try the method for the first time. This can be a further obstacle to the widespread use of the technique. We hope the appendix can help to address this potential limitation by concisely describing the methodological approaches that can be used for interpreting data from FFC measurements.

Acknowledgements

The authors thank Prof. Pellegrino Conte and Prof. Carmine D'Agostino for useful discussion and valuable suggestions for improvements.

References

1. F. Noack, Nuclear Magnetic Relaxation Spectroscopy, in *NMR*, ed. P. Diehl, E. Fluck and R. Kosfeld, Springer, Berlin, Heidelberg, 1971, vol. 3.
2. R. Kimmich and E. Anoardo, *Prog. Nucl. Magn. Reson. Spectrosc.*, 2004, **44**, 257–320.

3. F. Noack, St. Becker and J. Struppe, Applications of Field-Cycling NMR, in *Annual Report on NMR Spectroscopy*, ed. G. A. Webb, Academic Press, 1997, vol. 33, pp. 1–36.
4. E. Anoardo, G. Galli and G. Ferrante, *Appl. Magn. Reson.*, 2001, **20**, 365–404.
5. E. Anoardo, S. Kruber, G. O. Forte and G. A. Dominguez, New Trends in Field-cycling NMR Technology, in *Field-cycling NMR Relaxation: Instrumentation, Model Theories and Applications*, ed. R. Kimmich, Royal Society of Chemistry, 2019, pp. 67–87.
6. G. Ferrante and S. Sykora, *Advances in Inorganic Chemistry*, 2004, vol. 57, pp. 405–470.
7. R. J. S. Brown, The Earth's-Field NML development at Chevron, *Concepts Magn. Reson.*, 2001, **13**(6), 344–366.
8. N. F. Ramsey and R. V. Pound, *Phys. Rev.*, 1951, **81**, 278.
9. E. M. Purcell and R. V. Pound, *Phys. Rev.*, 1951, **81**, 279.
10. M. H. Levitt, *Spin Dynamics: basics of Nuclear Magnetic resonance*, Wiley, 2nd edn, 2007.
11. P. J. Hore, *Nuclear Magnetic Resonance*, Oxford, 2014.
12. J. Keller, *Understand NMR spectroscopy*, Willey, 2nd edn, 2010.
13. R. Kimmich, *Principles of Soft-Matter Dynamics: Basic Theories, Non-invasive Methods, Mesoscopic Aspects*, Springer, 2012.
14. A. Abragam, *Principles of Nuclear Magnetism*, Oxford University Press, 1960.
15. R. Kimmich, *NMR Tomography Diffusometry Relaxometry*, Springer, 1997.
16. C. P. Slichter, *Principle of Magnetic Resonance*, Harper&Row publisher, 1963.
17. T. C. Farrar and E. D. Becker, *Pulse and Fourier Transform NMR. Introduction to theory and Methods*, Academic Press Inc., 1971.
18. J.-P. Korb, *Prog. Nucl. Magn. Reson. Spectrosc.*, 2018, **104**, 12–55.
19. R. Kimmich and N. Fatkullin, *Prog. Nucl. Magn. Reson. Spectrosc.*, 2017, **101**, 18–50.
20. N. Bloembergen, E. M. Purcell and R. V. Pound, *Phys. Rev.*, 1948, **73**(7), 679.
21. J.-P. Korb, *New J. Phys.*, 2011, **13**, 035016.
22. D. A. Faux and P. J. McDonald, *Phys. Rev. E*, 2017, **95**, 033116.
23. D. A. Faux, P. J. McDonald and N. C. Howlet, *Microporous Mesoporous Mater.*, 2018, **269**, 39–42.
24. J.-P. Korb, G. Freiman, B. Nicot and P. Ligneul, *Phys. Rev. E: Stat., Nonlinear, Soft Matter Phys.*, 2009, **80**, 061601.
25. O. V. Bycbuk and B. O'Sbaugbnessy, *Phys. Rev. Lett.*, 1995, **74**(10), 1795.
26. O. V. Bycbuk and B. O'Sbaugbnessy, *J. Chem. Phys.*, 1994, **101**(1), 772.
27. P. E. Levitz and J.-P. Korb, *Europhys. Lett.*, 2005, **70**(5), 684–689.
28. P. E. Levitz, *J. Phys.: Condens. Matter*, 2005, **17**, s4059–s4074.
29. P. E. Levitz, J. P. Korb and D. Petit, *Eur. Phys. J. E: Soft Matter Biol. Phys.*, 2003, **12**, 29–33.
30. P. E. Levitz, *Field-cycling NMR Relaxometry: Instrumentation, Model Theories and Applications*, ed. R. Kimmich, Royal Society of Chemistry, 2019, ch. 12.

31. D. A. Faux, R. Kogon, V. Bortolotti and P. J. McDonald, *Molecules*, 2019, **24**, 3688.
32. D. A. Faux and P. J. McDonald, *Phys. Rev. E*, 2017, **95**, 033117.
33. D. A. Faux, S.-H. P. Cachia, P. J. McDonald, J. S. Bhatt and N. C. Howlett, *Phys. Rev. E: Stat., Nonlinear, Soft Matter Phys.*, 2015, **91**, 032311.
34. D. A. Faux, P. J. McDonald, N. C. Howlett, J. S. Bhatt and S. V. Churakov, *Phys. Rev. E: Stat., Nonlinear, Soft Matter Phys.*, 2013, **87**, 062309.
35. T. Zavada and R. Kimmich, *J. Chem. Phys.*, 1998, **109**(16), 6929.
36. T. Zavada, R. Kimmich, J. Grandjean and A. Kobelkov, *J. Chem. Phys.*, 1999, **110**(14), 6977.
37. B. Halle, H. Jòhannesson and K. Venu, *J. Magn. Reson.*, 1998, **135**(1), 1–13.
38. G. Lipari and A. Szabo, *J. Am. Chem. Soc.*, 1982, **104**, 4546–4559.
39. I. Bertini, M. Fragai, C. Luchinat and G. Parigi, *Magn. Reson. Chem.*, 2000, **38**(7), 543–550.
40. B. Halle, *J. Chem. Phys.*, 2009, **131**(22), 224507.
41. D. Bertsekas, *SIAM J. Control Optim.*, 1982, **20**(2), 221–246.
42. E. Gafni and D. Bertsekas, *SIAM J. Control Optim.*, 1984, **22**(6), 936–964.
43. G. Landi, G. V. Spinelli, F. Zama, D. C. Martino, P. Conte, P. Lo Meo and V. Bortolotti, *Applied Mathematics and Computation*, 2023, **444**, 127809.
44. G. Fellet, P. Conte, V. Bortolotti, F. Zama, G. Landi, D. F. Chillura Martino, V. Ferro, L. Marchiol and P. Lo Meo, *Agriculture*, 2022, **12**(320), 320–340.

APPENDIX C

Genesis of the Signal in MRI

G. RIVA* AND A. SPADARO

Casa di Cura Clinica del Mediterraneo, *via* Ettore Fieramosca, 100, Ragusa, Italy
*Email: Drgaetanoriva@gmail.com

C.1 Some General Information

The genesis of the MR image is based on complex physical phenomena that have been already described in Appendices A and B. Therefore, here only a very brief and practical overview is provided to the reader.

It is worth noting that MRI has been developed mainly for the purpose of medical diagnosis. For this reason, one has to account for the fact that hydrogen is the most abundant nucleus in the human body. Indeed, as a general remark, it is also the most abundant nucleus in nearly all kinds of systems, environmental compartments included. For this reason, we may state that MRI is addressed to monitor the behaviour of proton nuclei in any ^{1}H-rich system.

As previously reported (Appendix A), the motion of each nucleus is described by a vector referred to as the angular momentum that, in turn, is related to the nuclear spin. Being a moving electric charge, the nucleus can also be considered as a small magnetic dipole.

The nuclei having magnetic properties inside biological tissues are those of hydrogen (^{1}H), carbon-13 (^{13}C), nitrogen-14 (^{14}N), fluorine-19 (^{19}F), sodium-23 (^{23}Na), phosphorus-31 (^{31}P) and potassium-39 (^{39}K). All of these are also indicated as NMR *visible* nuclei as they have a nuclear spin number $(I) \neq 0$. Among these, as already stated, ^{1}H is the most abundant in the human organism. In fact, it is possible to find approximately an average of 10^{19} hydrogen atoms per 1 mm^{3} of each human tissue.

New Developments in NMR No. 32
The Environment in a Magnet: Applications of NMR Techniques to Environmental Problems
Edited by Pellegrino Conte, Delia Francesca Chillura Martino and Paolo Lo Meo

Published by the Royal Society of Chemistry, www.rsc.org

The MRI phenomenon applied to biological tissues can be summarized as the result of the changes induced by the application of radiofrequencies (RF) on hydrogen nuclei (of water) when they are placed in a static magnetic field ($\overrightarrow{B_0}$). Indeed, the magnetic moments describing the magnetic properties of the hydrogen nuclei lose their degeneration following $\overrightarrow{B_0}$ application. In particular, two different spin levels are generated. The first one is described by the magnetic moment precession movement along the $\overrightarrow{B_0}$ direction (traditionally, the $+z$-axis), while the second one is described by the magnetic moment rotating around the $-z$-axis. The former condition is indicated as the parallel orientation (it is also referred to as the most probable, or the lowest energetic level), whereas the latter is referred to as the antiparallel orientation. This is the less probable or the highest energetic level. In both cases, precession occurs without phase coherence with a frequency proportional to $\overrightarrow{B_0}$, according to Larmor's law (that is, $\omega_0 = \gamma B_0$). The sum of the z-components of all the nuclear magnetic moments results in a non-null vector referred to as magnetization ($\overrightarrow{M}$) that is oriented along the direction of $\overrightarrow{B_0}$. As reported in Appendix A, the behaviour of the magnetization after a radiofrequency pulse is applied along a direction that is perpendicular to the z-axis allows the so-called NMR phenomenon resulting in the graphs presented in Appendices A and B or the images presented in Chapter 9. The radiofrequency pulse, consisting in the application for a very short period of time of a magnetic field (usually indicated as $\overrightarrow{B_1}$) applied along the $-x$-axis, is also indicated as the flip angle pulse (FAP). This is because, following its application, the magnetization is moved away from the $+z$-axis to another position forming an angle β from the z-axis, the angle width depending on the pulse duration.

Following nuclear excitation due to $\overrightarrow{B_1}$ application, the magnetization relaxes (*i.e.*, the nuclei return to the lowest spin level) *via* two different mechanisms. One of the two is indicated as longitudinal (or spin–lattice) relaxation. It refers to the realignment of the magnetization along the direction of $\overrightarrow{B_0}$. The second mechanism is referred to as transversal (or spin–spin) relaxation. The latter is due to the dephasing (or coherence loss) of the $\overrightarrow{M}$ components in the xy-plane (see Appendix A for more details about both mechanisms). The time needed for the longitudinal relaxation is indicated as T_1 (or longitudinal relaxation time), while the time during which transversal relaxation occurs is named T_2 (or transversal relaxation time). The inverse values of both relaxation times define the longitudinal $\left(\frac{1}{T_1} = R_1\right)$ and the transversal $\left(\frac{1}{T_2} = R_2\right)$ relaxation rates, respectively. In Chapter 13, it is reported how T_1 and T_2 values change as the correlation time (τ_C) varies. Here, it must be reminded that the correlation time measures the time needed for a molecular system to rotate by 1 rad when it is assumed to have a spherical shape, or to move a distance equal to its gyration radius when the shape is not spherical. The longitudinal relaxation time shows a minimum as τ_C increases, whereas, in the same τ_C interval, the transversal relaxation time decreases. Three different motion regimes can be recognized (the reader is referred to Chapter 13 for more details). The fast motion regime, occurring when $(\omega_0\tau_C)^2 \ll 1$, is associated to very short correlation times (*i.e.*, $\leq 10^{-10}$ s) describing the motion of small molecules (*e.g.*, bulk water). When $(\omega_0\tau_C)^2 \gg 1$, the slow motion regime is achieved. It occurs when slowly tumbling molecules (*e.g.*, polymers such as proteins with a low hydration degree in fibrosis) with correlation times $>10^{-7}$ s are present. The condition $(\omega_0\tau_C)^2 \approx 1$, corresponding to the T_1 minimum, is indicated as the intermediate-motion regime. It occurs when correlation times range in the interval from $\sim 10^{-9}$ s to $\sim 10^{-6}$ s, typical for molecular systems such as fats and glyco-lipoproteins in human tissues. The molecular tumbling and rotations, monitored by the correlation times, induce small fluctuations in the applied magnetic field. As a consequence, a molecular system can produce a range of T_1 and T_2 values according to the chemical environment where it belongs. In other words, just as an example, water molecules surrounding a tumor tissue may show correlation times different from the same molecules surrounding healthy tissues, thereby generating different magnetic field fluctuations which, in turn, are responsible for the differences in the T_1 and T_2 values. These differences are used to generate the images typical for the MRI analyses (see below).

C.2 The Parameters to Account for the Generation of the Images[1]

The intrinsic features of biological tissues include the proton density (PD), the proton longitudinal, and the proton transversal relaxation times. These parameters can vary from one tissue to another so that the resulting images can differ in their tissue–tissue contrast, depending on which of the three aforementioned parameters is emphasized in a magnetic resonance sequence. This is the basis for the exceptional soft-tissue discrimination and diagnostic magnetic resonance imaging potential. As an example, due to their specific differences in terms of these three parameters, MRI can differentiate tissues appearing virtually indistinct on computed tomography (CT) scans without the need for contrast medium administration.

In the previous section, T_1 has been defined as the time needed for the realignment of the magnetization along the direction of $\overrightarrow{B_0}$, while T_2 was identified as the time needed for the coherence loss of the magnetization in the *xy*-plane. Practically speaking, on the one hand, the longitudinal relaxation time is associated to the recovery of all the nuclei to the ground state in order to have them available once again for the next excitation. On the other hand, the transversal relaxation time determines how quickly a signal fades after excitation.

Proton density refers to the number of excitable protons per volume unit in a given tissue. Notably, each tissue is characterized by its own PD which, in turn, determines the maximum signal intensity that can be acquired for that tissue. Images where the contrast among tissues is due to the different proton densities are referred to as either proton density-weighted or proton density images. PD can be maximized by minimizing T_1 and T_2. This is explained by considering that the lower the two relaxation time values, the higher the number of scans that can be applied, thereby improving NMR sensitivity (the reader is referred to Appendix A to have more information about NMR sensitivity and the scan number).

The three parameters described above can be experimentally modulated by the repetition or delay time (RT) and the echo time (ET).

The repetition time refers to the time interval between two successive scans. As indicated in Appendix A, reliable NMR spectra can be obtained only when the repetition time is $\geq 5T_1$. In fact, only when this condition is satisfied, signal saturation is avoided and quantitative assessment of NMR spectra can be achieved. The term "signal saturation" refers to the condition by which, after excitation, nuclei do not completely relax to the ground state because of the insufficient time between two successive pulses. When this occurs, the application of the second pulse "finds" most of the nuclei already in the excited state. Therefore, the number of nuclei that are excited by this second pulse is not representative of the real NMR observable nucleus population. As a consequence, quantitative assessment of NMR spectra cannot be achieved. However, when dealing with MRI, the unsatisfactory $RT \geq 5T_1$ is not a disadvantage. In fact, as an example, if a short repetition time (*e.g.*, $\leq$600 ms) is applied, only tissues containing quickly relaxing nuclei can be fully detected. This means that if a tissue with slowly relaxing ^1H nuclei stands by another tissue whose ^1H nuclei are quickly relaxing, the image of the latter is clearer than that of the former. By using the language of spectroscopy, one can say that the spin levels of the protons belonging to the first tissue are saturated because of the long proton relaxation. As a consequence, these protons cannot produce intense signals. Conversely, the spin levels of the quickly relaxing protons in the second tissue are not saturated and the protons can produce very intense signals.

When the repetition time is increased, the contrast between the images of the two tissues decreases. Let us, for example, consider RT = 2000 ms. As this condition is applied, the protons in both tissues have enough time to relax completely to the ground state. Therefore, no saturation is obtained and the maximum signal intensity for both tissues can be achieved.

The image contrast that can be obtained by applying short RT values is based on the saturation of the spin levels of the slowly relaxing nuclei and produce images that are referred to as T_1-weighted. Conversely, when sufficiently long RT values are used, the contrast between images is weighted only by proton density. Therefore, when long RT values are used, PD weighted images are obtained.

The echo time is the time interval between the moment when the pulse is applied and that when the maximum intensity of the NMR signal is achieved. In other words, let us consider the application of a 90° pulse. Due to this, the magnetization reverses along the *xy*-plane and starts to dephase as a consequence of the different chemical environments where the protons lay. After a time Δ, the application of a 180° pulse train allows refocusing of the magnetization. Therefore, the signal is acquired after waiting for another Δ time. By changing the Δ value, the signal intensity is modulated according to the relationship: $I(\tau) = I_0 \exp\left(-\frac{\tau}{T_2}\right)$, where $I(\tau)$ is the signal intensity when $\tau = 2\Delta$, while I_0 is the signal intensity in the absence of any perturbation. When the condition $\lim_{\tau \to 0}\left[I_0 \exp\left(-\frac{\tau}{T_2}\right)\right] = I_0$ is satisfied, it is possible to state that the images from different tissues cannot be differentiated. In other words, when the echo time is very short (that is, $\tau \to 0$, or practically speaking <30 ms), the dephasing is only slightly started, and all the tissues are approximately equally distinguishable. Conversely, the longer the echo time, the more contrast affects tissue images, thereby permitting a better differentiation among tissues. The image contrast filtered by lengthening the echo time provides the T_2-weighted images. As in the case of long RT values, also here it is possible to state that PD-weighted images are obtained when short ETs are applied.

C.3 The Conditions to Produce Reliable Images in a Reasonable Time Interval

In the previous section, we outlined that the contrast among tissue images can be obtained by operating on the repetition and echo times as summarized in Table C.1.

For the sake of clarity, it must be noted that, in the MRI literature, T_1, T_2, and PD are usually referred to as intrinsic parameters because they are related to the physical–chemical nature of the observed nuclei. Conversely, parameters such as the repetition time, the echo time, and the flip angle pulse are indicated as extrinsic parameters because they can be instrumentally modulated.

By looking at Table C.1, one must consider that the terms “short” and “long” have no meaning from a scientific point of view. The question arises:

Table C.1 Image contrast as affected by repetition (RT) and echo (ET) times. Reproduced from ref. 1 with permission from Springer Verlag, Copyright 2003.

Image contrast	RT	ET
T_1-weighted	Short	Short
T_2-weighted	Long	Long
PD-weighted	Long	Short

how long and how short should RT and ET be? Even if one decides to apply a repetition time of 600 ms (as in the case of the example reported in the previous section), this can be unreasonably long if one considers that the aforementioned RT value must be used in analyses involving patients that must not move during the experiment. Moreover, if an even shorter RT was applied while a 90° flip angle pulse is used, the saturation condition can also be achieved. This is because, in an interconnected complex system such as a tissue, the relaxation rate is not the same for all the nuclei. Therefore, only a small amount of the net magnetization can be transferred on the *xy*-plane, thereby causing a decrease in signal intensity.

Different *escamotages* can be used to reduce the experimental time and to avoid the problem of saturation. The first one is the application of a <90° flip angle pulse. In fact, just as an example, if a 30° FAP is applied, both T_1 and T_2 are strongly reduced (because a shorter time is needed both to recover the magnetization along the *z*-axis and to null the vectorial sum of the *xy*-components of the magnetization in the *xy*-plane). Thus, a shorter RT can be used and a reduction of the whole experimental time is achieved. As a rule of thumb, when RT $\geq$ 50 ms, a flip angle pulse of 50–60° can be applied.

In the case of echo sequences where T_2 is the main filter for image contrast, the application of a RT <100 ms results in a low (or poor) signal-to-noise ratio. As a general rule, it is possible to state that the relationship between FAP and ET is as follows:

- FAP $\geq$ 75° and ET $\leq$ 14 ms provide pseudo-T_1 contrast.
- FAP $\leq$ 45° and ET $\geq$ 18 ms provide pseudo-T_2 contrast.

In 3D sequences, where RT can be in the range 8–15 ms, the FAP applied to generate T_1-weighted images must be around 25°. However, if a FAP <25° is chosen in an echo sequence, together with long repetition times (*i.e.*, >50 ms), differences among T_1 values are nulled, thereby allowing the generation of T_2-weighted images modulated by ET. In summary, as a rule of thumb, the FAP in T_1-weighted images is >45°, while in the T_2-weighted ones it is <45°.

The second *escamotage* to modulate image contrast while reducing the experimental machine time is the presaturation. This consists in an initial 180° inverting pulse that is applied before the data for image generation are acquired. Comparatively, a 180° inverting pulse achieves a more pronounced T_1 effect than a 90° pulse because it completely flips the longitudinal magnetization. Consequently, T_1 relaxation starts from the −*y*-axis instead of the *xy*-plane, thereby resulting in the availability of twice the longitudinal magnetization. Moreover, the T_1 effect can be controlled by adjusting the time interval between the 180° inversion pulse and the excitation pulse. The latter time interval is also known as the inversion time (IT). By carefully selecting the IT, the signal contribution from a specific tissue can be eliminated by applying the excitation pulse when the tissue lacks magnetization. This allows the suppression of the signal from fat with a short IT and from myalgic encephalomyelitis with a long IT. Notably, this technique finds practical application in late-enhancement imaging for patients with myocardial infarction.

C.4 The Images: Acquisition of Information on Body Slices

Magnetic resonance imaging produces cross-sectional images of the human body or any other three-dimensional system. To achieve this, the excitation pulse is specifically targeted only to the desired slice for imaging, rather than applying it to the entire body. For instance, let us focus on a transverse (axial) slice or cross-section taken through the body. In most MR scanners, the magnetic field is aligned along the body axis of the system undergoing the imaging process. This direction is indicated as "z" to signify the main magnetic field's orientation, as previously reported.

In MRI, the so-called magnetic field gradients are applied. These can be considered as wedges, where the thicker side indicates higher field strength and the tip indicates lower field strength. Both the excitation of a specific slice and the identification of the signal's origin within that slice rely on the principle that the precessional or Larmor frequency is directly proportional to the magnetic field strength.

It is important to remind that protons are excited only when they are subjected to an RF pulse with a frequency that roughly matches their Larmor frequency. If a uniform magnetic field with the same strength were uniformly applied throughout the body, all protons would possess the same Larmor frequency and would be simultaneously excited by a single RF pulse. In order to achieve selective excitation of a specific slice, a gradient coil is used to generate an inhomogeneous magnetic field along the z-direction in a linear way. This produces a smooth gradient of magnetic field strength, which is weakest at the patient's head and strongest at the feet, for instance. Consequently, the Larmor frequencies vary gradually along the z-axis, resulting in each slice having its distinct frequency. By applying an RF pulse that corresponds to the Larmor frequency of the desired slice, only protons within that particular slice will be excited, whereas the rest of the body remains unaffected. This technique ensures targeted imaging of the desired slice without disturbing the surrounding areas.

As indicated above, gradients are extra magnetic fields generated by gradient coils. They are designed in a way that a magnetic field is either added to or subtracted from the main magnetic field. As protons move along the gradient, they experience temporary exposure to magnetic fields of varying strengths, resulting in distinct precessional frequencies for distinct positions.

The thickness of the generated slice depends on the gradient's steepness, where a shallow gradient produces a thicker slice and a steep gradient generates a thinner slice. To determine the slice position, the center frequency of the RF pulse is adjusted accordingly. Once the desired slice position and thickness are chosen through the appropriate slice-select gradient, the next step involves identifying the spatial position of an MR signal. This task, known as spatial encoding, is the most challenging aspect in the production of an MR image. It needs the application of additional gradients that modify the magnetic field strength along the *y*- and *x*-axes.

Spatial encoding consists of two main steps known as phase and frequency encoding.[1]

After the spins have been excited and are precessing in the *xy*-plane, a gradient in the *y*-direction (from top to bottom) is activated for phase encoding. This *y*-direction gradient causes the Larmor frequencies of the spins to vary based on their positions along the gradient. Consequently, the spins located higher up in the scanner are subjected to a stronger magnetic field. The latter affects magnetization by dephasing all the $\vec{M}$ components at different rates. In particular, the rates of the $\vec{M}$ components corresponding to the nuclei in the stronger magnetic field are faster than those of the nuclei located further down, and "feeling" a weaker magnetic field. Hence, a phase shift occurs among the spins relative to each other. The extent of this phase shift is determined by factors such as the duration and amplitude of the phase-encoding gradient and the specific physical locations of the precessing nuclei along its length. After a given time interval, the field gradients are turned off. Therefore, all the spins return to their initial precession rate, but they are, now, either ahead or behind in phase as compared to their state without the application of the field gradients. This results in a linear variation of phase along the *y*-axis, thereby allowing each line within the slice to be distinguished by its distinct phase characteristics.

The second spatial dimension of the MR signal is established by introducing frequency variations along the *x*-direction. To achieve this, a frequency-encoding gradient is applied specifically along the *x*-axis, which generates a magnetic field increasing in strength from right to left. Consequently, the spins on the left side precess at a slower rate than those on the right side. When the MR signal is acquired with the frequency-encoding gradient active, it results in an entire spectrum of frequencies. This spectrum consists of high frequencies originating from the right edge of the slice and low frequencies from the left edge. As a result, each column of the slice is characterized by a specific frequency.

Frequency and phase work together to provide distinctive spatial identification for each volume element which is referred to as a voxel. The MR signal obtained as indicated above contains two essential types of information. The frequency component identifies the position of the signal along the *x*-axis. To extract this spatial information accurately, a Fourier transform is applied (the reader is referred to Appendix A for more details about the Fourier transform). This mathematical operation allows differentiation of the signals based on their frequencies along the frequency encoding direction. Due to this, a precise spatial localization along the *x*-axis can be obtained.

The phase distribution within each frequency component provides valuable information about the signal's origin along the *y*-axis. However, the determination of the phases of individual spins cannot be obtained from a single signal. Rather, it requires a collection of signals. This situation is similar to that consist in solving a mathematical equation with many unknowns, for which the overall result is known, but none is clear about each single unknown. In order to calculate the unknowns, a set of different equations, whose number must be equal to that of the unknowns, must be accounted for. In the case of the MR signal, this involves repeating the sequence many times by varying the gradient strengths. The acquisition of a set of echoes with different phase encodings makes it possible to obtain the necessary phase-encoded spatial information through a second Fourier transform, this time along the *y*-axis. Consequently, for spatial encoding in two dimensions, the Fourier transform is applied twice, leading to the term "two-dimensional Fourier transform" (2D-FT). To perform such complex calculations – similar to solving a set of *m* equations with *m* unknowns – an MR scanner requires a dedicated computer, often referred to as an array processor. During the process, repeated measurements are taken with a specific temporal delay, that is, the repetition time. The number of phase-encoding steps performed depends on the desired image quality. By choosing more phase-encoding steps, resolution and image quality enhancement can be obtained. However, this also extends the scanning time.

C.5 The Images: The Achievement of a 3D Map of the Body

In the previous section, the achievement of images of body slices has been described. However, more than on slices, it can be important to have information about entire volumes of the body. The advantage of such a choice relies on the fact that the acquired source data set may need to undergo postprocessing, such as creating reconstructions in different planes. Moreover, when acquiring thin slices, there is the risk of overwhelming the MR signal with noise. Thin slices result in weaker MR signals due to fewer spins being excited. However, this limitation can be addressed by leveraging the stronger signal generated by imaging the entire volume and then extracting the individual slices afterward.

As compared to the acquisition of slice images, that of an entire volume needs an additional step to encode for the third dimension. This is achieved by the application of a phase-encoding gradient along the *z*-axis. The discussion above covers how encoding can be applied to both the *x*- and *y*-axes. The same approach can be extended when the third dimension (*i.e.*, the *z*-axis) is introduced. The sole difference consists in the fact that, instead of a 2D Fourier transform, a 3D one must be applied. The three-dimensional Fourier transform produces a 3D dataset of a volume without any gaps between slices. This dataset allows for the generation of reconstructions in any plane or projections with the help of appropriate reconstruction algorithms. The non-negligible disadvantage of a 3D scanning with Fourier transform is the much longer machine time required for the experimental session. As a consequence, a human in the machine may suffer not only for the prolonged noises, but also for the claustrophobic sensations.

C.6 The *K*-space[1]

The data collected as indicated in the last two sections are stored in a mathematical area referred to as the *K*-space. Simply speaking, the *K*-space is an $(m \times n \times t)$ three-dimensional matrix, where m, n, and t are the dimensions along which the field gradients and the encoding steps described in the sections above are applied. For the sake of simplicity, the three-dimensional matrix can also be indicated as $(K_x \times K_y \times K_z)$, so that the x-, y-, and z-dimensions are explicated. This matrix represents the set of data before the Fourier transform is applied in the three (x-, y-, and z-) dimensions. In the *K*-space, each line represents a single measurement, and a line is acquired for every phase-encoding step. The center line contains data that remain unaffected by the phase-encoding gradient. This is also recognized as the gradient isocenter. Once the scan is completed and the *K*-space is filled, the raw data are processed using 3D Fourier transform to generate an MR image. Noteworthily, the lines in the *K*-space do not directly correspond to the lines in the resulting MR image. Instead, the data in the center of the *K*-space play a primary role in determining the images' contrast, while the periphery (*i.e.*, the outer lines) primarily contains spatial information.

C.7 The New *K*-space Sampling Strategies

An already established technical innovation that is associated with the classical samples of the *K*-space described above is represented by parallel imaging. This *K*-space sampling technique is based on the so-called SMASH (simultaneous acquisition of spatial harmonic) and SENSE (sensitivity encoding). Both SMASH and SENSE involve the use of multichannel receiver coils to obtain additional spatial information used to complete a less dense sampling of phase coding lines of the *K*-space. For each sequence, the *K*-space is acquired with a reduced number of phase coding lines with the advantage of reducing acquisition times. Missing information is integrated through a sort of interpolation that uses sophisticated AI (artificial intelligence) systems.

C.8 From the *K*-space to the MR Image: The Fourier Transform

The Fourier transform decomposes signals or curves into a sum of sine and cosine waves each having different frequencies. To reconstruct the MR images, the Fourier discrete transform operation (also referred to as DFT) is used rather than a continuous transform. The data stored in the *K*-space are, actually, discrete being only a set of a finite number of figures. The DFT and its inverse are defined as a mathematical series (*i.e.*, a sum of terms) where the number of terms corresponds to that of samples of the curve to be transformed. The terms of the series are added together to calculate the value of a pixel (that is, a point; for a better definition of the pixel, the reader is referred to the next section) in the final MR image. The fast Fourier transform (FFT) is an algorithm recognized as the most efficient for the calculation of the DFT and is the one implemented nowadays in all the MR machines.

A two-dimensional DFT (DFT-2D) reconstructs a 2D image from the *K*-space. The 2D DFT is implemented as many separate FFTs, one for each row and column in the *K*-space. Therefore, if the *K*-space contains, *e.g.*, 256 rows and 256 columns, the reconstruction of the MR image requires a total of 512 FFTs. The volumetric Fourier transform (3D), used to reconstruct volumetric images, is an extension of the DFT-2D.

C.9 The Signal-to-noise Ratio and the Resolution

As reported in many chapters of this book, the resolution and the signal-to-noise ratio (S/N or SNR) are very important parameters to obtain reliable information from NMR investigations. Resolution refers to the ability to differentiate among different close peaks, in the case of NMR spectroscopy, or to distinguish one structure from another in the case of MRI. The signal-to-noise ratio is the measure of the NMR sensitivity. In other words, it refers to the ability to recognize a signal from a background noise. The simplest source of noise in all the NMR techniques is, for example, the current in the electric circuits. If the signal is not "high" enough as compared to the noise generated by the electric currents, no images can be obtained (in the case of spectroscopy – see Appendix A – no signal can be achieved). In other words, lack of sensitivity is obtained. Noise in MRI can arise from various factors, including imperfections in the MR system caused by magnetic field inhomogeneities, thermal noise originating from the RF coils, nonlinearity in signal amplifiers, aspects related to image processing, as well as patient-related factors resulting from body movement or respiratory motion. On the one hand, to ameliorate (*i.e.*, increasing) the SNR, one must operate on magnetic field strength, scan parameters (RT, ET, FAP), the number of scans, and selection of the transmit and receive coil (RF coil). On the other hand, resolution is increased when the size of the (image) matrix, slice thickness and receiver bandwidth, and field of view (FOV) are accounted for.

The first three parameters (*i.e.*, magnetic field strength, scan parameters, and the number of scans) have been already discussed in Appendices A and B, and in the previous sections of this appendix. Very briefly, it must be reminded that the stronger the magnetic field intensity, the larger the distance among the nuclear spin levels, thereby allowing a larger number of nuclei to be excited and, therefore, increasing sensitivity. How RT, ET, and FAP can be modulated to increase sensitivity is reported in Sections C.2 and C.3. Finally, SNR $\propto \sqrt{\text{NS}}$, where NS stands for the number of scans, that is, how many times the experiment is repeated under the very same conditions. As noise lines are randomly distributed, while the signal appears always in the same position, experiment repetition allows the collection of data that, once summed, provide a baseline where the background noise is reduced and the signal is magnified.

The amelioration of the signal-to-noise ratio can be obtained also by placing the RF coil as closest as possible to the tissue to be analysed, thus permitting the maximization of the transmitting and receiving signals by minimizing the possibility of signal distortion. In order to achieve this aim, different volume coil configurations have been developed, such as saddle, birdcage, and body coils,[1–3] together with surface,[1,3] intracavity,[1,3] and phased array coils.[1,3] The reader is referred to the specific literature for details on the aforementioned coils.[1–4]

For what concerns resolution, let us consider that after data are collected as previously discussed, they must be handled in a way that the body scanned image can be obtained. To this aim, each image coming from the acquired data is divided into many intersecting horizontal and vertical lines which provide a kind of "net" where the single net mesh is referred to as a pixel. The set of intersecting horizontal and vertical lines is indicated as the matrix. Pixels can be also considered as the two-dimensional representation of the volume elements (*i.e.*, voxels, which were defined above). Voxels, however, are 3D "objects", while pixels are 2D ones. For the reconstruction of the three-dimensional body "picture," the third dimension of the voxel coincides with the layer thickness. The matrix size is given by the pixel number, such as, for example, 64×64, 128×128, 256×256, or 512×512. In these cases, the matrix is square. However, rectangular matrices with, *e.g.*, 190×256 pixels can also be used. In general, the larger the number of pixels, the higher the image resolution. Currently, resolutions as high as 1024×1024 are possible.[4]

The choice of the matrix size has important consequences on the machine time. In fact, as the number of pixels is increased, a longer time is needed to acquire the data in all the dimensions where investigations are carried out. Therefore, the right resolution choice must account for the clinical problem and the type of tissue structure under examination. This choice is affected by the experience of the human operator.

The field of view and SNR are strongly connected. The former is the distance (usually expressed in mm) over which a magnetic resonance image is displayed. It is typically divided into several hundred picture elements which are referred to as pixels, as reported above. Pixel size in the frequency encoding direction is calculated by dividing the FOV by the matrix size in the frequency encoding direction. Moreover, the pixel size in the phase encoding direction is calculated by dividing the FOV and matrix size in the phase encoding direction. Based on this, it is possible to state that resolution increase can be obtained by tuning FOV size. A smaller FOV leads to a smaller pixel size, provided that the matrix remains unchanged. The pixel size is critical for the spatial resolution of the MR image. When the FOV remains the same, a higher matrix resolution (that is, more pixels) results in improved spatial resolution, while a lower matrix resolution (that is, fewer pixels) leads to poorer spatial resolution. From what has been reported so far, one might conclude that the matrix should be as large as possible to capture a maximum number of picture elements. This is true in terms of image resolution. However, the minimum pixel size is limited because, in general, the signal-to-noise ratio decreases with the size of the voxel.

A further factor limiting resolution is the image acquisition or scan time. It is directly proportional to the size of the matrix. In order to achieve a better resolution in a reasonable scan time, FOV reduction only in the phase encoding direction can be considered. This means that a rectangular FOV is achieved. The aforementioned reduction is possible because the spatial resolution is determined by the frequency encoding direction, while the scan time is dependent on the phase encoding direction. Filling only one-half the normal number of phase-encoding lines in the *K*-space reduces the imaging time and the FOV by 50%.[1] However, when a rectangular FOV is applied, artifacts can be obtained. In fact, mapping the signal from outside the FOV in the phase encoding direction back into the image results in misplacement at an incorrect location. According to Weishaupt *et al.*,[1] "*this kind of foldover can be suppressed by specific antialiasing options such as "no phase wrap". Moreover, reduction of the FOV in the phase-encoding direction is associated with a slight drop in SNR. A rectangular FOV is typically used to image the spine and extremities and for MR angiography. Scan time can be shortened further on state-of-the-art scanners that allow one to use rectangular fields of view in combination with rectangular pixels*".

References

1. D. Weishaupt, V. D. Köchli and B. Marincek, *How does MRI work?*, Springer Verlag, Berlin, Hiedelberg, 2003.
2. M. Kaiser, M. Detert, M. A. Rube, A. El-Tahir, O. J. Elle and A. Melzer, *et al.*, Resonant marker design and fabrication techniques for device visualization during interventional magnetic resonance imaging, *Biomed. Tech.*, 2015, **60**(2), 89–103.
3. B. Gruber, M. Froeling, T. Leiner and D. W. J. Klomp, RF coils: A practical guide for nonphysicists, *J. Magn. Reson. Imaging*, 2018, **48**(3), 590–604.
4. M. B. Dale, M. A. Brown and R. C. Semelka, *MRI Basic Principles and Applications*, John Wiley & Sons, Ltd, Chichester, UK, 2015.

Subject Index